Lecture Notes in Computer Science 5473

Commenced Publication in 1973
Founding and Former Series Editors:
Gerhard Goos, Juris Hartmanis, and Jan van Leeuwen

D1366830

Marc Fischlin (Ed.)

Topics in Cryptology – CT-RSA 2009

The Cryptographers' Track at the RSA Conference 2009
San Francisco, CA, USA, April 20-24, 2009
Proceedings

 Springer

Volume Editor

Marc Fischlin
TU Darmstadt, Theoretical Computer Science
Hochschulstrasse 10, 64289, Darmstadt, Germany
E-mail: marc.fischlin@gmail.com

Library of Congress Control Number: Applied for

CR Subject Classification (1998): E.3, D.4.6, K.6.5, C.2, K.4.4

LNCS Sublibrary: SL 4 – Security and Cryptology

ISSN	0302-9743
ISBN-10	3-642-00861-5 Springer Berlin Heidelberg New York
ISBN-13	978-3-642-00861-0 Springer Berlin Heidelberg New York

springer.com

© Springer-Verlag Berlin Heidelberg 2009
Printed in Germany

Typesetting: Camera-ready by author, data conversion by Scientific Publishing Services, Chennai, India
Printed on acid-free paper   SPIN: 12637942   06/3180   5 4 3 2 1 0

Preface

The 2009 RSA conference was held in San Francisco, USA, during April 20-24. The conference is devoted to security-related topics and, as part of this, hosts a distinguished track for cryptographic research. Since 2001 the proceedings of this Cryptographers' Track (CT-RSA) have been published in the series *Lecture Notes in Computer Science* of Springer.

The proceedings of CT-RSA 2009 contain 31 papers selected from 93 submissions, covering a wide variety of cryptographic areas. Each submission was anonymized for the reviewing process and was assigned to at least three of the 25 Program Committee members. Submissions co-authored by committee members were assigned to at least five members. After carefully considering more than 15,000 lines (more than 100,000 words) of reviews and online discussions, the committee selected 31 submissions for acceptance. The program also included an invited talk by Kenny Paterson entitled "Cryptography and Secure Channels."

I would like to thank all the authors who submitted papers. I am also indebted to the Program Committee members and all external reviewers for their voluntary work. The committee's work was tremendously simplified by Shai Halevi's submission software and his support. I would also like to thank the CT-RSA Steering Committee for electing me as Chair, and all the people from the RSA conference team for their support, especially Bree LaBollita.

January 2009   Marc Fischlin

CT-RSA 2009

RSA Conference 2009, Cryptographers' Track

Moscone Center, San Francisco, CA, USA
April 20–24, 2009

Program Chair

Marc Fischlin   Darmstadt University of Technology, Germany

Program Committee

Michel Abdalla	ENS & CNRS, France
Zuzana Beerliova-Trubiniova	ETH Zurich, Switzerland
Alex Biryukov	University of Luxembourg, Luxembourg
Melissa Chase	Microsoft Research, USA
Alex Dent	Royal Holloway, UK
Nelly Fazio	City University of New York, USA
Juan Garay	AT&T Labs - Research, USA
Amir Herzberg	Bar-Ilan University, Israel
Dennis Hofheinz	CWI, The Netherlands
Nick Howgrave-Graham	NTRU Cryptosystems, USA
Stanislaw Jarecki	UC Irvine, USA
Marc Joye	Thomson, France
Alexander May	Bochum University, Germany
Jesper Buus Nielsen	University of Aarhus, Denmark
Giuseppe Persiano	University of Salerno, Italy
Josef Pieprzyk	Macquarie University, Australia
Vincent Rijmen	K.U. Leuven, Belgium, and Graz University of Technology, Austria
Kazue Sako	NEC, Japan
Christian Schaffner	CWI, The Netherlands
Berry Schoenmakers	TU Eindhoven, The Netherlands
Willy Susilo	University of Wollongong, Australia
Pim Tuyls	Philips, The Netherlands
Jorge Villar	UPC Barcelona, Spain
Bogdan Warinschi	University of Bristol, UK

External Reviewers

Divesh Aggarwal	Dmitry Khovratovich	Thomas Popp
Toshinori Araki	Eike Kiltz	Dominik Raub
Giuseppe Ateniese	Ilya Kizhvatov	Thomas Ristenpart
Man Ho Au	Sandeep Kumar	Matt Robshaw
Roberto Avanzi	Alptekin Kupcu	Pankaj Rohatgi
Rikke Bendlin	Gregor Leander	Dries Schellekens
Johannes Blömer	Anja Lehmann	Martin Schlaeffer
Colin Boyd	Helger Lipmaa	Jacob Schuldt
Christoph de Canniere	Xiaomin Liu	Gautham Sekar
Srdjan Capkun	Jiqiang Lu	Haya Shulman
Rafik Chaabouni	Christoph Lucas	Nigel Smart
Donghoon Chang	Roel Maes	Martijn Stam
Sherman Chow	Mark Manulis	Ron Steinfeld
Christophe Clavier	Krystian Matusiewicz	Marc Stevens
Erik Dahmen	Sigurd Torkel Meldgaard	Bjoern Tackmann
Jean-François Dhem	Florian Mendel	Christophe Tartary
Orr Dunkelman	Nele Mentens	Tamir Tassa
Serge Fehr	Kazuhiko Mihematsu	Isamu Teranishi
Jun Feng	Gert Lassoe Mikkelsen	Stefano Tessaro
Pierre-Alain Fouque	Paul Morrissey	Elmar Tischhauser
Jakob Funder	Nicky Mouha	Nikos Triandopoulos
Steven Galbraith	Elke De Mulder	Michael Tunstall
Martin Geisler	Tomislav Nad	Mike Tunstall
Rosario Gennaro	Gregory Neven	Osman Ugus
Benedikt Gierlichs	Long Nguyen	Frederic Vercauteren
Aline Gouget	Phong Nguyen	Damien Vergnaud
Tim Gueneysu	Antonio Nicolosi	Peishun Wang
Carmiy Hazay	Ivica Nikolic	Benne de Weger
Martin Hirt	Svetla Nikova	Ralf-Philipp Weinmann
Qiong Huang	Satoshi Obana	Enav Weinreb
Xinyi Huang	Francis Olivier	Christopher Wolf
Sebastiaan Indesteege	Claudio Orlandi	Oliver Wuebbolt
Vicenzo Iovino	Elisabeth Oswald	Ng Ching Yu
Vincenzo Iovino	Dan Page	Tsz Hon Yuen
Toshiyuki Isshiki	Kenny Paterson	Hong-Sheng Zhou
Gene Itkis	Giuseppe Persiano	Vassilis Zikas
Emilia Kasper	Krzysztof Pietrzak	

Table of Contents

Multi-Party Protocols

Security of Encryption Schemes

Faults and Countermeasures

Countermeasures and Faults

Adaptive-ID Secure Revocable Identity-Based Encryption

Benoît Libert[1] and Damien Vergnaud[2],[⋆]

[1] Université Catholique de Louvain, Microelectronics Laboratory,
Crypto Group Place du Levant, 3 – 1348 Louvain-la-Neuve – Belgium
[2] Ecole Normale Supérieure – C.N.R.S. – I.N.R.I.A.
45, Rue d'Ulm – 75230 Paris CEDEX 05 – France

Abstract. Identity-Based Encryption (IBE) offers an interesting alternative to PKI-enabled encryption as it eliminates the need for digital certificates. While revocation has been thoroughly studied in PKIs, few revocation mechanisms are known in the IBE setting. Until quite recently, the most convenient one was to augment identities with period numbers at encryption. All non-revoked receivers were thus forced to obtain a new decryption key at discrete time intervals, which places a significant burden on the authority. A more efficient method was suggested by Boldyreva, Goyal and Kumar at CCS'08. In their *revocable* IBE scheme, key updates have logarithmic (instead of linear in the original method) complexity for the trusted authority. Unfortunately, security could only be proved in the selective-ID setting where adversaries have to declare which identity will be their prey at the very beginning of the attack game. In this work, we describe an adaptive-ID secure *revocable* IBE scheme and thus solve a problem left open by Boldyreva *et al.*.

Keywords. Identity-based encryption, revocation, provable security.

1 Introduction

Introduced by Shamir [32] and conveniently implemented by Boneh and Franklin [8], identity-based encryption (IBE) aims to simplify key management by using human-intelligible identifiers (e.g. email addresses) as public keys, from which corresponding private keys are derived by a trusted authority called Private Key Generator (PKG). Despite its many appealing advantages, it makes it difficult to accurately control users' decryption capabilities or revoke compromised identities. While IBE has been extensively studied using pairings (see [13] and references therein) or other mathematical tools [18,9], little attention has been paid to the efficient implementation of identity revocation until very recently [4].

⋆ The first author acknowledges the Belgian National Fund for Scientific Research (F.R.S.-F.N.R.S.) for their financial support and the BCRYPT Interuniversity Attraction Pole. The second author is supported by the European Commission through the IST Program under Contract ICT-2007-216646 ECRYPT II and by the French *Agence Nationale de la Recherche* through the PACE project.

M. Fischlin (Ed.): CT-RSA 2009, LNCS 5473, pp. 1–15, 2009.
© Springer-Verlag Berlin Heidelberg 2009

RELATED WORK. In public key infrastructures (PKI), revocation is taken care of either via certificate revocation lists (CRLs), by appending validity periods to certificates or using involved combinations of such techniques (e.g. [29,1,30,21,25]). However, the cumbersome management of certificates is precisely the burden that identity-based encryption strives to alleviate. Yet, the private key capabilities of misbehaving/compromised users should be promptly disabled after their detection. One of the cited reasons for the slow adoption of the IBE technology among standards is its lack of support for identity revocation. Since *only* the PKG's public key and the recipient's identity should be needed to encrypt, there is no way to notify senders that a specific identity was revoked.

To address this issue, Boneh and Franklin [8] suggested that users can periodically receive new private keys. Current validity periods are then appended to identities upon encryption so as to add timeliness to the decryption operation and provide automatic identity revocation: to revoke a specific user, the PKG simply stops issuing new keys for his identity. Unfortunately, this solution requires the PKG to perform linear work in the number of registered receivers and regularly generate fresh private keys for all users, which does not scale well as their number grows: each non-revoked user must obtain a new key at each period, which demands to prove his identity to the PKG and establish a secure channel to fetch the key.

Other solutions were suggested [7,20,27,3] to provide immediate revocation but they require the cooperation of an online semi-trusted party (called mediator) at each decryption, which is not totally satisfactory either since it necessarily incurs communication between users and the mediator.

Recently, Boldyreva, Goyal and Kumar [4] (BGK) significantly improved the technique suggested by Boneh and Franklin [8] and reduced the authority's periodic workload to be logarithmic (instead of linear) in the number of users while keeping the scheme efficient for senders and receivers. Their *revocable* IBE primitive (or R-IBE for short) uses a binary tree data structure and also builds on Fuzzy Identity-Based Encryption (FIBE) schemes that were introduced by Sahai and Waters [31]. Unfortunately, their R-IBE scheme only offers security guarantees in the relaxed selective-ID model [15,16] wherein adversaries must choose the target identity ahead of time (even before seeing the system-wide public key). The reason is that current FIBE systems are only secure in (a natural analogue of) the selective-ID model. Boldyreva *et al.* explicitly left open the problem of avoiding this limitation using their approach.

As noted in [5,6], selective-ID secure schemes can give rise to fully secure ones, but only under an exponential reduction in the size of the identity space. Also, while a random-oracle-using [2] transformation was reported [5] to turn any selective-ID secure IBE scheme into an adaptive-ID secure one, it entails a degradation factor of q_H (*i.e.*, the number of random oracle queries) in the reduction and additionally fails to provide "real-world" security guarantees [14]. In the standard model, it has even been shown [22] that the strongest flavor of selective-ID security (*i.e.*, the IND-sID-CCA one that captures chosen-ciphertext adversaries) does not even imply the weakest form of adaptive-ID security (which is the one-wayness against chosen-plaintext attacks).

OUR CONTRIBUTION. We describe an IBE scheme endowed with a similar and equally efficient revocation mechanism as in the BGK system while reaching security in the stronger *adaptive*-ID sense (as originally defined by Boneh and Franklin [8]), where adversaries choose the target identity in the challenge phase. We emphasize that, although relatively loose, the reduction is polynomial in the number of adversarial queries. Our construction uses the same binary tree structure as [4] and applies the same revocation technique. Instead of FIBE systems, we utilize a recently considered variant [28] of the Waters IBE [33]. To obtain a fairly simple security reduction, we use the property that the simulator is able to compute at least one private key for each identity. This notably brings out the fact that ordinary (as opposed to fuzzy) IBE systems can supersede the particular instance of FIBE scheme considered in [4] to achieve revocation. From an efficiency standpoint, our R-IBE performs essentially as well as the BGK construction.

ORGANIZATION. Section 2 first recalls the syntax and the security model of the R-IBE primitive. Section 3 explains the BGK revocation technique that we also use. Our scheme and its security analysis and then detailed in section 4.

2   Definitions

MODEL AND SECURITY DEFINITIONS. We recall the definition of R-IBE schemes and their security properties as defined in [4].

Definition 1. *An* identity-based encryption with efficient revocation, *or simply* Revocable IBE *(R-IBE) scheme is a 7-tuple* $(\mathcal{S}, \mathcal{SK}, \mathcal{KU}, \mathcal{DK}, \mathcal{E}, \mathcal{D}, \mathcal{R})$ *of efficient algorithms with associated message space* $\mathcal{M}$, *identity space* $\mathcal{I}$ *and time space* $\mathcal{T}$:

- *The* **Setup** *algorithm* $\mathcal{S}$ *is run by a key authority[1]. Given a security parameter* λ *and a maximal number of users* N, *it outputs a master public/secret key pair* $(\mathsf{mpk}, \mathsf{msk})$, *an initial state* st *and an empty revocation list* RL.
- *The stateful* **Private Key Generation** *algorithm* $\mathcal{SK}$ *is run by the key authority that takes as input the system master key pair* $(\mathsf{mpk}, \mathsf{msk})$, *an identity* $id \in \mathcal{I}$ *and state* st *and outputs a private key* d_{id} *and an updated state* st.
- *The* **Key Update Generation** *algorithm* $\mathcal{KU}$ *is used by the key authority. Given the master public and secret keys* $(\mathsf{mpk}, \mathsf{msk})$, *a key update time* $t \in \mathcal{T}$, *a revocation list* RL *and a state* st, *it publishes a key update* ku_t.
- *The* **Decryption Key Generation** *algorithm* $\mathcal{DK}$ *is run by the user. Given a private key* d_{id} *and a key update* ku_t, *it outputs a decryption key* $d_{id,t}$ *to be used during period* t *or a special symbol* $\perp$ *indicating that* id *was revoked.*
- *The randomized* **Encryption** *algorithm* $\mathcal{E}$ *takes as input the master public key* mpk, *an identity* $id \in \mathcal{I}$, *an encryption time* $t \in \mathcal{T}$, *and a message* $m \in \mathcal{M}$ *and outputs a ciphertext* c. *For simplicity and w.l.o.g. we assume that* id *and* t *are efficiently computable from* c.

---

[1] We follow [4] and call the trusted authority "key authority" instead of "PKG".

- *The deterministic **Decryption** algorithm $\mathcal{D}$ takes as input a decryption key $d_{id,t}$ and a ciphertext c, and outputs a message $m \in \mathcal{M}$ or a special symbol $\perp$ indicating that the ciphertext is invalid.*
- *The stateful **Revocation** algorithm $\mathcal{R}$ takes as input an identity to be revoked $id \in \mathcal{I}$, a revocation time $t \in \mathcal{T}$, a revocation list RL and state st, and outputs an updated revocation list RL.*

Correctness requires that, for any outputs $(\mathsf{mpk}, \mathsf{msk})$ of $\mathcal{S}$, any $m \in \mathcal{M}$, any $id \in \mathcal{I}$ and $t \in \mathcal{T}$, all possible states st and revocation lists RL, if id is not revoked by time t, then for $(d_{id}, st) \leftarrow \mathcal{SK}(\mathsf{mpk}, \mathsf{msk}, id, st)$, $ku_t \leftarrow \mathcal{KU}(\mathsf{mpk}, \mathsf{msk}, t, RL, st)$, $d_{id,t} \leftarrow \mathcal{DK}(d_{id}, ku_t)$ we have $\mathcal{D}(d_{id,t}, \mathcal{E}(\mathsf{mpk}, id, t, m)) = m$.

Boldyreva *et al.* formalized the *selective-revocable-ID* security property that captures the usual notion of selective-ID[2] security but also takes revocations into account. In addition to a private key generation oracle $\mathcal{SK}(.)$ that outputs private keys for identities of her choosing, the adversary is allowed to revoke users at will using a dedicated oracle $\mathcal{R}(.,.)$ (taking as input identities id and period numbers t) and can obtain key update information (which is assumed to be public) for any period t via queries $\mathcal{KU}(t)$ to another oracle. The following definition extends the security property expressed in [4] to the adaptive-ID setting.

Definition 2. *A R-IBE scheme is* revocable-ID *secure if any probabilistic polynomial time (PPT) adversary $\mathcal{A}$ has negligible advantage in this experiment:*

$$\boxed{\mathbf{Expt}_{\mathcal{A}}^{\text{IND-RID-CPA}}(\lambda)}$$

$(\mathsf{mpk}, \mathsf{msk}, RL, st) \leftarrow \mathcal{S}(\lambda, n)$
$(m_0, m_1, id^\star, t^\star, s) \leftarrow \mathcal{A}^{\mathcal{SK}(\cdot), \mathcal{KU}(\cdot), \mathcal{R}(\cdot, \cdot)}(\mathsf{find}, \mathsf{mpk})$
$d^\star \xleftarrow{R} \{0, 1\}$
$c^\star \leftarrow \mathcal{E}(\mathsf{mpk}, id^\star, t^\star, m_{d^\star})$
$d \leftarrow \mathcal{A}^{\mathcal{SK}(\cdot), \mathcal{KU}(\cdot), \mathcal{R}(\cdot, \cdot)}(\mathsf{guess}, s, c^\star)$
return 1 *if* $d = d^\star$ *and* 0 *otherwise.*

Beyond $m_0, m_1 \in \mathcal{M}$ and $|m_0| = |m_1|$, the following restrictions are made:

1. *$\mathcal{KU}(\cdot)$ and $\mathcal{R}(\cdot, \cdot)$ can be queried on time which is greater than or equal to the time of all previous queries i.e. the adversary is allowed to query only in non-decreasing order of time. Also, $\mathcal{R}(\cdot, \cdot)$ cannot[3] be queried on time t if $\mathcal{KU}(\cdot)$ was queried on t.*
2. *If $\mathcal{SK}(\cdot)$ was queried on identity $id^\star$ then $\mathcal{R}(\cdot, \cdot)$ must be queried on $(id^\star, t)$ for some $t \leq t^\star$.*

$\mathcal{A}$'s advantage is $\mathbf{Adv}_{\mathcal{A}}^{\text{IND-RID-CPA}}(\lambda) = \left| \Pr[\mathbf{Expt}_{\mathcal{A}}^{\text{IND-RID-CPA}}(\lambda) = 1] - \frac{1}{2} \right|$.

---

[2] Considered by Canetti, Halevi and Katz [15,16], this relaxed notion forces the adversary to choose the target identity before seeing the master public key.

[3] As in [4], we assume that revocations are made effective before that key updates are published at each time period. Otherwise, $\mathcal{A}$ could trivially win the game by corrupting and revoking $id^\star$ at period $t^\star$ but *after* having queried $\mathcal{KU}(t^\star)$.

This definition naturally extends to the *chosen-ciphertext* scenario where the adversary is further granted access to a decryption oracle $\mathcal{D}(\cdot)$ that, on input of a ciphertext c and a pair (id, t), runs $\mathcal{D}(d_{id,t}, c)$ to return some $m \in \mathcal{M}$ or $\perp$. Of course, $\mathcal{D}(\cdot)$ cannot be queried on the ciphertext $c^{\star}$ for the pair $(id^{\star}, t^{\star})$.

BILINEAR MAPS AND HARDNESS ASSUMPTIONS. We use prime order groups $(\mathbb{G}, \mathbb{G}_T)$ endowed with an efficiently computable map $e : \mathbb{G} \times \mathbb{G} \to \mathbb{G}_T$ such that:

1. $e(g^a, h^b) = e(g, h)^{ab}$ for any $(g, h) \in \mathbb{G} \times \mathbb{G}$ and $a, b \in \mathbb{Z}$;
2. $e(g, h) \neq 1_{\mathbb{G}_T}$ whenever $g, h \neq 1_{\mathbb{G}}$.

In such *bilinear groups*, we rely on a variant of the (now classical) Decision Bilinear Diffie-Hellman (DBDH) problem.

Definition 3. *Let $(\mathbb{G}, \mathbb{G}_T)$ be bilinear groups of prime order $p > 2^{\lambda}$ and $g \in \mathbb{G}$. The* **modified Decision Bilinear Diffie-Hellman Problem** *(mDBDH) is to distinguish the distributions $(g^a, g^b, g^c, e(g, g)^{bc/a})$ and $(g^a, g^b, g^c, e(g, g)^d)$ for random values $a, b, c, d \xleftarrow{R} \mathbb{Z}_p^*$. The advantage of a distinguisher $\mathcal{B}$ is*

$$\mathbf{Adv}_{\mathbb{G}, \mathbb{G}_T}^{\mathrm{mDBDH}}(\lambda) = \big| \Pr[a, b, c \xleftarrow{R} \mathbb{Z}_p^* : \mathcal{B}(g^a, g^b, g^c, e(g, g)^{bc/a}) = 1]$$
$$- \Pr[a, b, c, d \xleftarrow{R} \mathbb{Z}_p^* : \mathcal{B}(g^a, g^b, g^c, e(g, g)^d) = 1] \big|.$$

This problem is equivalent (see [17, Lemma 3.1] for a proof) to the original DBDH problem which is to tell apart $e(g, g)^{abc}$ from random given (g^a, g^b, g^c).

3   The BGK Construction

The idea of the scheme described by Boldyreva, Goyal and Kumar consists in assigning users to the leaves of a complete binary tree. Upon registration, the key authority provides them with a set of distinct private keys (all corresponding to their identity) for each node on the path from their associated leaf to the root of the tree. During period t, a given user's decryption key can be obtained by suitably combining any one of its node private keys with a key update for period t and associated with the *same* node of the tree.

At period t, the key authority publishes key updates for a set Y of nodes that contains no ancestors of revoked users and exactly one ancestor of any non-revoked one (so that, when no user is revoked, Y contains only the root node as illustrated on the figure where the nodes of Y are the squares). Then, a user assigned to leaf v is able to form an effective decryption key for period t if the set Y contains a node on the path from the root to v. By doing so, every update of the revocation list RL only requires the key authority to perform logarithmic work in the overall number of users. The size of users' private keys also logarithmically depends on the maximal number of users but, when the number of revoked users is reasonably small (as is likely to be the case in practice since one can simply re-initialize the whole system otherwise), the revocation method is much more efficient than the one initially suggested in [8].

Another attractive feature of this technique is that it can be used for temporary revocation. When a key is suspected of being compromised, the matching identity can be temporarily revoked while an investigation is conducted, and then reinstated if necessary.

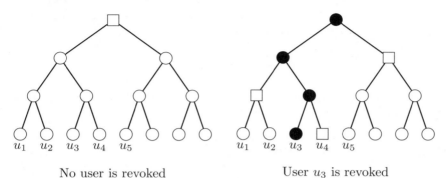

No user is revoked   User u_3 is revoked

The scheme of Boldyreva *et al.* builds on the fuzzy identity-based encryption (FIBE) primitive [31]. In FIBE systems, identities are seen as sets of descriptive attributes and users' keys can decrypt ciphertexts for which a certain threshold (called the "error tolerance") of attributes match between the ciphertext and the key. The private key of an identity (*i.e.*, a set of attributes) is generated using a new polynomial (of degree one less than the error tolerance) whose constant term is part of the master key of the scheme. The revocable IBE scheme of [4] uses a special kind of fuzzy IBE where ciphertexts are encrypted using the receiver's identity and the period number as "attributes". The decryption key of the receiver has to match both attributes to decrypt the ciphertext. For each node on the path from the root to its assigned leaf, the user is given a key attribute that is generated using a new polynomial of degree 1 for which the constant term is always the master secret. The same polynomials are used, for each node, to generate key updates. To compute a decryption key for period t, each user thus needs to combine two key attributes associated with the *same* node of the tree.

To date, existing FIBE schemes are only provably secure in the selective-ID sense and the construction of [4] has not been extended to the adaptive-ID model. As we will see, classical pairing-based IBE systems actually allow instantiating the same underlying revocation mechanism in the adaptive-ID setting.

4   An Adaptive-ID Secure Scheme

4.1   Intuition

We start from the same general idea as Boldyreva *et al.* but, instead of using fuzzy identity-based cryptosystems, we build on a recently suggested [28] variant of Waters' IBE [33] where, somewhat in the fashion of Gentry's IBE [24], the simulator is able to compute a decryption key for any identity in the security proof. In this variant, the master public key consists of $(X = g^x, Y, h) \in \mathbb{G}^3$ and a vector $\bar{u} = (u_0, u_1, \ldots, u_n) \in \mathbb{G}_1^{n+1}$ implementing Waters' "hashing" technique that maps strings $id = i_1 \ldots i_n \in \{0,1\}^n$ onto $F_u(id) = u_0 \cdot \prod_{j=1}^{n} u_i^{i_j}$. To derive a private key for the identity id, the authority picks $r, s \xleftarrow{R} \mathbb{Z}_p^*$ and sets

$$d_{id} = (d_1, d_2, d_3) = \left((Y \cdot h^r)^{1/x} \cdot F_u(id)^s, \ X^s, \ r\right)$$

so that $e(d_1, X) = e(Y, g) \cdot e(h, g)^{d_3} \cdot e(F(id), d_2)$. Ciphertexts are encrypted as

$$C = (C_0, C_1, C_2, C_3) = \left(m \cdot e(Y, g)^z, \ X^z, \ F_u(id)^z, \ e(g, h)^z\right)$$

and decrypted by computing $m = C_0 \cdot e(C_2, d_2) \cdot C_3^{d_3} / e(C_1, d_1)$ (the correctness can be checked by noting that $e(d_1, X)^z = e(Y, g)^z \cdot e(h, g)^{z d_3} \cdot e(F(id), d_2)^z)$.

We consider a two-level hierarchical extension of the above system where the second level identity is the period number. The shape of private keys thus becomes $(d_1, d_2, d_3, d_4) = \left((Y \cdot h^r)^{1/x} \cdot F_u(id)^{s_1} \cdot F_v(t)^{s_2}, \ X^{s_1}, \ X^{s_2}, \ r\right)$ for some function $F_v(t)$. Since we only need a polynomial number of time periods, we can settle for the Boneh-Boyen selective-ID secure identity hashing $F_v(id) = v_0^{id} \cdot v_1$ [5], for some $v_0, v_1 \in \mathbb{G}$, at level 2 (instead of Waters' technique).

Then, we also assign users to the leaves of a binary tree T. For each node $\theta \in \mathsf{T}$, the key authority splits $Y \in \mathbb{G}$ into new shares $Y_{1,\theta}, Y_{2,\theta}$ such that $Y = Y_{1,\theta} \cdot Y_{2,\theta}$. To derive users' private keys, the key authority computes a triple $(d_{1,\theta}, d_{2,\theta}, d_{3,\theta}) = \left((Y_{1,\theta} \cdot h^{r_{1,\theta}})^{1/x} \cdot F_u(id)^{s_{1,\theta}}, \ X^{s_{1,\theta}}, \ r_{1,\theta}\right)$ for each node θ on the path from the root to the leaf corresponding to the user. Key updates are triples $(ku_{1,\theta}, ku_{2,\theta}, ku_{3,\theta}) = \left((Y_{2,\theta} \cdot h^{r_{2,\theta}})^{1/x} \cdot F_v(t)^{s_{2,\theta}}, \ X^{s_{2,\theta}}, \ r_{2,\theta}\right)$ associated with non-revoked nodes $\theta \in \mathsf{T}$ during period t. Users' decryption keys can be obtained by combining any two such triples $(d_{1,\theta}, d_{2,\theta}, d_{3,\theta})$, $(ku_{1,\theta}, ku_{2,\theta}, ku_{3,\theta})$ for the same node θ. Revocation is handled as in [4], by having the key authority stop issuing key updates for nodes outside the set Y.

In the selective-ID sense, the binary tree technique of [4] can be applied to a 2-level extension of the Boneh-Boyen HIBE by sharing the master secret key in a two-out-of-two fashion, using new shares for each node. Directly extending the technique to the adaptive-ID setting with Waters' IBE is not that simple. In the security reduction of [33], the simulator does not know the master key or the private key of the target identity. The difficulty that we are faced with is that, at the first time that a tree node is involved in a private key query or a key update query, the simulator has to decide which one of the two master key shares it will have to know for that node. This is problematic when the target identity $id^\star$ is not known and has not been assigned a leaf yet: which share should be known actually depends on whether the considered node lies on the path connecting the target identity to the root of the tree. To address this issue, we used a variant of the Waters IBE where the simulator knows at least one valid decryption key for each identity[4] and can answer queries regardless of whether nodes are on the path from $id^\star$ to the root.

---

[4] After the completion of the paper, we noticed that a 2-level instance of the original Waters HIBE can be used and allows for shorter ciphertexts. As will be shown in an updated version of this work, it unfortunately ends up with an equally loose reduction since the simulator has to guess upfront which private key query (if any) will involve the target identity.

4.2   Description

The scheme uses the same binary tree structure as in [4] and we employ similar notations. Namely, root denotes the root node of the tree T. If v is a leaf node, we let Path(v) stand for the set of nodes on the path from v to root. Whenever θ is a non-leaf node, θ_l and θ_r respectively denote its left and right children.

In the description hereafter, we use the same node selection algorithm (called KUNodes) as in [4]. At each time period, this algorithm determines the smallest subset $\mathsf{Y} \subset \mathsf{T}$ of nodes that contains an ancestor of all leaves corresponding to non-revoked users. This minimal set precisely contains nodes for which key updates have to be publicized in such a way that only non-revoked users will be able to generate the appropriate decryption key for the matching period.

To identify the set Y, KUNodes takes as input the tree T, the revocation list RL and a period number t. It first marks (in black on the figure) all ancestors of users that were revoked by time t as revoked nodes. Then, it inserts in Y the non-revoked children of revoked nodes. Its formal specification is the following:

$$KUNodes(\mathsf{T}, RL, t)$$
$$\mathsf{X}, \mathsf{Y} \leftarrow \emptyset$$
$$\forall (v_i, t_i) \in RL$$
$$\quad \text{if } t_i \leq t \ \text{ then add Path}(v_i) \text{ to } \mathsf{X}$$
$$\forall \theta \in \mathsf{X}$$
$$\quad \text{if } \theta_l \notin \mathsf{X} \ \text{ then add } \theta_l \text{ to } \mathsf{Y}$$
$$\quad \text{if } \theta_r \notin \mathsf{X} \ \text{ then add } \theta_r \text{ to } \mathsf{Y}$$
$$\text{If } \mathsf{Y} = \emptyset \ \text{ then add root to } \mathsf{Y}$$
$$\text{Return } \mathsf{Y}$$

As in [4], we assume that the number of time periods t_{max} is polynomial in the security parameter λ, so that a degradation of $O(1/t_{max})$ in the security reduction is acceptable.

Setup $\mathcal{S}(\lambda, n, N)$: given security parameters $\lambda, n \in \mathbb{N}$ and a maximal number of users $N \in \mathbb{N}$ that the scheme must be prepared for, the key authority defines $\mathcal{I} = \{0,1\}^n$, $\mathcal{T} = \{1, \ldots, t_{max}\}$ and does the following.
  1. Select bilinear groups $(\mathbb{G}, \mathbb{G}_T)$ of prime order $p > 2^\lambda$ with $g \xleftarrow{R} \mathbb{G}^*$.
  2. Randomly choose $x \xleftarrow{R} \mathbb{Z}_p^*$, $h, Y \xleftarrow{R} \mathbb{G}^*$ as well as two random vectors $\overline{u} = (u_0, u_1, \ldots, u_n) \in \mathbb{G}^{*n+1}$ and $\overline{v} = (v_0, v_1) \in \mathbb{G}^{*2}$ that define functions $F_u : \mathcal{I} \rightarrow \mathbb{G}$, $F_v : \mathcal{T} \rightarrow \mathbb{G}$ such that, when $id = i_1 \ldots i_n \in \mathcal{I} = \{0,1\}^n$,

$$F_u(id) = u_0 \cdot \prod_{j=1}^{n} u_j^{i_j} \qquad\qquad F_v(t) = v_0^t \cdot v_1$$

  3. Set the master key as $\mathsf{msk} := x$ and initialize a revocation list $RL := \emptyset$ and a state $st = \mathsf{T}$ consisting of a binary tree T with $N < 2^n$ leaves.
  4. Define the master public key to be $\mathsf{mpk} := (X = g^x, Y, h, \overline{u}, \overline{v})$.

Private Key Generation $\mathcal{SK}(\mathsf{mpk}, \mathsf{msk}, id, st)$: Parse mpk as $(X, Y, h, \overline{u}, \overline{v})$, msk as x and st as T.

1. Choose an unassigned leaf v from T and associate it with $id \in \{0,1\}^n$.
2. For all nodes $\theta \in \mathsf{Path}(v)$ do the following:
   a. Retrieve $Y_{1,\theta}$ from T if it was defined[5]. Otherwise, choose it at random $Y_{1,\theta} \overset{R}{\leftarrow} \mathbb{G}$, set $Y_{2,\theta} = Y/Y_{1,\theta}$ and store the pair $(Y_{1,\theta}, Y_{2,\theta}) \in \mathbb{G}^2$ at node θ in $st = \mathsf{T}$.
   b. Pick $s_{1,\theta}, r_{1,\theta} \overset{R}{\leftarrow} \mathbb{Z}_p^*$ and set

$$d_{id,\theta} = (d_{1,\theta}, d_{2,\theta}, r_{1,\theta}) = \left((Y_{1,\theta} \cdot h^{r_{1,\theta}})^{1/x} \cdot F_u(id)^{s_{1,\theta}},\ X^{s_{1,\theta}},\ r_{1,\theta} \right).$$

3. Return $d_{id} = \{(\theta, d_{id,\theta})\}_{\theta \in \mathsf{Path}(v)}$ and the updated state $st = \mathsf{T}$.

Key Update Generation $\mathcal{KU}(\mathsf{mpk}, \mathsf{msk}, t, RL, st)$: Parse mpk as $(X, Y, h, \overline{u}, \overline{v})$, msk as x and st as T. For all nodes $\theta \in \mathsf{KUNodes}(\mathsf{T}, RL, t)$,

1. Fetch $Y_{2,\theta}$ from T if it was previously defined. If not, choose a fresh pair $(Y_{1,\theta}, Y_{2,\theta}) \in \mathbb{G}^2$ such that $Y = Y_{1,\theta} \cdot Y_{2,\theta}$ and store it in θ.
2. Choose $s_{2,\theta}, r_{2,\theta} \overset{R}{\leftarrow} \mathbb{Z}_p^*$ and compute

$$ku_{t,\theta} = (ku_{1,\theta}, ku_{2,\theta}, r_{2,\theta}) = \left((Y_{2,\theta} \cdot h^{r_{2,\theta}})^{1/x} \cdot F_v(t)^{s_{2,\theta}},\ X^{s_{2,\theta}},\ r_{2,\theta} \right).$$

Then, return $ku_t = \{(\theta, ku_{t,\theta})\}_{\theta \in \mathsf{KUNodes}(\mathsf{T}, RL, t)}$ and the updated $st = \mathsf{T}$.

Decryption Key Generation $\mathcal{DK}(\mathsf{mpk}, d_{id}, ku_t)$: Parse d_{id} into $\{(i, d_{id,i})\}_{i \in \mathsf{I}}$ and ku_t as $\{(j, ku_{t,j})\}_{j \in \mathsf{J}}$ for some sets of nodes $\mathsf{I}, \mathsf{J} \in \mathsf{T}$. If there exists no pair $(i,j) \in \mathsf{I} \times \mathsf{J}$ such that $i = j$, return $\bot$. Otherwise, choose an arbitrary such pair $i = j$, parse $d_{id,i} = (d_{1,i}, d_{2,i}, r_{1,i})$, $ku_{t,i} = (ku_{1,i}, ku_{2,i}, r_{2,i})$ and set the updated decryption key as

$$d_{id,t} = \left(d_{t,1},\ d_{t,2},\ d_{t,3},\ d_{t,4} \right) = \left(d_{1,i} \cdot ku_{1,i},\ d_{2,i},\ ku_{2,i},\ r_{1,i} + r_{2,i} \right)$$
$$= \left((Y \cdot h^{d_{t,4}})^{1/x} \cdot F_u(id)^{s_{1,i}} \cdot F_v(t)^{s_{2,i}},\ X^{s_{1,i}},\ X^{s_{2,i}}, d_{t,4} \right).$$

Finally, check that $d_{id,t}$ satisfies

$$e(d_{t,1}, X) = e(Y, g) \cdot e(g, h)^{d_{t,4}} \cdot e(F_u(id), d_{t,2}) \cdot e(F_v(t), d_{t,3}) \qquad (1)$$

and return $\bot$ if the above condition fails to hold. Otherwise return $d_{id,t}$.

Encryption $\mathcal{E}(\mathsf{mpk}, id, t, m)$: to encrypt $m \in \mathbb{G}_T$ for $id = i_1 \dots i_n \in \{0,1\}^n$ during period t, choose $z \overset{R}{\leftarrow} \mathbb{Z}_p^*$ and compute

$$C = \left(id, t, C_0, C_1, C_2, C_3, C_4 \right)$$
$$= \left(id,\ t,\ m \cdot e(g, Y)^z,\ X^z,\ F_u(id)^z,\ F_v(t)^z,\ e(g, h)^z \right).$$

---

[5] To avoid having to store $Y_{1,\theta}$ for each node, the authority can derive it from a pseudo-random function of θ using a shorter seed and re-compute it when necessary.

Decryption $\mathcal{D}(\text{mpk}, d_{id,t}, C)$: Parse C as $\big(id, t, C_0, C_1, C_2, C_3, C_4\big)$ and the decryption key $d_{id,t}$ as $(d_{t,1}, d_{t,2}, d_{t,3}, d_{t,4})$. Then, compute and return

$$m = C_0 \cdot \Big(\frac{e(C_1, d_{t,1})}{e(C_2, d_{t,2}) \cdot e(C_3, d_{t,3}) \cdot C_4^{d_{t,4}}} \Big)^{-1}. \tag{2}$$

Revocation $\mathcal{R}(\text{mpk}, id, t, RL, st)$: let v be the leaf node associated with id. To revoke the latter at period t, add (v, t) to RL and return the updated RL.

CORRECTNESS. We know that well-formed decryption keys always satisfy relation (1). If we raise both members of (1) to the power $z \in \mathbb{Z}_p^*$ (*i.e.*, the encryption exponent), we see that the quotient of pairings in (2) actually equals $e(g, Y)^z$.

EFFICIENCY. The efficiency of the scheme is comparable to that of the revocable IBE described in [4]: ciphertexts are only slightly longer (by an extra element of $\mathbb{G}_T$) and decryption is even slightly faster since it incurs the evaluation of a product of only 3 pairings (against 4 in [4]). Both schemes feature the same logarithmic complexity in the number of users in terms of private key size and space/computational cost for issuing key updates.

4.3   Security

The security proof is based on the one of [28] with the difference that we have to consider the case where the challenge identity is compromised at some point but revoked for the period during which the challenge ciphertext is created.

Theorem 1. *Let us assume that an IND-RID-CPA adversary $\mathcal{A}$ runs in time ζ and makes at most q private key queries over t_{max} time periods. Then, there exists an algorithm $\mathcal{B}$ solving the mDBDH problem with advantage $\mathbf{Adv}_{\mathcal{B}}^{\text{mDBDH}}(\lambda)$ and within running time $O(\zeta) + O(\varepsilon^{-2} \ln \delta^{-1})$ for sufficiently small ε and δ. The advantage of $\mathcal{A}$ is then bounded by*

$$\mathbf{Adv}_{\mathcal{A}}^{\text{IND-RID-CPA}}(\lambda) \leq 4 \cdot t_{max} \cdot q^2 \cdot (n+1) \cdot \Big(4 \cdot \mathbf{Adv}^{\text{mDBDH}}(\lambda) + \delta \Big). \tag{3}$$

Proof (sketch). The complete proof is deferred to the full version of the paper due to space limitation but we give its intuition here. We construct a simulator $\mathcal{B}$ that is given a tuple (g^a, g^b, g^c, T) and uses the adversary $\mathcal{A}$ to decide if $T = e(g, g)^{bc/a}$. The master public key is prepared as $X = g^a$, $h = g^b$ and $Y = X^\gamma \cdot h^{-r^\star}$ for random $\gamma, r^\star \xleftarrow{R} \mathbb{Z}_p^*$. The vector $\overline{v} = (v_0, v_1)$ is chosen so that $F_v(t) = g^{\beta(t-t^\star)} \cdot X^\alpha$ for random values $\alpha, \beta \xleftarrow{R} \mathbb{Z}_p^*$ and where $t^\star \xleftarrow{R} \{1, \ldots, t_{max}\}$ is chosen at random as a guess for the time period of the challenge phase. Finally, the $(n+1)$-vector $\overline{u}$ is chosen so as to have $F_u(id) = g^{J(id)} \cdot X^{K(id)}$ for some integer-valued functions $J, K : \{0, 1\}^n \to \mathbb{Z}$ chosen by the simulator according to Waters' technique [33].

To be successful, $\mathcal{B}$ needs to have $J(id^\star) = 0$ in the challenge phase and, by choosing $\overline{u}$ such that $J(.)$ is relatively small in absolute value, this will be the case with probability $O(1/q(n+1))$. The simulator also hopes that $\mathcal{A}$ will indeed make her challenge query for period $t^\star$, which occurs with probability $1/t_{max}$. The security proof relies on the fact that, with non-negligible probability, $\mathcal{B}$ can compute a valid decryption key for each identity $id \in \{0,1\}^n$. If $J(id) \neq 0$, $\mathcal{B}$ can do it using the Boneh-Boyen technique [5] while, in the case $J(id) = 0$, a valid key $d_{id,t}$ for period t is obtained by choosing $s_1, s_2 \xleftarrow{R} \mathbb{Z}_p^*$ and setting

$$(d_{t,1}, d_{t,2}, d_{t,3}, d_{t,4}) = \big(g^\gamma \cdot F_u(id)^{s_1} \cdot F_v(t)^{s_2}, \; X^{s_1}, \; X^{s_2}, \; r^\star\big), \qquad (4)$$

which has the required shape since $(Y \cdot h^{r^\star})^{1/a} = g^\gamma$.

In the challenge phase, when $\mathcal{A}$ hopefully comes up with a pair $(id^\star, t^\star)$ such that $J(id^\star) = 0$ and $t^\star$ is the expected time period, $\mathcal{B}$ flips a coin $d^\star \xleftarrow{R} \{0,1\}$ and constructs the ciphertext $C^\star$ as follows:

$$C_1^\star = g^c \qquad C_2^\star = (g^c)^{K(id^\star)} \qquad C_3^\star = (g^c)^\alpha \qquad C_4^\star = T$$

$$C_0^\star = m_{d^\star} \cdot \frac{e(C_1^\star, d_{t^\star,1})}{e(C_2^\star, d_{t^\star,2}) \cdot e(C_3^\star, d_{t^\star,3}) \cdot C_4^{\star d_{t^\star,4}}} \qquad (5)$$

where $d_{id^\star,t^\star} = (d_{t^\star,1}, d_{t^\star,2}, d_{t^\star,3}, d_{t^\star}, 4)$ is a valid decryption key calculated as per (4) for the pair $(id^\star, t^\star)$. If T actually equals $e(g,g)^{bc/a}$, $C^\star$ is easily seen to be a valid encryption of $m_{d^\star}$ using the encryption exponent $z = c/a$. If T is random on the other hand (say $T = e(g,h)^{z'}$ for a random $z' \in_R \mathbb{Z}_p^*$), we can check that $C_0^\star = m_{d^\star} \cdot e(Y,g)^z \cdot e(g,h)^{(z-z')r^\star}$, which means that $m_{d^\star}$ is perfectly hidden from $\mathcal{A}$'s view as long as $r^\star$ is so.

We now have to make sure that no information on $r^\star$ ever leaks during the game. To do so, we distinguish two kinds of adversaries:

- Type I adversaries choose to be challenged on an identity $id^\star$ that is corrupted at some point of the game but is revoked at period $t^\star$ or before.
- Type II adversaries do not corrupt the target identity $id^\star$ at any time.

At the outset of the game, the simulator $\mathcal{B}$ flips a coin $c_{mode} \xleftarrow{R} \{0,1\}$ as a guess for the type of adversarial behavior that it will be faced with. In the expectation of Type I adversary (i.e., $c_{mode} = 0$), $\mathcal{B}$ additionally has to guess which private key query will involve the identity $id^\star$ that $\mathcal{A}$ chooses to be challenged upon. If $c_{mode} = 0$, it thus draws $j^\star \xleftarrow{R} \{1, \ldots, q\}$ at the beginning of the game and the input id_j of the jth private key query happens to be $id^\star$ with probability $1/q$.

Regardless of the value of c_{mode}, for each tree node $\theta \in \mathsf{T}$, $\mathcal{B}$ splits the public key element $Y \in \mathbb{G}$ into two node-specific multiplicative shares $(Y_{1,\theta}, Y_{2,\theta})$ such that $Y = Y_{1,\theta} \cdot Y_{2,\theta}$. That is, at the first time that a node $\theta \in \mathsf{T}$ is involved in some query, $\mathcal{B}$ defines and stores exponents $\gamma_{1,\theta}, \gamma_{2,\theta}, r_{1,\theta}^\star, r_{2,\theta}^\star$ such that $\gamma = \gamma_{1,\theta} + \gamma_{2,\theta}$, $r^\star = r_{1,\theta}^\star + r_{2,\theta}^\star$ and defines $Y_{1,\theta} = X^{\gamma_{1,\theta}} \cdot h^{-r_{1,\theta}^\star}$, $Y_{2,\theta} = X^{\gamma_{2,\theta}} \cdot h^{-r_{2,\theta}^\star}$.

From here on, we assume that $\mathcal{B}$ is fortunate in the random guesses that it makes (i.e., $c_{mode} \in \{0,1\}$ and $j^\star \xleftarrow{R} \{1, \ldots, q\}$ if $c_{mode} = 0$). Then, the treatment of $\mathcal{A}$'s queries is the following. Revocation queries are dealt with by following the specification of the revocation algorithm that simply inserts the appropriate node trees in the revocation list RL. The way to answer other queries now depends on the bit c_{mode}.

- If $c_{mode} = 0$, $\mathcal{B}$ uses the following strategy.

  - $\mathcal{SK}(.)$ queries: let id_j be the input of the jth private key query and let $v \in \mathsf{T}$ be the node that $\mathcal{B}$ assigns to id_j.
    - If $j \neq j^\star$, for each node $\theta \in \mathsf{Path}(v)$, $\mathcal{B}$ re-computes $Y_{1,\theta} = X^{\gamma_{1,\theta}} \cdot h^{-r^\star_{1,\theta}}$ using the shares $(\gamma_{1,\theta}, \gamma_{2,\theta}, r^\star_{1,\theta}, r^\star_{2,\theta})$. It picks $r_{1,\theta}, s_{1,\theta} \xleftarrow{R} \mathbb{Z}_p^*$, defines $W = Y_{1,\theta} \cdot h^{r_{1,\theta}}$ and calculates

    $$(d_{1,\theta}, d_{2,\theta}) = \left(F_u(id_j)^{s_{1,\theta}} \cdot W^{-\frac{K(id_j)}{J(id_j)}}, \ X^{s_{1,\theta}} \cdot W^{-\frac{1}{J(id_j)}} \right) \qquad (6)$$

    which is well-defined since $J(id_j) \neq 0$ and can be checked to provide a correctly-shaped triple $d_{id_j,\theta} = (d_{1,\theta}, d_{2,\theta}, r_{1,\theta})$ for node θ if we set $s_{\tilde{1},\theta} = s_{1,\theta} - w/(aJ(id_j))$ where $w = \log_g(W)$. Indeed,

    $$W^{1/a} \cdot F_u(id_j)^{s_{\tilde{1},\theta}} = W^{1/a} \cdot F_u(id_j)^{s_{1,\theta}} \cdot (g^{J(id_j)} \cdot X^{K(id_j)})^{-\frac{w}{aJ(id_j)}}$$
    $$= F_u(id_j)^{s_{1,\theta}} \cdot W^{-\frac{K(id_j)}{J(id_j)}}$$

    and $X^{s_{\tilde{1},\theta}} = X^{s_{1,\theta}} \cdot W^{-\frac{1}{J(id_j)}}$. In this case, for all nodes $\theta \in \mathsf{Path}(v)$, the share $r^\star_{1,\theta}$ remains perfectly hidden from $\mathcal{A}$'s view.
    - If $j = j^\star$ (and thus $id_j = id^\star$ if $\mathcal{B}$ was lucky when choosing $j^\star$), for each node $\theta \in \mathsf{Path}(v)$, $\mathcal{B}$ picks a random $s_{1,\theta} \xleftarrow{R} \mathbb{Z}_p^*$ and uses the shares $(\gamma_{1,\theta}, \gamma_{2,\theta}, r^\star_{1,\theta}, r^\star_{2,\theta})$ to compute a triple $d_{id_j,\theta} = (d_{1,\theta}, d_{2,\theta}, r^\star_{1,\theta})$ where

    $$(d_{1,\theta}, d_{2,\theta}) = (g^{\gamma_{1,\theta}} \cdot F_u(id_j)^{s_{1,\theta}}, \ X^{s_{1,\theta}})$$

    We see that $d_{id_j,\theta}$ is well-formed since $(Y_{1,\theta} \cdot h^{r^\star_{1,\theta}})^{1/a} = g^{\gamma_{1,\theta}}$. In this case, the shares $\{r^\star_{1,\theta}\}_{\theta \in \mathsf{Path}(v)}$ are revealed to $\mathcal{A}$ as part of $d_{id_j,\theta}$.
  - $\mathcal{KU}(.)$ queries:
    - For periods $t \neq t^\star$, $\mathcal{B}$ runs $\mathsf{KUNodes}(\mathsf{T}, RL, t)$ to find the right set Y of non-revoked nodes. For each $\theta \in \mathsf{Y}$, $\mathcal{B}$ re-constructs $Y_{2,\theta} = X^{\gamma_{2,\theta}} \cdot h^{-r^\star_{2,\theta}}$ using the shares $(\gamma_{1,\theta}, \gamma_{2,\theta}, r^\star_{1,\theta}, r^\star_{2,\theta})$. It sets $W = Y_{2,\theta} \cdot h^{r_{2,\theta}}$ for a random $r_{2,\theta} \xleftarrow{R} \mathbb{Z}_p^*$. Then, it picks $s_{2,\theta} \xleftarrow{R} \mathbb{Z}_p^*$ and computes

    $$(ku_{1,\theta}, ku_{2,\theta}) = \left(F_v(t)^{s_{2,\theta}} \cdot W^{-\frac{\alpha}{\beta(t-t^\star)}}, \ X^{s_{2,\theta}} \cdot W^{-\frac{1}{\beta(t-t^\star)}} \right),$$

    which is well-defined since $F_v(t) = g^{\beta(t-t^\star)} \cdot X^\alpha$ and $t \neq t^\star$ and, if we define $s_{\tilde{2},\theta} = s_{2,\theta} - w/(\beta a(t - t^\star))$ (with $w = \log_g(W)$), we have

    $$W^{1/a} \cdot F_v(t)^{s_{\tilde{2},\theta}} = W^{1/a} \cdot F_v(t)^{s_{2,\theta}} \cdot (g^{\beta(t-t^\star)} \cdot X^\alpha)^{-\frac{w}{\beta a(t-t^\star)}}$$
    $$= F_v(t)^{s_{2,\theta}} \cdot W^{-\frac{\alpha}{\beta(t-t^\star)}}$$

    and $X^{s_{\tilde{2},\theta}} = X^{s_{2,\theta}} \cdot W^{-\frac{1}{\beta(t-t^\star)}}$. Finally, $\mathcal{B}$ returns $\{(ku_{1,\theta}, ku_{2,\theta}, r_{2,\theta})\}_{\theta \in \mathsf{Y}}$ and, for all nodes $\theta \in \mathsf{Y}$, the share $r^\star_{2,\theta}$ remains perfectly hidden.

. For period $t = t^\star$, $\mathcal{B}$ determines the set $\mathsf{Y} \in \mathsf{T}$ of non-revoked nodes using $\mathsf{KUNodes}(\mathsf{T}, RL, t)$. For each $\theta \in \mathsf{Y}$, $\mathcal{B}$ uses the shares $(\gamma_{1,\theta}, \gamma_{2,\theta}, r_{1,\theta}^\star, r_{2,\theta}^\star)$ to construct $ku_{t^\star,\theta}$ as the triple $ku_{t^\star,\theta} = (ku_{1,\theta}, ku_{2,\theta}, r_{2,\theta}^\star)$ where

$$(ku_{1,\theta}, ku_{2,\theta}) = \left(g^{\gamma_{2,\theta}} \cdot F_v(t^\star)^{s_{2,\theta}}, \; X^{s_{2,\theta}}\right)$$

for a random $s_{2,\theta} \xleftarrow{R} \mathbb{Z}_p^\star$. This pair has the correct distribution since $(Y_{2,\theta} \cdot h^{r_{2,\theta}^\star})^{1/a} = g^{\gamma_{2,\theta}}$. In this case, shares $\{r_{2,\theta}^\star\}_{\theta \in \mathsf{Y}}$ are given away.

By inspection, we check that, with non-negligible probability, $\mathcal{B}$ never has to reveal two complementary shares $r_{1,\theta}^\star, r_{2,\theta}^\star$ of $r^\star$ for any node $\theta \in \mathsf{T}$. Let $v^\star$ be the leaf that $\mathcal{B}$ assigns to the target identity $id^\star$ (which is also $id_{j^\star}$ with probability $1/q$). For all $\theta \in \mathsf{Path}(v^\star)$, $\mathcal{A}$ never sees both $r_{1,\theta}^\star$ and $r_{2,\theta}^\star$ because, according to the rules of definition 2, $id^\star$ must be revoked by period $t^\star$ if $\mathcal{A}$ decides to corrupt it at some point. Then, no ancestor of $v^\star$ lies in the set Y determined by $\mathsf{KUNodes}$ at period $t^\star$.

• The case $c_{mode} = 1$ is easier to handle. Recall that, if $\mathcal{A}$ indeed behaves as a Type II adversary, it does not query the private key of $id^\star$ at any time.

  - $\mathcal{SK}(.)$ queries: let id be the queried identity. We must have $J(id) \neq 0$ with non-negligible probability and $\mathcal{B}$ can compute a private key as suggested by relation (6) in the case $c_{mode} = 0$. In particular, the value $r_{1,\theta}^\star$ does not leak for any $\theta \in \mathsf{Path}(v)$ where v is the leaf associated with id.
  - $\mathcal{KU}(.)$ queries are processed exactly as in the case $c_{mode} = 0$. Namely, $\mathcal{B}$ distinguishes the same situations $t \neq t^\star$ and $t = t^\star$ and only reveals $r_{2,\theta}^\star$, for non-revoked nodes $\theta \in \mathsf{Y}$, when generating updates for period $t = t^\star$.

Again, the simulator does not reveal both $r_{1,\theta}^\star$ and $r_{2,\theta}^\star$ for any node since $r_{1,\theta}^\star$ is never used to answer private key queries.

With non-negligible probability, the value $r^\star$ thus remains independent of $\mathcal{A}$'s view for either value of $c_{mode} \in \{0, 1\}$. This completes the outline of the proof, which is more thoroughly detailed the full paper. $\square$

As we mentioned earlier, the reduction leaves room for improvement as its quadratic degradation factor q^2 becomes cubic since we must have $t_{max} \geq q$ to tolerate a polynomial number $O(q)$ of revocation queries. Although loose, the reduction is polynomial and thus solves the problem left open in [4].

Chosen-ciphertext security can be efficiently achieved using the usual techniques [16,11,12] or, since the outlined simulator knows a valid private key for each identity as in [24], in the fashion of Cramer-Shoup [19].

5   Conclusion

We showed that regular IBE schemes can be used to implement the efficient revocation mechanism suggested by Boldyreva $et\,al.$ and notably provide the first adaptive-ID secure revocable IBE. The latter was obtained by sharing the key generation process of a 2-level HIBE system from the "commutative-blinding family" (initiated with the first scheme of [5]). As another extension, the same ideas make it possible to construct revocable identity-based broadcast encryption schemes (using the recent Boneh-Hamburg constructions [10] for instance) in the selective-ID model.

An open problem is to devise adaptive-ID secure R-IBE systems with a tighter reduction than what we could obtain. It would also be interesting to see how revocation can be handled in the context of hierarchical IBE [26,23], where each entity of the hierarchy should be responsible for revoking its children.

Acknowledgements

We thank the reviewers for their comments.

References

1. Aiello, W., Lodha, S.P., Ostrovsky, R.: Fast Digital Identity Revocation. In: Krawczyk, H. (ed.) CRYPTO 1998. LNCS, vol. 1462, pp. 137–152. Springer, Heidelberg (1998)
2. Bellare, M., Rogaway, P.: Random oracles are practical: A paradigm for designing efficient protocols. In: ACM CCS, pp. 62–73 (1993)
3. Baek, J., Zheng, Y.: Identity-Based Threshold Decryption. In: Bao, F., Deng, R., Zhou, J. (eds.) PKC 2004. LNCS, vol. 2947, pp. 262–276. Springer, Heidelberg (2004)
4. Boldyreva, A., Goyal, V., Kumar, V.: Identity-Based Encryption with Efficient Revocation. In: ACM-CCS 2008 (2008)
5. Boneh, D., Boyen, X.: Efficient Selective-ID Secure Identity-Based Encryption Without Random Oracles. In: Cachin, C., Camenisch, J.L. (eds.) EUROCRYPT 2004. LNCS, vol. 3027, pp. 223–238. Springer, Heidelberg (2004)
6. Boneh, D., Boyen, X.: Secure Identity-Based Encryption Without Random Oracles. In: Franklin, M. (ed.) CRYPTO 2004. LNCS, vol. 3152, pp. 443–459. Springer, Heidelberg (2004)
7. Boneh, D., Ding, X., Tsudik, G., Wong, M.: A Method for Fast Revocation of Public Key Certificates and Security Capabilities. In: 10th USENIX Security Symposium, pp. 297–308 (2001)
8. Boneh, D., Franklin, M.: Identity-Based Encryption from the Weil Pairing. SIAM Journal of Computing 32(3), 586–615 (2003); earlier version in: Kilian, J. (ed.) CRYPTO 2001. LNCS, vol. 2139, p. 213–229. Springer, Heidelberg (2001)
9. Boneh, D., Gentry, C., Hamburg, M.: Space-Efficient Identity-Based Encryption Without Pairings. In: FOCS 2007, pp. 647–657 (2007)
10. Boneh, D., Hamburg, M.: Generalized Identity Based and Broadcast Encryption Schemes. In: Asiacrypt 2008. LNCS, vol. 5350, pp. 455–470 (2008)
11. Boneh, D., Katz, J.: Improved Efficiency for CCA-Secure Cryptosystems Built Using Identity-Based Encryption. In: Menezes, A. (ed.) CT-RSA 2005. LNCS, vol. 3376, pp. 87–103. Springer, Heidelberg (2005)
12. Boyen, X., Mei, Q., Waters, B.: Direct Chosen Ciphertext Security from Identity-Based Techniques. In: ACM CCS 2005, pp. 320–329 (2005)
13. Boyen, X.: A Tapestry of Identity-Based Encryption: Practical Frameworks Compared. Int. J. Applied Cryptography (IJACT) 1(1) (2008)
14. Canetti, R., Goldreich, O., Halevi, S.: The random oracle methodology, revisited. Journal of the ACM 51(4), 557–594 (2004); Earlier version in: STOC 1998 (1998)
15. Canetti, R., Halevi, S., Katz, J.: A Forward-Secure Public-Key Encryption Scheme. In: Biham, E. (ed.) EUROCRYPT 2003. LNCS, vol. 2656, pp. 254–271. Springer, Heidelberg (2003)

16. Canetti, R., Halevi, S., Katz, J.: Chosen-Ciphertext Security from Identity-Based Encryption. In: Cachin, C., Camenisch, J.L. (eds.) EUROCRYPT 2004. LNCS, vol. 3027, pp. 207–222. Springer, Heidelberg (2004)
17. Canetti, R., Hohenberger, S.: Chosen-Ciphertext Secure Proxy Re-Encryption. In: ACM CCS 2007, pp. 185–194. ACM Press, New York (2007)
18. Cocks, C.: An Identity-Based Encryption Scheme Based on Quadratic Residues. In: Honary, B. (ed.) Cryptography and Coding 2001. LNCS, vol. 2260, pp. 360–363. Springer, Heidelberg (2001)
19. Cramer, R., Shoup, V.: A Practical Public-Key Cryptosystem Provably Secure Against Adaptive Chosen Ciphertext Attack. In: Krawczyk, H. (ed.) CRYPTO 1998. LNCS, vol. 1462, pp. 13–25. Springer, Heidelberg (1998)
20. Ding, X., Tsudik, G.: Simple Identity-Based Cryptography with Mediated RSA. In: Joye, M. (ed.) CT-RSA 2003. LNCS, vol. 2612, pp. 193–210. Springer, Heidelberg (2003)
21. Elwailly, F., Gentry, C., Ramzan, Z.: QuasiModo: Efficient Certificate Validation and Revocation. In: Bao, F., Deng, R., Zhou, J. (eds.) PKC 2004. LNCS, vol. 2947, pp. 375–388. Springer, Heidelberg (2004)
22. Galindo, D.: A Separation Between Selective and Full-Identity Security Notions for Identity-Based Encryption. In: Gavrilova, M.L., Gervasi, O., Kumar, V., Tan, C.J.K., Taniar, D., Laganá, A., Mun, Y., Choo, H. (eds.) ICCSA 2006. LNCS, vol. 3982, pp. 318–326. Springer, Heidelberg (2006)
23. Gentry, C., Silverberg, A.: Hierarchical ID-Based Cryptography. In: Zheng, Y. (ed.) ASIACRYPT 2002. LNCS, vol. 2501, pp. 548–566. Springer, Heidelberg (2002)
24. Gentry, C.: Practical Identity-Based Encryption Without Random Oracles. In: Vaudenay, S. (ed.) EUROCRYPT 2006. LNCS, vol. 4004, pp. 445–464. Springer, Heidelberg (2006)
25. Goyal, V.: Certificate Revocation Using Fine Grained Certificate Space Partitioning. In: Dietrich, S., Dhamija, R. (eds.) FC 2007 and USEC 2007. LNCS, vol. 4886, pp. 247–259. Springer, Heidelberg (2007)
26. Horwitz, J., Lynn, B.: Toward hierarchical identity-based encryption. In: Knudsen, L.R. (ed.) EUROCRYPT 2002. LNCS, vol. 2332, pp. 466–481. Springer, Heidelberg (2002)
27. Libert, B., Quisquater, J.-J.: Efficient revocation and threshold pairing based cryptosystems. In: PODC 2003, pp. 163–171. ACM Press, New York (2003)
28. Libert, B., Vergnaud, D.: Towards Black-Box Accountable Authority IBE with Short Ciphertexts and Private Keys. In: Jarecki, S., Tsudik, G. (eds.) PKC 2009. LNCS, vol. 5443, pp. 235–255. Springer, Heidelberg (2009)
29. Micali, S.: Efficient Certificate Revocation. Technical Report MIT/LCS/TM-542-b (1996)
30. Micali, S.: Novomodo: Scalable Certificate Validation and Simplified PKI Management. In: PKI Research Workshop (2002)
31. Sahai, A., Waters, B.: Fuzzy Identity-Based Encryption. In: Cramer, R. (ed.) EUROCRYPT 2005. LNCS, vol. 3494, pp. 457–473. Springer, Heidelberg (2005)
32. Shamir, A.: Identity-Based Cryptosystems and Signature Schemes. In: Blakely, G.R., Chaum, D. (eds.) CRYPTO 1984. LNCS, vol. 196, pp. 47–53. Springer, Heidelberg (1985)
33. Waters, B.: Efficient Identity-Based Encryption Without Random Oracles. In: Cramer, R. (ed.) EUROCRYPT 2005. LNCS, vol. 3494, pp. 114–127. Springer, Heidelberg (2005)

An Efficient Encapsulation Scheme from Near Collision Resistant Pseudorandom Generators and Its Application to IBE-to-PKE Transformations

Takahiro Matsuda[1], Goichiro Hanaoka[2], Kanta Matsuura[1], and Hideki Imai[2,3]

[1] The University of Tokyo, Tokyo, Japan
{tmatsuda,kanta}@iis.u-tokyo.ac.jp
[2] National Institute of Advanced Industrial Science and Technology, Tokyo, Japan
{hanaoka-goichiro,h-imai}@aist.go.jp
[3] Chuo University, Tokyo, Japan

Abstract. In [14], Boneh and Katz introduced a primitive called *encapsulation* scheme, which is a special kind of commitment scheme. Using the encapsulation scheme, they improved the generic transformation by Canetti, Halevi, and Katz [17] which transforms any semantically secure identity-based encryption (IBE) scheme into a chosen-ciphertext secure public key encryption (PKE) scheme (we call the *BK transformation*). The ciphertext size of the transformed PKE scheme directly depends on the parameter sizes of the underlying encapsulation scheme. In this paper, by designing a size-efficient encapsulation scheme, we further improve the BK transformation. With our proposed encapsulation scheme, the ciphertext overhead of a transformed PKE scheme via the BK transformation can be that of the underlying IBE scheme plus 384-bit, while the original BK scheme yields that of the underlying IBE scheme plus at least 704-bit, for 128-bit security. Our encapsulation scheme is constructed from a pseudorandom generator (PRG) that has a special property called *near collision resistance*, which is a fairly weak primitive. As evidence of it, we also address how to generically construct a PRG with such a property from any one-way permutation.

Keywords: encapsulation, pseudorandom generator, public key encryption, identity-based encryption, IND-CCA security.

1 Introduction

1.1 Background

Studies on constructing and understanding efficient public key encryption (PKE) schemes secure against chosen ciphertext attacks (CCA) [38,22] are important research topics in the area of cryptography. Among several approaches towards CCA secure PKE schemes, one of the promising approaches is the "IBE-to-PKE" transformation paradigm [17], which is a method to obtain CCA secure PKE schemes from identity-based encryption (IBE) schemes [42,13].

M. Fischlin (Ed.): CT-RSA 2009, LNCS 5473, pp. 16–31, 2009.

In [17], Canetti, Halevi, and Katz showed a generic construction of CCA secure PKE schemes from any semantically secure IBE and a one-time signature (we call this IBE-to-PKE transformation the *CHK transformation*). This construction is fairly simple. Specifically, its ciphertext consists of (χ, vk, σ) where χ is a ciphertext of the underlying IBE scheme (under identity "*vk*"), vk is a verification of a one-time signature scheme, and σ is a valid signature of χ (under verification key vk). However, due to the use of a one-time signature, ciphertext size of the resulting scheme becomes larger than that of the underlying IBE scheme for $|vk|$ and $|\sigma|$, which might result in significantly large ciphertexts.

This method was later improved by Boneh and Katz [14] (we call the *BK transformation*) by replacing a one-time signature in the CHK transformation with an *encapsulation* scheme and a message authentication code (MAC) scheme, where an encapsulation scheme (the notion of which is introduced in the same paper [14]) is a special kind of commitment scheme that commits a random value. This method has a possibility of drastically reducing computation costs for encryption and decryption algorithms and ciphertext size of the transformed PKE scheme, compared to the CHK transformation. However, its ciphertext size directly depends on the size of parameters (commitment, decommitment, and the committed value) of the underlying encapsulation scheme, and thus an encapsulation scheme with large parameters still yields a large ciphertext for a transformed PKE scheme. Since the concrete encapsulation scheme that Boneh and Katz presented in [14] (we call the *BK encapsulation scheme*) had somewhat large parameters, PKE schemes transformed via the BK transformation could not be as size-efficient as existing practical CCA secure PKE schemes, e.g. [19,31].

Thus, there is still a room for further improvement for the BK transformation in terms of ciphertext size, by designing an encapsulation scheme with small parameter sizes.

1.2   Our Contribution

In this paper, focusing on the size-efficiency of the BK transformation, we present an efficient encapsulation scheme. Specifically, for 128-bit security, the ciphertext overhead (the difference of size between the whole ciphertext and its plaintext) of a PKE scheme obtained via the BK transformation with our encapsulation scheme can be that of the underlying IBE scheme plus 384-bit, while that of a PKE scheme via the BK transformation with their encapsulation scheme needs to be that of the underlying IBE scheme plus at least 704-bit.

The main building block used in the proposed encapsulation scheme is a pseudorandom generator (PRG) with a special property called *near collision resistance for predetermined parts of output* (NCR for short), which was first introduced and used by Boldyreva and Fischlin in [10]. Roughly speaking, NCR property is target collision resistance [34,7] for some part of output. In this paper, we only consider κ-least significant bits of output as the predetermined parts of NCR property, where κ is the security parameter. See Section 2.2 for more details.

We also show concrete instantiations of a PRG with NCR property. One construction is a slight modification of a practical PRG [1] used in practice which is based on cryptographic hash functions such as SHA-1. If we can assume that the hash functions used in the PRG satisfy target collision resistance, we immediately obtain a PRG with NCR property. Though we can provide only a heuristic analysis for this construction, we believe that it is fairly reasonable to assume that this practical PRG satisfies NCR property and we can use it in practical scenarios.

In order to confirm that a PRG with NCR property, though seemingly strong, is actually a fairly weak primitive, we also address how to generically construct such a PRG from any one-way permutation. Interestingly, the construction is the well-known one by Blum and Micali [9] and Yao [45] itself. Namely, the Blum-Micali-Yao PRG has NCR property as it is.

1.3   Related Works

CCA Security of PKE. The notion of CCA security of a PKE scheme was first introduced by Naor and Yung [35] and later extended by Rackoff and Simon [38] and Dolev, Dwork, and Naor [22]. Naor and Yung [35] proposed a generic construction of non-adaptive CCA secure PKE schemes from semantically secure PKE schemes, using *non-interactive zero knowledge proofs* which yield inefficient and complicated structure and are not practical. Based on the Naor-Yung paradigm, some generic constructions of fully CCA secure PKE schemes were also proposed, e.g., [22]. The first practical CCA secure PKE scheme is proposed by Cramer and Shoup [19], and they also generalized their method as *universal hash proof* technique [20] as well as some other instantiations of it. Kurosawa and Desmedt [31] further improved efficiency of the Cramer-Shoup scheme. Today, many practical CCA secure PKE schemes that pursue smaller ciphertext overhead and/or basing the security on weaker intractability assumptions are known, such as [15,29,41,27,18,25].

Canetti, Halevi, and Katz [17] proposed a novel methodology for achieving CCA security from IBE schemes. See Section 1.1 for more details.

Peikert and Waters [37] proposed a methodology to obtain CCA secure PKE schemes using a new primitive called *lossy trapdoor functions*. Recently, Rosen and Segev [40] proposed a generic paradigm for obtaining CCA secure PKE schemes from (injective) trapdoor functions that are one-way under *correlated products*. Hanaoka and Kurosawa [25] proposed yet another paradigm to achieve CCA secure PKE schemes from any CPA secure *broadcast encryption with verifiability*.

In the random oracle methodology [5], several generic methodologies (e.g., [6,23,36]) and concrete practical schemes are known. However, since the results from several papers, such as [16], have shown that this methodology has some problem, in this paper we focus only on the constructions in the standard model.

Other IBE-to-PKE Transformations and Tag-Based Encryption. The IBE-to-PKE transformations we focus on in this paper are the ones that are generically applicable to any IBE scheme (with semantic security). However, there are also several transformations that can be applied to only IBE schemes with special structures or properties. We review them below. Using ideas from [17,14] and specific algebraic properties of the underlying IBE schemes [11,44], Boyen, Mei, and Waters [15] proposed the currently best CCA secure PKE schemes in terms of ciphertext overhead. Using chameleon hash functions [30], Abe et al. [3] proposed several IBE-to-PKE transformations for *partitioned* identity-based key encapsulation mechanisms and constructed several CCA secure PKE schemes via the *Tag-KEM/DEM* paradigm [4]. Zhang [46] independently proposed two transformations that also use chameleon hash functions, where the first transformation is applicable to schemes with *separable* property which are similar to [3] and the second transformation is applicable generically but needs stronger security for a chameleon hash function. In this paper, we do not aim at a size-efficient IBE-to-PKE transformation at the cost of "generality" for the underlying IBE, so that the transformation can be widely applicable. Moreover, a chameleon hash function usually yields a computation of exponentiations, which is heavier compared to computation of "symmetric-key" primitives such as a computation of block ciphers. Recently, Matsuda et al. [33] proposed another generic and efficient IBE-to-PKE transformation. But it requires IBE to be non-malleable, and no concrete efficient IBE scheme (other than the CCA secure IBE schemes) is known so far.

Kiltz [28] showed that the IBE-to-PKE transformation paradigm can be generically applied to *tag-based encryption* schemes [32] of appropriate security, which are weaker primitives than IBE schemes.

2 Preliminaries

In this section, we review the definitions necessary for describing our result.

Notations. In this paper, "$x \leftarrow y$" denotes that x is chosen uniformly at random from y if y is a set or x is output from y if y is a function or an algorithm. "$x||y$" denotes a concatenation of x and y. "$|x|$" denotes the size of the set if x is a finite set or bit length of x if x is an element of some set. "x-LSB(y)" denotes x-least significant bits of y if y is a string. For simplicity, in most cases we drop the security parameter (1^κ) for input of the algorithms considered in this paper.

2.1 Encapsulation Scheme

Boneh and Katz [14] introduced the notion of an encapsulation scheme, which works as the main building block in the BK transformation. Roughly speaking, an encapsulation scheme is a kind of commitment scheme that commits a random value, so that it can be later recovered by using a decommitment.

Formally, an encapsulation scheme E consists of the following three (probabilistic) algorithms. A setup algorithm Encap.Setup takes 1^κ (security parameter κ) as input, and outputs a public parameter prm. A commitment algorithm Encap.Com takes a public parameter prm as input, and outputs a committed value $r \in \mathcal{V}$, a commitment $c \in \mathcal{C}$, and a decommitment $d \in \mathcal{D}$ (where $\mathcal{V}$, $\mathcal{C}$, and $\mathcal{D}$ are a committed value space, a commitment space, and a decommitment space of E, respectively). A recovery algorithm Encap.Rec takes a public parameter prm, a commitment $c \in \mathcal{C}$, and a decommitment $d \in \mathcal{D}$ as input, and outputs a committed value $r \in \mathcal{V} \cup \{\bot\}$. We require Encap.Rec(prm, c, d) = r for all prm output from Encap.Setup and all $(r, c, d) \in \mathcal{V} \times \mathcal{C} \times \mathcal{D}$ output from Encap.Com(prm).

Hiding Property. We define the advantage of an adversary $\mathcal{A}$ against hiding property of an encapsulation scheme E as follows:

$$\mathsf{Adv}_{E,\mathcal{A}}^{\mathrm{Hiding}} = \left| \Pr \left[\begin{array}{l} b_C \leftarrow \{0,1\}; \quad \mathsf{prm} \leftarrow \mathsf{Encap.Setup}(1^\kappa); \\ (r_1^*, c^*, d^*) \leftarrow \mathsf{Encap.Com}(\mathsf{prm}); \\ r_0^* \leftarrow \mathcal{V}; \quad b_A \leftarrow \mathcal{A}(\mathsf{prm}, r_{b_C}^*, c^*) \end{array} : b_A = b_C \right] - \frac{1}{2} \right|.$$

Definition 1. *We say that an encapsulation scheme E is (t, ϵ)-hiding if we have* $\mathsf{Adv}_{E,\mathcal{A}}^{\mathrm{Hiding}} \leq \epsilon$ *for any algorithm $\mathcal{A}$ running in time less than t.*

Binding Property. We define the advantage of an adversary $\mathcal{A}$ against binding property of an encapsulation scheme E as follows:

$$\mathsf{Adv}_{E,\mathcal{A}}^{\mathrm{Binding}} = \Pr \left[\begin{array}{l} \mathsf{prm} \leftarrow \mathsf{Encap.Setup}(1^\kappa); \\ (r^*, c^*, d^*) \leftarrow \mathsf{Encap.Com}(\mathsf{prm}); \\ d' \leftarrow \mathcal{A}(\mathsf{prm}, r^*, c^*, d^*) \end{array} : \begin{array}{l} \mathsf{Encap.Rec}(\mathsf{prm}, c^*, d') \\ \notin \{\bot, r^*\} \wedge \ d' \neq d^* \end{array} \right].$$

Definition 2. *We say that an encapsulation scheme E is (t, ϵ)-binding if we have* $\mathsf{Adv}_{E,\mathcal{A}}^{\mathrm{Binding}} \leq \epsilon$ *for any algorithm $\mathcal{A}$ running in time less than t.*

2.2   Pseudorandom Generator

Let $G : \mathcal{D} \rightarrow \mathcal{R}$ be a function with $|\mathcal{D}| \leq |\mathcal{R}|$. We define the advantage of an adversary $\mathcal{A}$ against pseudorandomness of G as follows:

$$\mathsf{Adv}_{G,\mathcal{A}}^{\mathrm{PR}} = \left| \Pr \left[\begin{array}{l} b_C \leftarrow \{0,1\}; \quad x^* \leftarrow \mathcal{D}; \quad y_1^* \leftarrow G(x^*); \\ y_0^* \leftarrow \mathcal{R}; \quad b_A \leftarrow \mathcal{A}(y_{b_C}^*) \end{array} : b_A = b_C \right] - \frac{1}{2} \right|.$$

Definition 3. *We say that a function G is (t, ϵ)-pseudorandom if we have* $\mathsf{Adv}_{G,\mathcal{A}}^{\mathrm{PR}} \leq \epsilon$ *for any algorithm $\mathcal{A}$ running in time less than t. We also say that G is a (t, ϵ)-pseudorandom generator (PRG).*

Near Collision Resistance (for Predetermined Parts of Output). Boldyreva and Fischlin [10] introduced the notion of near collision resistance (NCR) for predetermined parts of output of a PRG. Roughly speaking, NCR property ensures that given a randomly chosen input $x \in \mathcal{D}$, no adversary can efficiently find another input $x'(\neq x) \in \mathcal{D}$ such that the predetermined parts of output becomes identical. Since an adversary cannot have a control over one of the inputs, it is more related to target collision resistance [34,7] than ordinary (any) collision resistance [21]. According to the authors of [10] "near collision resistance" is named after [8].

In this paper, we only consider κ-least significant bits of output of G as the predetermined parts for NCR property, where κ is the security parameter. Formally, we define the advantage of an adversary $\mathcal{A}$ against NCR for κ-least significant bits of output of G as follows:

$$\mathsf{Adv}_{G,\mathcal{A}}^{\text{NCR-}\kappa\text{-LSB}} = \Pr[x^* \leftarrow \mathcal{D}; x' \leftarrow \mathcal{A}(x^*) : \kappa\text{-LSB}(G(x')) = \kappa\text{-LSB}(G(x^*)) \land x' \neq x^*].$$

Definition 4. *We say that a function (or pseudorandom generator) G is (t, ϵ)-near collision resistant for κ-least significant bits of output if we have $\mathsf{Adv}_{G,\mathcal{A}}^{\text{NCR-}\kappa\text{-LSB}} \leq \epsilon$ for any algorithm $\mathcal{A}$ running in time less than t. We also say that G is (t, ϵ)-NCR-κ-LSB.*

2.3 Target Collision Resistant Hashing

Let $H : \mathcal{D} \to \mathcal{R}$ be a hash function with $|\mathcal{D}| \geq |\mathcal{R}|$. We define the advantage of an adversary $\mathcal{A}$ against target collision resistance of H as follows:

$$\mathsf{Adv}_{H,\mathcal{A}}^{\text{TCR}} = \Pr[x^* \leftarrow \mathcal{D}; x' \leftarrow \mathcal{A}(x^*) : H(x') = H(x^*) \land x' \neq x^*].$$

Definition 5. *We say that a hash function H is (t, ϵ)-target collision resistant if we have $\mathsf{Adv}_{H,\mathcal{A}}^{\text{TCR}} \leq \epsilon$ for any algorithm $\mathcal{A}$ running in time less than t. We also say that H is a (t, ϵ)-target collision resistant hash function (TCRHF).*

2.4 Other Primitives

Here, we review only the algorithms of public key encryption, identity-based encryption, and a message authentication code (MAC) scheme. Refer to [14] for security definitions of these primitives.

A public key encryption (PKE) scheme Π consists of the following three (probabilistic) algorithms. A key generation algorithm PKE.KG takes 1^κ (security parameter κ) as input, and outputs a pair of a secret key sk and a public key pk. An encryption algorithm PKE.Enc takes a public key pk and a plaintext $m \in \mathcal{M}$ as input, and outputs a ciphertext χ (where $\mathcal{M}$ is a plaintext space of Π). A decryption algorithm PKE.Dec takes a secret key sk and a ciphertext χ as input, and outputs a plaintext $m \in \mathcal{M} \cup \{\bot\}$. We require PKE.Dec($sk$, PKE.Enc($pk, m$)) = m for all (sk, pk) output from PKE.KG and all $m \in \mathcal{M}$.

An identity-based encryption (IBE) scheme Π consists of the following four (probabilistic) algorithms. A setup algorithm IBE.Setup takes 1^κ (security parameter κ) as input, and outputs a pair of a master secret key msk and global parameters prm. A key extraction algorithm IBE.Ext takes global parameters prm, a master secret key msk, and an identity ID $\in \mathcal{I}$ as input, and outputs a decryption key dk_{ID} corresponding to ID (where $\mathcal{I}$ is an identity space of Π). An encryption algorithm IBE.Enc takes global parameters prm, an identity ID $\in \mathcal{I}$, and a plaintext $m \in \mathcal{M}$ as input, and outputs a ciphertext χ (where $\mathcal{M}$ is a plaintext space of Π). A decryption algorithm IBE.Dec takes a decryption key dk_{ID} and a ciphertext χ as input, and outputs a plaintext $m \in \mathcal{M} \cup \{\bot\}$. We require IBE.Dec(IBE.Ext(prm, msk, ID), IBE.Enc(prm, ID, m)) = m for all (msk, prm) output from IBE.Setup, all ID $\in \mathcal{I}$, and all $m \in \mathcal{M}$.

A message authentication code (MAC) scheme Σ consists of the following two algorithms. A MAC generating algorithm MAC.Mac takes a MAC key $k \in \mathcal{K}$ and a message $m \in \mathcal{M}$ as input, and outputs a valid MAC tag tag on m under k (where $\mathcal{K}$ and $\mathcal{M}$ are a MAC key space and a message space of Σ, respectively). A verification algorithm MAC.Verify takes a MAC key k, a message m, and a MAC tag tag as input, and outputs accept if tag is a valid MAC tag on m or reject otherwise. We require MAC.Verify$(k, m, \text{MAC.Mac}(k, m)) = $ accept for all $k \in \mathcal{K}$ and all $m \in \mathcal{M}$.

3   The Boneh-Katz Transformation

In this section, we briefly review the IBE-to-PKE transformation by Boneh and Katz [14]. Let $\Pi = $ (IBE.Setup, IBE.Ext, IBE.Enc, IBE.Dec) be an IBE scheme, E = (Encap.Setup, Encap.Com, Encap.Rec) be an encapsulation scheme, and $\Sigma = $ (MAC.Mac, MAC.Verify) be a MAC scheme. Then, a PKE scheme $\Pi' = $ (PKE.KG, PKE.Enc, PKE.Dec) obtained via the BK transformation is as shown in Fig. 1. CCA security of Π' was proved assuming that Π is IND-sID-CPA secure, E satisfies binding and hiding, and Σ is one-time secure. See [14] for details.

Notice that the overhead of ciphertext size from that of the underlying IBE scheme is caused by a commitment c, a decommitment d, and a MAC tag tag. Since the size of a MAC tag can be κ-bit for κ-bit security and is optimal, designing an encapsulation scheme such that the sizes of parameters (c, d) are small is desirable for obtaining a PKE scheme with a small ciphertext overhead.

In [14], the authors also showed a concrete construction of an encapsulation scheme. Here, we briefly review their encapsulation scheme. Encap.Setup(1^κ) picks a target collision resistant hash function (TCRHF) TCR and a pairwise-independent hash function (PIHF) h, and outputs prm $\leftarrow$ (TCR, h). Encap.Com(prm) picks a decommitment d randomly, computes $c \leftarrow$ TCR(d) and $r \leftarrow h(d)$, then outputs (r, c, d). Encap.Rec(prm, c, d) checks whether TCR$(d) = c$ or not, and outputs $r \leftarrow h(d)$ if this holds or $\perp$ otherwise.

Their scheme only uses a TCRHF and a PIHF for both the commitment and the recovery algorithms and thus is fairly efficient in terms of computation cost.

PKE.KG(1^κ) :	PKE.Enc(PK, m) :
(msk, prm$_I$) $\leftarrow$ IBE.Setup(1^κ)	$(r, c, d) \leftarrow$ Encap.Com(prm$_E$)
prm$_E$ $\leftarrow$ Encap.Setup(1^κ)	$y \leftarrow$ IBE.Enc(prm$_I$, c, $(m\|\|d)$)
$SK \leftarrow$ msk; $PK \leftarrow$ (prm$_I$, prm$_E$)	tag $\leftarrow$ MAC.Mac(r, y)
Output (SK, PK).	Output $\chi \leftarrow \langle c, y, \text{tag} \rangle$.
PKE.Dec(SK, χ) :	
Parse χ as $\langle c, y, \text{tag} \rangle$.; $dk_c \leftarrow$ IBE.Ext(prm$_I$, msk, c)	
$(m\|\|d) \leftarrow$ IBE.Dec(dk_c, y) (if this returns $\perp$ then output $\perp$ and stop.)	
$r \leftarrow$ Encap.Rec(prm$_E$, c, d) (if this returns $\perp$ then output $\perp$ and stop.)	
Output m if MAC.Verify$(r, y, \text{tag}) = $ accept. Otherwise output $\perp$.	

Fig. 1. The Boneh-Katz Transformation

However, due to the leftover hash lemma [26] used to show hiding property, we need to set d to be at least 448-bit for 128-bit security (it achieves hiding property in a statistical sense). Thus, even though we use an efficient IBE scheme such as [11] as the underlying IBE scheme in the BK transformation, it results in a PKE scheme with somewhat large ciphertext because of the size of d. However, as the authors of [14] pointed out, it is important to note that we do not need "statistical security" for neither hiding nor binding properties. We only need "computational security" for both. (Our proposed encapsulation scheme in the next section actually achieves them in computational sense.)

4   Proposed Encapsulation Scheme

As we have seen in Section 3, designing an encapsulation scheme with small parameter size is important for the size-efficiency of the BK transformation. In this section, we present an efficient encapsulation scheme using a PRG with NCR property and prove its security. We also show a concrete instantiation of the PRG with NCR property.

Let $G : \{0,1\}^\kappa \to \{0,1\}^{2\kappa}$ be a PRG (with NCR-κ-LSB property). Then we construct an encapsulation scheme $E = (\mathsf{Encap.Setup}, \mathsf{Encap.Com}, \mathsf{Encap.Rec})$ as follows.

$\mathsf{Encap.Setup}(1^\kappa)$: Set $\mathsf{prm} \leftarrow G$ and output prm.
$\mathsf{Encap.Com}(\mathsf{prm})$: Pick $d \in \{0,1\}^\kappa$ uniformly at random, compute $(r||c) \leftarrow G(d)$
  such that $|r| = |c| = \kappa$, and output (r, c, d).
$\mathsf{Encap.Rec}(\mathsf{prm}, c, d)$: Compute $(r||c') \leftarrow G(d)$ such that $|r| = |c'| = \kappa$, and
  output r if $c' = c$ or $\perp$ otherwise.

4.1   Security

In this subsection, we prove hiding and binding properties of the proposed scheme. The proofs for both properties are fairly intuitive and easy to see. Specifically, pseudorandomness of G provides hiding property and NCR-κ-LSB of G provides binding property of the proposed encapsulation scheme E.

Theorem 1. *If G is a (t, ϵ_{prg})-PRG, then the proposed encapsulation scheme E is $(t, 2\epsilon_{prg})$-hiding.*

Proof. Suppose $\mathcal{A}$ is an adversary that breaks (t, q, ϵ_{hide})-hiding property of E, which means that $\mathcal{A}$ with running time t wins the hiding game with probability $\frac{1}{2} + \epsilon_{hide}$. Then we construct a simulator $\mathcal{S}$ who can break $(t, \frac{1}{2}\epsilon_{hide})$-pseudorandomness of G. Our simulator $\mathcal{S}$, simulating the hiding game for $\mathcal{A}$, plays the PRG game with the PRG challenger $\mathcal{C}$ as follows.

Given a 2κ-bit string $y^*_{b_C}$, first $\mathcal{S}$ sets $\mathsf{prm} \leftarrow G$ and $(r^*_1||c^*) \leftarrow y^*_{b_C}$ such that $|r^*_1| = |c^*| = \kappa$, and then picks $b_S \in \{0,1\}$ and $r^*_0 \in \{0,1\}^\kappa$ uniformly at random. $\mathcal{S}$ gives $(\mathsf{prm}, r^*_{b_S}, c^*)$ to $\mathcal{A}$. After $\mathcal{A}$ outputs his guess b_A, $\mathcal{S}$ sets $b'_S \leftarrow 1$ if $b_A = b_S$ or $b'_S \leftarrow 0$ otherwise. Then $\mathcal{S}$ outputs b'_S as its guess.

Next, we estimate the advantage of $\mathcal{S}$. We have

$$\mathsf{Adv}_{G,\mathcal{S}}^{\mathrm{PR}} = |\Pr[b_S' = b_C] - \frac{1}{2}| = \frac{1}{2}|\Pr[b_S' = 1|b_C = 1] - \Pr[b_S' = 1|b_C = 0]|$$

$$= \frac{1}{2}|\Pr[b_A = b_S|b_C = 1] - \Pr[b_A = b_S|b_C = 0]|.$$

To complete the proof, we prove the following claims.

Claim 1. $\Pr[b_A = b_S|b_C = 1] = \frac{1}{2} + \epsilon_{hide}$

Proof of Claim 1. In the case $b_C = 1$, c^* and r_1^* are computed with G with a uniformly chosen input $d^* \in \{0,1\}^\kappa$ (i.e. $(r_1^*||c^*) = y_1^* = G(d^*)$). On the other hand, r_0^* is chosen uniformly by $\mathcal{S}$. Thus, the view of $\mathcal{A}$ is exactly the same as that in the hiding game (with the challenger's bit is b_S). Therefore, the probability that $b_A = b_S$ occurs is exactly the same as the probability that $\mathcal{A}$ succeeds in guessing in the hiding game, i.e., $\frac{1}{2} + \epsilon_{hide}$.   □

Claim 2. $\Pr[b_A = b_S|b_C = 0] = \frac{1}{2}$

Proof of Claim 2. In the case $b_C = 0$, since y_0^* given to $\mathcal{S}$ is a uniformly chosen 2κ-bit string, c^* and r_1^* are both uniformly and independently distributed in $\{0,1\}^\kappa$. Therefore, c^* may not necessarily be in the range of G and thus $\mathcal{S}$'s simulation for $\mathcal{A}$ may be imperfect. Thus, $\mathcal{A}$ may notice that he is in the simulated game and act unfavorably for $\mathcal{S}$. However, r_0^* is also uniformly chosen from $\{0,1\}^\kappa$ by $\mathcal{S}$. Since the distribution of the uniformly distributed value r_1^* and the distribution of a uniformly chosen value r_0^* are perfectly indistinguishable, it is information-theoretically impossible for $\mathcal{A}$ to distinguish r_1^* and r_0^*. Therefore, the probability that $b_A = b_S$ occurs is exactly $\frac{1}{2}$.   □

Above shows that if $\mathcal{A}$ wins the hiding game of E with advantage greater than ϵ_{hide}, then $\mathcal{S}$ breaks pseudorandomness of G with advantage greater than $\frac{1}{2}\epsilon_{hide}$, which completes the proof of Theorem 1.   □

Theorem 2. *If G is (t, ϵ_{ncr})-NCR-κ-LSB, then the proposed encapsulation scheme E is (t, ϵ_{ncr})-binding.*

Proof. Suppose $\mathcal{A}$ is an adversary that breaks (t, ϵ_{bind})-binding property of E, which means that $\mathcal{A}$ with running time t wins the binding game with probability ϵ_{bind}. Then we construct a simulator $\mathcal{S}$ who can break (t, ϵ_{bind})-NCR-κ-LSB property of G. The description of $\mathcal{S}$ is as follows.

Given $d^* \in \{0,1\}^\kappa$ which is chosen uniformly, first $\mathcal{S}$ sets $\mathsf{prm} \leftarrow G$ and computes $(r^*||c^*) \leftarrow G(d^*)$ such that $|r^*| = |c^*| = \kappa$. Then $\mathcal{S}$ gives $(\mathsf{prm}, r^*, c^*, d^*)$ to $\mathcal{A}$. After $\mathcal{A}$ outputs d', $\mathcal{S}$ simply outputs it as its output.

Note that $\mathcal{S}$'s simulation for $\mathcal{A}$ is perfect. Next, we estimate the advantage of $\mathcal{S}$. Let r' and c' be defined as $(r'||c') = G(d')$ such that $|r'| = |c'| = \kappa$. We have

$$\mathsf{Adv}_{G,\mathcal{S}}^{\mathrm{NCR}\text{-}\kappa\text{-}\mathrm{LSB}} = \Pr[\kappa\text{-}\mathsf{LSB}(G(d')) = \kappa\text{-}\mathsf{LSB}(G(d^*)) \wedge d' \neq d^*]$$

$$= \Pr[c' = c^* \wedge d' \neq d^*] = \Pr[\mathsf{Encap.Rec}(\mathsf{prm}, c^*, d') \neq \perp \wedge d' \neq d^*]$$

$$\geq \Pr[\mathsf{Encap.Rec}(\mathsf{prm}, c^*, d') \notin \{r^*, \perp\} \wedge d' \neq d^*] = \epsilon_{bind},$$

where the transition from the second to the third equalities is due to the definition of the recovery algorithm Encap.Rec of our encapsulation scheme E in this section. Above means that if $\mathcal{A}$ succeeds in breaking binding property of E with advantage greater than ϵ_{bind}, $\mathcal{S}$ also succeeds in breaking NCR-κ-LSB property with advantage greater than ϵ_{bind}, which completes the proof of Theorem 2. $\square$

4.2   Concrete Instantiation of PRG with NCR Property

In this subsection, we show a concrete construction of a PRG that has NCR property for practical scenarios. Specifically, we discuss that it is reasonable to assume that (a slight modification of) the PRG currently described in FIPS 186-2, Revised Appendix 3.1 [1] satisfies NCR property. Here, we briefly review the essential construction of the PRG in [1].

Let $H : \{0,1\}^* \rightarrow \{0,1\}^m$ be a cryptographic hash function. The construction of a PRG $\mathsf{FIPSPRG}_c^H : \{0,1\}^m \rightarrow \{0,1\}^{cm}$ for $c \geq 1$ is as follows:

Step 1. On input $x \in \{0,1\}^m$, set $x_0 \leftarrow x$.
Step 2. Compute $w_i \leftarrow H(x_{i-1})$ and $x_i \leftarrow (1 + x_{i-1} + w_i) \bmod 2^m$ for $1 \leq i \leq c$.
Step 3. Output $(w_1 \| w_2 \| \ldots \| w_c)$.

Then, we define our PRG G^H by interchanging the first and the second m-bit blocks of $\mathsf{FIPSPRG}_2^H$, i.e.,

$$G^H(x) = (\ H(\ (1 + x + H(x)) \bmod 2^m\ ) \ \| \ H(x)\ ).$$

Note that G^H is a PRG as long as $\mathsf{FIPSPRG}_2^H$ is. Moreover, since m-least significant bits of $G^H(x)$ is $H(x)$ itself, if we can assume that H satisfies target collision resistance [34,7], then we will obviously obtain a PRG with NCR-m-LSB.

Below, we address the above in a more formal manner.

Definition 6. *(FIPS186-2-PRG Assumption) We say that the (t, ϵ)-FIPS186-2-PRG assumption with regard to $\mathsf{FIPSPRG}_c^H$ holds if we can assume that the PRG $\mathsf{FIPSPRG}_c^H$ constructed using a hash function H as above is a (t, ϵ)-PRG.*

Theorem 3. *If the (t, ϵ_{fips})-FIPS186-2-PRG assumption with regard to $\mathsf{FIPSPRG}_2^H$ holds, then G^H constructed as above is a (t, ϵ_{fips})-PRG.*

Proof. Suppose $\mathcal{A}$ is an adversary that breaks the (t, ϵ_{pr})-pseudorandomness of G^H. Then we construct a simulator $\mathcal{S}$ who can break (t, ϵ_{pr})-FIPS186-2-PRG assumption with regard to $\mathsf{FIPSPRG}_2^H$, which means that $\mathcal{S}$ can break the (t, ϵ_{pr})-pseudorandomness of $\mathsf{FIPSPRG}_2^H$. The description of $\mathcal{S}$ is as follows.

Given a $2m$-bit string $y_{b_C}^*$, $\mathcal{S}$ sets z^* as a $2m$-bit string such that first and the second m-bit blocks of $y_{b_C}^*$ are interchanged. Then $\mathcal{S}$ gives z^* to $\mathcal{A}$. After $\mathcal{A}$ outputs its guess b_A, $\mathcal{S}$ sets $b_S \leftarrow b_A$ and output b_S as its guess.

Notice that $\mathcal{S}$ simulates the experiment of attacking pseudorandomness of G^H perfectly for $\mathcal{A}$. Namely, if $b_C = 1$, i.e., $y_{b_C}^* = \mathsf{FIPSPRG}_2^H(x)$ where $x \in \{0,1\}^m$ is chosen uniformly at random, z^* given to $\mathcal{A}$ is a $2m$-bit string that is $\mathsf{FIPSPRG}_2^H(x)$ for the uniformly random value x. On the other hand, if $b_C = 0$, i.e., $y_{b_C}^*$ is a uniformly chosen $2m$-bit string, then z^* given to $\mathcal{A}$ is also a uniformly random $2m$-bit string. Therefore, we have

$$\mathsf{Adv}^{\mathrm{PR}}_{\mathsf{FIPSPRG}_2^H,\mathcal{S}} = |\Pr[b_S = b_C] - \frac{1}{2}| = |\Pr[b_A = b_C] - \frac{1}{2}| = \epsilon_{pr}.$$

Above shows that if $\mathcal{A}$ breaks pseudorandomness of G^H with advantage greater than ϵ_{pr}, then $\mathcal{S}$ breaks pseudorandomness of $\mathsf{FIPSPRG}_2^H$ with advantage greater than ϵ_{pr}. This completes the proof of Theorem 3.   □

Theorem 4. *If a hash function H that is a building block of G^H is a (t, ϵ_{tcr})-TCRHF, then G^H is (t, ϵ_{tcr})-NCR-m-LSB.*

Proof. Suppose $\mathcal{A}$ is an adversary that breaks the (t, ϵ_{ncr})-NCR-m-LSB of G^H. Then we construct a simulator $\mathcal{S}$ who can break (t, ϵ_{ncr})-target collision resistance of H. The description of $\mathcal{S}$ is as follows.

Given $x^* \in \{0,1\}^m$ which is chosen uniformly at random, $\mathcal{S}$ gives x^* to $\mathcal{A}$. After $\mathcal{A}$ outputs x', $\mathcal{S}$ outputs x' as its own output.

It is easy to see that the $\mathcal{S}$'s simulation of the experiment attacking NCR-m-LSB of G^H for $\mathcal{A}$ is perfect. $\mathcal{S}$'s advantage is estimated as

$$\begin{aligned}
\mathsf{Adv}^{\mathrm{TCR}}_{H,\mathcal{S}} &= \Pr[H(x') = H(x^*) \wedge x' \neq x^*] \\
&= \Pr[m\text{-}\mathsf{LSB}(G^H(x')) = m\text{-}\mathsf{LSB}(G^H(x^*)) \wedge x' \neq x^*] = \epsilon_{ncr}.
\end{aligned}$$

Above shows that if $\mathcal{A}$ breaks NCR-m-LSB property of G^H with advantage greater than ϵ_{ncr}, then $\mathcal{S}$ breaks target collision resistance of H with advantage greater than ϵ_{ncr}. This completes the proof of Theorem 4.   □

As shown above, since we do not need full power of collision resistance [21] but target collision resistance, we can set $m = \kappa$ for κ-bit security. In practice, (an appropriate modification of) SHA-1 may be used as H. (Though SHA-1 is known to be already broken as a collision resistant hash function [43], it is still reasonable to assume that SHA-1 is target collision resistant.)

Although the FIPS186-2-PRG assumption with regard to $\mathsf{FIPSPRG}_2^H$ is somewhat heuristic (note that the FIPS186-2-PRG assumption with regard to $\mathsf{FIPSPRG}_1^H$ is the same assumption that H with m-bit input space is a PRG), we note that the PRG $\mathsf{FIPSPRG}_c^H$ we introduced here is used (recommended) for generating randomness for Digital Signature Standard (DSS) and is also listed in Recommended techniques of CRYPTREC [2], and thus, using the PRG G^H we presented above as a PRG with NCR-κ-LSB in our encapsulation scheme is fairly reasonable.

One might still think that a PRG with NCR-κ-LSB is a somewhat strong primitive. However, we can actually show that a PRG with NCR-κ-LSB can be constructed from a fairly weak assumption. As addressed in the next section, existence of a PRG with NCR property is generically implied by existence of a one-way permutation which is one of the most fundamental cryptographic primitives.

5 PRG with Near Collision Resistance from Any One-Way Permutation

The security of the PRG we show in Section 4.2 is somewhat heuristic (though we believe it to be fairly reasonable to use in practical scenarios). Here, we show an evidence that a PRG with NCR-κ-LSB is actually a very weak primitive. Specifically, we address that a PRG with NCR-κ-LSB can be generically constructed based on any one-way permutation, which is a fundamental and weak assumption in the area of cryptography. Actually, the construction we show here is the well-known and well-studied PRG by Blum and Micali [9] and Yao [45] (we call the *BMY-PRG*) itself. Namely, the BMY-PRG construction satisfies NCR-κ-LSB property as it is. We briefly review the construction below.

Let $g : \{0,1\}^\kappa \to \{0,1\}^\kappa$ be a one-way permutation and $h : \{0,1\}^\kappa \to \{0,1\}$ be a hardcore bit function of g (e.g. the Goldreich-Levin bit [24]). Then the BMY-PRG $G : \{0,1\}^\kappa \to \{0,1\}^{\kappa+l}$ for $l > 0$ is defined as follows:

$$G(x) = \Big(h(x) \parallel h(g(x)) \parallel h(g^{(2)}(x)) \parallel \dots \parallel h(g^{(l-1)}(x)) \parallel g^{(l)}(x) \Big),$$

where $g^{(i)}(x) = g(g^{(i-1)}(x))$ and $g^{(1)}(x) = g(x)$. Pseudorandomness of G constructed as above was proved assuming the one-wayness of the permutation g. See [9,45] for details.

As for NCR-κ-LSB property, it was already mentioned by Boldyreva and Fischlin in [10] that the BMY-PRG has the property. Here, however, we prove for completeness. The following shows that a PRG with NCR-κ-LSB can be actually constructed only from a one-way permutation.

Theorem 5. *([10]) If G is constructed as above, then G is $(t,0)$-NCR-κ-LSB for any t.*

Proof. According to the definition of the NCR-κ-LSB advantage, for an adversary $\mathcal{A}$, we have

$$\mathsf{Adv}_{G,\mathcal{A}}^{\text{NCR-}\kappa\text{-LSB}}$$
$$= \Pr[x^* \leftarrow \{0,1\}^\kappa; x' \leftarrow \mathcal{A}(G, x^*) : \kappa\text{-LSB}(G(x')) = \kappa\text{-LSB}(G(x^*)) \wedge x' \neq x^*]$$
$$= \Pr[x^* \leftarrow \{0,1\}^\kappa; x' \leftarrow \mathcal{A}(G, x^*) : g^{(l)}(x') = g^{(l)}(x^*) \wedge x' \neq x^*].$$

Since g is a permutation, for any $x, x'(\neq x) \in \{0,1\}^\kappa$ and any $i \geq 1$, we have $g^{(i)}(x) \neq g^{(i)}(x')$. Therefore, of course we have $g^{(l)}(x) \neq g^{(l)}(x')$ for any $x, x'(\neq x) \in \{0,1\}^\kappa$ and any $l > 0$, and thus the above probability equals to zero for any adversary $\mathcal{A}$ with any running time.   □

6 Comparison

Table 1 shows the comparison among generic IBE-to-PKE transformations, including the CHK transformation (CHK) [17] where the one-time signature is instantiated with no stronger primitive tool than one implied by a one-way function (e.g. [39]) (CHK), the CHK transformation where the one-time signature is instantiated with the strongly unforgeable signature by Boneh and Boyen [12] (known as one of the best signature schemes in terms of signature size) (CHK w. BB), the BK transformation where the BK encapsulation scheme is used (BK), and the BK transformation where the encapsulation scheme is instantiated with ours in Section 4 (BK w. Ours).

In Table 1, the column "Overhead by Transformation" denotes how much the ciphertext size increases from that of the underlying IBE scheme (typical sizes for 128-bit security are given as numerical examples), and the column "Required Size for $\mathcal{M}_{IBE}$" denotes how much size is necessary for the plaintext space of the underlying IBE scheme.

Ciphertext Overhead. If the BK transformation is instantiated with our encapsulation scheme, overhead from the ciphertext of the underlying IBE scheme is caused by two κ-bit strings and one MAC. Thus, if we require 128-bit security, we can set each to be 128-bit and thus we will have 384-bit overhead in total. In the original BK scheme, on the other hand, the overhead is caused by a TCRHF (TCR), a MAC, and a large randomness d which is a decommitment of the BK encapsulation scheme (as already noted in Section 3, we need large size for d due to the leftover hash lemma [26]). Because of d, though size of the image TCR(d) with the TCRHF TCR and the MAC tag can be 128-bit, we need at least 448-bit for d, and the overhead in total needs to be at least 704-bit. Compared to the CHK scheme, even though the one-time signature is instantiated with the Boneh-Boyen scheme, the most size-efficient scheme in the standard model so far, it cannot provide a smaller overhead than ours.

Computation Overhead. If we use the PRG in Section 4.2 for our encapsulation scheme, then the essential efficiency of computations (two computations of a

Table 1. Comparison among Generic IBE-to-PKE Transformations

	Overhead by Transformation (Numerical Example (bit)‡)		Required Size for $\mathcal{M}_{IBE}$
CHK [17]	$\|vk\| + \|sig\|$	(> 10000)	$\|m_{PKE}\|$
CHK w. BB [12]†	$(2\|g_2\|) + (\|g_1\| + \|p\|)$	(≥ 1024)	$\|m_{PKE}\|$
BK [14]	$\|TCR(d)\| + \|d\| + \|MAC\|$	(704)	$\|m_{PKE}\| + \|d\|$
BK w. Ours	$2\kappa + \|MAC\|$	(384)	$\|m_{PKE}\| + \kappa$

vk and sig denote the verification key and the signature of the one-time signature in the CHK transformation. d is a randomness (decommitment) used in the BK encapsulation scheme. $\|m_{PKE}\|$ is a plaintext size of a transformed PKE scheme.

† We assume that the Boneh-Boyen signature scheme is implemented using a bilinear groups $(\mathbb{G}_1, \mathbb{G}_2)$ of prime order p with an asymmetric pairing $e : \mathbb{G}_1 \times \mathbb{G}_2 \to \mathbb{G}_T$ and generators $g_1 \in \mathbb{G}_1$ and $g_2 \in \mathbb{G}_2$.

‡ We assume that generators (g_1, g_2) of the Boneh-Boyen scheme can be removed from the size of the verification key. We set $\|g_2\| = \|p\| = 256$, $\|g_1\| \geq 256$, and $\|d\| = 448$.

cryptographic hash function) of the encapsulation scheme is comparable to the BK encapsulation scheme (one computation of a cryptographic hash function and one computation of a PIHF, the latter of which is usually a cheap arithmetic computation over some finite field). If we use the PRG in Section 5 for our encapsulation scheme, then, because of the computation of the BMY-PRG, our encapsulation scheme requires heavier computations for both commitment and recovery algorithms compared to the BK encapsulation scheme. Specifically, for obtaining a 2κ-bit pseudorandom string from κ-bit string with the BMY-PRG, we have to compute a one-way permutation κ times (though this can be reduced to $\kappa/(\log \kappa)$ times by taking not just one bit but $\log \kappa$ bits of hardcore bits in each iteration of the computation of a one-way permutation in the BMY-PRG, this is still far worse than the BK encapsulation scheme). However, since the computations in these encapsulation schemes are all "symmetric-key" computations, in most cases they are not so significant compared to the computations done in encryption and decryption algorithms of the IBE scheme, which usually include computations of exponentiations and/or pairings.

Acknowledgement

The authors would like to thank Kazuo Ohta for his helpful comments and suggestions. The authors also would like to thank anonymous reviewers of CT-RSA'09 for their invaluable comments.

References

1. Digital Signature Standard (DSS). FIPS 186-2 (+ Change Notice), Revised Appendix 3.1 (2000), http://csrc.nist.gov/publications/fips/fips186-2/fips186-2-change1.pdf
2. CRYPTREC Report 2007 (in Japanese), p. 31 (2007), http://www.cryptrec.go.jp/report/c07_wat_final.pdf, Older but English version is also available. CRYPTREC Report 2002, p. 23 (2002), http://www.ipa.go.jp/security/enc/CRYPTREC/fy15/doc/c02e_report2.pdf
3. Abe, M., Cui, Y., Imai, H., Kiltz, E.: Efficient hybrid encryption from ID-based encryption (2007), eprint.iacr.org/2007/023
4. Abe, M., Gennaro, R., Kurosawa, K., Shoup, V.: Tag-KEM/DEM: A new framework for hybrid encryption and a new analysis of Kurosawa-Desmedt KEM. In: Cramer, R. (ed.) EUROCRYPT 2005. LNCS, vol. 3494, pp. 128–146. Springer, Heidelberg (2005)
5. Bellare, M., Rogaway, P.: Random oracles are practical: A paradigm for designing efficient protocols. In: Proc. of CCS 1993, pp. 62–73. ACM, New York (1993)
6. Bellare, M., Rogaway, P.: Optimal asymmetric encryption. In: De Santis, A. (ed.) EUROCRYPT 1994. LNCS, vol. 950, pp. 92–111. Springer, Heidelberg (1995)
7. Bellare, M., Rogaway, P.: Collision-resistant hashing: Towards making UOWHFs practical. In: Kaliski Jr., B.S. (ed.) CRYPTO 1997. LNCS, vol. 1294, pp. 470–484. Springer, Heidelberg (1997)
8. Biham, E., Chen, R.: Near-collisions of SHA-0. In: Franklin, M. (ed.) CRYPTO 2004. LNCS, vol. 3152, pp. 290–305. Springer, Heidelberg (2004)
9. Blum, M., Micali, S.: How to generate cryptographically strong sequences of pseudo-random bits. SIAM J. Computing 13(4), 850–864 (1984)

10. Boldyreva, A., Fischlin, M.: On the security of OAEP. In: Lai, X., Chen, K. (eds.) ASIACRYPT 2006. LNCS, vol. 4284, pp. 210–225. Springer, Heidelberg (2006)

11. Boneh, D., Boyen, X.: Efficient selective-ID secure identity-based encryption without random oracles. In: Cachin, C., Camenisch, J.L. (eds.) EUROCRYPT 2004. LNCS, vol. 3027, pp. 223–238. Springer, Heidelberg (2004)

12. Boneh, D., Boyen, X.: Short signatures without random oracles. In: Cachin, C., Camenisch, J.L. (eds.) EUROCRYPT 2004. LNCS, vol. 3027, pp. 56–73. Springer, Heidelberg (2004)

13. Boneh, D., Franklin, M.: Identity-based encryption from the Weil pairing. In: Kilian, J. (ed.) CRYPTO 2001. LNCS, vol. 2139, pp. 213–229. Springer, Heidelberg (2001)

14. Boneh, D., Katz, J.: Improved efficiency for CCA-secure cryptosystems built using identity-based encryption. In: Menezes, A. (ed.) CT-RSA 2005. LNCS, vol. 3376, pp. 87–103. Springer, Heidelberg (2005)

15. Boyen, X., Mei, Q., Waters, B.: Direct chosen ciphertext security from identity-based techniques. In: Proc. of CCS 2005, pp. 320–329. ACM Press, New York (2005)

16. Canetti, R., Goldreich, O., Halevi, S.: The random oracle methodology, revisited. In: Proc. of STOC 1998, pp. 209–218. ACM Press, New York (1998)

17. Canetti, R., Halevi, S., Katz, J.: Chosen-ciphertext security from identity-based encryption. In: Cachin, C., Camenisch, J.L. (eds.) EUROCRYPT 2004. LNCS, vol. 3027, pp. 207–222. Springer, Heidelberg (2004)

18. Cash, D., Kiltz, E., Shoup, V.: The twin Diffie-Hellman problem and applications. In: Smart, N.P. (ed.) EUROCRYPT 2008. LNCS, vol. 4965, pp. 127–145. Springer, Heidelberg (2008)

19. Cramer, R., Shoup, V.: A practical public key cryptosystem provably secure against adaptive chosen ciphertext attack. In: Krawczyk, H. (ed.) CRYPTO 1998. LNCS, vol. 1462, pp. 13–25. Springer, Heidelberg (1998)

20. Cramer, R., Shoup, V.: Universal hash proofs and a paradigm for adaptive chosen ciphertext secure public-key encryption. In: Knudsen, L.R. (ed.) EUROCRYPT 2002. LNCS, vol. 2332, pp. 45–64. Springer, Heidelberg (2002)

21. Damgård, I.B.: Collision free hash functions and public key signature schemes. In: Price, W.L., Chaum, D. (eds.) EUROCRYPT 1987. LNCS, vol. 304, pp. 203–216. Springer, Heidelberg (1988)

22. Dolev, D., Dwork, C., Naor, M.: Non-malleable cryptography. In: Proc. of STOC 1991, pp. 542–552. ACM Press, New York (1991)

23. Fujisaki, E., Okamoto, T.: How to enhance the security of public-key encryption at minimum cost. In: Imai, H., Zheng, Y. (eds.) PKC 1999. LNCS, vol. 1560, pp. 53–68. Springer, Heidelberg (1999)

24. Goldreich, O., Levin, L.A.: Hardcore predicate for all one-way functions. In: Proc. of STOC 1989, pp. 25–32. ACM Press, New York (1989)

25. Hanaoka, G., Kurosawa, K.: Efficient chosen ciphertext secure public key encryption under the computational Diffie-Hellman assumption. In: Pieprzyk, J. (ed.) ASIACRYPT 2008. LNCS, vol. 5350, pp. 308–325. Springer, Heidelberg (2008)

26. Håstad, J., Impagliazzo, R., Levin, L., Luby, M.: Construction of a pseudorandom generator from any one-way function. SIAM J. Computing 28(4), 1364–1396 (1999)

27. Hofheinz, D., Kiltz, E.: Secure hybrid encryption from weakened key encapsulation. In: Menezes, A. (ed.) CRYPTO 2007. LNCS, vol. 4622, pp. 553–571. Springer, Heidelberg (2007)

28. Kiltz, E.: Chosen-ciphertext security from tag-based encryption. In: Halevi, S., Rabin, T. (eds.) TCC 2006. LNCS, vol. 3876, pp. 581–600. Springer, Heidelberg (2006)

29. Kiltz, E.: Chosen-ciphertext secure key-encapsulation based on gap hashed diffie-hellman. In: Okamoto, T., Wang, X. (eds.) PKC 2007. LNCS, vol. 4450, pp. 282–297. Springer, Heidelberg (2007)

30. Krawczyk, H., Rabin, T.: Chameleon hashing and signatures. In: Proc. of NDSS 2000, pp. 143–154. Internet Society (2000)

31. Kurosawa, K., Desmedt, Y.: A new paradigm of hybrid encryption scheme. In: Franklin, M. (ed.) CRYPTO 2004. LNCS, vol. 3152, pp. 426–442. Springer, Heidelberg (2004)

32. MacKenzie, P., Reiter, M.K., Yang, K.: Alternatives to non-malleability: Definitions, constructions and applications. In: Naor, M. (ed.) TCC 2004. LNCS, vol. 2951, pp. 171–190. Springer, Heidelberg (2004)

33. Matsuda, T., Hanaoka, G., Matsuura, K., Imai, H.: Simple CCA-secure public key encryption from any non-malleable identity-based encryption. In: The proceedings of ICISC 2008 (to appear, 2008)

34. Naor, M., Yung, M.: Universal one-way hash functions and their cryptographic applications. In: Proc. of STOC 1989, pp. 33–43. ACM Press, New York (1989)

35. Naor, M., Yung, M.: Public-key cryptosystems provably secure against chosen ciphertext attacks. In: Proc. of STOC 1990, pp. 427–437. ACM, New York (1990)

36. Okamoto, T., Pointcheval, D.: REACT: Rapid enhanced-security asymmetric cryptosystem transform. In: Naccache, D. (ed.) CT-RSA 2001. LNCS, vol. 2020, pp. 159–174. Springer, Heidelberg (2001)

37. Peikert, C., Waters, B.: Lossy trapdoor functions and their applications. In: Proc. of STOC 2008, pp. 187–196. ACM Press, New York (2008)

38. Rackoff, C., Simon, D.R.: Non-interactive zero-knowledge proof of knowledge and chosen ciphertext attack. In: Feigenbaum, J. (ed.) CRYPTO 1991. LNCS, vol. 576, pp. 433–444. Springer, Heidelberg (1992)

39. Reyzin, L., Reyzin, N.: Better than BiBa: Short one-time signatures with fast signing and verifying. In: Batten, L.M., Seberry, J. (eds.) ACISP 2002. LNCS, vol. 2384, pp. 144–153. Springer, Heidelberg (2002)

40. Rosen, A., Segev, G.: Chosen-ciphertext security via correlated products. In: Reingold, O. (ed.) TCC 2009. LNCS, vol. 5444, pp. 419–436. Springer, Heidelberg (2009), eprint.iacr.org/2008/116/

41. Shacham, H.: A Cramer-Shoup encryption scheme from the linear assumption and from progressively weaker linear variants (2007), eprint.iacr.org/2007/074/

42. Shamir, A.: Identity-based cryptosystems and signature schemes. In: Blakely, G.R., Chaum, D. (eds.) CRYPTO 1984. LNCS, vol. 196, pp. 47–53. Springer, Heidelberg (1985)

43. Wang, X., Yin, Y.L., Yu, H.: Finding collisions in the full SHA-1. In: Shoup, V. (ed.) CRYPTO 2005. LNCS, vol. 3621, pp. 17–36. Springer, Heidelberg (2005)

44. Waters, B.: Efficient identity-based encryption without random oracles. In: Cramer, R. (ed.) EUROCRYPT 2005. LNCS, vol. 3494, pp. 114–127. Springer, Heidelberg (2005)

45. Yao, A.C.: Theory and application of trapdoor functions. In: Proc. of FOCS 1982, pp. 80–91. IEEE Computer Society Press, Los Alamitos (1982)

46. Zhang, R.: Tweaking TBE/IBE to PKE transforms with chameleon hash functions. In: Katz, J., Yung, M. (eds.) ACNS 2007. LNCS, vol. 4521, pp. 323–339. Springer, Heidelberg (2007)

Universally Anonymous IBE Based on the Quadratic Residuosity Assumption

Giuseppe Ateniese[1] and Paolo Gasti[2,*]

[1] The Johns Hopkins University
[2] University of Genova
ateniese@cs.jhu.edu, gasti@disi.unige.it

Abstract. We introduce the first universally anonymous, thus key-private, IBE whose security is based on the standard quadratic residuosity assumption. Our scheme is a variant of Cocks IBE (which is not anonymous) and is efficient and highly parallelizable.

1 Introduction

Identity-based encryption was introduced by Shamir in 1984 [21]. He asked whether it was possible to encrypt a message by just using the identity of the intended recipient. Several partial and inefficient solutions were proposed after Shamir's initial challenge but it was only in 2000 that Sakai et al. [19], Boneh and Franklin [8], and Cocks [11] came up with very practical solutions.

The Boneh-Franklin work has been the most influential of all: it did not just introduce the first practical IBE scheme but, more importantly, it provided appropriate assumptions and definitions and showed how to pick the right curves, how to encode and map elements into points, etc.

Cocks' scheme is *per se* revolutionary: it is the first IBE that does not use pairings but rather it works in standard RSA groups and its security relies on the standard quadratic residuosity assumption (within the random oracle model). Cocks IBE, however, encrypts the message bit by bit and thus it is considered very bandwidth consuming. On the other end, Cocks [11] observes that his scheme can be used in practice to encrypt short session keys in which case the scheme becomes very attractive. We may add that the importance of relying on such a standard assumption should not be underestimated. In fact, this is what motivated the recent work of Boneh, Gentry, and Hamburg [9] where a new space-efficient IBE scheme is introduced whose security is also based on the quadratic residuosity assumption. Unfortunately, as the authors point out [9], their scheme is not efficient, it is more expensive than Cocks IBE and in fact it is more expensive than all standard IBE and public-key encryption schemes since its complexity is quartic in the security parameter (in particular, the encryption algorithm may take several seconds to complete even on a fast machine).

However, the scheme of Boneh et al. [9] has an important advantage over the scheme of Cocks: it provides anonymity, i.e., nobody can tell who the intended

---

* Work done while visiting The Johns Hopkins University.

M. Fischlin (Ed.): CT-RSA 2009, LNCS 5473, pp. 32–47, 2009.

recipient is by just looking at the ciphertext. Anonymity, or key-privacy, is a very important property that was first studied by Bellare et al. [4]. Recipient anonymity can be used, for example, to thwart traffic analysis, to enable searching on encrypted data [7], or to anonymously broadcast messages [1]. Several IBE schemes provide anonymity, for instance the Boneh-Franklin scheme is anonymous. Other schemes that do not originally provide anonymity can be either properly modified [10] or adapted to work in the XDH setting [2,3,6,20].

At this point, it is natural to ask whether it is possible to enhance Cocks IBE and come up with a variant that provides anonymity and that, unlike Boneh et al.'s scheme [9], is as efficient as the original scheme of Cocks.

The first attempt in this direction has been proposed recently by Di Crescenzo and Saraswat [14]. They provide the first public-key encryption with keyword search (PEKS) that is not based on pairings. Although their scheme is suitable for PEKS, we note that when used as an IBE it becomes quite impractical: it uses four times the amount of bandwidth required by Cocks and it requires each user to store and use a very large number of secret keys (four keys per each bit of the plaintext). In addition, the security of their scheme is based on a new assumption they introduce but we can show that their assumption is equivalent to the standard quadratic residuosity one.

Universal anonymity is a new and exciting notion introduced at Asiacrypt 2005 by Hayashi and Tanaka [17]. An encryption scheme is universally anonymous if ciphertexts can be made anonymous by anyone and not just by whoever created the ciphertexts. Specifically, a universally anonymizable public-key encryption scheme consists of a standard public-key encryption scheme and two additional algorithms: one used to anonymize ciphertexts, which takes as input only the public key of the recipient, and the other is used by the recipient to decrypt anonymized ciphertexts.

The following observations are obvious but worth emphasizing: (1) A universally anonymous scheme is also key-private in the sense of Bellare et al. [4]. What makes universally anonymity interesting and unique is that anyone can anonymize ciphertexts using just the public key of the recipient. (2) Key-private schemes can be more expensive than their non-private counterparts. For instance, RSA-OAEP can be made key-private as shown in [4] but the new anonymous variant is more expensive. (3) The concept of universal anonymity makes sense also for schemes that are already key-private. For instance, ElGamal is key-private only by assuming that all keys are generated in the same group and participants share the same public parameters. But in many scenarios this is not the case. In PGP, for instance, parameters for each user are selected in distinct groups. Evidently, ElGamal applied in different algebraic groups is not anonymous anymore as one can test whether a given ciphertext is in a group or not.

Our contributions are
(1) We enhance Cocks IBE and make it universally anonymous, and thus key-private in the sense of Bellare et al. [4]. Our variant of Cocks IBE can be seen as the most efficient anonymous IBE whose security is based on the quadratic residuosity assumption. The efficiency of our scheme is comparable to that of

Cocks IBE. In fact, it is substantially more efficient than the recent scheme of Boneh et al. [9] and the IBE that derives from the PEKS construction by Di Crescenzo et al. [14]. In addition, the ciphertext expansion of our scheme is comparable to that of Cocks IBE.

(2) We implemented our variant and measured its performance. We show that in practice the efficiency of Cocks IBE and the variant we propose in this paper compare favorably even with that of the Boneh-Franklin scheme.

(3) Incidentally, our solutions and techniques can be used to simplify the PEKS construction in [14] and our Lemma 2 in Section 2.3 can be used to show that the new security assumption introduced in [14] is actually equivalent to the standard quadratic residuosity assumption, thus making the elegant Di Crescenzo-Saraswat PEKS scheme the first one whose security is based solely on such a standard assumption (which was left as an open problem in the area). However, we will not elaborate on this point any further for lack of space.

Hybrid Encryption and CCA-security. It is well-known that in order to encrypt long messages, asymmetric encryption can be used in combination with symmetric encryption for improved efficiency. This simple and well-known paradigm has been formalized only recently by Cramer and Shoup [13,12] and Shoup [23]. It is introduced as the KEM-DEM construction which consists of two parts: the key encapsulation mechanism (KEM), used to encrypt a symmetric key, and the data encapsulation mechanism (DEM) that is used to encrypt the plaintext via a symmetric cipher.

The focus of this paper is on variants of Cocks IBE which can be proven secure only in the random oracle model. Thus, it makes sense to consider KEM-DEM constructions that are CCA-secure in such a model. It is possible to show (see, e.g., Bentahar et al. [5]) that if a KEM returns $(\text{Encrypt}_{\text{UAnonIBE}}(K), F(K))$, where $\text{Encrypt}_{\text{UAnonIBE}}(K)$ is a one-way encryption for an identity and F is a hash function modeled as a random oracle, then the combination of this KEM with a CCA-secure DEM results in a CCA-secure hybrid encryption. (Note that one-way encryption is implied by CPA-security.) Since our scheme UAnonIBE and its efficient variants are CPA-secure in the random oracle model, the resulting hybrid encryption that follows from the paradigm above is a CCA-secure encryption in the random oracle model.

2   Preliminaries

In this section, we recall first the IBE scheme proposed by Cocks [11]. Then we show that Cocks IBE is not anonymous due to a test proposed by Galbraith, as reported in [7]. Finally, we show that Galbraith's test is the "best test" possible against the anonymity of Cocks IBE. We assume that N is a large-enough RSA-type modulus. Hence, throughout the paper, we will omit to consider cases where randomly picked elements are in $\mathbb{Z}_N$ but not in $\mathbb{Z}_N^*$ or, analogously, have Jacobi symbol over N equal to 0 since these cases occur only with negligible probability[1]. Therefore, for consistency, we always assume to work in $\mathbb{Z}_N^*$ rather than in $\mathbb{Z}_N$ even though $\mathbb{Z}_N^*$ is not closed under modular addition.

We will denote with $\mathbb{Z}_N^*[+1]$ ($\mathbb{Z}_N^*[-1]$) the set of elements in $\mathbb{Z}_N^*$ with Jacobi symbol $+1$ (-1, resp.) and with $\mathbb{QR}(N)$ the set of quadratic residues (or squares) in $\mathbb{Z}_N^*$. The security of Cocks IBE (and our variants) relies on the standard quadratic residuosity assumption which simply states that the two distributions $DQR(n) = \{(c,N) : (N,p,q) \overset{R}{\leftarrow} Gen(1^n), \ c \overset{R}{\leftarrow} \mathbb{QR}(N)\}$ and $DQRN(n) = \{(c,N) : (N,p,q) \overset{R}{\leftarrow} Gen(1^n), \ c \overset{R}{\leftarrow} \mathbb{Z}_N^*[+1] \setminus \mathbb{QR}(N)\}$ are computationally indistinguishable, where n is a security parameter and $Gen(\cdot)$ generates a RSA-type n-bit Blum modulus and its two prime factors.

2.1 Cocks' IBE Scheme

Let $N = pq$ be a Blum integer, i.e., where p and q are primes each congruent to 3 modulo 4. In addition, we consider $H : \{0,1\}^* \to \mathbb{Z}_N^*[+1]$ a full-domain hash which will be modeled as a random oracle in the security analysis.

Master Key: The secret key of the trusted authority is (p,q) while its public key is $N = pq$.

Key Generation: Given the identity ID, the authority generates $a = H(ID)$ (thus the Jacobi symbol $\left(\frac{a}{N}\right)$ is $+1$). The secret key for the identity ID is a value r randomly chosen in $\mathbb{Z}_N^*$ such that $r^2 \equiv a \bmod N$ or $r^2 \equiv -a \bmod N$. This value r is stored and returned systematically.

Encryption: To encrypt a bit $b \in \{-1,+1\}$ for identity ID, choose uniformly at random two values $t, v \in \mathbb{Z}_N^*$, such that $\left(\frac{t}{N}\right) = \left(\frac{v}{N}\right) = b$, and compute:

$$(c,d) = \left(t + \frac{a}{t} \bmod N, v - \frac{a}{v} \bmod N\right)$$

Decryption: Given a ciphertext (c,d), first set $s = c$ if $r^2 \equiv a \bmod N$ or $s = d$ otherwise. Then, decrypt by computing:

$$\left(\frac{s + 2r}{N}\right) = b$$

Notice that $s + 2r \equiv w(1 + r/w)^2 \bmod N$, thus the Jacobi symbol of $s + 2r$ is equal to that of w, where w is either t or v.

2.2 Galbraith's Test (GT)

As mentioned in the paper by Boneh et al. [7], Galbraith showed that Cocks' scheme is not anonymous. Indeed, let $a \in \mathbb{Z}_N^*[+1]$ be a public key and consider the following set:

---

[1] The Jacobi symbol of $a \in \mathbb{Z}_N$ is denoted as $\left(\frac{a}{N}\right)$ and is either -1, 0, or $+1$. However, $\left(\frac{a}{N}\right) = 0$ if and only if $\gcd(a,N) \neq 1$, thus this case happens only with negligible probability since the value $\gcd(a,N)$ would be a non-trivial factor of N.

$$S_a[N] = \left\{ t + \frac{a}{t} \bmod N \mid t \in \mathbb{Z}_N^* \right\} \subset \mathbb{Z}_N^*$$

Given two random public keys $a, b \in \mathbb{Z}_N^*[+1]$, Galbraith's test (which we will denote with "$GT(\cdot)$") allows us to distinguish the uniform distribution on the set $S_a[N]$ from the uniform distribution on the set $S_b[N]$. Given $c \in \mathbb{Z}_N^*$, the test over the public key a is defined as the Jacobi symbol of $c^2 - 4a$ over N, that is:

$$GT(a, c, N) = \left(\frac{c^2 - 4a}{N} \right)$$

Notice that when c is sampled from $S_a[N]$, the test $GT(a, c, N)$ will always return $+1$ given that $c^2 - 4a = (t - (a/t))^2$ is a square. However, if c is sampled from $S_b[N]$ the test is expected to return $+1$ with probability negligibly close to $1/2$ since, in this case, the distribution of the Jacobi symbol of the element $c^2 - 4a$ in $\mathbb{Z}_N^*$ follows the uniform distribution on $\{-1, +1\}$.

It is mentioned in [7] that since Cocks ciphertext is composed of several values sampled from either $S_a[N]$ (and $S_{-a}[N]$) or $S_b[N]$ (and $S_{-b}[N]$, respectively), then an adversary can repeatedly apply Galbraith's test to determine with overwhelming probability whether a given ciphertext is intended for a or b. However, one must first prove some meaningful results about the distribution of Jacobi symbols of elements of the form $c^2 - 4b$ in $\mathbb{Z}_N^*$, for *fixed* random elements $a, b \in \mathbb{Z}_N^*[+1]$ and for $c \in S_a[N]$. These results are reported in the next section.

2.3   Relevant Lemmata and Remarks

Damgård in [15] studied the distribution of Jacobi symbols of elements in $\mathbb{Z}_N^*$ in order to build pseudo-random number generators. In his paper, Damgård reports of a study performed in the 50s by Perron in which it is proven that for a prime p and for any a, the set $a + \mathbb{QR}(p)$ contains as many squares as non squares in $\mathbb{Z}_p^*$ when $p \equiv 1 \mod 4$, or the difference is just 1 when $p \equiv 3 \mod 4$. It is possible to generalize Perron's result to study the properties of the set $a + \mathbb{QR}(N)$ in $\mathbb{Z}_N^*$ but we also point out that the security of Cocks IBE implicitly depends on the following Lemma:

Lemma 1. *Let (a, N) be a pair such that $(N, p, q) \xleftarrow{R} Gen(1^n)$ and $a \xleftarrow{R} \mathbb{Z}_N^*[+1]$. The distribution $\left\{ \left(\frac{t^2 + a}{N} \right) : t \xleftarrow{R} \mathbb{Z}_N^* \right\}$ is computationally indistinguishable from the uniform distribution on $\{-1, +1\}$ under the quadratic residuosity assumption.*

To prove the Lemma above it is enough to observe that if we compute the Jacobi symbol of a value $c \in S_a[N]$ we obtain:

$$\left(\frac{c}{N} \right) = \left(\frac{(t^2 + a)/t}{N} \right) = \left(\frac{t^2 + a}{N} \right) \left(\frac{t}{N} \right)$$

However the Jacobi symbol of t over N is the plaintext in Cocks IBE and thus Lemma 1 must follow otherwise the CPA-security of Cocks IBE would not hold.

Remark. Let's pick c randomly in $\mathbb{Z}_N^*$. If $GT(a, c, N) = -1$, we can clearly conclude that $c \notin S_a[N]$. However, if $GT(a, c, N) = +1$, what is the probability that $c \in S_a[N]$? The answer is $1/2$ since a t exists such that $c = t + a/t$ whenever $c^2 - 4a$ is a square and this happens only half of the times (clearly $GT(a, c, N)$ is equal to 0 with negligible probability hence we do not consider this case). To summarize:

$$GT(a, c, N) = \begin{cases} +1 \Longrightarrow c \in S_a[N] \text{ with prob. } 1/2 \\ -1 \Longrightarrow c \notin S_a[N] \end{cases}$$

We will argue that there is no *better* test against anonymity over an encrypted bit. That is, we show that a test that returns $+1$ to imply that $c \in S_a[N]$ with probability $1/2 + \delta$ (for a non-negligible $\delta > 0$) cannot exist under the quadratic residuosity assumption. We first notice that $c \in S_a[N]$ if and only if $\Delta = c^2 - 4a$ is a square. Indeed, if $c = t + a/t$ then $\Delta = (t - a/t)^2$. If Δ is a square then the quadratic equation $c = t + a/t$ has solutions for t in $\mathbb{Z}_N^*$ with overwhelming probability. Thus $S_a[N]$ can alternatively be defined as the set of all $c \in \mathbb{Z}_N^*$ such that $c^2 - 4a$ is a square.

Intuitively, we can see Galbraith's test as an algorithm that checks whether the discriminant Δ has Jacobi symbol $+1$ or -1, and this is clearly *the best it can do* since the factors of the modulus N are unknown. (Remember that we do not consider cases where the Jacobi symbol is 0 since they occur with negligible probability.) Indeed, if $\pm x$ and $\pm y$ are the four distinct square roots modulo N of Δ, then $t^2 - ct + a$ is congruent to 0 modulo N whenever t is congruent modulo N to any of the following four distinct values:

$$\frac{c \pm x}{2} \text{ and } \frac{c \pm y}{2}$$

We denote with $GT_a^N[+1]$ the set $\{c \in \mathbb{Z}_N^* \mid GT(a, c, N) = +1\}$. Analogously, we define $GT_a^N[-1]$ as the set $\{c \in \mathbb{Z}_N^* \mid GT(a, c, N) = -1\}$. We prove the following Lemma:

Lemma 2. [VQR–Variable Quadratic Residuosity] *The distributions* $D_0(n) = \{(a, c, N) : (N, p, q) \xleftarrow{R} Gen(1^n),\ a \xleftarrow{R} \mathbb{Z}_N^*[+1],\ c \xleftarrow{R} S_a[N]\}$ *and* $D_1(n) = \{(a, c, N) : (N, p, q) \xleftarrow{R} Gen(1^n),\ a \xleftarrow{R} \mathbb{Z}_N^*[+1],\ c \xleftarrow{R} GT_a^N[+1] \setminus S_a[N]\}$ *are computationally indistinguishable under the quadratic residuosity assumption.*

Proof. We assume there is a PPT adversary A that can distinguish between $D_0(n)$ and $D_1(n)$ with non-negligible advantage and we use it to solve a random instance of the quadratic residuosity problem. We make no assumptions on how A operates and we evaluate it via oracle access.

The simulator is given a random tuple (N, x) where $(N, p, q) \xleftarrow{R} Gen(1^n)$ and $x \in \mathbb{Z}_N^*[+1]$. The simulation proceeds as follows:

1. Find a random $r \in \mathbb{Z}_N^*$ such that $a = (r^2 - x)/4$ has Jacobi symbol $+1$ (see Lemma 1). A receives as input (a, r, N), where $a \in \mathbb{Z}_N^*[+1]$ and $r \in GT_a^N[+1]$. Notice that releasing the public key a effectively provides A with the ability to generate several values in $S_a[N]$ (and $S_{-a}[N]$);

2. If A responds that $r \in S_a[N]$ then output *"x is a square"* otherwise output *"x is not a square"*.

The value a is distributed properly and it has already been established that $r \in S_a[N]$ if and only if $r^2 - 4a$ is a square. But $r^2 - 4a = x$, therefore A cannot have non-negligible advantage under the quadratic residuosity assumption. □

The next Lemma easily follows from Lemma 1 since $c^2 - 4a$ can be written as $c^2 + h$ for a fixed $h \in \mathbb{Z}_N^*[+1]$.

Lemma 3. *Let (a, N) be a pair such that $(N, p, q) \xleftarrow{R} Gen(1^n)$ and a $\xleftarrow{R} \mathbb{Z}_N^*[+1]$. The distribution $\{GT(a, c, N) : c \xleftarrow{R} \mathbb{Z}_N^*\}$ is computationally indistinguishable from the uniform distribution on $\{-1, +1\}$.*

3 Our Basic Construction and Its Efficient Variants

We extend Cocks' scheme to support anonymity. Unlike previous proposals, our scheme UAnonIBE has efficiency, storage, and bandwidth requirements similar to those of the original scheme by Cocks (which is not anonymous). Our scheme is also the first universally anonymous IBE, according to the definition in [17] (although we do not include the extra algorithms as in [17] to keep the presentation simple).

3.1 The Basic Scheme

Let $H : \{0, 1\}^* \rightarrow \mathbb{Z}_N^*[+1]$ be a full-domain hash modeled as a random oracle. Let n and m be two security parameters. The algorithms which form UAnonIBE are defined as follows (all operations are performed modulo N):

Master Key: The public key of the trusted authority is the n-bit Blum integer $N = pq$, where p and q are $n/2$-bit primes each congruent to 3 modulo 4.

Key Generation: Given the identity ID, the authority generates $a = H(ID)$ (thus the Jacobi symbol $\left(\frac{a}{N}\right)$ is $+1$). The secret key for the identity ID is a value r randomly chosen in $\mathbb{Z}_N^*$ such that $r^2 \equiv a \mod N$ or $r^2 \equiv -a \mod N$. This value r is stored and returned systematically.

Encryption: To encrypt a bit $b \in \{-1, +1\}$ for identity ID, choose uniformly at random two values $t, v \in \mathbb{Z}_N^*$, such that $\left(\frac{t}{N}\right) = \left(\frac{v}{N}\right) = b$, and compute $(c, d) = \left(t + \frac{a}{t}, v - \frac{a}{v}\right)$.

Then, compute the *mask* to anonymize the ciphertext (c, d) as follows:

1. Pick two indices k_1 and k_2 independently from the geometric distribution[2] D with probability parameter $1/2$;
2. Select random T, V in $\mathbb{Z}_N^*$ and set $Z_1 = c + T$ and $Z_2 = d + V$;
3. For $1 \leq i < k_1$, select random values $T_i \in \mathbb{Z}_N^*$ s.t. $GT(a, Z_1 - T_i, N) = -1$;
4. For $1 \leq i < k_2$, select random values $V_i \in \mathbb{Z}_N^*$ s.t. $GT(-a, Z_2 - V_i, N) = -1$;
5. Set $T_{k_1} = T$ and $V_{k_2} = V$;
6. For $k_1 < i \leq m$, select random values $T_i \in \mathbb{Z}_N^*$;
7. For $k_2 < i \leq m$, select random values $V_i \in \mathbb{Z}_N^*$;

Finally, output $(Z_1, T_1, \ldots, T_m)$ and $(Z_2, V_1, \ldots, V_m)$.[3]

Decryption: Given a ciphertext $(Z_1, T_1, \ldots, T_m)$ and $(Z_2, V_1, \ldots, V_m)$, first discard one of the two tuples based on whether a or $-a$ is a square. Let's assume we keep the tuple $(Z_1, T_1, \ldots, T_m)$ and we discard the other. In order to decrypt, find the smallest index $1 \leq i \leq m$ s.t. $GT(a, Z_1 - T_i, N) = +1$ and output:

$$\left(\frac{Z_1 - T_i + 2r}{N} \right) = b$$

We run the same procedure above if the second tuple is actually selected and the first is discarded. It is enough to replace a with $-a$, Z_1 with Z_2, and T_i with V_i.

3.2 Security Analysis

We need to show that our scheme, UAnonIBE, is ANON-IND-ID-CPA-secure [1,9], that is, the ciphertext does not reveal any information about the plaintext and an adversary cannot determine the identity under which an encryption is computed, even thought the adversary selects the identities and the plaintext.

In [16], Halevi provides a sufficient condition for a CPA public-key encryption scheme to meet the notion of key-privacy, or anonymity, as defined by Bellare et al. in [4]. In [1], Abdalla et al. extend Halevi's condition to identity-based encryption. In addition, their notion is defined within the random oracle model and Halevi's statistical requirement is weakened to a computational one. Informally, it was observed that if an IBE scheme is already IND-ID-CPA-secure then the oracle does not have to encrypt the message chosen by the adversary but can encrypt a random message of the same length. The game where the oracle replies with an encryption on a random message is called ANON-RE-CPA. In [1], it was shown that if a scheme is IND-ID-CPA-secure and ANON-RE-CPA-secure then it is also ANON-IND-ID-CPA-secure.

---

[2] The geometric distribution is a discrete memoryless random distribution for $k = 1, 2, 3, \ldots$ having probability function $\Pr[k] = p(1-p)^{k-1}$ where $0 < p < 1$. Therefore, for $p = 1/2$ the probability that $k_1 = k$ is 2^{-k}. For more details see, e.g., [24].

[3] Note that if we build a sequence $c_1, \ldots, c_m$ by selecting a k from D and setting $c_i = -1$ for $1 \leq i < k$, $c_k = 1$, and $c_i \in_R \{-1, 1\}$ for $k < i \leq m$, we have that $\Pr[c_i = 1] = 2^{-i} + \sum_{j=2}^{i} 2^{-j} = 1/2$.

ANON-RE-CPA game. We briefly describe the security game introduced by Abdalla et al. in [1]. MPK represents the set of public parameters of the trusted authority. The adversary A has access to a random oracle H and to an oracle $KeyDer$ that given an identity ID returns the private key for ID according to the IBE scheme.

Experiment $\mathbf{Exp}_{\mathbf{IBE},\mathbf{A}}^{\mathbf{anon-re-cpa-b}}(n)$:
  pick random oracle H
  $(ID_0, ID_1, msg, state) \leftarrow A^{KeyDer(\cdot),H}(find, MPK)$
  $W \xleftarrow{R} \{0,1\}^{|msg|}; C \leftarrow Enc^H(MPK, ID_b, W)$
  $b' \leftarrow A^{KeyDer(\cdot),H}(guess, C, state)$
  return b'

The adversary cannot request the private key for ID_0 or ID_1 and the message msg must be in the message space associated with the scheme. The ANON-RE-CPA-advantage of an adversary A in violating the anonymity of the scheme IBE is defined as:

$$\mathbf{Adv}_{\mathbf{IBE},\mathbf{A}}^{\mathbf{anon-re-cpa}}(n) = \Pr\left[\mathbf{Exp}_{\mathbf{IBE},\mathbf{A}}^{\mathbf{anon-re-cpa-1}}(n) = 1\right] - \Pr\left[\mathbf{Exp}_{\mathbf{IBE},\mathbf{A}}^{\mathbf{anon-re-cpa-0}}(n) = 1\right]$$

A scheme is said to be ANON-RE-CPA-secure if the above advantage is negligible in n.

Theorem 1. UAnonIBE *is* ANON-IND-ID-CPA-*secure in the random oracle model under the quadratic residuosity assumption.*

In order to simplify the proof of theorem 1, we make and prove an important claim first. We will show that a ciphertext for a random $a \in \mathbb{Z}_N^*[+1]$ is indistinguishable from a sequence of random elements in $\mathbb{Z}_N$ to a PPT distinguisher $\mathcal{DS}$. In particular, let $\mathcal{O}\{S_a[N], GT_a^N[-1], GT_a^N[+1]\}$ be an oracle that returns UAnonIBE encryptions of random messages under the public key a and let $\mathcal{O}^*$ an oracle that returns a $(m+1)$-tuple of elements picked uniformly at random from $\mathbb{Z}_N^*$. We prove the following:

Claim. The distinguisher $\mathcal{DS}$ has only negligible advantage in distinguishing the outputs of the oracles $\mathcal{O}\{S_a[N], GT_a^N[-1], GT_a^N[+1]\}$ and $\mathcal{O}^*$ under the quadratic residuosity assumption.

Proof. Let $(Z_1, T_1, \ldots, T_m)$ be the output of $\mathcal{O}\{S_a[N], GT_a^N[-1], GT_a^N[+1]\}$. In particular, a $c \in S_a[N]$ is randomly picked and Z_1 is set to $c + T_k$, where k is chosen according to the geometric distribution D defined in the UAnonIBE encryption algorithm. Let $(U_0, U_1, \ldots, U_m)$ be the output of $\mathcal{O}^*$.

First, notice that there exists a minimal index k' such that $GT(a, U_0 - U_{k'}, N) = +1$. Such an index exists with probability $1 - 2^{-m}$ because of Lemma 3. Second, it is easy to see that the distribution induced by the index k' is indistinguishable from the distribution D. This still follows from Lemma 3 since we know that the probability that $GT(a, U_0 - U_i, N) = +1$ is negligibly close to $1/2$, for $1 \leq i \leq m$. Hence, the probability that $k' = v$, for a positive integer $v \in \mathbb{N}$, is negligibly close to 2^{-v}. Thus, both indices k and k' determine the same distribution except with negligible probability (to account for the cases where Galbraith's test returns 0). Finally, because of Lemma 2, $\mathcal{DS}$ cannot determine whether $U_0 - U_{k'} \in GT_a^N[+1]$ is in $S_a[N]$ or not. $\qquad\square$

Remark. We point out an insightful analogy between the oracle $\mathcal{O}\{S_a[N], GT_a^N[-1], GT_a^N[+1]\}$ in the claim above and the oracle $\mathcal{O}\{\mathbb{QR}(N), \mathbb{Z}_N^*[-1], \mathbb{Z}_N^*[+1]\}$ which picks elements $c \in \mathbb{QR}(N)$ (rather than in $S_a[N]$) and sets $Z_1 = c + T_k$, where k is chosen according to D. Then, it generates elements $T_1, \ldots, T_m$ such that: (1) $Z_1 - T_i \in \mathbb{Z}_N^*[-1]$, for $1 \leq i < k$, (2) $Z_1 - T_k \in \mathbb{QR}(N)$, and (3) $Z_1 - T_i \in \mathbb{Z}_N^*$, for $k < i \leq m$. Evidently, even the outputs of this oracle are indistinguishable from the outputs of $\mathcal{O}^*$ under the quadratic residuosity assumption.

Proof of Theorem 1. It must be clear that UAnonIBE is IND-ID-CPA-secure since Cocks IBE is IND-ID-CPA-secure in the random oracle model under the quadratic residuosity assumption and the mask is computed without knowing the plaintext or the secret key of the intended recipient. Thus, we only need to show that UAnonIBE is ANON-RE-CPA-secure. But, because of the Claim above, a PPT adversary A must have negligible advantage in determining whether the ciphertext C returned by $ENC^H(\cdot)$ is for ID_0 or ID_1 because C is (with overwhelming probability) a proper encryption for both ID_0 and ID_1 on two *random* bits. (It is equivalent to respond to A with two $(m+1)$-tuples of random elements in $\mathbb{Z}_N^*$.) $\qquad\square$

3.3   A First Efficient Variant: Reducing Ciphertext Expansion

The obvious drawback of the basic scheme is its ciphertext expansion. Indeed, for each bit of the plaintext $2 \cdot (m+1)$ values in $\mathbb{Z}_N^*$ must be sent while in Cocks IBE each bit of the plaintext requires two values in $\mathbb{Z}_N^*$. Therefore, we need a total of $2 \cdot (m+1) \cdot n$ bits for a single bit in the plaintext, where n and m are the security parameters (e.g., $n = 1024$ and $m = 128$). However, this issue is easy to fix. Intuitively, since our scheme requires the random oracle model for its security, we could use another random oracle that expands a short seed into a value selected uniformly and independently in $\mathbb{Z}_N^*$.

Specifically, a function $G : \{0,1\}^* \rightarrow \mathbb{Z}_N^*$ is used, which we model as a random oracle, that maps a e-bit string α to a random value in $\mathbb{Z}_N^*$. The parameter e must be large enough, e.g., $e = 160$.

It is tempting to use the oracle G and a single short seed α plus a counter to generate all values $T_1, \ldots, T_m$ and $V_1, \ldots, V_m$. This first solution would provide minimal ciphertext expansion, since only the seed α must be sent, however it may turn out to be computationally expensive. To see this, consider that an α must be found such that $GT(a, Z_1 - T_i, N) = -1$ for $1 \leq i < k_1$. Now, if k_1 happens to be large, say $k_1 = 20$, then clearly finding a suitable α could be computationally intensive. Nevertheless, we prove that this scheme is secure as long as the basic UAnonIBE scheme is secure. More importantly, we emphasize that the proof of security of all other schemes proposed after this first one can easily be derived from the proof of the following theorem.

Theorem 2. *The first efficient variant of* UAnonIBE *is* ANON-IND-ID-CPA-*secure in the random oracle model under the quadratic residuosity assumption.*

Proof. We let the simulator S play the role of a man-in-the-middle attacker between two ANON-RE-CPA games: the first game is against the basic UAnonIBE and the second game is against an adversary A that has non-negligible advantage in breaking the first variant of UAnonIBE. We show that S can use A to win in the first ANON-RE-CPA game, thus violating the quadratic residuosity assumption. The simulation is straightforward: S forwards the H-queries and $KeyDer$-queries to the respective oracles. When A challenges for identities ID_0 and ID_1, S challenges on the same identities in the first ANON-RE-CPA game. Then S receives the ciphertext $(Z_1, T_1, \ldots, T_m)$, $(Z_2, V_1, \ldots, V_m)$. S sends to A, $(Z_1, \alpha), (Z_2, \beta)$ where α and β are chosen uniformly at random in $\{0, 1\}^e$. At this point, the simulator responds to the G-queries as follows:

$$G(\alpha \parallel i) = T_i \text{ and } G(\beta \parallel i) = V_i \text{ , for } 1 \leq i \leq m,$$

and with random values in $\mathbb{Z}_N^*$ in any other cases. The adversary A eventually returns its guess which S uses in the first game in order to win with non-negligible advantage.   □

The obvious next-best solution is to use a single seed per value. Thus, rather than sending the ciphertext as per our basic scheme, that is $(Z_1, T_1, \ldots, T_m)$ and $(Z_2, V_1, \ldots, V_m)$, the following values could be sent:

$$(Z_1, \alpha_1, \ldots, \alpha_m) \text{ and } (Z_2, \beta_1, \ldots, \beta_m),$$

where α_i, β_i are chosen uniformly at random in $\{0, 1\}^e$ until the conditions in steps 3. and 4. of the encryption algorithm of the basic scheme are satisfied. The recipient would then derive the intended ciphertext by computing $T_i = G(\alpha_i)$ and $V_i = G(\beta_i)$, for $1 \leq i \leq m$. If we set e to be large enough, say $e = 160$, then clearly the security of this variant is equivalent to the one of the basic scheme in the random oracle model and a single bit of the plaintext would require $2 \cdot (m \cdot e + n)$ bits rather than $2 \cdot (m \cdot n + n)$, where $e < n$. Hence, for $n = 1024$, $m = 128$ and $e = 160$, we need to send $2 \cdot (160 \cdot 128 + 1024)$ bits while Cocks' scheme requires only $2 \cdot (1024)$ bits.

On a closer look however, it is easy to see that since G is a random oracle we just need to ensure that its inputs are repeated only with negligible probability. Let $X = x^{(1)} x^{(2)} \ldots x^{(t)}$ be the plaintext of t bits. For each plaintext X, the sender selects a random message identifier $MID_X \in \{0, 1\}^{e_1}$ which is sent along with the ciphertext. For bit $x^{(j)}$, the sender computes:

$$(Z_1^{(j)}, \alpha_1^{(j)}, \ldots, \alpha_m^{(j)}) \text{ and } (Z_2^{(j)}, \beta_1^{(j)}, \ldots, \beta_m^{(j)}),$$

where the coefficients $\alpha_i^{(j)}$, $\beta_i^{(j)}$ are chosen uniformly at random in $\{0,1\}^e$ until the conditions in steps 3. and 4. of the encryption algorithm of the basic scheme are satisfied (thus notice that e can be small but still big enough to be able to find those values $T_i^{(j)}$ and $V_i^{(j)}$ that satisfy such conditions). The recipient will derive the intended ciphertext by computing:

$$T_i^{(j)} = G(MID_X \,||\, 0 \,||\, \alpha_i^{(j)} \,||\, i \,||\, j) \text{ or } V_i^{(j)} = G(MID_X \,||\, 1 \,||\, \beta_i^{(j)} \,||\, i \,||\, j),$$

where $i \in \{1,\ldots,m\}$ and $j \in \{1,\ldots,t\}$. As an example, we can set $m = 128$, $e_1 = 160$, and $e = 8$. In this case the ciphertext expansion per single bit of the plaintext is only $2 \cdot (1024 + 1024)$ bits which is twice the amount required by Cocks IBE for $n = 1024$. (In addition, extra 160 bits are needed for MID_X but these bits are transmitted only once per message.)

3.4   A Second Efficient Variant: Trade-Off between Ciphertext Expansion and Performance

We propose a second variant of UAnonIBE which provides an optimal trade-off between efficiency and ciphertext expansion. Our performance tests show that this variant is in practice as efficient as any of the previous variants and at the same time it provides the smallest ciphertext expansion (thus we recommend this version for practical systems).

We fix a new global parameter ℓ which is a small positive integer. Let $X = x^{(1)}x^{(2)}\ldots x^{(t)}$ be the plaintext of t bits. For each plaintext X, the sender selects a random identifier $MID_X \in \{0,1\}^{e_1}$ which is sent along with the ciphertext. For bit $x^{(j)}$, the sender computes:

$$(Z_1^{(j)},\alpha_1^{(j)},\ldots,\alpha_\ell^{(j)}) \text{ and } (Z_2^{(j)},\beta_1^{(j)},\ldots,\beta_\ell^{(j)})$$

where $\alpha_i^{(j)}$, $\beta_i^{(j)}$ are in $\{0,1\}^e$, when $i < \ell$, and $\alpha_\ell^{(j)}$, $\beta_\ell^{(j)}$ are in $\{0,1\}^{e'}$, for some $e' > e$. The intended ciphertext is derived by the recipient by computing:

$$T_i^{(j)} = G(MID_X \,||\, 0 \,||\, \alpha_i^{(j)} \,||\, i \,||\, j) \text{ or } V_i^{(j)} = G(MID_X \,||\, 1 \,||\, \beta_i^{(j)} \,||\, i \,||\, j)$$

for $i < l$, and

$$T_i^{(j)} = G(MID_X \,||\, 0 \,||\, \alpha_\ell^{(j)} \,||\, i \,||\, j) \text{ or } V_i^{(j)} = G(MID_X \,||\, 1 \,||\, \beta_\ell^{(j)} \,||\, i \,||\, j)$$

for $i \geq \ell$. Note that in this variant of our basic scheme an arbitrary number of $T_i^{(j)}$ and $V_i^{(j)}$ can be generated (i.e., there is no fixed global parameter m).

Given the distribution of k_1, k_2, for a large enough ℓ, we expect $k_1 \leq \ell$ or $k_2 \leq \ell$ with high probability. When $k_1 \leq \ell$ or $k_2 \leq \ell$ the scheme is as efficient as the first variant. When $k_1 > \ell$ (or $k_2 > \ell$) the computational cost of finding a value for $\alpha_\ell^{(j)}$ (or $\beta_\ell^{(j)}$) is exponential in $k_1 - \ell$ ($k_2 - \ell$, respectively).

As an example, we set the global parameter $\ell = 6$ and then $e_1 = 160$, $e = 8$, $e' = 80$, and $n = 1024$. The ciphertext expansion of this variant of UAnonIBE is $2 \cdot ((\ell - 1) \cdot e + e' + n)$, therefore, the ciphertext size for a single bit of the plaintext is now only $2 \cdot (120 + 1024)$ bits which is very close to the number of bits $(2 \cdot (1024))$ required by Cocks IBE (which is not anonymous). Note that for each message, the sender also transmits the random message identifier MID_X.

4   Optimizations and Implementation

An important aspect that should be considered in order to implement UAnonIBE efficiently is the value of the parameters ℓ, e (the size of $\alpha_1^{(j)}, \ldots, \alpha_{\ell-1}^{(j)}$) and e' (the size of $\alpha_\ell^{(j)}$). These values affect both the ciphertext expansion and the encryption time significantly, therefore they must be selected carefully. Choosing e or e' to be too small can reduce the probability of encrypting to an unacceptable level. Choosing ℓ to be too small can make the encryption process very slow. If we set $e = 8$ and $e' = 80$, we can find a suitable value for each $\alpha_i^{(j)}$, and therefore encrypt, with a probability of at least $1 - 2^{-80}$. We found that the value $\ell = 6$ is the best compromise between encryption time and ciphertext expansion. If we set $e = 8$, $e' = 80$ and $\ell = 6$, the ciphertext expansion for a 128-bit message is 3840 bytes more than a Cocks encryption for both $+a$ and $-a$: for a 1024-bit modulus N the encrypted message size is about 36KB instead of 32KB with Cocks IBE.

Size of e	2	4	6	8	10	Cocks IBE
Ciphertext size in bytes	33748	34708	35668	36628	37588	32768

We implemented the second efficient variant of UAnonIBE and compared it with the original Cocks IBE [11] and the scheme proposed by Boneh and Franklin based on pairings [8]. We have two goals in mind. The first is to show that Cocks IBE and our schemes are practical when used as hybrid encryption algorithms (following the KEM-DEM paradigm), even when compared with the Boneh-Franklin IBE, with the clear advantage compared to other IBE schemes of relying

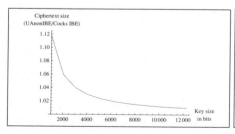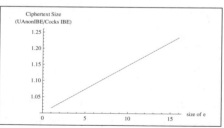

Fig. 1. The two graphs show UAnonIBE's ciphertext size relative to Cocks' scheme. The first shows how the relative bandwidth overhead introduced by our solution decreases with the size of the master parameter, while the second shows how the size of the ciphertext increases, varying the size of e and fixing $e' = 10 \cdot e$, compared to Cocks' ciphertext.

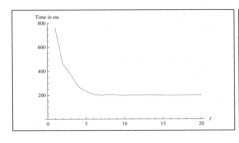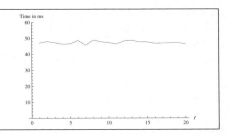

Fig. 2. The two graphs show the average time required respectively to anonymize and de-anonymize a 128-bit message encrypted with Cocks algorithm varying ℓ

on a well-established assumption. Our second goal is to show that the efficiency of our scheme is comparable to that of the original scheme by Cocks.

For our performance analysis, we set the size of the values $\alpha_i^{(j)}$ and $\beta_i^{(j)}$ with $1 \leq i < \ell$ to 8 bits and the size of $\alpha_\ell^{(j)}$ and $\beta_\ell^{(j)}$ to 80 bits. However, our tests showed that the size of those parameters have no measurable impact on the performance of the scheme. In order to calculate the optimal value for ℓ, we measured the time required to anonymize a key of 128 bit (for a total of 256 encrypted bits, considering both cases $+a$ and $-a$). Figure 2 summarizes our results. The value $\ell = 6$ seems to be optimal, since further increasing ℓ does not noticeably affect the time required to anonymize or decrypt a message.

Experimental Setup. We employed the MIRACL software package, developed by Shamus Software [22], to run our tests. MIRACL is a comprehensive library often used to implement cryptographic systems based on pairings. We used the optimized implementation of the Boneh-Franklin IBE provided by the library and we implemented Cocks IBE and our scheme with an RSA modulus of 1024 bits. The implementation of the Boneh-Franklin IBE uses a 512-bit prime p, Tate pairing and a small 160-bit subgroup q. The curve used is $y^2 = x^3 + x$ instead of $y^2 = x^3 + 1$ because it allows for a faster implementation. Those two settings should provide the same level of security according to NIST [18]. The tests were run on a machine that consisted of an Intel Pentium 4 2.8GHz with 512MB RAM. The system was running Linux kernel 2.6.20 with the compiler GCC 4.1.2. We implemented the cryptographic primitives using version 5.2.3 of the MIRACL library. Every source file was compiled with optimization '-02', as suggested in the MIRACL documentation. The table below shows average times over 1000 runs of the cryptographic operations specified in the first column. The symmetric key encrypted in our tests is of 128 bits.

	Extract	Encrypt	Decrypt	Anonymous	Universally Anonymous
Boneh-Franklin	9.1 ms	91.6 ms	85.4 ms	YES	NO
Cocks IBE	14.2 ms	115.3 ms	35.0 ms	NO	NO
Our Scheme	14.2 ms	319.4 ms	78.1 ms	YES	YES

In the table we also indicate whether a scheme is anonymous or not. Cocks IBE is not anonymous while Boneh-Franklin IBE is anonymous but not universally anonymous. One could try to turn Boneh-Franklin IBE into a universally anonymous scheme using for example the techniques in [17]. But, even assuming that this is possible, the new scheme would be different and more expensive than the original one and still depending on pairing-based assumptions.

5 Conclusions

We proposed UAnonIBE: the first IBE providing universal anonymity (thus key-privacy) and secure under the standard quadratic residuosity assumption. The efficiency and ciphertext expansion of our scheme are comparable to those of Cocks IBE. We showed that Cocks IBE and our anonymous variant are suitable in practice whenever hybrid encryption (KEM-DEM paradigm) is employed. We believe our schemes are valid alternatives to decidedly more expensive schemes introduced in [9] (which, in addition, are anonymous but not universally anonymous).

Acknowledgments. We are grateful to Giovanni Di Crescenzo, Marc Joye, Ivano Repetto, and the anonymous reviewers for their insightful comments.

References

1. Abdalla, M., Bellare, M., Catalano, D., Kiltz, E., Kohno, T., Lange, T., MaloneLee, J., Neven, G., Paillier, P., Shi, H.: Searchable Encryption Revisited: Consistency Properties, Relation to Anonymous IBE, and Extensions. In: Shoup, V. (ed.) CRYPTO 2005. LNCS, vol. 3621, pp. 205–222. Springer, Heidelberg (2005)
2. Ateniese, G., Camenisch, J., de Medeiros, B.: Untraceable RFID Tags via Insubvertible Encryption. In: CCS 2005: Proceedings of the 12th ACM conference on Computer and communications security, pp. 92–101. ACM, New York (2005)
3. Ballard, L., Green, M., de Medeiros, B., Monrose, F.: Correlation-Resistant Storage via KeywordSearchable Encryption. In: Cryptology ePrint Archive, Report 2005/417 (2005), http://eprint.iacr.org/2005/417
4. Bellare, M., Boldyreva, A., Desai, A., Pointcheval, D.: Key-Privacy in Public-Key Encryption. In: Boyd, C. (ed.) ASIACRYPT 2001. LNCS, vol. 2248, pp. 566–582. Springer, Heidelberg (2001)
5. Bentahar, K., Farshim, P., Malone-Lee, J., Smart, N.: Generic Constructions of Identity-Based and Certificateless KEMs. In: Cryptology ePrint Archive, Report 2005/058 (2005), http://eprint.iacr.org/2005/058
6. Boneh, D., Boyen, X., Shacham, H.: Short Group Signatures. In: Franklin, M. (ed.) CRYPTO 2004. LNCS, vol. 3152, pp. 41–55. Springer, Heidelberg (2004)
7. Boneh, D., Di Crescenzo, G., Ostrovsky, R., Persiano, G.: Public Key Encryption with Keyword Search. In: Cachin, C., Camenisch, J.L. (eds.) EUROCRYPT 2004. LNCS, vol. 3027, pp. 506–522. Springer, Heidelberg (2004)
8. Boneh, D., Franklin, M.: Identity-Based Encryption from the Weil Pairing. SIAM Journal on Computing 32(3), 586–615 (2003)
9. Boneh, D., Gentry, C., Hamburg, M.: Space-Efficient Identity Based Encryption Without Pairings. In: FOCS 2007: Proceedings of the 48th Annual IEEE Symposium on Foundations of Computer Science, pp. 647–657. IEEE Computer Society, Washington (2007)

10. Boyen, X., Waters, B.: Anonymous Hierarchical Identity-Based Encryption (Without Random Oracles). In: Dwork, C. (ed.) CRYPTO 2006. LNCS, vol. 4117, pp. 290–307. Springer, Heidelberg (2006)
11. Cocks, C.: An Identity Based Encryption Scheme Based on Quadratic Residues. In: Honary, B. (ed.) Cryptography and Coding 2001. LNCS, vol. 2260, pp. 360–363. Springer, Heidelberg (2001)
12. Cramer, R., Shoup, V.: A Practical Public Key Cryptosystem Provably Secure against Adaptive Chosen Ciphertext Attack. In: Krawczyk, H. (ed.) CRYPTO 1998. LNCS, vol. 1462, pp. 13–25. Springer, Heidelberg (1998)
13. Cramer, R., Shoup, V.: Design and Analysis of Practical Public-Key Encryption Schemes Secure against Adaptive Chosen Ciphertext Attack. SIAM Journal on Computing 33(1), 167–226 (2004)
14. Di Crescenzo, G., Saraswat, V.: Public Key Encryption with Searchable Keywords Based on Jacobi Symbols. In: Srinathan, K., Rangan, C.P., Yung, M. (eds.) INDOCRYPT 2007. LNCS, vol. 4859, pp. 282–296. Springer, Heidelberg (2007)
15. Damgård, I.B.: On the Randomness of Legendre and Jacobi Sequences. In: Goldwasser, S. (ed.) CRYPTO 1988. LNCS, vol. 403, pp. 163–172. Springer, Heidelberg (1990)
16. Halevi, S.: A Sufficient Condition for Key-Privacy. In: Cryptology ePrint Archive, Report 2005/05 (2005), http://eprint.iacr.org/2005/005
17. Hayashi, R., Tanaka, K.: Universally Anonymizable Public-Key Encryption. In: Roy, B. (ed.) ASIACRYPT 2005. LNCS, vol. 3788, pp. 293–312. Springer, Heidelberg (2005)
18. NIST. The Case for Elliptic Curve Cryptography, http://www.nsa.gov/ia/industry/crypto_elliptic_curve.cfm
19. Sakai, R., Ohgishi, K., Kasahara, M.: Cryptosystems Based on Pairing. In: Symposium on Cryptography and Information Security (SCIS 2000), Okinawa, Japan (2000)
20. Scott, M.: Authenticated ID-based Key Exchange and Remote Log-in With Insecure Token and PIN Number. In: Cryptology ePrint Archive, Report 2002/164 (2002), http://eprint.iacr.org/2002/164
21. Shamir, A.: Identity-Based Cryptosystems and Signature Schemes. In: Blakely, G.R., Chaum, D. (eds.) CRYPTO 1984. LNCS, vol. 196, pp. 47–53. Springer, Heidelberg (1985)
22. Shamus Software. The MIRACL library, http://www.shamus.ie
23. Shoup, V.: A Proposal for an ISO Standard for Public Key Encryption (Version 2.1) (manuscript) (December 20, 2001), http://www.shoup.net/papers/iso-2_1.pdf
24. Spiegel, M.R.: Theory and Problems of Probability and Statistics. McGraw-Hill, New York (1992)

Attacks on the DECT Authentication Mechanisms

Stefan Lucks[1], Andreas Schuler[2], Erik Tews[3], Ralf-Philipp Weinmann[4],
and Matthias Wenzel[5]

[1] Bauhaus-University Weimar, Germany
[2] Chaos Computer Club Trier, Germany
[3] FB Informatik, TU Darmstadt, Germany
[4] FSTC, University of Luxembourg
[5] Chaos Computer Club München, Germany
Stefan.Lucks@uni-weimar.de, krater@ccc-trier.de,
e_tews@cdc.informatik.tu-darmstadt.de, ralf-philipp-weinmann@uni.lu,
mazzoo@mazzoo.de

Abstract. Digital Enhanced Cordless Telecommunications (DECT) is
a standard for connecting cordless telephones to a fixed telecommunications network over a short range. The cryptographic algorithms used in
DECT are not publicly available. In this paper we reveal one of the two
algorithms used by DECT, the DECT Standard Authentication Algorithm (DSAA). We give a very detailed security analysis of the DSAA
including some very effective attacks on the building blocks used for
DSAA as well as a common implementation error that can practically
lead to a total break of DECT security. We also present a low cost attack on the DECT protocol, which allows an attacker to impersonate a
base station and therefore listen to and reroute all phone calls made by
a handset.

1 Introduction

Digital Enhanced Cordless Telecommunications (DECT) is a standard for
connecting cordless telephones to a fixed telecommunications network. It was
standardized in 1992 by the CEPT, a predecessor of the ETSI (European
Telecommunications Standards Institute) and is the de-facto standard for cordless telephony in Europe today. For authentication and privacy, two proprietary
algorithms are used: The DECT Standard Authentication Algorithm (DSAA)
and the DECT Standard Cipher (DSC). These algorithms have thus far only
been available under a Non-Disclosure Agreement from the ETSI and have not
been subject to academic scrutiny. Recently, the DSAA algorithm was reverse
engineered by the authors of this paper to develop an open-source driver for a
PCMCIA DECT card.

Our contribution: This paper gives the first public description of the DSAA as
well as cryptanalytic results on its components. Furthermore we show two types

M. Fischlin (Ed.): CT-RSA 2009, LNCS 5473, pp. 48–65, 2009.

of flaws that result in practical attacks against DECT implementations. One is a protocol flaw in the authentication mechanism, the other is a combination of a common implementation error combined with a brittle protocol deisgn.

The paper is structured as follows: Section 2 describes the authentication methods used in DECT. Section 3 details how easily a DECT base station can be impersonated in practice and outlines the consequences. Section 4 describes the DECT Standard Authentication Algorithm and explains how a weak PRNG can lead to a total break of DECT security. Section 5 presents the first public analysis of the DSAA. We conclude the paper in Section 6.

1.1   Notation and Conventions

We use **bold font** for variable names in algorithm descriptions as well as for input and output parameters. Hexadecimal constants are denoted with their least significant byte first in a `typewriter font`. For example, if all bits of the variable **b** are 0 except for a single bit, we write 0100 if $\mathbf{b}[0] = 1$, 0200 if $\mathbf{b}[1] = 1$, 0001 if $\mathbf{b}[8] = 1$ and 0080 if $\mathbf{b}[15] = 1$. Function names are typeset with a sans-serif font.

Function names written in capital letters like A11 are functions that can be found in the public DECT standard [5]. Conversely function names written in lowercase letters like step1 have been introduced by the authors of this paper. Functions always have a return value and never modify their arguments.

To access a bit of an array, the [·] notation is used. For example foo[0] denotes the first bit of the array foo. If more than a single bit, for example a byte should be extracted, the [· ... ·] notation is used. For example foo[0...7] extracts the first 8 bits in foo, which is the least significant byte in foo.

To assign a value in pseudocode, the ← operator is used. Whenever the operators + and ∗ are used in pseudocode, they denote addition and multiplication modulo 256. For example foo[0...7] ← bar[0...7] ∗ barn[0...7] multiplies the first byte in bar with the first byte in barn, reduces this value modulo 256 and stores the result in the first byte of foo.

If a bit or byte pattern is repeated, the (·)˙ notation can be used. For example instead of writing `aabbaabbaabb`, we can write $(\mathtt{aabb})^3$. For concatenating two values, the || operator is used. For example `aa`||`bb` results in `aabb`.

1.2   Additional Terminology

In the following we will use the terminology of the DECT standards [5,4]. To make this paper self-contained, we briefly explain the most important terms: A FT is fixed terminal, also called base station. A PT is a portable terminal, e.g. a telephone or handset. The Radio Fixed Party Identity (RFPI) is a 40-bit identifier of an FT. A PT is identified by a International Portable User Identity (IPUI), a value similar to the RFPI of variable length. In challenge/response authentications, responses are named RES1 or RES2. The value received during the authentication is called SRES1 or SRES2, and the value calcluated by the station (expected as a response) is called XRES1 or XRES2.

2   Authentication in DECT

The public standard describing the security features of DECT specifies four different authentication processes A11, A12, A21 and A22. These four processes are used for both authentication and key derivation and make use of an authentication algorithm A. DECT equipment conforming to the GAP standard [4] must support the DSAA to achieve vendor interoperability.

The algorithms A11 and A12 are used during the authentication of a PT. They are also used to derive a key for the DSC and to generate keying material during the initial pairing of a PT with a FT. The algorithms A21 and A22 are only used during the authentication of a FT. Furthermore the processes A11, A12, A21 and A22 are used to pair a PT with a FT.

2.1   Keys Used in DECT

In most cases of residential DECT usage, the user buys a DECT FT, and one or more DECT PTs. The first step then is to pair the PTs with the FT, unless they have been bought as a bundle and the pairing was already completed by the manufacturer. This procedure is called key allocation in the DECT standards and described in more detail in Section 2.5. After this process, every DECT PT shares a 128 bit secret key with the FT, called the UAK.

In all scenarios we have seen so far, the UAK was the only key used to derive any other keys, but the DECT standard allows two alternative options:

– The UAK is used together with a UPI, a short 16-32 bit secret, manually entered by the user of the PT. The UAK und UPI are then used to derive a key.
– No UAK is used. Instead a short 16-32 bit secret called AC is entered by the user, which forms the key. The DECT standard suggests that the AC should only be used, if a short term coupling between PT and FT is required.

For the rest of this paper we will mainly focus on the first case, where the only the UAK is used.

2.2   Authentication of a PT and Derivation of the DSC Key

When a PT needs to authenticate itself against a FT, the procedures A11 and A12 are used. The FT generates two random 64 bit values RS and RAND_F and sends them to the PT.

The PT uses the A11 algorithm, which takes a 128 bit key UAK and a 64 bit value RS as input, and generates a 128 bit output KS, which is used as an intermediate key. The PT then uses the A12 algorithm, which takes KS and

RAND_F as input and produces two outputs: a 32 bit response called SRES1 and a 64 bit key DCK, which can be used for the DSC. SRES1 then is sent to the FT.

The same computation is done on the FT too, except here the first output of A12 is called XRES1 instead of SRES1. The FT receives the value SRES1 from the PT and compares it with his own value XRES1. If both are equal, the PT is authenticated.

2.3   Authentication of a FT

When a FT needs to authenticate itself against a PT, a similar procedure is used.

First the PT generates a 64 bit value RAND_P and sends it to a FT. Then the FT generates a 64 bit value RS and uses A21 to compute a 128 bit intermediate key KS from UAK and RS. Now A22 is used to compute a 32 bit response SRES2 from KS and RAND_P. A22 only generates SRES2 and no key for the DSC. The FT sends SRES2 and KS to the PT.

After having received KS, the PT can do the same computations. Here the output of A22 is called XRES2. The PT now compares XRES2 with the received value SRES2. If both are equal, the FT is authenticated.

This protocol might seem odd at the first look. As far as we know, the design goal was to build a protocol where a PT can be used in a roaming scenario, similar to GSM. Here, the home network provider could hand over a couple of (RS, KS) pairs to the partner network, which can then allow a PT to operate without having to know the UAK.

2.4   Mutual Authentication

The standard specifies different methods of mutual authentication:

  - A direct method, which simply consists of executing A11 and A12 to authenticate the PT first followed by A21 and A22 to authenticate the FT.
  - Indirect methods which involve a one-sided authentication of the PT together with a cipher key derivation that is used for data confidentiality.

However, even though the indirect methods are recommended for all applications except for local loop installations (see the reference configurations in Appendix F.2 of [5]), they are inherently flawed as they do not provide a mutual authentication at all. This indirect method is reminiscent of the case of GSM [1]. A derived cipher key does not necessarily have to be used, a FT may simply send a message indicating that it does not support encryption – it is an optional feature in the GAP standard. Moreover, even if encryption is enabled, being able to transmit encrypted messages under a derived key does not proof possession of this key: The FT may just replay authentication challenges, subsequently replaying encrypted messages that were previously recorded.

2.5   Key Allocation

Most DECT systems allow an automatic pairing process. To initiate pairing the user switches both a PT and a FT to a dedicated pairing mode and enters the same PIN number on both devices[1]. This step needs to be repeated with all DECT PT devices. Each PT performs a handshake with the FT and a mutual authentication using the **DSAA** algorithm and the PIN as a shared secret is performed. During this handshake, three 64-bit random numbers are generated. However only a single 64 bit random number RS sent by the FT is used together with the PIN to generate the 128 bit UAK. For a 4 digit PIN number, there are only $2^{77.288}$ possible values for the UAK. If a flawed random number generator is used on the FT for which an attacker can predict the subset of random numbers generated during key allocation, the number of possibe UAKs shrinks accordingly. This can be exploited by sniffing challenge-response pairs ((RAND_F, RS), SRES1) at any time after key allocation that can be used as 32-bit filters. In practice we did indeed find weak PRNGs implemented in the firmware of several base stations – across a variety of vendors – in one specific case only providing 24 bits of entropy for the 64 bit value RS. This leads to a very practical and devastating attack against DECT PTs using vulnerable DECT stacks.

3   Impersonating a Base Station

As described in the previous section, in most cases authentication of the FT is optional. This makes DECT telephones vulnerable to a very simple, yet effective and practical attack: An attacker impersonates a DECT FT by spoofing its RFPI and faking the authentication of the PT. This is done by sending out random values RAND_F and RS for which any response SRES1 is accepted. Subsequently the impersonating FT simply rejects any attempts to do cipher mode switching. This technique is significantly simpler to implement than a protocol reflection attack and has been verified to work in practice by us.

We implemented this attack in practice by modifying the driver of a PCMCIA DECT card. The drivers and firmware for this card do not support the DECT Standard Cipher. Furthermore, the frames are completely generated in software which allows us to easily spoof the RFPI of another base station. Upon initialization of the card, the RFPI was read from the card and written to a structure in memory. We patched the driver such that the RFPI field in this structure was overwritten with an assumed RFPI value of our choosing directly after the original value was written there. Then we modified the routine comparing the RES values returned by the PT with the computed XRES values. We verified that we were indeed broadcasting a fake RFPI with a USRP [3] and a DECT sniffer that was written for the GNURadio framework by the authors.

---

[1] Some DECT FTs are shipped with a fixed default PIN number – usually specified in the manual – which user has to enter as given on the DECT PT.

For our lab setup, we used an ordinary consumer DECT handset paired to a consumer base station. We set the modified driver of our PCMCIA card to broadcast the RFPI of this base station and added the IPUI of the phone to the database of registered handsets of the card. The device key was set to an arbitrary value. After jamming the DECT over-the-air communication for a short time, the handset switched to our faked base station with a probability of about 50%. From this point on, every call made by the phone was handled by our PCMCIA hard, and we where completely able to trace all communications and reroute all calls. No warning or error message was displayed on the telephone. Both the handset and the base station where purchased in 2008, which shows that even current DECT phones do not authenticate base stations and also do not force encrypted communication.

This attack shows that it is possible to intercept, record and reroute DECT phonecalls with equipment as expensive as a wireless LAN card, making attacks on DECT as cheap as on wireless LANs. Subsequently we also succeeded in converting this card to a passive sniffing device with a custom-written Linux and firmware[2].

4   The DECT Standard Authentication Algorithm

The algorithms A11, A12, A21, and A22 can be seen as wrappers around an algorithm, we call DSAA. The algorithm DSAA accepts an 128 bit key and a 64 bit random as input and produces a 128 bit output. This output is now modified as follows:

- A11 just returns the whole output of DSAA, without any further modification.
- A21 behaves similar to A11, but here, every second bit of the output is inverted, starting with the first bit of the output. For example if the first byte of output of DSAA is ca, then the first byte of output of A21 is 60.
- A22 just returns the last 4 bytes of output of DSAA as RES.
- A12 is similar to A22, except here, the middle 8 bytes of DSAA are returned too, as DCK.

5   Security Analysis of the DECT Authentication

DSAA is surprisingly insecure. The middle 64 bits of the output of DSAA only depend on the middle 64 bits of the key. This allows trivial attacks against DSAA, which allow the recovery of all 128 secret key bits with an effort in the magnitude of about 2^{64} evaluations of DSAA. Even if attacks against the DSAA cannot improved past this bound, keep in mind the entropy problems of the random number generators that we found and described in Section 2.5.

---

[2] This software is available at http://www.dedected.org

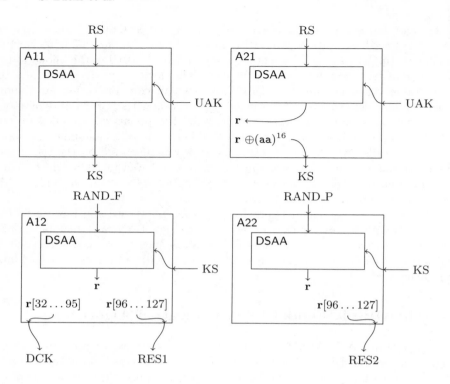

Fig. 1. The four DSAA algorithms

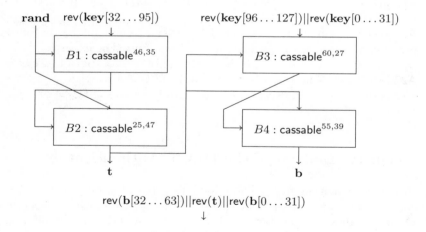

Fig. 2. DSAA overview

Table 1. The DSAA S-Box (**sbox**)

	0	1	2	3	4	5	6	7	8	9	a	b	c	d	e	f
00	b0	68	6f	f6	7d	e8	16	85	39	7c	7f	de	43	f0	59	a9
10	fb	80	32	ae	5f	25	8c	f5	94	6b	d8	ea	88	98	c2	29
20	cf	3a	50	96	1c	08	95	f4	82	37	0a	56	2c	ff	4f	c4
30	60	a5	83	21	30	f8	f3	28	fa	93	49	34	42	78	bf	fc
40	61	c6	f1	a7	1a	53	03	4d	86	d3	04	87	7e	8f	a0	b7
50	31	b3	e7	0e	2f	cc	69	c3	c0	d9	c8	13	dc	8b	01	52
60	c1	48	ef	af	73	dd	5c	2e	19	91	df	22	d5	3d	0d	a3
70	58	81	3e	fd	62	44	24	2d	b6	8d	5a	05	17	be	27	54
80	5d	9d	d6	ad	6c	ed	64	ce	f2	72	3f	d4	46	a4	10	a2
90	3b	89	97	4c	6e	74	99	e4	e3	bb	ee	70	00	bd	65	20
a0	0f	7a	e9	9e	9b	c7	b5	63	e6	aa	e1	8a	c5	07	06	1e
b0	5e	1d	35	38	77	14	11	e2	b9	84	18	9f	2a	cb	da	f7
c0	a6	b2	66	7b	b1	9c	6d	6a	f9	fe	ca	c9	a8	41	bc	79
d0	db	b8	67	ba	ac	36	ab	92	4b	d7	e5	9a	76	cd	15	1f
e0	4e	4a	57	71	1b	55	09	51	33	0c	b4	8e	2b	e0	d0	5b
f0	47	75	45	40	02	d1	3c	ec	23	eb	0b	d2	a1	90	26	12

Besides that, the security of DSAA mainly relies on the security of the **cassable** block cipher. Our analysis of **cassable** showed that **cassable** is surprisingly weak too.

5.1   The **Cassable** Block Cipher

The DSAA can be interpreted as a cascade of four very similarly constructed block ciphers. We shall call this family of block ciphers **cassable**. A member of this family is a substitution-linear network parametrized by two parameters that only slightly change the key scheduling. The block cipher uses 6 rounds, each of the rounds performing of a key addition, applying a bricklayer of S-Boxes and a mixing step in sequence. The last round is not followed by a final key addition, so that the last round is completely invertible, besides the key addition. This reduces the effective number of rounds to 5.

In the following we will describe how the **cassable** block ciphers are constructed. The functions $\sigma_i : GF(2)^{64} \rightarrow GF(2)^{64}$ with $1 \leq i \leq 4$ denoting bit permutations that are used to derive the round keys from the cipher key. The function $\lambda_i : (\mathbb{Z}/256\mathbb{Z})^8 \rightarrow (\mathbb{Z}/256\mathbb{Z})^8$ denotes the mixing functions used in the block ciphers, the function $\gamma : GF(2)^{64} \rightarrow GF(2)^{64}$ is a bricklayer transform that is defined as:

$$\gamma(A||B||\cdots||A) = \rho(A)||\rho(B)||\cdots||\rho(H)$$

with $A, B, \ldots, H \in GF(2)^8$ and $\rho : GF(2)^8 \rightarrow GF(2)^8$ denoting the application of the invertible S-Box that is given in Table 1. The linear transforms perform butterfly-style mixing:

$$\lambda_1 : (A,\ldots,H) \mapsto (2A+E, 2B+F, 2C+G, 2D+H, 3E+A, 3F+B, 3G+C, 3H+D)$$
$$\lambda_2 : (A,\ldots,H) \mapsto (2A+C, 2B+D, 3C+A, 3D+B, 2E+G, 2F+H, 3G+E, 3H+F)$$
$$\lambda_3 : (A,\ldots,H) \mapsto (2A+B, 3B+A, 2C+D, 3D+C, 2E+F, 3F+E, 2G+H, 3H+G)$$

The round keys $K_i \in GF(2)^{64}$ with $1 \le i \le 6$ are iteratively derived from the cipher key $K_0 \in GF(2)^{64}$ using the following parametrized function $\sigma_{(m,l)}$:

$$\sigma_{(m,l)} : (k_0, \ldots, k_{63}) \mapsto (k_m, k_{(m+l) \bmod 64}, k_{(m+2l) \bmod 64}, \ldots, k_{(m+63l) \bmod 64})$$

by simply applying a $\sigma_{(m,l)}$ to the cipher key i times:

$$K_i = \sigma^i_{(l,m)}(K)$$

The individual bytes of the key K_i can be accessed by $K_{i,A}$ to $K_{i,H}$.

To be able to compose the round function, we identify the elements of the vector space $GF(2)^8$ with the elements of the ring $\mathbb{Z}/256\mathbb{Z}$ using the canonical embedding. Given a fixed σ, the round function f_r for round r with $1 \le r \le 6$ that transforms a cipher key K and a state X into the state of the next round then looks as follows:

$$f_r : (X, K) \mapsto \lambda_{(((r-1) \bmod 3)+1)}(X \oplus \sigma^r(K))$$

The Mixing Layer. To diffuse local changes in the state bits widely, the functions λ_i with $1 \le i \le 3$ (lambda1, lambda2, and lambda3 in the pseudo-code) are used. These form a butterfly network. At first look, it seems that full diffusion is achieved after the third round, because every byte of the state depends on every other byte of the input at this point. However, we made an interesting observation: The λ_i functions only multiply the inputs with either the constant 2 or 3. This means that for the components of the output vector formed as

$$c = (a * 2 + b) \bmod 256$$

the lowestmost bit of c will be equal to the lowestmost bit of b and not depend on a at all. This observation will be used in Section 5.2.

The S-Box. The DSAA S-Box has a tendency towards flipping the lowest bit. If a random input is chosen, the lowest bit of the output will equal to the lowest bit of the input with a probability of $\frac{120}{256}$. For up to three rounds we were able to find exploitable linear approximations depending on the lowest bits of the input bytes, the lowest bits of the state and various bits of the key. Although this sounds promising, the linear and differential properties of the S-Box are optimal. Interpolating the S-Box over $GF(2^8)$ yields dense polynomials of degree 254, interpolation over $GF(2)$ results in equations of maximum degree.

The Key Scheduling. The key bit permutation used in the key scheduling is not optimal for the cassable ciphers used in DSAA. Although the bit permutation could have a maximum order of 64, a lower order was observed for the cassable ciphers instantiated, namely 8 and 16.

5.2   A Practical Attack on Cassable

The individual block ciphers used within the DSAA can be fully broken using differential cryptanalysis [2] with only a very small number of chosen plaintexts. Assume that we have an input $m = m_A||m_B||m_C||m_D||m_E||m_F||m_G||m_H$ with $m_i \in \{0,1\}^8$ and a second input $m' = m'_A||m'_B||m'_C||m'_D||m'_E||m'_F||m'_G||m'_H$ where every second byte is the same, i. e. $m_B = m'_B, m_D = m'_D, m_F = m'_F$, and $m_H = m'_H$. Now both inputs are encrypted. Let $s = s_{i,A}||\ldots||s_{i,H}$ and $s' = s'_{i,A}||\ldots||s'_{i,H}$ be the states after i rounds of cassable. After the first round, $s_{1,B} = s'_{1,B}, s_{1,D} = s'_{1,D}, s_{1,F} = s'_{1,F}$, and $s_{1,H} = s'_{1,H}$ holds. This equality still holds after the second round. After the third round, the equality is destroyed, but $s_{3,A} \equiv s'_{3,A} \bmod 2, s_{3,C} \equiv s'_{3,C} \bmod 2, s_{3,E} \equiv s'_{3,E}$, and $s_{3,G} \equiv s'_{3,G} \bmod 2$ holds. The key addition in round four preserves this property, with only the fourth application of the S-Box $\rho_{4,j}$ destroying it.

An attacker can use this to recover the secret key of the cipher. Assume the attacker is able to encrypt two such messages m and m' with the same secret key and see the output. He can invert the lambda3 and gamma steps of the last round, because they are not key-dependent. To recover the value of $s_{3,A} \oplus K_{4,A}$ and $s_{3,E} \oplus K_{4,E}$, he only needs 32 round key bits of round key 6 which are added to $s_{5,A}, s_{5,C}, s_{5,E}$, and $s_{5,G}$, and 16 round key bits of round key 5, which are added to $s_{4,A}$ and $s_{4,E}$. Due to overlaps in the round key bits these are only 38 different bits for cassable[46,35], 36 different bits for cassable[25,47], 42 different bits for cassable[60,27], and 40 different bits for cassable[55,39]. After the attacker has recovered $s_{3,A} \oplus K_{4,A}, s_{3,E} \oplus K_{4,E}, s'_{3,A} \oplus K_{4,A}$ and $s'_{3,E} \oplus K_{4,A}$, he checks whether $s_{3,A} \oplus K_{4,A} = s'_{3,A} \oplus K_{4,A} \bmod 2$ and $s_{3,E} \oplus K_{4,E} = s'_{3,E} \oplus K_{4,E} \bmod 2$ holds. If at least one of the conditions is not satisfied, he can be sure that his guess for the round key bits was wrong. Checking all possible values for these round key bits will eliminate about $\frac{3}{4}$ of the key space with computational costs of about 2^k invocations of cassable, if there are k different key bits for the required round key parts of round key 5 and 6.

After having eliminated 75% of the key space, an attacker can repeat this with another pair on the remaining key space and eliminate 75% of the remaining key space again. Iterating this procedure with a total of 15 pairs, only about 2^{34} possible keys are expected to remain. These can then be checked using exhaustive search. The total workload amounts to $2^k + \frac{1}{4}2^k + \frac{1}{16}2^k + \frac{1}{64}2^k + \ldots + 2^{34}$ block cipher invocations which is bounded by $1.5834 \cdot 2^k$ for $k \geq 36$. For cassable[25,47], this would be about $2^{36.7}$.

An efficient implementation needs only negligible memory when every possible value of the k round key bits is enumerated and every combination is checked against all available message pairs. Only the combinations which pass their tests against all available pairs are saved, which should be about 2^{k-30}.

If the attacker can choose the input for cassable, he can choose 16 different inputs, where every second byte is set to an arbitrary constant. If the attacker can only observe random inputs, he can expect to find a pair in whcih every second byte is the same after 2^{16} random inputs. After $4 \cdot 2^{16}$ inputs, the expected number of pairs is about $4 \cdot 4 = 16$, which is sufficient for the attack. If not enough pairs are available to the attacker, the attack is still possible, however with increased computational effort and memory usage.

5.3   A Known-Plaintext Attack on Three Rounds Using a Single Plaintext/Ciphertext Pair

Three rounds of the cassable block cipher can be attacked using a single plaintext/ciphertext pair. This is of relevance as attacking B_4 or B_2 allows us to invert the preceding ciphers B_1 and B_3.

Assume a plaintext $m = m_A || m_B || m_C || m_D || m_E || m_F || m_G || m_H$ encrypted over three rounds. The output after the third round then is $S_3 = s_{3,A}, \ldots, s_{3,H}$. As in the previous attack, we can invert the diffusion layer λ_3 and the S-Box layer ρ without knowing any key bits, obtaining $(z_0, \ldots, z_7) := S_2 \oplus K_3$ with $z_i \in GF(2)^8$ for $0 \leq i < 8$. At this point the diffusion is not yet complete. For instance, the following relation holds for z_0:

$$z_0 = \rho((2 \cdot \rho(m_0 \oplus K_{1,A}) + \rho(m_4 \oplus K_{1,E})) \oplus K_{2,A}) + \\ \rho((2 \cdot \rho(m_2 \oplus K_{1,C}) + \rho(m_6 \oplus K_{1,G})) \oplus K_{2,C}) \oplus K_{3,A}$$

Due to overlaps in the key bits, for the block cipher B_1 the value z_0 then only depends on 41 key bits, for B_2 on 36 key bits, for B_3 on 44 key bits and for B_4 on 46 key bits.

We can use the equations for the z_i as a filter which discard $\frac{255}{256}$ of the searched key bit subspace.

In the following, we give an example of how the attacks works for B2: Starting with z_0, we expect 2^{28} key bit combinations after the filtering step. Interestingly, the key bits involved in z_0 for B2 are the same as for z_2, so we can use this byte to filter down to about 2^{20} combinations. Another filtering step using both z_4 and z_6 will just cost us an additional 4 key bits, meaning we can filter about 2^{24} combinations down to about 2^8. All of these filtering steps can be chained without storing intermediate results in memory, making the memory complexity negligible.

For the remaining combinations we can exhaustively search through the remaining 24 key bits, giving a 2^{32} work factor. The overall cost of the attack is dominated by the first filtering step however, which means the attack costs about 2^{36} cassable invocations.

For B4, the key bit permutations work against our favor: After filtering with z_0 we expect 2^{38} key bit combinations to remain. Subsequently we filter with z_2, which causes another 6 key bits to be involved (z_4 and z_6 would involve 10 more key bits). This yields 2^{36} key bit combinations. Subsequently filtering with z_4 involves 8 more key bits, causing the number of combinations to stay at 2^{36}. Finally we can filter with z_6, which adds 4 more key bits, bringing the number of combinations down to 2^{32}. As there are no more unused key bits left, we can test all of the 2^{32} key candidates. The total cost for this attack is again dominated by the first filtering step which requires 2^{46} cassable invocations. Again the attack can be completed using negligible memory by chaining the filtering conditions.

The attacks on B2 and B4 can be used to attack a reduced version of the DSAA where B1 and B3 are 6 round versions of cassable and B2 and B4 are reduced to three rounds. An attack on this reduced version costs approximately 2^{44} invocations of the reduced DSAA since approximately three 6 round cassable invocations are used per DSAA operation.

6   Conclusion

We have shown the first public description of the DSAA algorithm, which clearly shows that the algorithm only provides at most 64 bit of symmetric security. An analysis using only the official documents published by ETSI would not have revealed these information.

We could also show that the building blocks used for the DSAA have some serious design flaws, which might allow attacks with a complexity below 2^{64}. Especially the block cipher used in DSAA seems to be weak and can be completely broken using differential cryptanalysis.

Although 64 bit of symmetric security might be sufficient to hold off unmotivated attackers, most of the currently deployed DECT systems might be much easier attackable, because encryption and an authentication of the base station is not always required. This allows an attacker spending about 30$ for a PCMCIA card to intercept most DECT phone calls and totally breach the security architecture of DECT.

Currently, we see two possible countermeasures. First, all DECT installations should be upgraded to require mutual authentication and encryption of all phone calls. This should only be seen as a temporary fix until a better solution is available.

A possible long term solution would be an upgrade of the DECT security architecture to use *public* well analyzed methods and algorithms for key exchange and traffic encryption and integrity protection. A possible alternative could be IEEE 802.11 based Voice over IP phone systems, where networks can be encrypted using WPA2. These systems are currently more costly than DECT installations and still more difficult to configure than DECT phones for a novice user, but encrypt and protect all calls and signaling informations using AES-CCMP and allow a variety of different protocols for the key exchange. However it is open to debate whether these systems can provide a viable alternative to DECT systems because of their different properties in term of power consumption, radio spectrum and quality of service provided.

We would like to thank all the people who supported and helped us with this paper, especially those, whose names are not mentioned in this document.

References

1. Barkan, E., Biham, E., Keller, N.: Instant ciphertext-only cryptanalysis of GSM encrypted communication. In: Boneh, D. (ed.) CRYPTO 2003. LNCS, vol. 2729, pp. 600–616. Springer, Heidelberg (2003)
2. Biham, E., Shamir, A.: Differential cryptanalysis of DES-like cryptosystems. In: Menezes, A., Vanstone, S.A. (eds.) CRYPTO 1990. LNCS, vol. 537, pp. 2–21. Springer, Heidelberg (1991)
3. Ettus, M.: USRP user's and developer's guide. Ettus Research LLC (February 2005)
4. European Telecommunications Standards Institute. ETSI EN 300 444 V1.4.2 (2003-02): Digital Enhanced Cordless Telecommunications (DECT); Generic Access Profile (February 2003)
5. European Telecommunications Standards Institute. ETSI EN 300 175-7 V2.1.1: Digital Enhanced Cordless Telecommunications (DECT); Common Interface (CI); Part 7: Security Features (August. 2007)

A   Pseudocode for the DSAA

The DSAA (see Algorithm 1) uses four different 64 bit block cipher like functions as building blocks. DSAA takes a random value **rand** $\in \{0,1\}^{64}$ and a key **key** $\in \{0,1\}^{128}$ as input and splits the 128 bit key into two parts of 64 bit. The first part of the key are the 64 middle bits of the key. DSAA calls the step1 function with the random value and the first part of the key to produce the first 64 bits of output, which only depend on the middle 64 bits of the key. Then the output of step1 is used to produce the second 64 bits of output using the step2 function and the second half of the key. Please note that the second half of the output only depends on the first half of the output and the second part of the key.

Algorithm 1. DSAA (**rand** $\in \{0,1\}^{64}$, **key** $\in \{0,1\}^{128}$)

1: $\mathbf{t} \leftarrow$ step1(rev(**rand**), rev(**key**[32 . . . 95]))
2: $\mathbf{b} \leftarrow$ step2(**t**, rev(**key**[96 . . . 127])||rev(**key**[0 . . . 31]))
3: **return**   rev(**b**[32 . . . 63])||rev(**t**)||rev(**b**[0 . . . 31]))

We will now have a closer look at the functions step1 and step2. Both are very similar and each one uses two block cipher like functions as building blocks.

Algorithm 2. step1(**rand** $\in \{0,1\}^{64}$, **key** $\in \{0,1\}^{64}$)

1: $\mathbf{k} = \text{cassable}_{\mathbf{rand}}^{46,35}(\mathbf{key})$
2: **return** $\text{cassable}_{\mathbf{k}}^{25,47}(\mathbf{rand})$

---

step1 takes a 64 bit key **key** and a 64 bit random value **rand** as input and uses two block ciphers to produce its output. The key is used as a key for the first cipher and the random value as a plaintext. The value **rand** then is used as an input to the second block cipher and is encrypted with the output of the first block cipher as the key.

Algorithm 3. cassable$_{\textbf{key}}^{\textbf{start,step}}(\textbf{m} \in \{0,1\}^{64})$

1: $\textbf{t} \leftarrow \textbf{key}$
2: $\textbf{s} \leftarrow \textbf{m}$
3: **for** $\textbf{i} = 0$ to 1 **do**
4:     $\textbf{t} \leftarrow$ sigma($\textbf{start}, \textbf{step}, \textbf{t}$)
5:     $\textbf{s} \leftarrow$ lambda1(gamma($\textbf{s} \oplus \textbf{t}$))
6:     $\textbf{t} \leftarrow$ sigma($\textbf{start}, \textbf{step}, \textbf{t}$)
7:     $\textbf{s} \leftarrow$ lambda2(gamma($\textbf{s} \oplus \textbf{t}$))
8:     $\textbf{t} \leftarrow$ sigma($\textbf{start}, \textbf{step}, \textbf{t}$)
9:     $\textbf{s} \leftarrow$ lambda3(gamma($\textbf{s} \oplus \textbf{t}$))
10: **end for**
11: **return s**

To describe the block ciphers, we introduce a family of block ciphers we call cassable. These block ciphers differ only in their key schedule, where round keys are always bit permutations of the input key. All bit permutations used by cassable can be described by two numbers **start** and **step**.

The block cipher cassable itself is a substitution linear network. To mix the round key into the state, a simple XOR is used. Additionally, $\mathbb{Z}_{256}$-linear mixing is used for diffusion and an 8×8 S-Box for non-linearity of the round function.

Algorithm 4. step2($\textbf{rand} \in \{0,1\}^{64}, \textbf{key} \in \{0,1\}^{64}$)

1: $\textbf{k} =$ cassable$_{\textbf{rand}}^{60,27}(\textbf{key})$
2: **return cassable**$_{\textbf{k}}^{55,39}(\textbf{rand})$

---

step2 is similar to step1, just two other bit permutations are used. The function rev simply reverses the order of the bytes of its input.

Algorithm 5. rev($\textbf{in} \in \{0,1\}^{i*8}$)

Ensure: Byte-reverses the input **in**
    for $\textbf{j} = 0$ to $i - 1$ **do**
        $\textbf{k} \leftarrow i - j - 1$
        $\textbf{out}[j * 8 \ldots j * 8 + 7] \leftarrow \textbf{in}[k * 8 \ldots k * 8 + 7]$
    end for
    return out

Algorithm 6. lambda1(in $\in \{0,1\}^{64}$)

1: **out**$[0\ldots 7] \leftarrow$ **in**$[32\ldots 39] + 2 *$ **in**$[0\ldots 7]$
2: **out**$[32\ldots 39] \leftarrow$ **in**$[0\ldots 7] + 3 *$ **in**$[32\ldots 39]$
3: **out**$[8\ldots 15] \leftarrow$ **in**$[40\ldots 47] + 2 *$ **in**$[8\ldots 15]$
4: **out**$[40\ldots 47] \leftarrow$ **in**$[8\ldots 15] + 3 *$ **in**$[40\ldots 47]$
5: **out**$[16\ldots 23] \leftarrow$ **in**$[48\ldots 55] + 2 *$ **in**$[16\ldots 23]$
6: **out**$[48\ldots 55] \leftarrow$ **in**$[16\ldots 23] + 3 *$ **in**$[48\ldots 55]$
7: **out**$[24\ldots 31] \leftarrow$ **in**$[56\ldots 63] + 2 *$ **in**$[24\ldots 31]$
8: **out**$[56\ldots 63] \leftarrow$ **in**$[24\ldots 31] + 3 *$ **in**$[56\ldots 63]$
9: **return out**

Algorithm 7. lambda2(in $\in \{0,1\}^{64}$)

1: **out**$[0\ldots 7] \leftarrow$ **in**$[16\ldots 23] + 2 *$ **in**$[0\ldots 7]$
2: **out**$[16\ldots 23] \leftarrow$ **in**$[0\ldots 7] + 3 *$ **in**$[16\ldots 23]$
3: **out**$[8\ldots 15] \leftarrow$ **in**$[24\ldots 31] + 2 *$ **in**$[8\ldots 15]$
4: **out**$[24\ldots 31] \leftarrow$ **in**$[8\ldots 15] + 3 *$ **in**$[24\ldots 31]$
5: **out**$[32\ldots 39] \leftarrow$ **in**$[48\ldots 55] + 2 *$ **in**$[32\ldots 39]$
6: **out**$[48\ldots 55] \leftarrow$ **in**$[32\ldots 39] + 3 *$ **in**$[48\ldots 55]$
7: **out**$[40\ldots 47] \leftarrow$ **in**$[56\ldots 63] + 2 *$ **in**$[40\ldots 47]$
8: **out**$[56\ldots 63] \leftarrow$ **in**$[40\ldots 47] + 3 *$ **in**$[56\ldots 63]$
9: **return out**

Algorithm 8. lambda3(in $\in \{0,1\}^{64}$)

1: **out**$[0\ldots 7] \leftarrow$ **in**$[8\ldots 15] + 2 *$ **in**$[0\ldots 7]$
2: **out**$[8\ldots 15] \leftarrow$ **in**$[0\ldots 7] + 3 *$ **in**$[8\ldots 15]$
3: **out**$[16\ldots 23] \leftarrow$ **in**$[24\ldots 31] + 2 *$ **in**$[16\ldots 23]$
4: **out**$[24\ldots 31] \leftarrow$ **in**$[16\ldots 23] + 3 *$ **in**$[24\ldots 31]$
5: **out**$[32\ldots 39] \leftarrow$ **in**$[40\ldots 47] + 2 *$ **in**$[32\ldots 39]$
6: **out**$[40\ldots 47] \leftarrow$ **in**$[32\ldots 39] + 3 *$ **in**$[40\ldots 47]$
7: **out**$[48\ldots 55] \leftarrow$ **in**$[56\ldots 63] + 2 *$ **in**$[48\ldots 55]$
8: **out**$[56\ldots 63] \leftarrow$ **in**$[48\ldots 55] + 3 *$ **in**$[56\ldots 63]$
9: **return out**

Algorithm 9. sigma(**start, step, in** $\in \{0,1\}^{64}$)

1: **out** $\leftarrow (00)^8$
2: **for** $i = 0$ to 63 **do**
3:     **out**[**start**] $\leftarrow$ **in**[i]
4:     **start** $\leftarrow$ (**start** + **step**) mod 64
5: **end for**
6: **return out**

Algorithm 10. gamma(in $\in \{0,1\}^{64}$)

1: **for** $i = 0$ to 7 **do**
2:     **out**$[i * 8\ldots i * 8 + 7] \leftarrow$ **sbox**[**in**$[i * 8\ldots i * 8 + 7]$]
3: **end for**
4: **return out**

B   Test Vectors for **DSAA**

To make implementation of these algorithms easier, we decided to provide some test vectors. Let us assume that A11 is called with the key K=ffff9124ffff9124ffff9124ffff9124 and the RS=0000000000000000 as in [5] Annex K. These values will be passed directly to the DSAA algorithm. Now, step1(0000000000000000, 2491ffff2491ffff) will be called. While processing the input, the internal variables will be updated according to Table 2. The final result after step2(ca41f5f250ea57d0, 2491ffff2491ffff) has been calculated is 93638b457afd40fa585feb6030d572a2, which is the UAK. The internal states of step2 can be found in Table 3.

Table 2. Trace of step1(0000000000000000, 2491ffff2491ffff)

algorithm	after line	i	t	s
cassable^{46,35}	5	0	0000000000000000	549b363670244848
cassable^{46,35}	7	0	0000000000000000	51d3084936beeaae
cassable^{46,35}	9	0	0000000000000000	20e145b2c0816ec6
cassable^{46,35}	5	1	0000000000000000	4431b3d7c1217a7c
cassable^{46,35}	7	1	0000000000000000	6cdcc25bbe8bc07f
cassable^{46,35}	9	1	0000000000000000	2037df9f8856a0a2
cassable^{25,47}	5	0	77fe578089a40531	cce76e5f83f77b4c
cassable^{25,47}	7	0	f5b720768a8a8817	c69973d6388f3cf7
cassable^{25,47}	9	0	552023ae0791ddf4	1cd81853ba428a2c
cassable^{25,47}	5	1	8856a0a22037df9f	ca643e2238dc1d1d
cassable^{25,47}	7	1	89a4053177fe5780	82fa43b0725dc387
cassable^{25,47}	9	1	8a8a8817f5b72076	ca41f5f250ea57d0

Table 3. Trace of step2(ca41f5f250ea57d0, 2491ffff2491ffff)

algorithm	after line	i	t	s
cassable^{60,27}	5	0	66f9d1c1c6524b4b	39ad15f5f68ab424
cassable^{60,27}	7	0	5bd0d66bf152e4c0	59e160ed3bb1189c
cassable^{60,27}	9	0	d5ebead34f434050	0bc33d7c093128b8
cassable^{60,27}	5	1	d2c057d860e3dd72	3f538f008a2b52f9
cassable^{60,27}	7	1	c6d2e3614c5953cb	ab826a7542ffa5c7
cassable^{60,27}	9	1	f158c640d3f27cc3	757782ad02592b4e
cassable^{55,39}	5	0	b0ec588246ea9577	40be7413fe173981
cassable^{55,39}	7	0	df212e1b790245e6	087978cbb37813af
cassable^{55,39}	9	0	e671b9d44296ee08	d97b8d2dbae583b9
cassable^{55,39}	5	1	2a0f207383ec575d	1340ba1df9d60b52
cassable^{55,39}	7	1	d022e4e81dd712ee	f7af7e62a1fa5ce6
cassable^{55,39}	9	1	04b3db206f4e7d03	08d87f9aef21c939

C   Example of a Weak PRNG Used in DECT Stacks

Algorithm 11 is a typical example for the quality of pseudo random-number generators (PRNGs) used in DECT stacks. Although it is supposed to provide 64-bit of randomness per nonce output, it only manages to use 24 bits of entropy. Moreover, the total number of distinct 64-bit **rand** values of this PRNG is only 2^{22} since outputs collide.

Algorithm 11. vendor_A_PRNG($\textbf{xorval} \in \{0,1\}^8, \textbf{counter} \in \{0,1\}^{16}$)

1: **for** $\textbf{i} = 0$ to 7 **do**
2:     **out**$[(\textbf{i} * 8)) \ldots (\textbf{i} * 8 + 7)] \leftarrow \lfloor \textbf{counter}/2^i \rfloor \oplus \textbf{xorval}$
3: **end for**
4: **return out**

The values produced by this particular PRNG can be easily stored in ASCII representation in a file just 68 Megabytes big. This means that to identify vulnerable implementations, an attacker or evaluator simply has to search for an intercepted **rand** in this text file.

D   Structure of the **cassable** Block Cipher

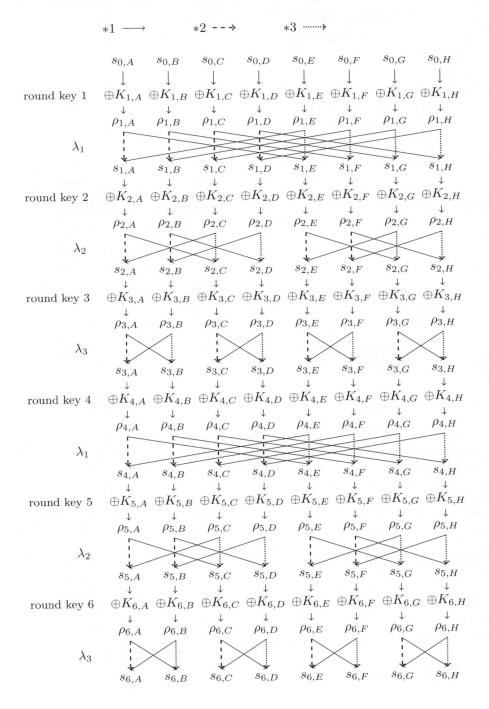

Comparison-Based Key Exchange and the Security of the Numeric Comparison Mode in Bluetooth v2.1

Andrew Y. Lindell

Aladdin Knowledge Systems and Bar-Ilan University, Israel
andrew.lindell@aladdin.com, lindell@cs.biu.ac.il

Abstract. In this paper we study key exchange protocols in a model where the key exchange takes place between devices with limited displays that can be compared by a human user. If the devices display the same value then the human user is convinced that the key exchange terminated successfully and securely, and if they do not then the user knows that it came under attack. The main result of this paper is a rigorous proof that the numeric comparison mode for device pairing in Bluetooth version 2.1 is secure, under appropriate assumptions regarding the cryptographic functions used. Our proof is in the standard model and in particular does not model any of the functions as random oracles. In order to prove our main result, we present formal definitions for key exchange in this model and show our definition to be equivalent to a simpler definition. This is a useful result of independent interest that facilitates an easier security analysis of protocols in this model.

1 Introduction

A central problem in cryptography is that of enabling parties to communicate secretly and reliably in the presence of an adversary. This is often achieved by having the parties run a protocol for generating a mutual and secret session key. This session key can then be used for secure communication using known techniques (e.g., applying encryption and message authentication codes to all communication). Two important parameters to define regarding this problem relate to the strength of the adversary and the communication model and/or initial setup for the parties. The problem of session-key generation was initially studied by Diffie and Hellman [8] who considered a passive adversary that can eavesdrop on the communication of the parties, but cannot actively modify messages on the communication line. Thus, the parties are assumed to be connected by reliable, albeit non-private, channels. Many efficient and secure protocols are known for this scenario. In contrast, in this paper, we consider a far more powerful adversary who can modify and delete messages sent between the parties, as well as insert messages of its own choice. It is well known that in the presence of such a powerful adversary, it is impossible for the parties to generate a secret session key if they have no initial secrets and can only communicate over the

M. Fischlin (Ed.): CT-RSA 2009, LNCS 5473, pp. 66–83, 2009.
© Springer-Verlag Berlin Heidelberg 2009

adversarially controlled channel. This is due to the fact that the adversary can carry out a separate execution with each of the parties, where in each execution it impersonates the other. Since there is no initial secret (like a password or public-key infrastructure), there is nothing that prevents the adversary from succeeding in its impersonation.

The common solution to the above problem is to indeed introduce a shared secret, like a password, or to assume a public-key infrastructure. However, these solutions are not always possible nor always desired (a user cannot memorize a long private-key and short human-memorizable passwords are notoriously problematic). Another option is therefore to assume that the parties have an additional *authenticated communication channel* that cannot be tampered with by the adversary and can be used to send a short message [10,16]. There are a number of ways that such a channel can be implemented in reality. In this paper, we consider the case that the parties running the key exchange protocol (or, more accurately, the *devices*) each have a screen upon which they can display a short (say, 6 digit) number. The human user then compares to make sure that both devices display the same number, and if they do, is convinced that the key exchange terminated securely. We remark that although this does not seem to be an authenticated communication channel, it is essentially equivalent to one. This is because one party can send a short message to the other party (using the insecure channel), and then they can both display the message on their screens. If the adversary modifies the message en route, then this will be detected by the human user who will reject the result. Thus, the screens can be used to communicate a single short number from one party to the other (for usability reasons, it is required that only a single value be displayed).

Our results. Our main result is a *rigorous proof of security* of the numeric comparison mode in the simple pairing protocol of Bluetooth version 2.1 [1]. The importance of this result is due to the popularity of Bluetooth, and the unfortunate historic fact that vulnerabilities have often been found in unproven key exchange protocols, sometimes many years after they were released. We stress that our analysis focuses solely on the numeric comparison mode and says nothing about the security of the entire standard (and in particular, nothing about the security regarding the interplay between the different modes and backward compatibility with version 2.0). We prove the security of the protocol in the standard model, by appropriately modeling the functions used in the Bluetooth protocol as standard cryptographic primitives. We stress that we do not model any of the functions as ideal primitives (like random oracles), although this would have made the proof of security much easier.

In order to prove our results, we present a formal definition of comparison-based key exchange that is based on the definitions of key exchange of [3,4]. Our definition is similar in spirit to that of [16], except that we focus specifically on the problem of key exchange, whereas [16] considered a more general setting of message authentication. As is standard for definitions of security for key exchange protocols, we consider a complex setting where many different protocol instances are run concurrently. Since it is difficult to analyze the security of protocols in complex settings, we present an alternative definition that implies our main definition. The alternative definition is slightly more restrictive but seems to capture the way protocols typically work in this setting. This definition is easier to work with, and to demonstrate this further, we show that it is equivalent to a definition whereby only a single protocol execution takes place. We believe that this alternative definition and its equivalence to the simpler setting is of independent interest as it facilitates significantly easier proofs of security of protocols in this model.

Related work. The problem of secure key exchange has achieved a huge amount of attention, whether it be in the plain model with an eavesdropping adversary, or whether it considers an active adversary and assumes the existence of a full public-key infrastructure, shared high quality secrets or low quality passwords. The comparison-based model that we consider here was first studied in [11,12,10], with a more general treatment appearing in [16]. Tight bounds for achieving information-theoretic security in this model were shown in [15]. The MA-DH protocol of [13] has many similarities to the Bluetooth v2.1 numeric comparison protocol analyzed in this paper. Nevertheless, it has significant differences, making it necessary to provide a separate security analysis and proof.

2   Comparison-Based Secure Key-Exchange – Definitions

Preliminaries. We denote the security parameter by n. A function $f : \mathbb{N} \to [0, 1]$ is negligible if for every polynomial $p(\cdot)$ there exists an integer N such that for every $n > N$ it holds that $f(n) < 1/p(n)$. We denote an arbitrary negligible function by negl.

Background. In this section, we adapt the definition of secure key exchange of [3,4] to our setting. Although the basic ideas are similar, there are a number of fundamental differences between this model and the classic model of key exchange. First and foremost, the parties do not only interact via regular communication channels. In particular, the parties are able to carry out a numeric comparison between two short numbers of length ℓ, and this can be used to prevent the adversary from carrying out a successful man-in-the-middle attack. We formally model the comparison as part of the protocol in the following simple way: each entity participating in a key exchange holds a local public "comparison variable" (the variable is public in the sense that the adversary can read its value whenever it wishes). The comparison variable can be set only once in any instance (i.e., it is write-once only); this rules out protocols that use multiple comparisons (arguably, such protocols have more limited use in practice). Another fundamental difference between this setting and the classic model of key exchange is that it is *not* enough for the adversary to learn the secret key that one of the parties obtains at the end of a protocol execution (it can always succeed in doing this by just interacting with the party). Rather, the adversary only succeeds if it manages to learn the secret key that a pair of parties obtain in an execution in which the *parties' comparison variables are equal*. A third difference is that there is no public-key infrastructure or secret setup information and thus all instances of the protocol are identical. This is in contrast to the shared secret setting where each pair of parties hold a shared secret key, and every protocol instance run by a party is initialized with the party's secret key. Despite this, the protocol is supposed to be secure in the presence of an active adversary, and not just an eavesdropping one.

We remark that in our setting here, it makes no sense to allow a single party to run many instances of the protocol concurrently. This is because each party has only one interface for displaying the comparison variable, and so more than one execution cannot be run at the same time. In addition, since there is no shared setup between different executions, allowing more than one execution would be equivalent in any case (when there is no shared setup, a number of executions by a single party is equivalent to a number of parties running a single execution each). Of course, the different parties running different executions may be running concurrently. We could additionally allow each party to run many executions sequentially, but this clearly makes no difference and thus for simplicity we just assume that each party runs one execution.

The definition. A protocol for secure key exchange assumes that there is a set of *principals* which are the parties (clients, servers or others) who will engage in the protocol. We denote by Π_i the instance of the protocol that is run by user P_i (recall that in contrast to [3,4] each party runs one execution only). The adversary is given oracle access to these instances and may also control some of the instances itself. We remark that unlike the standard notion of an "oracle", in this model instances maintain state which is updated as the protocol progresses. In addition to information regarding the protocol execution, the state of an instance Π_i includes the following variables (initialized as *null*):

- sid_i: the *session identifier* of this particular instance;
- comp_i: the aforementioned write-once *comparison variable* of the instance; we denote the length of comp_i by ℓ;
- pid_i: the *partner identifier* which is the name of the principal P_j with whom P_i's comparison variable is compared (we note that pid_i can never equal i); in our setting here, it is always the case that if $\mathsf{pid}_i = j$ then $\mathsf{pid}_j = i$ because the human comparing the variables will always work in this way;[1]
- acc_i: a boolean variable set to **true** or **false** denoting whether Π_i accepts or rejects at the end of the execution.

Partnering. We say that two instances Π_i and Π_j are *partnered* if the following properties hold: (1) $\mathsf{pid}_i = j$ (and thus by our requirement $\mathsf{pid}_j = i$); and (2) $\mathsf{sid}_i = \mathsf{sid}_j \neq null$. The notion of partnering is important for defining security, as we will see.

The adversary model. The adversary is given total control of the external network (i.e., the network connecting clients to servers). In particular we assume that the adversary has the ability to not only listen to the messages exchanged by

---

[1] This is in contrast to the standard setting of key exchange where P_1 may think that it's interacting with P_2 who in turn thinks that it's interacting with P_3.

players, but also to interject messages of its choice and modify or delete messages sent by the parties.[2] The above-described adversarial power is modeled by giving the adversary oracle access to the instances of the protocol that are run by the principals. Notice that this means that the parties actually only communicate through the adversary. The oracles provided to the adversary are as follows:

- Execute(i, j): When this oracle is called, pid_i is set to j and pid_j is set to i, and then a complete protocol execution between instances Π_i and Π_j is run. The oracle-output is the protocol transcript (i.e., the complete series of messages exchanged by the instances throughout the execution). These oracle calls reflect the adversary's ability to passively eavesdrop on protocol executions. As we shall see, the adversary should learn nothing from such oracle calls. If an Init call has already been made including i or j, then Execute(i, j) is ignored.
- Init(i, j): This call initializes $pid_i = j$ and $pid_j = i$. If pid_i or pid_j is already set, then this call does nothing. In addition, it returns the first message that Π_i sends to Π_j in a protocol execution.
- Send(i, M): This call sends the message M to the instance Π_i. The output of the oracle is whatever message the instance Π_i would send after receiving the message M (given its current state). This oracle allows the adversary to carry out an active man-in-the-middle attack on the protocol executions.
- Reveal(i): This call outputs the secret key sk_i that instance Π_i outputs at the end of the protocol execution. This oracle allows the adversary to learn session keys from previous and concurrent executions, modeling improper exposure of past session keys and ensuring independence of different session keys in different executions.
- Test(i): This call is needed for the definition of security and does not model any real adversarial ability. The adversary is only allowed to query it once, and the output is either the private session key of Π_i, denoted sk_i, or a random key sk that is chosen independently of the protocol executions (each case happens with probability $1/2$). The adversary's aim is to distinguish these two cases. We let b^i_{test} denote the bit chosen by Test(i) to determine whether to output sk_i or a random sk.

The security of key exchange protocols is composed of three components: non-triviality, correctness and privacy. We begin by stating the non-triviality requirement (this is different from the definition in [3,4] because we also require that $comp_i = comp_j$ so that a human user will accept the result):

Non-triviality. If two instances Π_i and Π_j that hold each other's partner identifier communicate without adversarial interference (as in an Execute call), then Π_i and Π_j are partnered, $comp_i = comp_j$ and they both accept.

---

[2] In principle, the adversary should also be given control over a subset of the oracles, modeling the case of an "inside attacker". However, in our setting, there are no initial secrets and thus this makes no difference.

Correctness. If two *partnered* instances Π_i and Π_j accept (i.e., $\mathsf{acc}_i = \mathsf{acc}_j = 1$) and $\mathsf{comp}_i = \mathsf{comp}_j$, then they must both conclude with the same session key (i.e., $sk_i = sk_j$).

Privacy. We now define what it means for a protocol to be private. Intuitively, a protocol achieves privacy if the adversary cannot distinguish real session keys from random ones. (This then implies that the parties can use their generated session keys in order to establish secure channels; see [5] for more discussion on this issue.) Of course, the adversary can always correctly guess the bit in a $\mathsf{Test}(i)$ query if it queried $\mathsf{Reveal}(i)$ or $\mathsf{Reveal}(j)$ when Π_i and Π_j are partnered. Therefore, $\mathcal{A}$ is only said to have succeeded if these oracles were not queried. In addition, we are only interested in the case that $\mathcal{A}$ correctly guesses the key when $\mathsf{comp}_i = \mathsf{comp}_j$ and both instances accept. This is due to the fact that if $\mathsf{comp}_i \neq \mathsf{comp}_j$ then the human user will not accept the result, and if one of the instances does not accept then no session-key will be output by that instance. This yields the following definition of adversarial success. Formally, we say that an adversary $\mathcal{A}$ **succeeds** if the following conditions are *all* fulfilled:

1. $\mathcal{A}$ outputs b_{test}^i
2. $\mathsf{comp}_i = \mathsf{comp}_j$ and $\mathsf{acc}_i = \mathsf{acc}_j = \mathsf{true}$
3. If Π_i and Π_j are partnered then $\mathcal{A}$ did not query $\mathsf{Reveal}(i)$ or $\mathsf{Reveal}(j)$.

Now, the adversary's **advantage** is formally defined by:

$$\mathsf{Adv}(\mathcal{A}) = |2 \cdot \mathrm{Prob}[\mathcal{A} \text{ succeeds }] - 1| \,.$$

We reiterate that an adversary is only considered to have succeeded if it correctly guesses the bit used by the $\mathsf{Test}(i)$ oracle, $\mathsf{comp}_i = \mathsf{comp}_j$, and the adversary did *not* query $\mathsf{Reveal}(i)$ or $\mathsf{Reveal}(j)$ when Π_i and Π_j are partnered. We stress that if $\mathsf{pid}_i = j$ but $\mathsf{sid}_i \neq \mathsf{sid}_j$, then Π_i and Π_j are not partnered and thus the adversary succeeds if $\mathsf{comp}_i = \mathsf{comp}_j$ and it correctly guesses the bit used by the $\mathsf{Test}(i)$ oracle, even if it queried $\mathsf{Reveal}(j)$.

An important observation here is that when there is no initial setup and only a short comparison channel of length ℓ is used, the adversary can gain an advantage of $2^{-\ell}$ for every pair of instances by just running two separate executions with two instances and hoping that their comparison variables will end up being equal. A protocol is therefore called private if it is limited to random success of this fashion. Notice that in $\mathsf{Execute}$ oracle calls, the adversary is passive and thus it should only have a negligible advantage in guessing the secret key of such an instance, irrespective of the value of ℓ. We do not explicitly require this, but rather provide it with a $2^{-\ell}$ advantage only when it queries the Send oracle. This is reminiscent of the definition for password-based key exchange of [2]. In order to define this, we define Q_{send} to be the number of protocol instances without common partner identifiers to which the adversary made Send oracle queries. We stress that if $\mathcal{A}$ makes multiple Send queries to Π_i and to Π_j, and $\mathsf{pid}_i = j$, then this is counted as 1 in Q_{send}. Formally, a protocol is said to be **private** if the advantage of the adversary is at most negligibly more than $Q_{\text{send}}/2^\ell$, where ℓ is the length of the comparison variable. In summary,

Definition 1 (comparison-based key exchange). *A comparison-based key exchange protocol with a comparison variable of length $\ell \in \mathsf{N}$ is said to be* secure *if for every probabilistic polynomial-time adversary $\mathcal{A}$ that makes at most Q_{send} queries of type* Send *to different protocol instances without common partner identifiers, there exists a negligible function* negl *such that*

$$\mathsf{Adv}(\mathcal{A}) < \frac{Q_{\mathrm{send}}}{2^\ell} + \mathsf{negl}(n).$$

Furthermore, the probability that the non-triviality or correctness requirement is violated is at most negligible in the security parameter n.

We note that the bound of $Q_{\mathrm{send}}/2^\ell$ for $\mathcal{A}$'s advantage is optimal. Specifically, one can construct an adversary $\mathcal{A}$ who obtains this exact advantage by separately interacting with two protocol instances Π_i and Π_j for which $\mathsf{pid}_i = j$ and $\mathsf{pid}_j = i$. At the end of the execution, $\mathcal{A}$ will know both sk_i and sk_j and will succeed if $\mathsf{comp}_i = \mathsf{comp}_j$. If this does not hold, then $\mathcal{A}$ can just invoke an Execute oracle call for two other instances, query the Test oracle for one of those instances, and then just randomly guess the test result, succeeding with probability one half. The advantage of this adversary $\mathcal{A}$ is as follows. First, under the assumption that an honest protocol execution yields a uniformly distributed comparison variable, we have that $\mathsf{comp}_i = \mathsf{comp}_j$ with probability exactly $2^{-\ell}$. In this case, $\mathcal{A}$ succeeds with probability 1. Noting further that if $\mathsf{comp}_i \neq \mathsf{comp}_j$ then $\mathcal{A}$ succeeds with probability $1/2$, we have:

$$\Pr[\mathcal{A} \text{ succeeds}] = \Pr[\mathcal{A} \text{ succeeds} \mid \mathsf{comp}_i = \mathsf{comp}_j] \cdot \Pr[\mathsf{comp}_i = \mathsf{comp}_j]$$
$$+ \Pr[\mathcal{A} \text{ succeeds} \mid \mathsf{comp}_i \neq \mathsf{comp}_j] \cdot \Pr[\mathsf{comp}_i \neq \mathsf{comp}_j]$$
$$= 1 \cdot \frac{1}{2^\ell} + \frac{1}{2} \cdot \left(1 - \frac{1}{2^\ell}\right)$$
$$= \frac{1}{2^\ell} + \frac{1}{2} - \frac{1}{2^{\ell+1}} = \frac{1}{2} + \frac{1}{2^{\ell+1}}$$

implying that $\mathcal{A}$'s advantage is $1/2^\ell$. Noting finally that $Q_{\mathrm{send}} = 1$ in this case, we have that $\mathcal{A}$ achieves the upper bound of $Q_{\mathrm{send}}/2^\ell$ on the advantage as stated in Definition 1. The above argument holds for any value of Q_{send} (and not just the special case that $Q_{\mathrm{send}} = 1$). In this case, $\mathcal{A}$ interacts separately with Q_{send} pairs and succeeds if for any of the pairs it holds that $\mathsf{comp}_i = \mathsf{comp}_j$ (or with probability $1/2$ otherwise, as above). Since the probability that $\mathsf{comp}_i = \mathsf{comp}_j$ in at least one of the executions is $Q_{\mathrm{send}}/2^\ell$ we have that $\mathcal{A}$ succeeds with probability $Q_{\mathrm{send}}/2^\ell + \frac{1}{2} \cdot (1 - Q_{\mathrm{send}}/2^\ell)$. As above, this results in an advantage of $Q_{\mathrm{send}}/2^\ell$, as required.

An alternative definition. In the full version of this paper [14] we present an alternative definition that is easier to work with. We prove that security under the alternative definition implies security under Definition 1 and thus it suffices to use the alternative definition. In addition, we prove that when considering the alternative definition, security in the concurrent setting with many protocols instances is equivalent to security in a one-time setting where an adversary interacts once with a protocol instance P_1 and once with a protocol instance P_2 (and where $\mathsf{pid}_1 = 2$ and $\mathsf{pid}_2 = 1$). Since there is only one instance of each type, from here on we just refer to the adversary interacting with parties P_1 and P_2. We use the alternative definition to prove the security of the Bluetooth protocol since it is significantly easier to work with. In order to facilitate reading the proof below, we briefly describe the alternative definition. We define two events referring to the adversary's success:

1. An adversary $\mathcal{A}$ succeeds in a **guess attack**, denoted $\mathsf{succ}_{\mathcal{A}}^{\mathrm{guess}}$, if it outputs the correct b_{test}^i after querying $\mathsf{Test}(i)$ for an instance Π_i that is partnered with some other instance Π_j, and $\mathsf{Reveal}(i)$ or $\mathsf{Reveal}(j)$ were not queried.
2. An adversary $\mathcal{A}$ succeeds in a **comparison attack**, denoted $\mathsf{succ}_{\mathcal{A}}^{\mathrm{comp}}$, if there exist two accepting instances Π_i and Π_j with $\mathsf{pid}_i = j$ and $\mathsf{pid}_j = i$ that are *not* partnered and yet $\mathsf{comp}_i = \mathsf{comp}_j$.

We say that a protocol is secure if for every probabilistic polynomial-time $\mathcal{A}$, the probability of $\mathsf{succ}_{\mathcal{A}}^{\mathrm{guess}}$ is at most negligibly greater than $1/2$, and the probability of $\mathsf{succ}_{\mathcal{A}}^{\mathrm{comp}}$ is at most negligibly greater than $2^{-\ell}$. The justification for this definition can be found in the full version of this paper, as well as the fact that it suffices to analyze the security of a protocol in a restricted setting where the adversary interacts with a single pair of parties P_1 and P_2.

3   Bluetooth Pairing in Numeric Comparison Mode

In this section, we describe the Bluetooth pairing protocol in the numeric comparison mode. We also describe the cryptographic functions that are used by the protocol, and state the assumptions that are needed regarding each one in order to prove the security of the protocol. The Bluetooth specification refers to devices A and B; in order to be consistent with our definitional notations, we refer to parties P_1 and P_2 instead.

3.1   Cryptographic Tools and Functions

The numeric comparison mode in the Bluetooth simple pairing protocol uses the following tools:

- An Elliptic curve group in which it is assumed that the Decisional Diffie-Hellman problem is hard. We denote the group by $\mathcal{G}$, the generator by g, and the group order by q.

- A non-interactive computationally binding and *non-malleable* commitment scheme C. We denote a commitment to a string x using coins r by $C(x; r)$. The computational binding property means that it is infeasible for any polynomial-time adversary $\mathcal{A}$ to find x, r, x', r' where $x \neq x'$ but $C(x; r) = C(x'; r')$. Informally speaking, non-malleability [9] means that given a commitment $c = C(x; r)$ it is infeasible for a polynomial-time adversary to generate a commitment c' so that later given (x, r) it can produce (x', r') such that $c' = C(x'; r')$ and x, x' are related via a predetermined polynomial-time computable relation; this is typically called **non-malleability with respect to opening** [7]. Our formal definition can be found in Appendix A and is adapted from the definition in [6] with some minor changes.

The commitment scheme is instantiated as follows: in order to commit to a string x, r^a where r^a is uniformly distributed and half of the length of the key for HMAC-SHA256, choose a random string r^b which is also half of the length of the key for HMAC-SHA256, compute HMAC-SHA256$_r(x)$ where $r = (r^a, r^b)$, and set the commitment value to be the 128 most significant bits of the result. We remark that it may appear to be more natural to use randomness that is the entire length of the HMAC key and then let x be the entire string that is committed to. Indeed, this would have been more natural. However, in the Bluetooth protocol, part of r must be considered to remain secret and we therefore take it to be part of the value being committed to. We remark that the computational binding of this commitment scheme follows directly from the assumption that it is hard to find a collision in SHA256. The assumption on non-malleability is less studied, but seems reasonable given the chaotic behavior of such functions. We remark also that it follows trivially from any random-oracle type assumption.

- A function $g : \{0, 1\}^* \to \{0, 1\}^\ell$ with the property that when any long-enough part of the input is uniformly distributed, then the output is close to uniform. We formalize this by allowing an adversary to choose two values α and β and then asking what the probability is that $g(\alpha, r) = \beta$ when $r \in_R \{0, 1\}^{n/2}$ is uniformly distributed. We call this *computational 2-universal hashing*. In order to be consistent with the exact use of g in the protocol, we first introduce the following notation: For an arbitrary string α, we denote by $\alpha[r]$ the string derived by combining α and r in a predetermined way. (In our use, $\alpha[r]$ will either be the concatenation of r after α, or it involves parsing α into α_1 and α_2 where $|\alpha_2| = n/2$ and then setting $\alpha[r] = (\alpha_1, r, \alpha_2)$.) Then:

Definition 2. *A function* $g : \{0, 1\}^* \to \{0, 1\}^\ell$ *is a* computational 2-universal hash function *if for every probabilistic polynomial-time machine $\mathcal{A}$ there exists a negligible function* negl *such that*

$$\Pr_{(\alpha,\beta)\leftarrow\mathcal{A}(1^n); r\leftarrow\{0,1\}^{n/2}}[g(\alpha[r]) = \beta] < \frac{1}{2^\ell} + \mathsf{negl}(n)$$

We stress that $r \in_R \{0, 1\}^{n/2}$ is *uniformly distributed* and thus chosen independently of α and β output by $\mathcal{A}$. The function g in Bluetooth is defined by $g(x) = \text{SHA256}(x) \bmod 2^{32}$. It seems very reasonable to assume that SHA256 fulfills this property.

- A pseudorandom function F keyed with keys output from Diffie-Hellman key exchange over Elliptic curve groups. This is implemented using HMAC-SHA256 and taking the 128 most significant bits. Formally, we say that a function F is **pseudorandom when keyed with** $\mathcal{G}$ if it is a pseudorandom function when the key is a random element of $\mathcal{G}$. It is easy to show that if F is pseudorandom when keyed with $\mathcal{G}$ and the Decisional Diffie-Hellman (DDH) assumption holds in $\mathcal{G}$, then it is pseudorandom when keyed with the result of a Diffie-Hellman key exchange. This follows directly from DDH which states that the result of a Diffie-Hellman key exchange is indistinguishable from a random element in $\mathcal{G}$. For simplicity, we state this directly in the Definition below:

Definition 3. *Let* gen(1^n) *be an algorithm that outputs the description of a group $\mathcal{G}$, its generator g, and its order q. A function ensemble $F = \{F_k\}$ is* pseudorandom when DDH-keyed with gen *if for every probabilistic polynomial-time distinguisher D there exists a negligible function* negl *such that*

$$\left| \Pr\left[D^{F_{g^{ab}}}(1^n, g^a, g^b) = 1 \right] - \Pr\left[D^H(1^n, g^a, g^b) = 1 \right] \right| < \mathsf{negl}(n)$$

where $(\mathcal{G}, g, q) \leftarrow$ gen(1^n), a, b are randomly chosen in $\{1, \ldots, q\}$, and H is a truly random function ensemble.

As we have mentioned, any function ensemble that is pseudorandom when keyed with a random element from $\mathcal{G}$ is also pseudorandom when DDH-keyed with gen, under the assumption that the DDH assumption holds relative to gen. Note that a "standard" pseudorandom function receives a uniformly distributed bit string. Therefore, this does not necessarily suffice (a random element of $\mathcal{G}$ is not necessarily a uniformly distributed bit string).

3.2   The Protocol and Correctness

The Bluetooth simple pairing protocol in numeric comparison mode appears in Figure 1 and is denoted Π. It is easy to see that Protocol Π is non-trivial. The proof that Π fulfills correctness is also not difficult and appears in the full version of this paper [14].

4   The Proof of Security

We now prove that Π is a secure comparison-based key-exchange protocol. The structure of our proof demonstrates the usefulness of the alternative definition as a tool; a proof of security that works directly with Definition 1 would be much more complex.

Theorem 4. *Assume that the Decisional Diffie-Hellman assumption holds relative to* gen, *that F is a pseudorandom function when DDH-keyed with* gen, *that C is a computationally-binding non-malleable commitment scheme and that g is a computational 2-universal hash function. Then, Protocol Π is a secure comparison-based key-exchange protocol.*

Protocol Π

- **Pre-protocol exchange:** Parties P_1 and P_2 exchange party identifiers 1 and 2 (in Bluetooth, these are their respective Bluetooth addresses) as well as additional auxiliary information α_1 and α_2 (we will ignore the content of this information here). P_1 sets $\mathsf{pid}_1 = 2$, and P_2 sets $\mathsf{pid}_2 = 1$.

- **Phase 1 – Public-Key Exchange:**
  1. The initiating party P_1 generates a Diffie-Hellman value by choosing a random $a \in \{1, \ldots, q\}$ and computing $pk_1 = g^a$. P_1 sends pk_1 to P_2.

  2. Upon receiving pk_1 from P_1, party P_2 chooses a random $b \in \{1, \ldots, q\}$, computes $pk_2 = g^b$, and sends pk_2 to P_1.

  3. Party P_1 sets $\mathsf{sid}_1 = (pk_1, pk_2)$ and party P_2 sets $\mathsf{sid}_2 = (pk_1, pk_2)$.

- **Phase 2 – Authentication Stage 1:**
  1. P_2 chooses a random string $r_2 \in_R \{0,1\}^n$ and sends $c_2 = C(pk_2, pk_1, 0; r_2)$ to P_1.

  2. P_1 chooses a random string $r_1 \in_R \{0,1\}^n$ and sends r_1 to P_2.

  3. P_2 sends r_2 to P_1. Upon receiving r_2, party P_1 checks that $c_2 = C(pk_2, pk_1, 0; r_2)$, where c_2 is the value it received above and pk_1, pk_2 are as exchanged in phase 1.

  4. P_1 sets $\mathsf{comp}_1 = g(pk_1, pk_2, r_1, r_2)$ and P_2 sets $\mathsf{comp}_2 = g(pk_1, pk_2, r_1, r_2)$.

- **Phase 3 – Authentication Stage 2:**
  1. P_1 computes $k = (pk_2)^a$ and P_2 computes $k = (pk_1)^b$, where the computation is in the group $\mathcal{G}$.

  2. P_1 computes $e_1 = F_k(r_1, r_2, 0, \alpha_1, 1, 2)$ and sends e_1 to P_2. (Note that 1 and 2 here, and below, are the parties identifiers and not constants. Thus, they are actually the parties' Bluetooth addresses and more.)

  3. P_2 checks that $e_1 = F_k(r_1, r_2, 0, \alpha_2, 1, 2)$; if yes it sends P_1 the value $e_2 = F_k(r_1, r_2, 0, \alpha_1, 2, 1)$ and sets $\mathsf{acc}_2 = \mathsf{true}$; otherwise P_2 sets $\mathsf{acc}_2 = \mathsf{false}$ and aborts.

  4. P_1 checks that $e_2 = F_k(r_1, r_2, 0, \alpha_1, 2, 1)$; if yes it sets $\mathsf{acc}_1 = \mathsf{true}$, and if not it sets $\mathsf{acc}_1 = \mathsf{false}$ and aborts.

- **Phase 4 – Link-Key Calculation:**
  1. Party P_1 outputs $sk_1 = F_k(r_1, r_2, \beta, 2, 1)$ where β is a fixed string.

  2. Party P_2 outputs $sk_2 = F_k(r_1, r_2, \beta, 2, 1)$.

Fig. 1. Bluetooth 2.1 Pairing – Numeric Comparison Mode

Proof. We prove the security of Π in two stages. First, we prove that $\mathsf{succ}_{\mathcal{A}}^{\text{guess}}$ occurs with probability at most negligibly greater than $1/2$. Intuitively, this holds because if P_1 and P_2 are partnered, then this implies that they both have the same Diffie-Hellman values and so essentially have completed a Diffie-Hellman key exchange *undisturbed*, with the adversary only eavesdropping. This in turn implies that F_k is a pseudorandom function and thus the session keys that are output are pseudorandom. We then proceed to prove that $\mathsf{succ}_{\mathcal{A}}^{\text{comp}}$ occurs with probability at most negligibly greater than $2^{-\ell}$. This follows from the security of the commitment scheme C and the 2-universality of g. Specifically, phase 2 of the protocol can be viewed as a method of choosing two random strings r_1 and r_2 that are (computationally) independent of each other. In order for this to hold even if $\mathcal{A}$ carries out a man-in-the-middle attack, the commitment scheme C must be non-malleable (see Appendix A). This forces $\mathcal{A}$ to either just copy the commitment sent by P_2 or modify it, in which case it will contain an independent r_2 value. If $\mathcal{A}$ copies the commitment, then it will contain the parties' public keys. However, by the assumption that they are not partnered, these keys do not match with those that the parties received in the protocol. $\mathcal{A}$ must therefore modify the commitment, resulting in r_1 and r_2 being independent of each other. Once this is given, it is possible to apply the 2-universality of g stating that whichever is chosen last causes the comparison value to be almost uniformly distributed. We proceed now to the formal proof.

As stated, we prove the protocol using the alternative definition (which is proven in the full version to imply security under Definition 1). We begin by proving that for every probabilistic polynomial-time $\mathcal{A}$ interacting with P_1 and P_2, it holds that

$$\Pr[\mathsf{succ}_{\mathcal{A}}^{\text{guess}}] < \frac{1}{2} + \mathsf{negl}(n)$$

Recall that $\mathsf{succ}_{\mathcal{A}}^{\text{guess}}$ occurs if $\mathcal{A}$ outputs the correct b_{test} value after querying $\mathsf{Test}(1)$ or $\mathsf{Test}(2)$ *and* P_1 is partnered with P_2. Now, by the protocol description, the session identifiers are defined to be (pk_1, pk_2) and thus if $\mathsf{sid}_1 = \mathsf{sid}_2$ it follows that P_1 and P_2 hold the same Diffie-Hellman values. Intuitively, this means that if $\mathcal{A}$ can guess the correct b_{test} value with non-negligible probability, then it can solve the DDH problem in $\mathcal{G}$ with non-negligible advantage. The formal reduction follows. Let $\mathcal{A}$ be a probabilistic polynomial-time adversary and let ϵ be a function such that $\Pr[\mathsf{succ}_{\mathcal{A}}^{\text{guess}}] = \frac{1}{2} + \epsilon(n)$. We show that ϵ must be negligible by presenting a distinguisher D that solves the DDH problem in $\mathcal{G}$ with advantage ϵ. Distinguisher D receives (g^a, g^b, k) and attempts to determine if $k = g^{ab}$ or if $k \in_R \mathcal{G}$. D invokes $\mathcal{A}$ and when it sends a Send oracle query to which P_1 is supposed to reply with its public-key exchange message, then D replies with $pk_1 = g^a$. Likewise, when $\mathcal{A}$ sends an analogous message for P_2 then D replies with $pk_2 = g^b$. If $\mathcal{A}$ does not forward the same pk_1, pk_2 messages unmodified (and so P_1 and P_2 are not partnered) then D outputs a random bit and halts. Otherwise, it proceeds. (Note that D may need to proceed with the simulation before knowing if they are partnered. In this case, it assumes that they will be, and if it turns out to be incorrect it immediately outputs a random bit and halts.) Now, from this step on, D acts exactly like the honest P_1 and P_2

would. In particular, when D reaches the authentication stage 2 of the protocol, it uses the value k that it received in its input to compute e_1 and e_2. Likewise, it uses k to compute sk_1 and sk_2. Now, when $\mathcal{A}$ queries $\mathsf{Test}(1)$ or $\mathsf{Test}(2)$, D chooses a random $b \in_R \{0,1\}$ and replies with sk_1 (or sk_2 respectively) if $b = 0$ and with a random value $\tilde{sk} \in_R \{0,1\}^{|sk_1|}$ otherwise. Finally, D outputs 1 if and only if $\mathcal{A}$ outputs $b_{\text{test}} = b$.

If $k = g^{ab}$ then the simulation above by D is exactly what $\mathcal{A}$ would see in a real protocol execution. Therefore,

$$\Pr[D(g^a, g^b, g^{ab}) = 1] = \frac{1}{2} + \epsilon(n)$$

In contrast, when k is a random value, the simulation by D is "wrong". In particular, the e_1, e_2, sk_1, sk_2 values are computed using a random key k independent of pk_1, pk_2, instead of using g^{ab}. We would like to claim that $\mathcal{A}$ outputs $b_{\text{test}} = b$ with probability $1/2$ in this case, but this may not be true because k has been used to compute e_1, e_2 which are seen by $\mathcal{A}$. Thus, if $\mathcal{A}$ was not computationally bounded it could determine $b_{\text{test}} = b$. Nevertheless, we prove that if F is indeed a pseudorandom function, then $\mathcal{A}$ can output $b_{\text{test}} = b$ with probability at most $1/2 + \mathsf{negl}(n)$. Let δ be a function such that $\mathcal{A}$ outputs $b_{\text{test}} = b$ in this case with probability $1/2 + \delta(n)$. We first prove that δ is a negligible function. Specifically, we construct a distinguisher D_F who receives an oracle that is either the pseudorandom function F_k or a truly random function. D_F invokes $\mathcal{A}$ and works in the same way as D with the following differences. First, D_F generates random pk_1 and pk_2 values itself and uses them. Second, it computes e_1, e_2, sk_1, sk_2 using its *function oracle*. If D_F is given a random function oracle, then sk_1, sk_2 are completely random and independent of everything that $\mathcal{A}$ has seen so far. Thus, information-theoretically, $\mathcal{A}$ outputs $b_{\text{test}} = b$ with probability exactly $1/2$. In contrast, if D_F is given F_k as an oracle, then it generates exactly the same distribution as D when $k \in_R \mathcal{G}$ is a random value. It follows that in this case $\mathcal{A}$ outputs $b_{\text{test}} = b$ with probability $1/2 + \delta(n)$. This implies that

$$\left| \Pr[D_F^{F_k}(1^n) = 1] - \Pr[D_F^H(1^n) = 1] \right| = \delta(n)$$

and so δ must be a negligible function, by the assumption that F_k is a pseudorandom function. Combining the above, we have that

$$\left| \Pr[D(g^a, g^b, g^{ab}) = 1] - \Pr[D(g^a, g^b, k) = 1] \right|$$
$$= \left| \frac{1}{2} + \epsilon(n) - \frac{1}{2} - \delta(n) \right| = |\epsilon(n) - \delta(n)|$$

and so ϵ must also be a negligible function, proving that $\mathsf{succ}_{\mathcal{A}}^{\text{guess}}$ occurs with probability that is at most negligibly greater than $1/2$, as required.

We now prove that for every probabilistic polynomial-time $\mathcal{A}$ interacting only with P_1 and P_2, it holds that

$$\Pr[\mathsf{succ}_{\mathcal{A}}^{\text{comp}}] < \frac{1}{2^{\ell}} + \mathsf{negl}(n)$$

Recall that $\mathsf{succ}_{\mathcal{A}}^{\mathrm{comp}}$ holds if P_1 and P_2 are *not* partnered, and yet $\mathsf{comp}_1 = \mathsf{comp}_2$. Since the session identifier in Π is defined to be the pair of public keys (pk_1, pk_2) exchanged in the first phase, we have that $\mathsf{succ}_{\mathcal{A}}^{\mathrm{comp}}$ can only hold if P_1 and P_2 hold different public keys. This occurs if at least one of the keys sent by an instance was not received as-is by the other instance, but was rather "modified" en route by $\mathcal{A}$.

We introduce the following notation that will be helpful in the proof below. If one instance sends a message α, then we denote by α' the message received by the other instance. Thus, the public key sent by P_1 is denoted pk_1 and the public key received by P_2 is denoted pk_1'. Using this notation, we have that P_1 and P_2 are not partnered if P_1 has $\mathsf{sid}_1 = (pk_1, pk_2')$ and P_2 has $\mathsf{sid}_2 = (pk_1', pk_2)$, and either $pk_1 \neq pk_1'$ or $pk_2 \neq pk_2'$ or both.

Now, the first authentication stage involves P_2 sending $c_2 = C(pk_2, pk_1', 0; r_2)$ and P_1 receiving some c_2'. Then, P_1 sends r_1 and P_2 receives r_1'. Finally, P_2 returns r_2 and P_1 receives some string r_2'. Using the above notation, we have that $\mathsf{succ}_{\mathcal{A}}^{\mathrm{comp}}$ occurs if and only if

$$(pk_1, pk_2') \neq (pk_1', pk_2) \quad \text{and} \quad g(pk_1, pk_2', r_1, r_2') = g(pk_1', pk_2, r_1', r_2) \quad (1)$$

(Note that $\mathsf{comp}_1 = g(pk_1, pk_2', r_1, r_2')$ and $\mathsf{comp}_2 = g(pk_1', pk_2, r_1', r_2)$.) Without loss of generality, we assume that $\mathcal{A}$ always causes P_1 and P_2 to be not partnered (otherwise it always fails so this does not make any difference), and so $(pk_1, pk_2') \neq (pk_1', pk_2)$ always. We analyze the probability that Eq. (1) holds in two disjoint cases related to the possible schedulings of messages by $\mathcal{A}$:

1. *Case 1 – P_2 sends r_2 after P_1 has received c_2':* The main difficulty in the proof here is due to the fact that it is theoretically possible that $\mathcal{A}$ can make c_2' depend on c_2 (and likewise r_2' can depend on r_2). Therefore, the inability of $\mathcal{A}$ to succeed depends on the *non-malleability* of the commitment scheme C; see Definition 5 in Appendix A (familiarity with the exact definition is needed for the proof below). Let $\mathcal{A}$ be a probabilistic polynomial-time adversary. We prove that $\mathsf{succ}_{\mathcal{A}}^{\mathrm{comp}}$ occurs in this case with probability that is at most negligibly greater than $2^{-\ell}$. First, we show that there exists an adversary $\hat{\mathcal{A}}$, a relation $\hat{R}$ and a distribution $\hat{D}$ for the non-malleability experiment $\mathsf{Expt}_{\hat{\mathcal{A}}, \hat{R}, \hat{D}}^{\mathrm{real}}(1^n)$ such that

$$\Pr[\mathsf{Expt}_{\hat{\mathcal{A}}, \hat{R}, \hat{D}}^{\mathrm{real}}(1^n) = 1] = \Pr[\mathsf{succ}_{\mathcal{A}}^{\mathrm{comp}}] \quad (2)$$

Adversary $\hat{\mathcal{A}}$ for the non-malleability experiment begins by invoking $\mathcal{A}$ (the adversary for the key exchange protocol) and emulating the parties P_1 and P_2 until the point that P_2 is supposed to send c_2. Note that at this point, the keys pk_1' and pk_2 are fully defined. Then, $\hat{\mathcal{A}}$ outputs $z = (pk_2, pk_1')$. The distribution $\hat{D}$ receives z, chooses a random $r_2^a \in_R \{0,1\}^{n/2}$ and outputs $m_1 = (pk_2, pk_1', 0, r_2^a)$. Adversary $\hat{\mathcal{A}}$ then receives com_1 (by the definition of C, com_1 is a commitment to m_1 using random coins r_2^b of length $n/2$), and hands it to $\mathcal{A}$ as if it is the commitment c_2 sent by P_2 in the key exchange protocol. When $\mathcal{A}$ sends a commitment c_2' to P_1, then $\hat{\mathcal{A}}$ defines this to be com_2

and outputs it. Following this, as defined in the non-malleability experiment, $\hat{\mathcal{A}}$ receives dec_1 which is the string $(pk_2, pk_1', 0, r_2^a, r_2^b)$. $\hat{\mathcal{A}}$ defines $r_2 = (r_2^a, r_2^b)$ and hands it to $\mathcal{A}$ as if coming from P_2. Finally, when $\mathcal{A}$ wishes to send r_2' to P_1, $\hat{\mathcal{A}}$ defines $\mathsf{dec}_2 = (pk_2', pk_1, 0, r_2^{a'}, r_2^{b'})$ and $\sigma = (pk_1, pk_2', r_1, r_1', r_2^b, r_2^{b'})$ where these are the appropriate strings sent in the emulation carried out by $\hat{\mathcal{A}}$ ($\hat{\mathcal{A}}$ needs to include r_2^b and $r_2^{b'}$ because these are not part of the messages m_1, m_2 but randomness used to generate the commitments). Finally, R outputs 1 if and only if $g(pk_1, pk_2', r_1, r_2') = g(pk_1', pk_2, r_1', r_2)$, where the values input to g are parsed from m_1, m_2 and σ. Eq. (2) follows from the observation that $\hat{\mathcal{A}}$'s emulation of an execution of Π for $\mathcal{A}$ is perfect, and from the fact that R outputs 1 if and only if $\mathsf{succ}_{\mathcal{A}}^{\mathsf{comp}}$ occurs.

Now, by the assumption that the commitment scheme C is non-malleable with respect to opening, we have that there exists an adversary $\hat{\mathcal{A}}'$ such that

$$\Pr[\mathsf{Expt}_{\hat{\mathcal{A}}, \hat{R}, \hat{D}}^{\mathrm{real}}(1^n) = 1] < \Pr[\mathsf{Expt}_{\hat{\mathcal{A}}', \hat{R}, \hat{D}}^{\mathrm{sim}}(1^n) = 1] + \mathsf{negl}(n)$$

We don't know how $\hat{\mathcal{A}}'$ works, but we do know that it first outputs a string z and then a pair (σ, m_2). The output of the experiment is then equal to 1 if and only if $g(pk_1, pk_2', r_1, r_2') = g(pk_1', pk_2, r_1', r_2)$, where $pk_1, pk_2', pk_1', pk_2, r_1, r_1', r_2'$ are all derived from z, σ and m_2, and r_2^a is *uniformly distributed* and independent of all other values. We stress that r_2^a is random and independent since $r_2^a \in_R \{0,1\}^{n/2}$ is chosen randomly by $\hat{D}$ and not given to $\hat{\mathcal{A}}'$. (Note that we cannot say anything about r_2^b because this is chosen by $\hat{\mathcal{A}}'$ as part of σ.) We conclude this case by using the computational 2-universality of g. That is, letting $\beta = \mathsf{comp}_1$ (which is fully defined by z, σ and m_2) and $\alpha_1 = (pk_1', pk_2, r_1'), \alpha_2 = r_2^b$ (again, fully defined by z and σ), we have that

$$\Pr_{r_2^a \leftarrow \{0,1\}^{n/2}} [g(\alpha_1, r_2^a, \alpha_2) = \beta] < \frac{1}{2^\ell} + \mathsf{negl}(n).$$

Thus

$$\Pr[\mathsf{Expt}_{\hat{\mathcal{A}}', \hat{R}, \hat{D}}^{\mathrm{sim}}(1^n) = 1] < \frac{1}{2^\ell} + \mathsf{negl}(n),$$

implying that

$$\Pr[\mathsf{succ}_{\mathcal{A}}^{\mathsf{comp}}] = \Pr[\mathsf{Expt}_{\hat{\mathcal{A}}, \hat{R}, \hat{D}}^{\mathrm{real}}(1^n) = 1]$$

$$< \Pr[\mathsf{Expt}_{\hat{\mathcal{A}}', \hat{R}, \hat{D}}^{\mathrm{sim}}(1^n) = 1] + \mathsf{negl}(n) < \frac{1}{2^\ell} + \mathsf{negl}'(n)$$

proving that the probability that $\mathsf{succ}_{\mathcal{A}}^{\mathsf{comp}}$ is at most negligibly greater than $2^{-\ell}$, as required. (We remark that the above proof only works in the scheduling case where $\mathcal{A}$ sends c_2' to P_1 before receiving r_2 from P_2, because in the non-malleability experiment com_2 must be output by the adversary before it receives dec_1.)

2. *Case 2 – P_2 sends r_2 before P_1 has received c_2':* Observe that phase 2 involves P_2 sending c_2, P_1 sending r_1 and then P_2 replying with r_2. Thus, in this case, $\mathcal{A}$ effectively runs the executions with P_1 and P_2 *sequentially*. That is, $\mathcal{A}$

concludes phase 2 with P_2 before beginning phase 2 with P_1. Intuitively, in this case, $\mathsf{succ}_{\mathcal{A}}^{\mathsf{comp}}$ can only occur with probability $2^{-\ell}$ because comp_2 is *fixed* before r_1 is chosen by P_1. Thus, the computational 2-universality of g suffices to show that $\mathsf{comp}_1 = \mathsf{comp}_2$ with probability at most negligibly greater than $2^{-\ell}$. More formally, let β be the comp_2 value of P_2. By this scheduling case, this is fixed before P_1 receives c_2' and so, in particular, before it chooses r_1. However, if $\mathcal{A}$ can choose r_2' after receiving r_1 from P_1, then the property of g no longer holds (recall that α and β must be independent of r). Intuitively this is not a problem due to the computational binding property of C.

Formally, let $\mathcal{A}$ be an adversary for the key exchange protocol; we assume that $\mathcal{A}$ always sends a valid r_2' to P_1 (otherwise P_1 rejects). We construct α and β as required for g as follows. Invoke $\mathcal{A}$ and emulate an execution with P_1 and P_2 until the end of phase 2 with P_1. Since phase 2 has finished, the strings c_2' and r_2' are fully defined, as are $pk_1, pk_1', pk_2, pk_2', r_2, r_1'$ (recall that phase 2 with P_2 concluded before it even started with P_1). These values therefore define α and β as follows: $\alpha = (pk_1, pk_2', r_2')$ and $\beta = \mathsf{comp}_2 = g(pk_1', pk_2, r_1', r_2)$. Now, we argue that

$$\Pr[\mathsf{succ}_{\mathcal{A}}^{\mathsf{comp}}] < \Pr_{r \leftarrow \{0,1\}^n}[g(\alpha, r) = \beta] + \mathsf{negl}(n) \tag{3}$$

In order to see that this holds, after $\mathcal{A}$ sends r_2' at the end of phase 2 (in the above procedure for determining α and β), rewind $\mathcal{A}$ to the point before r_1 is sent by P_1. Then, replace it with the random string r in Eq. (3). The value r_1 sent by P_1 in the process of determining α and β is identically distributed to the value r from Eq. (3). Now, there are two possibilities: $\mathcal{A}$ sends the same r_2' as when determining α and β, or $\mathcal{A}$ sends a different r_2'. In the first case, we have that $\mathsf{succ}_{\mathcal{A}}^{\mathsf{comp}}$ occurs *if and only if* $g(\alpha, r) = \beta$. In the second case, we have that $\mathcal{A}$ can be used to contradict the binding property of C (the formal reduction of this fact is straightforward and thus omitted). Thus, this case can occur with at most negligible probability. Eq. (3) therefore follows. By the security of g, we have that $\mathsf{succ}_{\mathcal{A}}^{\mathsf{comp}}$ occurs with probability at most negligibly greater than $2^{-\ell} + \mathsf{negl}(n)$, as required.

This completes the proof of security. ∎

References

1. Specification of the Bluetooth system. Covered Core Package version 2.1 + EDR (July 26, 2007)
2. Bellare, M., Pointcheval, D., Rogaway, P.: Authenticated Key Exchange Secure Against Dictionary Attacks. In: Preneel, B. (ed.) EUROCRYPT 2000. LNCS, vol. 1807, pp. 139–155. Springer, Heidelberg (2000)
3. Bellare, M., Rogaway, P.: Entity Authentication and Key Distribution. In: Stinson, D.R. (ed.) CRYPTO 1993. LNCS, vol. 773, pp. 232–249. Springer, Heidelberg (1994)
4. Bellare, M., Rogaway, P.: Provably Secure Session Key Distribution: the Three Party Case. In: The 27th STOC 1995, pp. 57–66 (1995)

5. Canetti, R., Krawczyk, H.: Analysis of Key-Exchange Protocols and Their Use for Building Secure Channels. In: Pfitzmann, B. (ed.) EUROCRYPT 2001. LNCS, vol. 2045, pp. 453–474. Springer, Heidelberg (2001)
6. Di Crescenzo, G., Katz, J., Ostrovsky, R., Smith, A.: Efficient and Non-interactive Non-malleable Commitment. In: Pfitzmann, B. (ed.) EUROCRYPT 2001. LNCS, vol. 2045, pp. 40–59. Springer, Heidelberg (2001)
7. Di Crescenzo, G., Ishai, Y., Ostrovsky, R.: Non-Interactive and Non-Malleable Commitment. In: 30th STOC, pp. 141–150 (1998)
8. Diffie, W., Hellman, M.E.: New Directions in Cryptography. IEEE Transactions on Information Theory IT-22, 644–654 (1976)
9. Dolev, D., Dwork, C., Naor, M.: Non-Malleable Cryptography. SIAM Journal on Computing 30(2), 391–437 (2000)
10. Gehrmann, C., Mitchell, C., Nyberg, K.: Manual Authentication for Wireless Devices. RSA Cryptobytes 7, 29–37 (2004)
11. Hoepman, J.H.: The ephemeral pairing problem. In: Juels, A. (ed.) FC 2004. LNCS, vol. 3110, pp. 212–226. Springer, Heidelberg (2004)
12. Hoepman, J.H.: Ephemeral Pairing on Anonymous Networks. In: Hutter, D., Ullmann, M. (eds.) SPC 2005. LNCS, vol. 3450, pp. 101–116. Springer, Heidelberg (2005)
13. Laur, S., Nyberg, K.: Efficient Mutual Data Authentication Using Manually Authenticated Strings. In: Pointcheval, D., Mu, Y., Chen, K. (eds.) CANS 2006. LNCS, vol. 4301, pp. 90–107. Springer, Heidelberg (2006)
14. Lindell, Y.: Comparison-Based Key Exchange and the Security of the Numeric Comparison Mode in Bluetooth v2.1 (full version). ePrint Cryptology Archive, Report 2009/013 (2009)
15. Naor, M., Segev, G., Smith, A.: Tight Bounds for Unconditional Authentication Protocols in the Manual Channel and Shared Key Models. In: Dwork, C. (ed.) CRYPTO 2006. LNCS, vol. 4117, pp. 214–231. Springer, Heidelberg (2006)
16. Vaudenay, S.: Secure Communications over Insecure Channels Based on Short Authenticated Strings. In: Shoup, V. (ed.) CRYPTO 2005. LNCS, vol. 3621, pp. 309–326. Springer, Heidelberg (2005)

A   Non-malleable Commitments – Definition

Informally speaking, a commitment scheme is non-malleable if given a commitment c it is computationally hard to generate a commitment c' that is "related" by some predefined relation R. When considering computationally binding commitments, it is possible that c' can actually be a commitment to any value. Therefore, it is not clear what it means that the value committed to in c' is related to the value committed to in c. This problem is solved by defining the notion *with respect to opening* [7]. This means that given a decommitment for c to some value x, it is hard for the adversary (who generated c' after being given c) to generate a decommitment for c' to some x' so that x' is related to x. Of course, the probability of success depends on the relation (some are "easier" than others). Therefore, the requirement is that it is possible to generate a related commitment c' given c with the same probability as it is possible to generate a related x' without even being given x. This is formalized by defining two experiments: a real experiment in which the adversary is given c, and a simulation experiment where the adversary just outputs a message and hopes that it's related. Our formal definition is adapted from [6] with two minor changes. First, we allow the adversary $\mathcal{A}$ to provide input to the distribution machine that generates the value to be committed to. Second, we allow the adversary to output state information which is used by the relation. Both of these changes do not seem to make it particularly easier for the adversary, but they make the definition much more useful for proving the security of protocols which rely on non malleability. The experiments relate to a probabilistic polynomial-time adversary $\mathcal{A}$, a polynomial-time computable relation R and a probabilistic polynomial-time samplable distribution D. We also denote the committer/sender algorithm by P_1 and the receiver algorithm by P_2 (the receiver takes for input a commitment string and a decommitment value and output a string that represents the value that was committed to). The experiments are defined as follows:

Experiment $\mathsf{Expt}^{\mathrm{real}}_{\mathcal{A},R,D}(1^n)$:

1. $z \leftarrow \mathcal{A}(1^n)$
2. $m_1 \leftarrow D(1^n, z)$
3. $(\mathsf{com}_1, \mathsf{dec}_1) \leftarrow P_1(m_1)$
4. $\mathsf{com}_2 \leftarrow \mathcal{A}(1^n, \mathsf{com}_1)$
5. $(\sigma, \mathsf{dec}_2) \leftarrow \mathcal{A}(1^n, \mathsf{com}_1, \mathsf{dec}_1)$
6. $m_2 \leftarrow P_2(\mathsf{com}_2, \mathsf{dec}_2)$
7. Output 1 if and only if $\mathsf{com}_1 \neq \mathsf{com}_2$ and $R(\sigma, m_1, m_2) = 1$

Experiment $\mathsf{Expt}^{\mathrm{sim}}_{\mathcal{A}',R,D}(1^n)$:

1. $z \leftarrow \mathcal{A}'(1^n)$
2. $m_1 \leftarrow D(1^n, z)$
3. $(\sigma, m_2) \leftarrow \mathcal{A}'(1^n)$
4. Output 1 if and only if $R(\sigma, m_1, m_2) = 1$

We now define security by stating that for every $\mathcal{A}$ in the real experiment there exists an $\mathcal{A}'$ who succeeds with almost the same probability in the simulation experiment. We allow the machine $\mathcal{A}'$ to know the distribution machine D and relation R (unlike [6]); this suffices for our proof of security and is possibly a weaker requirement.

Definition 5. *A non-interactive commitment scheme C with sender/receiver algorithms (P_1, P_2) is non-malleable with respect to opening if for every probabilistic polynomial-time $\mathcal{A}$, every probabilistic polynomial-time samplable distribution D and every polynomial-time computable ternary relation R, there exists a probabilistic polynomial-time $\mathcal{A}'$ and a negligible function negl such that:*

$$\Pr\left[\mathsf{Expt}^{\mathrm{real}}_{\mathcal{A},R,D}(1^n) = 1\right] < \Pr\left[\mathsf{Expt}^{\mathrm{sim}}_{\mathcal{A}',R,D}(1^n) = 1\right] + \mathsf{negl}(n)$$

Key Insulation and Intrusion Resilience
over a Public Channel

Mihir Bellare[1], Shanshan Duan[2], and Adriana Palacio[3]

[1] Dept. of Computer Science & Engineering, University of California, San Diego
9500 Gilman Drive, La Jolla, CA 92093, USA
http://www-cse.ucsd.edu/users/mihir
[2] Dept. of Computer Science & Engineering, University of California, San Diego
9500 Gilman Drive, La Jolla, CA 92093, USA
http://www-cse.ucsd.edu/users/shduan
[3] Computer Science Department, Bowdoin College,
8650 College Station, Brunswick, ME 04011-8486, USA
http://academic.bowdoin.edu/faculty/A/apalacio/

Abstract. Key insulation (KI) and Intrusion resilience (IR) are methods
to protect a user's key against exposure by utilizing periodic communications with an auxiliary helper. But existing work assumes a secure channel
between user and helper. If we want to realize KI or IR in practice we must
realize this secure channel. This paper looks at the question of how to do
this when the communication is over what we are more likely to have in
practice, namely a public channel such as the Internet or a wireless network. We explain why this problem is not trivial, introduce models and
definitions that capture the desired security in a public channel setting,
and provide a complete (and surprising) answer to the question of when
KI and IR are possible over a public channel. The information we provide
is important to guide practitioners with regard to the usage of KI and IR
and also to guide future research in this area.

1   Introduction

Key Insulation (KI) [15,16] and Intrusion Resilience (IR) [13,20] are technologies
to protect against key exposure. They have been extensively researched in the
cryptographic community and we have lots of schemes, variations and extensions [13,14,15,16,18,19,20]. However, all this work assumes a secure communication channel between the parties. If we want to realize KI or IR in practice we
must realize this secure channel. How can this be done? Surprisingly, this fundamental question has received no attention until now. We address it and turn up
some surprising answers which have important implications for the realizability
of KI and IR in practice.

1.1   Background

An important threat to the security of cryptography-using applications is exposure of the secret key due to viruses, worms or other break-ins allowed by

M. Fischlin (Ed.): CT-RSA 2009, LNCS 5473, pp. 84–99, 2009.

operating-system holes. Forward security [1,2,7,10] is one way to counter this, or at least mitigate the damage caused. Here the user has a single, fixed public key pk whose lifetime is divided into stages $1, \ldots, N$. The secret (signing or decryption) key evolves with time: at the start of stage i, the user computes its stage i secret key usk_i as a function of its stage $i-1$ secret key usk_{i-1} and then discards the latter. The security condition is that for $j < i$, a break-in during stage i (resulting in exposure of usk_i) does not allow the adversary to compute usk_j or compromise its uses. (Meaning that forgery of documents with date j or decryption of ciphertexts sent in stage j remains hard.) Once usk_i is exposed, however, $usk_{i+1}, \ldots, usk_N$ are automatically compromised (they can be computed from usk_i), and the best the user can hope to do about this is detect the break-in and revoke the public key.

Key-insulated (KI) security as introduced by Dodis, Katz, Xu, and Yung [15,16] and refined by [4] attempts to provide both forward and backward security, meaning a break-in during stage i leaves usk_j uncompromised for all $j \neq i$. More generally, break-ins for all stages $i \in I$ leave usk_j and its uses secure for all $j \in [N] \setminus I$, where $[N] = \{1, \ldots, N\}$. To accomplish this, an auxiliary party, called a helper, is introduced. The secret key usk_i of stage i is now computed by the user not merely as a function of usk_{i-1}, but also of a key hsk_i sent by the helper to the user at the start of stage i. The advantage of this system (over a merely forward-secure one) is that the public key is *never* revoked. Intrusion resilience (IR) [20,13] is an extension where forward and backword security are provided even if both user and helper are compromised as long as the compromise is not simultaneous and, even in the latter case, forward security is assured. Further extensions and variants include KI with parallel helpers [18] and KI (hierarchical) identity-based encryption [19]. Our discussion below will focus on the simpler KI case. We will discuss the extension to IR later.

KI security requires that the communication channel between user and helper is secure. Indeed, if not, meaning if an adversary could obtain the helper keys $hsk_1, \ldots, hsk_N$ sent over the channel, a single break-in in a stage i would allow it to compute *all* subsequent user secret keys by simply using the key-update process of the user, and KI would end up providing no more than forward security, which does not even need a helper. In previous works, this secure-channel assumption is built into the model, which denies the adversary hsk_j unless it has broken in during stage j.

1.2 Realizing the Secure Channel

To deploy KI in practice we must have some way to realize the secure channel. In some settings it may be possible to do this through physical means, but such settings are rare. The range of application for KI would be greatly increased if the communication between user and helper could flow over a public channel such as the Internet or a wireless network. This would allow the helper to be, for example, a server on the Internet. Alternatively, the helper could be your cell phone with the user being your laptop. (In this case, even though the devices may be in close proximity, the communication would be over a public wireless phone network.)

While definitely important for applications, enabling KI over public channels looks at first to be something trivial. This is because we would appear to know very well how to implement a secure channel over a public one. After all, isn't this the main task of basic cryptography? Specifically, let us just use encryption and authentication, either under a symmetric key shared by the parties, or under public keys.

However, we make the important observation that this standard solution runs into an inherent problem here, where the name of the game is break-in and key exposure. Namely, if the adversary breaks in during some stage i, one should realistically assume it exposes not just usk_i but also any keys used to secure the channel. (Meaning either the shared key or the user's decryption key.) This renders the channel insecure from then on, and key-insulated security vanishes (more accurately, one has only forward security) as explained above.

The above indicates that realizing KI over a public channel is nontrivial but not (yet) that it is impossible. The reason is that we have not yet exploited the full power of the model. Specifically there are two capabilities one can offer the parties. First, since we are already in a setting where keys evolve, instead of trying to secure the channel with static keys, we could allow channel-securing keys to evolve as well. Second, we could allow the update process to be an interactive protocol rather than merely a single flow.

1.3 Our Model

What the above reflects is that we need a new model to formally investigate the possibility of KI over a public channel. Providing such a model is the first contribution to our paper. In our model, the user in stage i has (in addition to usk_i) a stage i channel-securing key uck_i, while the helper has a corresponding hck_i. At the start of stage $i + 1$, the parties engage in an arbitrary interactive *channel-update* protocol. This protocol uses —and aims to get its security from— the current channel keys uck_i, hck_i. Its goal is two-fold: to (securely) communicate hsk_{i+1} from helper to user, and to "refresh" the channel keys, meaning deal the helper with a new key hck_{i+1} and the user with a corresponding new key uck_{i+1}. Once the protocol terminates, the user can update usk_i to usk_{i+1} using hsk_{i+1} as before, install uck_{i+1} as its new channel key, and discard both usk_i and uck_i. As an example, the protocol could begin with an authenticated session-key exchange based on its current channel keys and then use the session key to securely transfer hsk_{i+1} and fresh channel keys. But now, a break-in during period i exposes not only usk_i and hsk_i but also uck_i. Actually we go further, allowing the adversary to even obtain the user coins underlying the stage i channel-update protocol execution. This is realistic because the intruder could be on the system when the protocol executes, but this added adversary capability will make our proofs harder. While the core elements of the new model are natural and clear, there are subtle details. In Section 4, we describe our model and provide a formal definition of KI security over a public channel.

1.4   Our Results

Now that we have a model, we ask whether it is possible to design KI schemes secure in this public channel model. Interestingly, the answer turns out to depend on whether the adversary is active or merely passive. Specifically, the answer is "no" in the first case and "yes" in the second. Let us now elaborate on these results.

ACTIVE SECURITY. The communication security model cryptographers prefer to consider is that of an active adversary who has full control of the channel. It can not only see all transmissions, but stop, inject or alter any transmission. This is the model adopted, for example, in the work of Canetti and Krawczyk defining notions of secure channels [11,12], and also in work on session-key exchange [5,11]. It would be desirable to achieve public-channel KI security in the face of such an adversary. We show that this is impossible. That is, even in our above-described model, which allows an interactive channel-update protocol and evolving channel-security keys, an active adversary can always succeed in breaking the scheme. The reason is that after it breaks in, it obtains the user's channel-security key and can thus impersonate the user. We note that authentication (such as an authenticated session-key exchange) does not prevent this since the adversary acquires all the user's credentials via the break-in. This negative result is particularly strong because our public-channel KI model is as generous as one can get, while keeping in the spirit of KI.

There seem to be only two ways to circumvent the negative result. The first is to revoke the public key upon break-in discovery, but if one is willing to do this, one may as well just use forward security and avoid the helper altogether. Indeed, the whole point of the helper and KI is to never have to revoke the public key. The other possibility is to use an out-of-band method to redistribute channel-securing keys after break-in discovery such as a physically secure channel. But this is just an assumed secure channel under another name, exactly what we are trying to avoid. In conclusion, our result suggests that it would be inadvisable to implement any form of KI when the channel may be open to active attack.

PASSIVE SECURITY. On the positive side, we show that public-channel KI is possible against an adversary that is allowed only a passive attack on the communication channel. (Meaning it can eavesdrop, but not inject messages.) Our method is general, meaning it yields a compiler that can take any KI scheme secure in the secure-channel model and turn it into a KI scheme secure in our public-channel model under passive attack. The transformation is simple. Our channel-update protocol begins with a secure key exchange (e.g., Diffie-Hellman) to get a session key under which the helper encrypts the data it needs to transmit. The key exchange is not authenticated: this is not necessary for security against passive attack and, given the above, would not help to achieve security against active attack. We clarify that our choice of channel-update protocol is purely illustrative. The reader can surely think of others that will work.

This positive result is significant for two reasons. First, it shows that KI is at least possible over a channel where the adversary may be able to eavesdrop but finds it hard to inject or corrupt transmissions. Second, the result shows that our new method, allowing an interactive channel-update protocol, has borne fruit. Indeed, even KI under passive attack is not possible when the communication consists of a single transmission from helper to user.

Although the protocol is simple, there are subtleties in the proof arising from the strength of our model which allows the adversary to obtain the user coins from the channel-update protocol execution in any stage in which it breaks in. A consequence of this is that the starting secure-channel KI scheme needs to have optimal threshold, meaning be secure even if there are break ins in all but one stage. Some early secure-channel KI encryption schemes [15] were threshold and did not have this property, and, in this case, we cannot offer security over a public channel even in the presence of a passive adversary. Luckily, secure-channel KI schemes with optimal threshold exist for both encryption [4] and signatures [16].

PRACTICAL IMPLICATIONS. Our results imply that KI will only work if one has a channel whose physical properties preclude active attack. Anyone contemplating actual usage of KI needs to be aware of this limitation and the need to be careful about the choice of channel.

1.5 Extensions

The intrusion resilience (IR) setting of [20,13] continues to make the secure-channel assumption, and our results extend to it. However the model is considerably more complex due to the presence of both refreshes and updates and again there are subtle details to be careful about in creating the public channel analog. In the full version of this paper [3], we recall the secure-channel IR model and then provide a detailed description of our public-channel IR model. When this is done, the negative result, showing the impossibility of IR over a public channel in the presence of an active adversary, carries over easily from the KI case since IR includes KI as a special case. We need to extend the previous positive result, however. We are able to show that secure key exchange can still be used for both refresh and update to transform any secure-channel IR scheme into a public-channel IR scheme secure against passive adversaries. The proof is, however, more complex than in the KI case and is given in [3]. Similar extensions hold for the many variant notions in this area, including strong KI security [15,16] and KI with parallel helpers [18].

1.6 Discussion

Cryptographic protocols commonly make the assumption that parties are connected by secure channels. This abstraction would seem both natural and convenient; after all, isn't this exactly what standard cryptography (encryption and authentication) gives us? Yet there are settings where secure channels are surprisingly difficult to realize. One example is secure computation, where a secure channel between each pair of parties is a standard assumption [8]. Yet this channel is astonishingly difficult to realize, at least in the public-key setting, due in part to the selective-decryption problem [17]. Solutions were finally given by [9]. Our work provides another example.

2   Definitions

We let $\mathbb{N} = \{1, 2, \ldots\}$ be the set of positive integers, and for $N \in \mathbb{N}$ we let $[N] = \{1, \ldots, N\}$. The empty string is denoted ε. The notation $x \xleftarrow{\$} S$ denotes that x is selected randomly from set S. Unless otherwise indicated, an algorithm may be randomized. An adversary is an algorithm. If A is an algorithm, then the notation $x \xleftarrow{\$} A(a_1, a_2, \ldots)$ denotes that x is assigned the outcome of the experiment of running A on inputs $a_1, a_2, \ldots$, with fresh coins. If A is deterministic, we might write $x \leftarrow A(a_1, a_2, \ldots)$ instead.

GAMES. We will use code-based games [6] in definitions and some proofs. We recall some background here. A game —see Figure 1 for an example— has an **Initialize** procedure, procedures to respond to adversary oracle queries, and a **Finalize** procedure. A game G is executed with an adversary A as follows. First, **Initialize** executes and its outputs are the inputs to A. Then, A executes, its oracle queries being answered by the corresponding procedures of G. When A terminates, its output becomes the input to the **Finalize** procedure. The output of the latter, denoted G^A, is called the output of the game, and we let "$G^A \Rightarrow y$" denote the event that this game output takes value y. Variables not explicitly initialized or assigned are assumed to have value $\perp$, except for booleans which are assumed initialized to false.

INTERACTIVE ALGORITHMS. We will model each party in a two-party protocol as an *interactive algorithm*. Such an algorithm I takes as input an *incoming message* M_{in}, a *current state* St, and a *decision* d which can be acc, rej or $\perp$. Its output, denoted $I(M_{\text{in}}, St, d)$, is a triple $(M_{\text{out}}, St', d')$ consisting of an *outgoing message*, an updated state, and an updated decision. We require that if $d \neq \perp$ then $M_{\text{out}} = \perp$, $St' = St$, and $d' = d$. Our convention is that the initial state provided to an interactive algorithm is its local input and random coins. Given a pair of interactive algorithms (I, J), we assume that the first move in the interaction always belongs to I. The first incoming message for I is set to ε. An interactive algorithm terminates when its decision becomes acc or rej. Once it terminates, it outputs $\perp$ as its outgoing message in response to any incoming message and its state and decision stay the same. The *local output* of an interactive algorithm is its final state.

Given a pair of interactive algorithms (I, J) with local inputs x_I, x_J and coins ω^I, ω^J respectively, we define $\mathbf{Run}(I, x_I, J, x_J; \omega^I, \omega^J)$ to be the quintuple (Conv, St_I, d_I, St_J, d_J) consisting of the conversation transcript (meaning the sequence of messages exchanged between the parties), I's local output, I's decision, J's local output, and J's decision, respectively, after an interaction in which I has local input x_I and random coins ω^I and J has local input x_J and random coins ω^J. We let $\mathbf{Run}(I, x_I, J, x_J)$ be the random variable whose value is $\mathbf{Run}(I, x_I, J, x_J; \omega^I, \omega^J)$ when ω^I, ω^J are chosen at random.

3 Key Insulation in the Secure-Channel Model

We will take a modular approach to KI over a public channel, where a public-channel KI scheme consists of a (standard) secure-channel KI scheme —meaning one in the model of an assumed-secure channel— together with a channel-key-update protocol. We will then be able to give "compiler" style results which transform any secure-channel KI scheme into a public-channel KI scheme for suitable channel-key-update protocols. (Of course, this is only for passive adversaries since in the active case we will show that KI over public channels is impossible.) To enable this we first recall a definition of secure-channel KI. The latter has been defined for both encryption [15,4] and signatures [16]. For simplicity, we will treat the case of signatures. The case of encryption is entirely analogous and all our results carry over. Our definition below differs from that of [16] in some details, but this does not affect the results.

A *key-updating signature scheme* KUS = (KG, HKU, UKU, Sig, Ver) is specified by five algorithms with the following functionality. The randomized *key-generation algorithm* KG returns (pk, usk_0, hsk), where pk is the user public key, usk_0 is the *stage 0 user secret key*, and hsk is the *master helper key*. The user is initialized with pk, usk_0, while the helper is initialized with pk, hsk. At the start of stage $a \geq 1$, the helper applies the deterministic *helper key-update algorithm* HKU to a, pk, hsk to obtain a *stage a helper key* hsk_a, which is then assumed to be conveyed to the user via a secure channel. The user receives hsk_a from the helper and then applies the deterministic *user key-update algorithm* UKU to a, pk, hsk_a, usk_{a-1} to obtain the *stage a user secret key* usk_a. The user then discards (erases) usk_{a-1}. In stage a the user can apply the *signing algorithm* Sig to a, its stage a secret key usk_a, and a message $M \in \{0,1\}^*$ to obtain a pair (a, σ), consisting of the stage number a and a signature σ. During stage a anyone can apply the deterministic *verification algorithm* Ver to pk, a message M, and a pair (i, σ) to obtain either 1, indicating acceptance, or 0, indicating rejection. We require that if (i, σ), where $1 \leq i \leq a$, was produced by applying the signing algorithm to i, usk_i, M then $\text{Ver}(pk, M, (i, \sigma)) = 1$.

SECURITY. Consider game KIS of Figure **??**. The **Initialize** procedure provides adversary A with input pk. A can call its **Next** oracle to move the system into the next stage. It may break in during the current stage by calling its **Expose** oracle and getting back the user and helper keys for that stage. A may obtain signatures for messages of its choice during the current stage by calling its **Sign** oracle. To win, A must output a message M and a signature (j, σ) such that j is an unexposed stage, $\text{Ver}(pk, M, (j, \sigma)) = 1$, and M was not queried to **Sign** during stage j. A's advantage is

$$\mathbf{Adv}^{\text{ki}}_{\text{KUS}}(A) = \Pr\left[\text{KIS}^A \Rightarrow \text{true}\right].$$

We adopt the convention that the *running time* of an adversary A is the execution time of the entire game, including the time taken for initialization, the time taken by the oracles to compute replies to the adversary's queries, and the time taken for finalization.

---

procedure Initialize
$(pk, usk_0, hsk) \overset{\$}{\leftarrow} \mathsf{KG}$; $a \leftarrow 0$; $S \leftarrow \emptyset$; $E \leftarrow \emptyset$
Return pk

procedure Next()
$a \leftarrow a + 1$
$hsk_a \leftarrow \mathsf{HKU}(a, pk, hsk)$
$usk_a \leftarrow \mathsf{UKU}(a, pk, hsk_a, usk_{a-1})$

procedure Finalize$(M, (j, \sigma))$
Return $(j \notin E \land (j, M) \notin S \land \mathsf{Ver}(pk, M, (j, \sigma)) = 1)$

procedure Expose()
$E \leftarrow E \cup \{a\}$
Return (usk_a, hsk_a)

procedure Sign(M)
$(a, \sigma) \overset{\$}{\leftarrow} \mathsf{Sig}(a, usk_a, M)$
$S \leftarrow S \cup \{(a, M)\}$
Return (a, σ)

Fig. 1. Game KIS used to define KI signatures in the secure-channel model

THE IMPLICIT SECURE-CHANNEL ASSUMPTION. As discussed in Section 1, the secure-channel assumption is implicit in the above model. This is due to the fact that A is *not* given hsk_a for stages a in which it did not make an **Expose** query. Also note that the assumption is necessary, for if A had an additional oracle **Get** that returned hsk_a, but the rest of the game was the same, it could win via

$\mathbf{Next}()$; $(hsk_1, usk_1) \leftarrow \mathbf{Expose}()$; $\mathbf{Next}()$
$hsk_2 \leftarrow \mathbf{Get}()$; $usk_2 \leftarrow \mathsf{UKU}(2, pk, hsk_2, usk_1)$
$(2, \sigma) \leftarrow \mathsf{Sig}(2, usk_2, 0)$; return $(0, (2, \sigma))$

4   Key Insulation in the Public-Channel Model

We saw above that a secure channel between helper and user is both assumed and necessary in the existing notion of KI. Here we consider how the channel can be implemented. Let us first discuss how key exposure implies failure of the obvious way to secure the channel.

STATIC KEYS WON'T SECURE THE CHANNEL. The obvious solution is to use standard cryptography. Let the helper have a signing key sk whose corresponding verification key vk is held by the user, and correspondingly, let the user have a decryption key dk whose corresponding encryption key ek is held by the helper. (These keys are generated and distributed honestly and securely along with usk_0, hsk when the system is initialized. The cryptography could be symmetric or asymmetric. In the first case, the signature is a MAC and encryption is symmetric, so that $sk = vk$ and $dk = ek$. In the second case, the signature and encryption are public-key based.) Now in stage a, the helper sends (C, σ) to the user, where C is an encryption of hsk_a under ek and σ is a signature of C under sk. The user verifies the signature using vk and decrypts C using dk to get hsk_a. This, however, fails completely to provide security in the key-exposure setting, even for an adversary that is merely passive with regard to channel access. (That is, it can eavesdrop the communication but not send messages itself.) This is because one must realistically assume that a break-in in a stage a exposes all information the user has, which includes not only usk_a but also dk. Equipped with usk_a, dk via the break-in, the adversary can now obtain the stage $a + 1$ channel transmission (C_{a+1}, σ_{a+1}) via its channel access, decrypt C_{a+1} using dk to get hsk_{a+1}, and compute $usk_{a+1} = \mathsf{UKU}(a + 1, pk, hsk_{a+1}, usk_a)$. Continuing in this fashion, it can obtain usk_i for all $i \geq a$.

EVOLVING CHANNEL-SECURING KEYS. The above is already something of which potential implementers should be aware, but not yet enough to give up hope of obtaining KI, for there is an obvious next step, which we take. Namely, let us allow the channel to be secured not under keys that are static but which themselves evolve, so that a break-in exposes only the current keys. This section introduces and formalizes a very general model to this end, where an interactive protocol (such as a secure key exchange) may be used in each step to provide a secure channel and also update the channel keys.

PUBLIC-CHANNEL KEY UPDATING SIGNATURE SCHEMES. A *public-channel key-updating signature scheme* is a triple $\mathsf{PCKUS} = (\mathsf{KUS}, \mathsf{CKG}, (\mathsf{U}, \mathsf{H}))$, where $\mathsf{KUS} = (\mathsf{KG}, \mathsf{HKU}, \mathsf{UKU}, \mathsf{Sig}, \mathsf{Ver})$ is a key-updating signature scheme, CKG is the channel-key-generation algorithm, and the channel-key-update protocol (U, H) is a pair of interactive algorithms to be run by user and helper, respectively. Let us now explain how the system runs.

Algorithm CKG returns (uck_0, hck_0), where uck_0 is the *stage 0 user channel key* and hck_0 is the *stage 0 helper channel key*. When the user is initialized, in addition to the public key pk and stage 0 user secret key usk_0 produced by KG, the user is given uck_0. When the helper is initialized, in addition to pk and the master helper key hsk (also generated by KG), the helper is given hck_0.

In any stage a ($a \geq 0$), the user holds not only its stage a user secret key usk_a, but also a stage a user channel key uck_a. The helper holds hsk and a stage a helper channel key hck_a. At the start of stage $a + 1$, the helper computes $hsk_{a+1} = \mathsf{HKU}(a + 1, pk, hsk)$. The parties then engage in the channel-key-update protocol (U, H). The local input of U is the stage a user secret key usk_a, the stage a user channel key uck_a and some random coins ω_a^U, while the local input of H is the stage $a + 1$ helper key hsk_{a+1}, the stage a helper channel key hck_a and some random coins ω_a^H. After the interaction, the expected local output of U is hsk_{a+1} plus the stage $a + 1$ user channel key uck_{a+1}, while the expected local output of H is the stage $a+1$ helper channel key hck_{a+1}. Once the protocol has completed, the user can update its key as before, namely it computes $usk_{a+1} = \mathsf{UKU}(a + 1, pk, hsk_{a+1}, usk_a)$. It then discards not only usk_a but also its previous channel key uck_a. We require the natural correctness condition, namely that the stage $a + 1$ helper key produced by U in the interaction in which U has input usk_a, uck_a, ω_a^U and H has input $hsk_{a+1}, hck_a, \omega_a^H$, is hsk_{a+1} with probability one. In addition, we require that at the end of the interaction, U's decision d_{a+1}^U and H's decision d_{a+1}^H are both acc.

SECURITY. We proceed to formalize two notions of security for public-channel key-updating signature schemes: key insulation under active and passive attacks. We

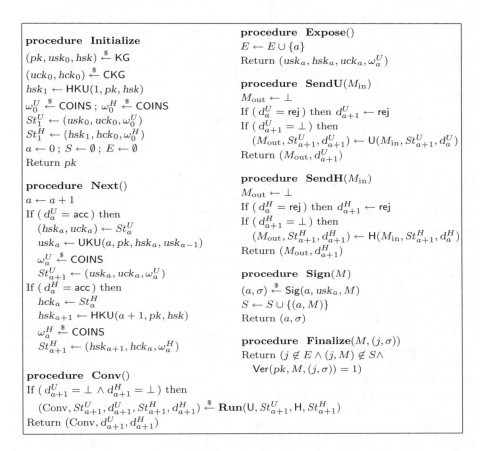

Fig. 2. Games used to define public-channel key insulation under active and passive attack. Game PCKI-aa includes all of the procedures except **Conv**, while game PCKI-pa includes all except **SendU** and **SendH**.

first provide definitions and then explanations. Let $\mathsf{PCKUS} = ((\mathsf{KG}, \mathsf{HKU}, \mathsf{UKU}, \mathsf{Sig}, \mathsf{Ver}), \mathsf{CKG}, (\mathsf{U}, \mathsf{H}))$ be a public-channel key-updating signature scheme. We consider an adversary A interacting with the games of Figure 2. The **Initialize** procedure gives A input pk. In an *active attack*, A is provided with oracles **Next**, **Expose**, **SendU**, **SendH**, and **Sign**, while in a *passive attack* it is provided with oracles **Next**, **Expose**, **Conv**, and **Sign**. It may query the oracles adaptively, in any order it wants, with the following restriction: In the case of an active adversary, as soon as **SendU** returns $d_{a+1}^U = \mathsf{acc}$ and **SendH** returns $d_{a+1}^H = \mathsf{acc}$, A makes a query to oracle **Next**. In the case of a passive adversary, every query to oracle **Conv** is immediately followed by a query to oracle **Next**. Eventually, A outputs a message M and a signature (j, σ) and halts. An active (resp., passive) adversary is said to win if game PCKI-aa (resp., PCKI-pa) returns true, meaning j is an unexposed stage, $\mathsf{Ver}(pk, M, (j, \sigma)) = 1$, and M was not queried to oracle **Sign** during stage j. For $\mathsf{atk} \in \{\mathsf{aa}, \mathsf{pa}\}$, A's atk-advantage is

$$\mathbf{Adv}_{\mathsf{PCKUS}}^{\mathrm{pcki\text{-}atk}}(A) \;=\; \Pr\Big[\, \mathsf{PCKI\text{-}atk}^A \Rightarrow \mathsf{true}\,\Big].$$

Again, we adopt the convention that the *running time* of an adversary A is the execution time of the entire game, including the time taken for initialization, the time taken by the oracles to compute replies to the adversary's queries, and the time taken for finalization.

EXPLANATION. An active adversary has full control over the communication between the helper and the user. It can deliver messages out of order, modify messages or inject messages of its own choosing. This is modeled by providing the adversary access to oracles **SendU** and **SendH**, which represent the user and helper, respectively, running the channel-key-update protocol. Once this protocol terminates, the adversary is required to call its **Next** oracle to move the system into the next stage. This models the user updating his keys as soon as he obtains the helper secret key for the next stage. As in the case of key insulation in the secure-channel model, the adversary may break in during the current stage by calling its **Expose** oracle, but here it gets back the user secret key, the helper key, the user channel key, and the user's coins, for that stage. As before, the adversary may obtain signatures for messages of its choice during the current stage by calling its **Sign** oracle. To win, it must output a valid forgery for an unexposed stage.

A passive adversary cannot modify or inject messages, but it can eavesdrop on the communication channel, obtaining transcripts of conversations between the user and the helper. We model this by providing the adversary access to oracle **Conv** which runs the channel-key-update protocol and returns the conversation transcript and the decisions of U and H. In all other respects, a passive adversary is like an active adversary: as soon as the channel-key-update protocol terminates, the adversary is required to call its **Next** oracle to move the system into the next stage, the adversary can break in during the current stage, it can obtain signatures for messages of its choice during the current stage, and its goal is to produce a valid forgery for an unexposed stage.

5   Impossibility of Public-Channel KI Under Active Attack

We show that the notion of public-channel key insulation under active attack is unachievable, meaning all public-channel key-updating signature schemes are vulnerable to an active attack. The precise statement of our result is the following.

Theorem 1. *Let* $\mathsf{PCKUS} = ((\mathsf{KG}, \mathsf{HKU}, \mathsf{UKU}, \mathsf{Sig}, \mathsf{Ver}), \mathsf{CKG}, (\mathsf{U}, \mathsf{H}))$ *be a public-channel key-updating signature scheme. Let* t_{KG}, t_{HKU}, t_{UKU}, t_{Sig}, t_{Ver}, *and* t_{CKG} *denote the running times of the corresponding algorithms, and* $t_{(\mathsf{U},\mathsf{H})}$ *denote the running time of protocol* (U, H). *Let* m *be the maximum number of moves in this protocol. Then there exists an adversary A against* PCKUS *that makes one query to oracle* **Next**, *one* **Expose** *query, at most* $\lceil m/2 \rceil$ **SendU** *queries, at most*

$2\lceil m/2 \rceil$ **SendH** *queries, and no* **Sign** *queries, such that*

$$\mathbf{Adv}^{\text{pcki-aa}}_{\text{PCKUS}}(A) = 1 .$$

Furthermore, the running time of A is $t_{\text{KG}} + t_{\text{CKG}} + 2t_{\text{HKU}} + 2t_{\text{UKU}} + 2t_{(\text{U,H})} + t_{\text{Sig}}.$

The proof of the above theorem, which is given in [3], is simple, as is not uncommon for impossibility results, where the key insights are in the development of the model and the question posed.

6   Possibility of Public-Channel KI Under Passive Attack

Given a KI signature scheme in the secure-channel model, we show in this section how to transform it into a KI signature scheme secure against passive attack in the public-channel model. We first discuss the primitives we use, namely an arbitrary secret-key-exchange protocol and an arbitrary one-time symmetric encryption scheme.

SECRET-KEY-EXCHANGE (SKE) PROTOCOL. An SKE protocol with key length k is a pair of interactive algorithms (I, J) each of which has local output a k bit string. We require that

$$\Pr\left[(K_I = K_J) \wedge (d^I = d^J = \mathsf{acc}) \right] = 1 ,$$

where $(\text{Conv}, K_I, d^I, K_J, d^J) \xleftarrow{\$} \mathbf{Run}(\mathsf{I}, \varepsilon, \mathsf{J}, \varepsilon)$ meaning the parties agree on a common key. For security we require that the common key be computationally indistinguishable from random. This is captured by defining the ske-advantage of an adversary A as

$$\mathbf{Adv}^{\text{ske}}_{(\mathsf{I},\mathsf{J})}(A) = 2 \cdot \Pr\left[\mathsf{SKE}^A_{(\mathsf{I},\mathsf{J})} \Rightarrow \mathsf{true} \right] - 1 ,$$

where game $\mathsf{SKE}_{(\mathsf{I},\mathsf{J})}$ is defined in Figure 3.

One example of a suitable SKE protocol is a Diffie-Hellman key exchange. (The DH key needs to be suitably hashed to a k bit string.) Another possibility, based on any asymmetric encryption scheme $(\mathsf{AKg}, \mathsf{AEnc}, \mathsf{ADec})$, works as follows. I picks a public/secret key pair (pk, sk) by running AKg and sends pk to J. The latter selects a random k-bit string K, encrypts it under pk using AEnc and sends the ciphertext to I. I decrypts the ciphertext with sk using ADec to obtain K.

SYMMETRIC ENCRYPTION. A symmetric encryption scheme $\mathsf{SE} = (\mathsf{SEnc}, \mathsf{SDec})$ with key length k consists of two algorithms. The encryption algorithm SEnc takes a k bit key K and plaintext $M \in \{0,1\}^*$ to return a ciphertext C. The decryption algorithm SDec takes K and C to return either a plaintext M or the symbol $\perp$. We require

$$\Pr\left[K \xleftarrow{\$} \{0,1\}^k \ : \ \mathsf{SDec}(K, \mathsf{SEnc}(K, M)) = M \right] = 1$$

for all $M \in \{0,1\}^*$. We also require standard IND-CPA security except that it need only be one-time. This is captured by letting

$$\mathbf{Adv}^{\text{ind-cpa}}_{\mathsf{SE}}(A) = 2 \cdot \Pr\left[\mathsf{INDCPA}^A_{\mathsf{SE}} \Rightarrow \mathsf{true} \right] - 1 ,$$

procedure Initialize	**procedure Initialize**
$b \xleftarrow{\$} \{0,1\}$	$K \xleftarrow{\$} \{0,1\}^k$
procedure Conv()	$b \xleftarrow{\$} \{0,1\}$
$(\mathrm{Conv}, K_1, d^I, K_1, d^J) \xleftarrow{\$} \mathbf{Run}(\mathsf{I}, \varepsilon, \mathsf{J}, \varepsilon)$	**procedure LR(M_0, M_1)**
$K_0 \xleftarrow{\$} \{0,1\}^k$	$C \xleftarrow{\$} \mathsf{SEnc}(K, M_b)$
Return $(\mathrm{Conv}, d^I, d^J, K_b)$	Return C
procedure Finalize(d)	**procedure Finalize(d)**
Return $(b = d)$	Return $(b = d)$

Fig. 3. Game $\mathsf{SKE}_{(\mathsf{I},\mathsf{J})}$ on the left is used to define security of SKE protocol (I, J) and game $\mathsf{INDCPA}_{\mathsf{SE}}$ on the right is used to define security of symmetric encryption scheme $\mathsf{SE} = (\mathsf{SEnc}, \mathsf{SDec})$. In both cases, the key length is k.

where game $\mathsf{INDCPA}_{\mathsf{SE}}$ is in Figure 3 and A is required to make only one LR query (this is how the one-time requirement is captured), consisting of a pair of equal-length messages.

CONSTRUCTION. Let $\mathsf{KUS} = (\mathsf{KG}, \mathsf{HKU}, \mathsf{UKU}, \mathsf{Sig}, \mathsf{Ver})$ be a key-updating signature scheme. We transform it into a public-channel key-updating signature scheme $\mathsf{PCKUS} = (\mathsf{KUS}, \mathsf{CKG}, (\mathsf{U}, \mathsf{H}))$, where CKG always returns $(\varepsilon, \varepsilon)$, by defining the channel-key-update protocol (U, H) in terms of any secret-key-exchange protocol (I, J) and symmetric encryption scheme $\mathsf{SE} = (\mathsf{SEnc}, \mathsf{SDec})$, both with the same key length k, as follows. The parties first run the secret-key-exchange protocol, with U playing the role of I and H playing the role of J, to agree on a common key K. The helper then encrypts hsk_i under K using SEnc to obtain a ciphertext C which it sends to the user. The latter decrypts C under K using SDec to obtain hsk_i.

We clarify that this particular channel-update protocol is chosen for illustrative purposes. Many others are possible, as the reader will probably see. However, it does include several different instantiations, arising from the different available choices of SKE protocols mentioned above.

SECURITY OF OUR CONSTRUCTION. We prove that if the given key-updating signature scheme is KI in the secure-channel model and the secret-key-exchange protocol as well as the symmetric encryption scheme are secure, then the public-channel key-updating signature scheme obtained using our construction is KI under passive attack in the public-channel model.

Theorem 2. *Let* $\mathsf{KUS} = (\mathsf{KG}, \mathsf{HKU}, \mathsf{UKU}, \mathsf{Sig}, \mathsf{Ver})$ *be a key-updating signature scheme. Let* $\mathsf{PCKUS} = ((\mathsf{KG}, \mathsf{HKU}, \mathsf{UKU}, \mathsf{Sig}, \mathsf{Ver}), \mathsf{CKG}, (\mathsf{U}, \mathsf{H}))$ *be the public-channel key-updating signature scheme constructed from* KUS, *secret-key-exchange protocol* (I, J) *and symmetric encryption scheme* $\mathsf{SE} = (\mathsf{SEnc}, \mathsf{SDec})$ *as described above. Let* t_{KG}, t_{HKU}, t_{UKU}, t_{Sig}, t_{Ver}, *and* t_{CKG} *denote the running times of the corresponding algorithms, and* $t_{(\mathsf{U},\mathsf{H})}$ *denote the running time of protocol* (U, H). *Let* A *be an adversary against* PCKUS, *making* q *queries to oracles*

Conv *and* **Next**, q_E *queries to* **Expose**, *and* q_S *queries to* **Sign**. *Then there exist adversaries E, B, S such that*

$$\mathbf{Adv}^{\text{pcki-pa}}_{\text{PCKUS}}(A) \leq q \cdot \mathbf{Adv}^{\text{ske}}_{(I,J)}(E) + q \cdot \mathbf{Adv}^{\text{ind-cpa}}_{\text{SE}}(B) + q \cdot \mathbf{Adv}^{\text{ki}}_{\text{KUS}}(S) \ . \quad (1)$$

Furthermore, the running times of E, B are both $t_{\text{KG}} + t_{\text{CKG}} + q_N \cdot t_{\text{HKU}} + q_N \cdot t_{\text{UKU}} + q_S \cdot t_{\text{Sig}} + q \cdot t_{(U,H)}$, *and the running time of S is* $t_{\text{CKG}} + \mathcal{O}(q + q_S + q_E) + q \cdot t_{(U,H)}$. *Also S makes* $q - 1$ *queries to its* **Expose** *oracle and* q *queries to its* **Next** *oracle.*

PROOF OVERVIEW. The proof that our construction achieves public-channel KI security under passive attack seems easy at first, but there are subtle difficulties arising from the fact that our model allows the adversary to obtain the user coins from the channel-update protocol execution in any stage in which it breaks in. This means that the adversary obtains the session key, and can check whether the ciphertext transmitted by the helper decrypts to the helper secret key for the stages in question, a value it also has from its break-in. Of course, in the real protocol, this will always be true. But the natural simulation is to consider a protocol in which, rather than encrypting the helper key under the session key yielded by the session-key exchange protocol, the helper encrypts a constant under a new, random key. The security of the session-key exchange protocol and the encryption scheme should imply that this makes no difference. However, the adversary can in fact detect the difference between the simulation and the real game because, as we said above, it can obtain the real session key and decrypt the ciphertext under it. To get around this, we guess a stage in which the adversary does not break in, and switch to the simulated key and message only in this stage, using the real key and real message in other stages. But to do this, our simulation needs to know the real message, which is the helper secret key, and the only way to get this is to break in. Luckily, it can do so by consequence of the assumed security of the underlying secure-channel KI scheme, but the result is a discrepancy in resources: even if the adversary against the public channel protocol does very few break-ins, the adversary against the secure-channel protocol breaks-in to $N - 1$ out of N stages. Therefore, it is required that the given secure-channel scheme be secure against $N - 1$ break-ins. Luckily, we have such schemes. For signatures, the schemes of [16] have the desired property. For encryption, some of the original schemes of [15] are threshold and don't have the property, but the scheme of [4] does. Due to space limitations, we defer the full proof of Theorem 2 to [3].

7   Intrusion Resilience in the Public-Channel Model

As in the case of key insulation, we take a modular approach to IR over a public channel, where a public-channel IR scheme consists of a secure-channel IR scheme—meaning one in the model of an assumed-secure channel—together with a channel-key-update protocol and a channel-key-refresh protocol. We will then show how to transform any secure-channel IR scheme into a public-channel IR scheme secure against passive adversaries for suitable channel-key-update and channel-key-refresh protocols. IR has been defined for signatures [20] and encryption [13,14]. For simplicity, we treat the case of signatures. The case of encryption is entirely analogous. In [3] we recall a definition of secure-channel IR, present our model for public-channel IR signatures, and construct a public-channel IR signature scheme secure against passive attack based on any secure-channel IR signature scheme, any secret-key-exchange protocol and any symmetric encryption scheme.

Acknowledgments

The first author was supported in part by NSF grant CNS 0524765 and CNS 0627779 and a gift from Intel corporation. The second author was supported in part by the grants of the first author.

References

1. Anderson, R.: Two Remarks on Public-Key Cryptology. In: 2000, and Invited Lecture at the Fourth Annual Conference on Computer and Communications Security, Zurich, Switzerland (April 1997) (manuscript)
2. Bellare, M., Miner, S.K.: A forward-secure digital signature scheme. In: Wiener, M. (ed.) CRYPTO 1999. LNCS, vol. 1666, p. 431. Springer, Heidelberg (1999)
3. Bellare, M., Duan, S., Palacio, A.: Key Insulation and Intrusion Resilience Over a Public Channel. IACR Eprint Archive (2009)
4. Bellare, M., Palacio, A.: Protecting against Key Exposure: Strongly Key-Insulated Encryption with Optimal Threshold. Applicable Algebra in Engineering, Communication and Computing 16(6), 379–396 (2006)
5. Bellare, M., Rogaway, P.: Authentication and key distribution. In: Stinson, D.R. (ed.) CRYPTO 1993. LNCS, vol. 773, pp. 232–249. Springer, Heidelberg (1994)
6. Bellare, M., Rogaway, P.: The Security of Triple Encryption and a Framework for Code-Based Game-Playing Proofs. In: Vaudenay, S. (ed.) EUROCRYPT 2006. LNCS, vol. 4004, pp. 409–426. Springer, Heidelberg (2006)
7. Bellare, M., Yee, B.S.: Forward-security in private-key cryptography. In: Joye, M. (ed.) CT-RSA 2003. LNCS, vol. 2612, pp. 1–18. Springer, Heidelberg (2003)
8. Ben-Or, M., Goldwasser, S., Wigderson, A.: Completeness Theorems for Non-Cryptographic Fault-Tolerant Distributed Computation. In: Proceedings of the 30th Annual Symposium on the Theory of Computing. ACM, New York (1998)
9. Canetti, R., Feige, U., Goldreich, O., Naor, M.: Adaptively Secure Multi- Party Computation. In: Proceedings of the 28th Annual Symposium on the Theory of Computing. ACM, New York (1996)
10. Canetti, R., Halevi, S., Katz, J.: A Forward-Secure Public-Key Encryption Scheme. In: Biham, E. (ed.) EUROCRYPT 2003. LNCS, vol. 2656. Springer, Heidelberg (2003)
11. Canetti, R., Krawczyk, H.: Analysis of Key-Exchange Protocols and Their Use for Building Secure Channels. In: Pfitzmann, B. (ed.) EUROCRYPT 2001. LNCS, vol. 2045, p. 453. Springer, Heidelberg (2001)

12. Canetti, R., Krawczyk, H.: Universally Composable Notions of Key Exchange and Secure Channels. In: Knudsen, L.R. (ed.) EUROCRYPT 2002. LNCS, vol. 2332, p. 337. Springer, Heidelberg (2002)
13. Dodis, Y., Franklin, M., Katz, J., Miyaji, A., Yung, M.: Intrusion-Resilient Public-Key Encryption. In: Joye, M. (ed.) CT-RSA 2003. LNCS, vol. 2612, pp. 19–32. Springer, Heidelberg (2003)
14. Dodis, Y., Franklin, M., Katz, J., Miyaji, A., Yung, M.: A Generic Construction for Intrusion-Resilient Public-Key Encryption. In: Okamoto, T. (ed.) CT-RSA 2004. LNCS, vol. 2964, pp. 81–98. Springer, Heidelberg (2004)
15. Dodis, Y., Katz, J., Xu, S., Yung, M.: Key-Insulated Public Key Cryptosystems. In: Knudsen, L.R. (ed.) EUROCRYPT 2002. LNCS, vol. 2332, p. 65. Springer, Heidelberg (2002)
16. Dodis, Y., Katz, J., Xu, S., Yung, M.: Strong Key-Insulated Signature Schemes. In: Desmedt, Y.G. (ed.) PKC 2003. LNCS, vol. 2567, pp. 130–144. Springer, Heidelberg (2002)
17. Dwork, C., Naor, M., Reingold, O., Stockmeyer, L.: Magic Functions. In: Proceedings of the 40th Symposium on Foundations of Computer Science. IEEE, Los Alamitos (1999)
18. Hanaoka, G., Hanaoka, Y., Imai, H.: Parallel Key-Insulated Public Key Encryption. In: Yung, M., Dodis, Y., Kiayias, A., Malkin, T.G. (eds.) PKC 2006. LNCS, vol. 3958, pp. 105–122. Springer, Heidelberg (2006)
19. Hanaoka, Y., Hanaoka, G., Shikata, J., Imai, H.: Identity-based Heirarchical Strongly Key-Insulated Encryption and its Application. In: Roy, B. (ed.) ASIACRYPT 2005. LNCS, vol. 3788, pp. 495–514. Springer, Heidelberg (2005)
20. Itkis, G., Reyzin, L.: SiBIR: Signer-Base -Resilient Signatures. In: Yung, M. (ed.) CRYPTO 2002. LNCS, vol. 2442, p. 499. Springer, Heidelberg (2002)

Statistically Hiding Sets

Manoj Prabhakaran[1,*] and Rui Xue[2,**]

[1] Dept. of Computer Science
University of Illinois, Urbana-Champaign
mmp@cs.uiuc.edu
[2] State Key Laboratory of Information Security
Institute of Software, Chinese Academy of Sciences
rxue@is.iscas.ac.cn

Abstract. Zero-knowledge set is a primitive introduced by Micali, Rabin, and Kilian (FOCS 2003) which enables a prover to commit a set to a verifier, without revealing even the size of the set. Later the prover can give zero-knowledge proofs to convince the verifier of membership/non-membership of elements in/not in the committed set. We present a new primitive called *Statistically Hiding Sets* (SHS), similar to zero-knowledge sets, but providing an information theoretic hiding guarantee, rather than one based on efficient simulation. Then we present a new scheme for statistically hiding sets, which does not fit into the "Merkle-tree/mercurial-commitment" paradigm that has been used for *all* zero-knowledge set constructions so far. This not only provides efficiency gains compared to the best schemes in that paradigm, but also lets us provide *statistical* hiding; previous approaches required the prover to maintain growing amounts of state with each new proof for such a statistical security.

Our construction is based on an algebraic tool called *trapdoor DDH groups* (TDG), introduced recently by Dent and Galbraith (ANTS 2006). However the specific hardness assumptions we associate with TDG are different, and of a strong nature — strong RSA and a knowledge-of-exponent assumption. Our new knowledge-of-exponent assumption may be of independent interest. We prove this assumption in the generic group model.

1 Introduction

Zero-knowledge set is a fascinating cryptographic primitive introduced by Micali, Rabin, and Kilian [22], which has generated much interest since then [24,21,10,8,17,9]. It enables a party (the prover) to commit a set — without revealing even its size — to another party (the verifier). Later the verifier can make membership queries with respect to the committed set; the prover can

---

* Supported in part by NSF grants CNS 07-16626 and CNS 07-47027.
** Work done mostly while visiting UIUC. Supported by the China Scholarship Council, the 973 Program (No. 2007CB311202), the 863 program (No. 2008AA01Z347), and NSFC grants (No. 60873260, 60773029).

M. Fischlin (Ed.): CT-RSA 2009, LNCS 5473, pp. 100–116, 2009.
© Springer-Verlag Berlin Heidelberg 2009

answer these queries and give proofs to convince the verifier of the correctness of the answers, without revealing anything further about the set.

In this paper, we revisit the notion of zero-knowledge sets. We provide an alternate notion — which we call statistically hiding sets (SHS) — that replaces the zero-knowledge property by the slightly weaker requirement of "statistical hiding." Statistical hiding, unlike zero-knowledge, does not require *efficient* simulatability; this relaxation is comparable to how witness-independence is a weakening of zero-knowledge property for zero-knowledge proofs. But the intuitive security guarantees provided by SHS is the same as that provided by zero-knowledge sets. (In particular, the informal description in the previous paragraph is applicable to both.)

Then we present a novel scheme for this new primitive, significantly departing from previous approaches for building zero-knowledge sets. While *all* previous approaches for zero-knowledge sets used a tree-based construction (along with a primitive called mercurial commitments), ours is a direct algebraic construction. To the best of our knowledge, this is the first construction for this kind of primitive that does not fit into the Merkle-tree/mercurial-commitment paradigm. This construction (a) provides *statistical* zero-knowledge, without the prover having to maintain growing amounts of state with each new proof[1] and (b) provides efficiency gains compared to previous constructions of zero-knowledge sets. Further, since the techniques used are different, our construction opens up the possibility of building zero-knowledge sets (or SHS) with certain features that are not amenable to the Merkle-tree/mercurial-commitment based approach.

Our construction is based on *trapdoor DDH groups* (TDG), a primitive introduced recently by Dent and Galbraith [13]. Ours is perhaps the first non-trivial application of this cryptographic primitive, illustrating its potential and versatility. The specific hardness assumptions we associate with TDG are different from those in [13], and of a strong nature (strong RSA and a Knowledge-of-Exponent assumption). While we believe these assumptions are reasonable given the state-of-the-art in algorithmic number theory, we do hope that our approach leads to newer efficient constructions of statistically hiding sets and similar tools *based on more standard assumptions*. We also hope that our work will help in better understanding certain powerful assumptions. In particular, the new simple and powerful knowledge-of-exponent assumption that we introduce (and prove to hold in a generic group model) may be of independent interest. See Section 1.4 below for more details.

1.1   Our Contributions

We briefly point out the highlights of this work, and discuss the tradeoffs we achieve.

---

[1] The Merkle-tree based approach requires the prover to use a pseudorandom function to eliminate the need for maintaining state that grows with each new proof. This makes the resulting zero-knowledge computational rather than perfect or statistical.

– Prior constructions for ZK sets either required the prover to accumulate more state information with each query, or guaranteed only computational hiding (which was based on the security of pseudorandom functions). In contrast, our construction for SHS provides unconditional statistical hiding (without growing state information). In particular, this makes our scheme unconditionally forward secure: even an unbounded adversary cannot break the security after the commitments and proofs are over.

However, our soundness guarantee depends on new complexity assumptions (see Section 1.4). But as we explain below, complexity assumptions are more justified when used for soundness than when used for hiding.

– Compared to previous ZK set constructions, we obtain efficiency gains in communication complexity and in the amount of private storage that the prover has to maintain after the commitment phase. The computational complexity of verification of the proofs is also better, depending on the specifics of the mercurial commitments and the group operations involved. However, the computational complexity of generating the proofs is higher in our case. In [25] we provide a detailed comparison.

– Since all previous constructions of ZK sets use a Merkle-tree/mercurial commitmet based approach, we consider it an important contribution to provide an alternate methodology. We hope that this can lead to constructions with features that could not be achieved previously. In particular, our construction suggests the possibility of achieving a notion of *updateability* with better privacy than obtained in [21]. We do not investigate this here.

– The definition of SHS is also an important contribution of this work. It differs from the definition of ZK sets in certain technical aspects (see Section 1.3) which might be more suitable in some situations. But more importantly, it provides a technically *relaxed* definition of security, retaining the conceptual security guarantees of ZK sets.[2] This technical relaxation has already helped us achieve a fixed-state construction with statistical security. Going further, we believe SHS could lead to better *composability* than ZK sets, because the hiding guarantee is formulated as a statistical guarantee and not in terms of efficient simulation. Again, we leave this for future investigation.

– Finally, in this work we introduce a new "knowledge-of-exponent" assumption (called KEA-DDH), closely related to the standard DDH assumption. We prove that the assumption holds in the generic group model. Due to its naturality, the new assumption could provide a useful abstraction for constructing and analysing new cryptographic schemes.

On the use of computational assumptions for soundness. The main disadvantage of our construction is the use of non-standard computational assumptions. However, we point out an arguably desirable trade-off it achieves, compared to existing constructions of ZK sets. When maintaining growing state information is not an

---

[2] The relaxation is in that we do not require efficient simulation. Note that compared to *computational* ZK sets, SHS' security is stronger in some aspects. We remark that though one could define *computationally* hiding sets, such a primitive does not have the above mentioned advantages that SHS has over ZK sets.

option, existing ZK set constructions offer only computational security for the prover, based on the security of the pseudorandom function being used. If the pseudorandom function gets broken eventually (due to advances in cryptanalysis, or better computational technology), then the prover's security is lost.

In contrast, our SHS construction provides *unconditional and everlasting security for the prover*. Further, security guarantee for the verifier depends only on assumptions on prover's computational ability *during the protocol execution*. So, in future, if the assumptions we make turn out to be false and even if an explicit algorithm is found to violate them, this causes no concern to a verifier who accepted a proof earlier.

In short, our use of stronger complexity assumptions is offset by the fact that they are required to hold only against adversaries operating during the protocol execution. In return, we obtain unconditional security after the protocol execution.

Statistically Hiding Databases. Merkle-tree based constructions of zero-knowledge sets naturally extend to zero-knowledge database with little or no overhead. Our construction does not extend in this manner. However in the full version [25] we point out a simple way to use zero-knowledge sets (or statistically hiding sets) in a black-box manner to implement zero-knowledge databases (or statistically hiding databases, respectively).

1.2 Related Work

Micali, Rabin and Kilian introduced the concept of zero-knowledge sets, and provided a construction based on a Merkle-tree based approach [22]. All subsequent constructions have followed this essential idea, until now. Chase et al. abstracted the properties of the commitments used in [22] and successfully formalized a general notion of commitments named *mercurial commitments* [10]. Catalano et al. [8] further clarified the notion of mercurial commitments used in these constructions. More recently Catalano et al. [9] introduced a variant of mercurial commitments that allowed using q-ary Merkle-trees in the above construction to obtain a constant factor reduction in the proof sizes (under stronger assumptions on groups with bilinear pairing).

Liskov [21] augmented the original construction of Micali et al. to be updatable. Gennaro and Micali [17] introduced a non-malleability-like requirement called independence and constructed mercurial commitments with appropriate properties which when used in the Merkle-tree based construction resulted in a zero-knowledge set scheme with the independence property. Ostrovsky et al. [24] extended the Merkle-tree based approach to handle more general datastructures (directed acyclic graphs); however in their notion of privacy the size of the data-structure is allowed to be publicly known, and as such they do not require mercurial commitments.

Our construction is motivated by prior constructions of *accumulators*. The notion of accumulators was first presented in [5] to allow rolling a set of values into one value, such that there is a short proof for each value that went into it. Barić and Pfitzmann [3] proposed a construction of a collision-resistant accumulator under the strong RSA assumption. Camenisch and Lysyanskaya [6] further developed a dynamic accumulator so that accumulated values can be added or removed from the accumulator. The most important difference between an accumulator and a zero-knowledge set is that the former does not require the prover to provide a proof of non-membership for elements not accumulated. Our scheme bears resemblance to the "universal accumulators" proposed in Li et al. [20], which does allow proofs of non-membership, but does not have the zero-knowledge or hiding property. Subsequent to our work Xue et al. [27] have proposed a more efficient scheme, secure in the random oracle model.

Trapdoor DDH groups were introduced by [13], and to the best of our knowledge has not been employed in any cryptographic application (except a simple illustrative example in [13]). They also gave a candidate for this primitive based on elliptic curve groups with composite order, using "hidden pairings." Indeed, Galbraith and McKee [15] had pointed out that if the pairing operation is not hidden, typical hardness assumptions like the RSA assumption may not be justified in those groups. [13] also defined another primitive called Trapdoor Discrete Logarithm groups; however the candidate they proposed for this – with some reservations – was subsequently shown to be insecure [23].

Our hardness assumptions on TDG are different from those in [13]. The first assumption we use, namely the Strong RSA assumption, was introduced by Barić and Pfitzmann in their work on accumulators [3] mentioned above, as well as by Fujisaki and Okamoto [14]. Subsequently it has been studied extensively and used in a variety of cryptographic applications (for e.g. [16,11,7,1,6,2,26]). The second assumption we make is a Knowledge-of-Exponent Assumption (KEA). The first KEA, now called KEA1, was introduced by [12], to construct an efficient public key cryptosystem secure against chosen ciphertext attacks. Hada and Tanaka [19] employed it together with another assumption (called KEA2) to propose 3-round, negligible-error zero-knowledge arguments for NP. Bellare and Palacio in [4] falsified the KEA2 assumption and used another extension of KEA1, called KEA3, to restore the results in [19].

1.3   Differences with the Original Definition

Our definition differs from that of Micali, Rabin and Kilian [22] in several technical aspects. The original definition was in the setting of the trusted setup of a common reference string; it also required proofs to be publicly verifiable; the zero-knowledge property was required for PPT verifiers and was defined in terms of a PPT simulator (the indistinguishability could be computational, statistical or perfect).

In contrast, we define an SHS scheme as a two-party protocol, with no requirement of public verifiability of the proofs; but we do not allow a trusted setup. We require the hiding property to hold even when the verifier is computationally unbounded; but we do not require an efficient simulation.

As mentioned before, we consider these differences to be of a technical nature: the basic utility provided by SHS is the same as that by ZK sets. However we do expect qualitative differences to show up when considering composition issues (cf. parallel composition of zero-knowledge proofs and witness indistinguishable proofs). We leave this for future investigation.

Our definition of the statistical hiding property is formulated using a computationally unbounded simulation. It is instructive to cast our security definition (when the verifier is corrupt) in the "real-world/ideal-world" paradigm of definitions that are conventional for multi-party computation. In the ideal world the corrupt verifier (simulator) can be computationally unbounded, but gets access only to a blackbox to answer the membership queries. We require a statistically indistinguishable simulation — effectively requiring security even in a computationally unbounded "environment." (However our definition is not in the Universal Composition framework, as we do not allow the environment to interact with the adversary during the protocol.)

In [25] we include a further discussion of the new definition, comparing and contrasting it with the definition of zero-knowledge *proofs*.

1.4   Assumptions Used

The hardness asssumptions used in this work are of a strong nature. We use a combination of a strong RSA assumption and a knowledge-of-exponent assumption. Further these assumptions are applied to a relatively new family of groups, namely, trapdoor DDH groups. Therefore we advise caution in using our protocol before gaining further confidence in these assumptions. Nevertheless we point out that *the assumptions are used only for soundness*. The statistical hiding property is *unconditional*. This means that even if an adversary manages to violate our assumptions *after* one finishes using the scheme, it cannot violate the security guarantee at that point.

The knowledge of exponent assumption we use — called KEA-DH — is a new proposal, but similar to KEA1 introduced in 1991 by Damgard [12]. We believe this powerful assumption could prove very useful in constructing efficient cryptographic schemes, yet is reasonable enough to be safely assumed in different families of groups. In this work we combine KEA-DH with another (more standard) assumption called the strong RSA assumption. In Section 3 we descibe these assumptions, and in the full version [25] we provide some further preliminary observations about them (including a proof that KEA-DH holds in a generic group model).

Our construction depends on the idea of trapdoor DDH groups by Dent and Galbraith [13]. (But our use of this primitive does not require exactly the same features as a trapdoor DDH scheme offers; in particular, we do not use the DDH assumption.)

2   Statistically Hiding Sets

In this section we present our definition of statistically hiding sets (SHS). It is slightly different from the original definition of zero-knowledge sets of Micali et al. [22], but offers the same intuitive guarantees.

Notation. We write Pr[experiment : condition] to denote the probability that a condition holds after an experiment. An experiment is a probabilistic computation (typically described as a sequence of simpler computations); the condition will be in terms of variables set during the experiment. We write {experiment : random variable} to denote the distribution in which a random variable will be distributed after an experiment.

The security parameter will be denoted by k. By a PPT algorithm we mean a non-uniform probabilistic algorithm which terminates within $\text{poly}(k)$ time for some polynomial poly. We say two distributions are *almost the same* if the statistical difference between them is negligible in k.

Statistically Hiding Set Scheme: Syntax. A (non-interactive) SHS scheme consists of four PPT algorithms: setup, commit, prove and verify.

- setup: This is run by the verifier to produce a public-key/private-key pair: $(\mathsf{PK}, \mathsf{VK}) \leftarrow \mathsf{setup}(1^k)$. Here k is the security parameter; the public-key PK is to be used by a prover who makes a commitment to the verifier, and the private (verification) key VK is used by the verifier to verify proofs of membership and non-membership.
- commit: Algorithm used by the prover to generate a commitment σ of a finite set S, along with private information ρ used for generating proofs: it takes as input $(\sigma, \rho) \leftarrow \mathsf{commit}(S, \mathsf{PK})$.
- prove: Algorithm used by the prover to compute non-interactive proofs of memberships or non-memberships of queried elements: $\pi \leftarrow \mathsf{prove}(x, \rho, \mathsf{PK})$. Here π is a proof of membership or non-membership of x in a set S that was committed using commit; ρ is the the private information computed by commit.
- verify: Algorithm used by the verifier to verify a given proof of membership or non-membership: $b \leftarrow \mathsf{verify}(\pi, \sigma, x, \mathsf{VK})$. The output b is either a bit 0 or 1 (corresponding to accepting a proof of non-membership, or a proof of membership, respectively), or $\bot$ (corresponding to rejecting the proof).

Statistically Hiding Set Scheme: Security. Let U_k stand for the universal set of elements allowed by the scheme for security parameter k (i.e., the sets committed using the scheme are subsets of U_k). U_k can be finite (e.g., $\{0,1\}^k$) or infinite (e.g. $\{0,1\}^*$). We say that PPT algorithms (setup, commit, prove, verify) form a secure SHS scheme if they satisfy the properties stated below.

- *Perfect completeness:* For any finite set $S \subseteq U_k$, and $x \in U_k$,

$$\Pr\Big[(\mathsf{PK}, \mathsf{VK}) \leftarrow \mathsf{setup}(1^k); (\sigma, \rho) \leftarrow \mathsf{commit}(\mathsf{PK}, S); \pi \leftarrow \mathsf{prove}(\mathsf{PK}, \rho, x);$$

$$b \leftarrow \mathsf{verify}(\mathsf{VK}, \pi, \sigma, x) : x \in S \implies b = 0 \text{ and } x \notin S \implies b = 1\Big] = 1$$

That is, if the prover and verifier honestly follow the protocol, then the verifier is always convinced by the proofs given by the prover (it never outputs $\bot$).

- *Computational Soundness:* For any PPT adversary $\mathcal{A}$, there is a negligible function ν s.t.

$$\Pr\Big[(\mathsf{PK}, \mathsf{VK}) \leftarrow \mathsf{setup}(1^k); (\sigma, x, \pi_0, \pi_1) \leftarrow \mathcal{A}(\mathsf{PK}); b_0 \leftarrow \mathsf{verify}(\mathsf{VK}, \pi_0, \sigma, x);$$

$$b_1 \leftarrow \mathsf{verify}(\mathsf{VK}, \pi_1, \sigma, x) : b_0 = 0 \text{ and } b_1 = 1\Big] \leq \nu(k)$$

That is, except with negligible probability, an adversary cannot produce a commitment σ and two proofs π_0, π_1 which will be accepted by the verifier respectively as a proof of non-membership and of membership of an element x in the committed set.

- *Statistical Hiding:* There exists a distribution $\mathsf{simcommit}(\mathsf{PK})$, and two distributions $\mathsf{simprove}_0(\mathsf{PK}, \rho, x)$ and $\mathsf{simprove}_1(\mathsf{PK}, \rho, x)$, such that for every adversary $\mathcal{A}$ (not necessarily PPT), and every finite $S \subseteq U_k$ and any polynomial $t = t(k)$, the following two distributions have a negligible statistical distance between them.

$$\left\{\begin{array}{l} (\mathsf{PK}, s_0) \leftarrow \mathcal{A}(1^k; \mathsf{aux}(S)); \\ (\sigma, \rho) \leftarrow \mathsf{commit}(\mathsf{PK}, S); \\ \pi_0 := \sigma; \\ \mathbf{for}\ i = 1, \ldots, t \\ \quad (x_i, s_i) \leftarrow \mathcal{A}(s_{i-1}, \pi_{i-1}); \\ \quad \pi_i \leftarrow \mathsf{prove}(\mathsf{PK}, \rho, x_i); \\ \mathbf{endfor} \\ \qquad : (s_t, \pi_t) \end{array}\right\} \qquad \left\{\begin{array}{l} (\mathsf{PK}, s_0) \leftarrow \mathcal{A}(1^k; \mathsf{aux}(S)); \\ (\sigma, \rho) \leftarrow \mathsf{simcommit}(\mathsf{PK}); \\ \pi_0 := \sigma; \\ \mathbf{for}\ i = 1, \ldots, t \\ \quad (x_i, s_i) \leftarrow \mathcal{A}(s_{i-1}, \pi_{i-1}); \\ \quad \pi_i \leftarrow \mathsf{simprove}_{\chi_{x_i}^S}(\mathsf{PK}, \rho, x_i); \\ \mathbf{endfor} \\ \qquad : (s_t, \pi_t) \end{array}\right\}$$

Here $\mathsf{aux}(S)$ denotes arbitrary auxiliary information regarding the set S being committed to.[3] χ_x^S is a bit indicating $x \in S$ or not, which is the only bit of information that each invocation of $\mathsf{simprove}$ uses. In the outcome of the experiment, the state s_t may include all information that $\mathcal{A}$ received so far, including σ and all the previous proofs π_i.

Note that the second of these two distributions does not depend on S, but only on whether $x_i \in S$ for $i = 1, \ldots, t$. We *do not* require the distributions $\mathsf{simcommit}(\cdot)$ and $\mathsf{simprove}(\cdot)$ to be efficiently sampleable. An alternate (but equivalent) definition of the hiding property is in terms of a (computationally unbounded) simulator which can statistically simulate an adversary's view, with access to only an oracle which tells it whether $x \in S$ for each x for which it has to produce a proof.

---

[3] We include aux only for clarity, because by virtue of the order of quantifiers ($\forall \mathcal{A}$ and S) we are allowing $\mathcal{A}$ auxiliary information about the set S.

3   Trapdoor DDH Group

Following Dent and Galbraith[13], we define a *trapdoor DDH group* (TDG, for short) as follows. The hardness assumptions we make on such a group are different.

A TDG is defined by two PPT algorithms: a group generation algorithm Gen and a trapdoor DDH algorithm TDDH.

- Gen takes as input the security parameter k (in unary), and outputs a description (i.e., an algorithm for the group operation) for a group $\mathbb{G}$, an element $g \in \mathbb{G}$, an integer $n = 2^{O(k)}$ as an upperbound on the order of g in $\mathbb{G}$, and a trapdoor τ. The representation of this output should be such that given a tuple $(\mathbb{G}, g, n)$ (purportedly) produced by Gen(1^k), it should be possible to efficiently verify that $\mathbb{G}$ is indeed a group and $g \in \mathbb{G}$ has order at most n.[4]
- TDDH takes as input $(\mathbb{G}, g, n, \tau)$ produced by Gen, as well as elements A, B, C and outputs 1 if and only if $A, B, C \in \mathbb{G}$ and there exist integers a, b, c such that $A = g^a, B = g^b, C = g^{ab}$.

We make the following two hardness assumptions on a TDG.

Assumption 1 (Strong RSA Assumption). *For every PPT algorithm $\mathcal{A}$, there is a negligible function ν such that* $\Pr\Big[(\mathbb{G}, g, n, \tau) \leftarrow$ Gen$(1^k); (x, e) \leftarrow \mathcal{A}(\mathbb{G}, g, n) : x \in \mathbb{G}, e > 1$ *and* $x^e = g\Big] < \nu(k)$. *The probability is over the coin tosses of* Gen *and* $\mathcal{A}$.

Assumption 2 (Diffie-Hellman Knowledge of Exponent Assumption). (KEA-DH) *For every PPT adversary $\mathcal{A}$, there is a PPT extractor $\mathcal{E}$ and a negligible function ν such that*

$$\Pr\Big[(\mathbb{G}, g, n, \tau) \leftarrow \text{Gen}(1^k); (A, B, C) \leftarrow \mathcal{A}(\mathbb{G}, g, n; r); z \leftarrow \mathcal{E}(\mathbb{G}, g, n; r)$$

$$: \big(\exists a, b : A = g^a, B = g^b, C = g^{ab}\big), C \neq A^z, C \neq B^z\Big] < \nu(k).$$

Here r stands for the coin tosses of $\mathcal{A}$; $\mathcal{E}$ may toss additional coins. The probability is over the coin tosses of Gen*, $\mathcal{A}$ (namely r), and any extra coins used by $\mathcal{E}$.*

Further this holds even if $\mathcal{A}$ is given oracle access to a DDH oracle for the group $\mathbb{G}$. The extractor $\mathcal{E}$ does not access this oracle.

Informally, KEA-DH says that if an adversary takes g generated by Gen and outputs a DDH tuple (g^a, g^b, g^{ab}), then it must know either a or b. However, since $\mathcal{A}$ may not know the order of g in $\mathbb{G}$, the integers a, b are not unique.

---

[4] The upperbound on the order of g can often be the order of $\mathbb{G}$ itself, which could be part of the description of $\mathbb{G}$. We include this upperbound explicitly in the output of Gen as it plays an important role in the construction and the proof of security. However we stress that the exact order of g will be (assumed to be) hard to compute from $(\mathbb{G}, g, n)$.

For our use later, we remark that the extractor $\mathcal{E}$ can be modified to output its inputs as well as the output of $\mathcal{A}$, along with z. Then KEA-DH asserts that if there exists an adversary that takes $(\mathbb{G}, g, n)$ as input and produces (α, A, B, C) which satisfies some prescribed property and in which (A, B, C) is a DDH tuple, with non-negligible probability, then there is a machine $\mathcal{E}$ which takes $(\mathbb{G}, g, n)$ as input, and with non-negligible probability outputs $(\mathbb{G}, g, \alpha, A, B, C, z)$ such that (α, A, B, C) satisfies the prescribed property and either $C = A^z$ or $C = B^z$,

In [25] we discuss the trapdoor DDH group proposed by Dent and Galbraith [13], and also make preliminary observations about our assumptions above. In particular we prove that KEA-DH holds in a generic group with a bilinear pairing operation.

Note that in Assumption 2 we allow $\mathcal{A}$ access to a DDH oracle in $\mathbb{G}$, but requires an extractor which does not have this access. This captures the intuition that $\mathcal{A}$ cannot effectively make use of such an oracle: either a query to such an oracle can be seen to be a DDH tuple by considering how the tuple was derived, or if not, it is highly unlikely that it will be a DDH tuple. Indeed, this is the case in the generic group. Finally, we point out that in the KEA-DH assumption the adversary does not obtain any auxiliary information (from Gen, for instance). For proving the security of our construction it will be sufficient to restrict to such adversaries (even though we allow auxiliary information in the security definition).

4   Our Construction

Our SHS construction somewhat resembles the construction of accumulators in [3,16,18,6,20]. In the following we require the elements in the universe (a finite subset of which will be committed to) to be represented by sufficiently large prime numbers. In [25] we show that such a representation is easily achieved (under standard number theoretic conjectures on dsitribution of prime numbers).

Construction 1 (Statistically Hiding Sets). The construction consists of the following four algorithms:

1. setup The verifier takes the security parameter k as the input and runs a trapdoor DDH group generation algorithm to obtain $(\mathbb{G}, g, n, \tau)$. The verifier then sends $(\mathbb{G}, g, n)$ to the prover. The prover verifies that $\mathbb{G}$ is indeed a group and $g \in \mathbb{G}$ has order at most n. (Recall that an upper bound on the order of g is explicitly included as part of description of $\mathbb{G}$. See Footnote 4.) In the following let $N = 2^k n$.
2. commit To commit to a set $S = \{p_1, p_2, \ldots, p_m\}$, the prover chooses an integer $v \in \{0, \ldots, N-1\}$ at random and computes

$$u = \prod_{i=1}^{i=m} p_i \qquad\qquad C = g^{uv}$$

and outputs $\sigma = C$ as the public commitment and $\rho = uv$ as the private information for generating proofs in future.

3. prove When the prover receives a query about an element p, it will respond depending on whether $p \in S$ or not:
   - $p \in S$: In this case the prover computes $u_0 = u/p$ and $C_0 = g^{vu_0}$ and sends $\pi = (\text{yes}; C_0)$ to the verifier.
   - $p \notin S$: Since $v < p$ and $\gcd(p, u) = 1$, we have $\gcd(p, uv) = 1$. The extended Euclidean algorithm allows the prover to compute integers (a_0, b_0) such that $a_0 uv + b_0 p = 1$. The prover then chooses a number $\gamma \in \{0, \ldots, N-1\}$ at random, and forms the pair of integers $(a, b) := (a_0 + \gamma p, b_0 - \gamma uv)$. Finally, the prover evaluates $A = g^a$, $B = g^b$, $D = C^a$ and sends $\pi = (\text{no}; A, B, D)$ to the verifier.
4. verify When the verifier queries about an element p and receives a proof π from the prover, it does the following:
   - If π is of the form (yes, C_0), i.e., the prover asserts that $p \in S$, then the verifier checks if $C_0^p = C$ holds. It outputs 1 if the preceding equation holds, and $\perp$ otherwise.
   - If π is of the form (no, A, B, D), i.e., the prover claims $p \notin S$, then the verifier checks if $\mathsf{TDDH}(A, C, D; \mathbb{G}, g, \tau) = 1$ (i.e., whether (g, A, C, D) is a DDH tuple in $\mathbb{G}$) and if $D \cdot B^p = g$ holds in the group. It outputs 0 if both the checks are satisfied, and $\perp$ otherwise.

5   Security of Our Scheme

Theorem 1. *Under the strong RSA and the KEA-DH assumptions, Construction 1 is a secure statistically hiding set scheme.*

In the rest of this section we prove this. The soundness of the construction depends on the computational assumptions (Lemma 2), but the statistical hiding property is unconditional (Lemma 6 which is proven using Lemma 3, Lemma 4 and Lemma 5).

Completeness. For the completeness property, we need to show that an honest prover can always convince the verifier about the membership of the queried element.

Lemma 1. *Construction 1 satisfies the completeness property.*

Proof. For any finite set S and any element p, we show that an honest prover's proof of membership or non-membership will be accepted by an honest verifier. Let $S = \{p_1, \ldots, p_m\}$, let u, v be computed as in the commitment phase of our construction and p be an element queried by the verifier.

If $p \in S$, then $p \mid uv$ and so uv/p is an integer. So the prover can indeed compute $C_0 = g^{uv/p}$ and send it to the verifier. The verifier will accept this proof since $C_0^p = C$.

If $p \notin S$, then $p \nmid u$. Note that since we require $p \geq N$ and $0 \leq v < N$, we have $p > v$. So $p \nmid uv$ and p being prime, $\gcd(p, uv) = 1$. So the prover can run the extended Euclidean algorithm and find $(a_0, b_0) \in \mathbb{Z} \times \mathbb{Z}$ such that

$a_0 uv + b_0 p = 1$. For any integer γ (and in particular for $\gamma \in \{0, \ldots, N-1\}$), for $(a, b) = (a_0 + \gamma p, b_0 - \gamma uv)$, it holds

$$auv + bp = (a_0 + \gamma p)uv + (b_0 - \gamma uv)p = a_0 uv + b_0 p = 1.$$

The prover evaluates $A = g^a$, $B = g^b$, $D = C^a = g^{uva}$ and sends (A, B, D) to the verifier. Clearly the tuple $(g, A, C, D) = (g, g^a, C, C^a)$ is a DDH tuple and hence $\mathsf{TDDH}(A, C, D; \mathbb{G}, g, \tau) = 1$. Also

$$D \cdot B^p = C^a \cdot B^p = g^{(uv)a} \cdot g^{pb} = g^{auv+bp} = g. \tag{1}$$

Hence in this case also the verifier accepts the proof. □

Soundness. The soundness property guarantees that the scheme does not allow a malicious prover to equivocate about the membership of any element p in the committed set.

Lemma 2. *Under the strong RSA and the KEA-DH assumptions, construction 1 has the soundness property.*

Proof. To violate the soundness of our construction, a corrupt prover $\mathcal{A}$ must make a commitment C and subsequently for some prime p give out valid proofs for membership and non-membership of p. That is, on input $(\mathbb{G}, g, n)$, $\mathcal{A}$ outputs (p, C, C_1, A, B, D) satisfying

$$C_1^p = C \tag{2}$$
$$\exists a \text{ s.t. } A = g^a, D = C^a \tag{3}$$
$$D \cdot B^p = g \tag{4}$$

We will show that the probability of this is negligible.

First note that from KEA-DH there is an extractor $\mathcal{E}$ corresponding to $\mathcal{A}$, such that with almost the same probability as $\mathcal{A}$ satisfies the above conditions $\mathcal{E}$ outputs $(\mathbb{G}, g, p, C, C_1, A, B, D, z)$ such that in addition to the above conditions, either $A = g^z$ or $C = g^z$.

1. In case z is such that $C = g^z$, we consider two cases.

– If $\gcd(p, z) = 1$: Then using extended Euclidean algorithm on input (p, z) one can efficiently compute integers (α, β) such that $\alpha z + bp = 1$. That is, $g^{az+bp} = g$. Using equation (2), $C_1^p = g^z$ and so this can be rewritten as $C_1^{ap} g^{bp} = g$. That is by setting $x = C_1^\alpha g^\beta$, we have $x^p = g$.

– If $\gcd(p, z) \neq 1$, then $p \mid z$ as p is a prime. Then z/p is an integer. Let $x = A^{z/p} \cdot B$; note that by substituting $C = g^z$ into equation (3) we get $D = A^z$, and hence by equation (4), $x^p = g$.

2. In case $A = g^z$, then in the subgroup generated by g, we have $x^a = x^z$ for all x (that is, a and z are congruent modulo the order of g). From equations (2) (3) and (4), we have

$$g = D \cdot B^p = C^a \cdot B^p = (C_1^p)^a \cdot B^p = (C_1^a \cdot B)^p.$$

Setting $x = C_1^a \cdot B = C_1^z \cdot B$ we have $x^p = g$.

Thus in all cases from the output of $\mathcal{E}$, one can efficiently find (x, p) such that $x^p = g$ and $p > 1$. This is possible with the same probability (up to negligible difference) as $\mathcal{A}$ succeeding in breaking the soundness. But by the strong RSA assumption this probability must be negligible, Hence the probability that $\mathcal{A}$ violates soundness is negligible.   □

Statistical Hiding. The hiding property is statistical and unconditional. We shall show that the commitment as well as the proofs given by an honest prover are distributed (almost) independent of the set that is committed to. The only guarantee that is required from the setup information $(\mathbb{G}, g, n)$ sent out by the verifier is that the order of $g \in \mathbb{G}$ is bounded by n (as will be verified by the prover).

First we prove that the commitment is almost uniformly distributed in the subgroup $\langle g \rangle$. We use the following observation.

Lemma 3. *Let $\mathcal{U}_m$ be the uniform distribution over $\mathbb{Z}_m$. Let $\mathcal{U}_{N|m}$ be the distribution over $\mathbb{Z}_m$ obtained by sampling an element uniformly from $\mathbb{Z}_N$ and reducing it modulo m. Then the statistical distance between $\mathcal{U}_m$ and $\mathcal{U}_{N|m}$ is at most m/N.*

Proof. $\mathcal{U}_{N|m}$ can be written as a convex combination of $\mathcal{U}_m$ and of the distribution of $v \bmod m$ when v is chosen uniformly at random from $\{tm, \ldots, N-1\}$, where $t = \lfloor N/m \rfloor$. The weights used in the convex combination are tm/N and $(N - tm)/N$ respectively. So the statistical difference is at most $(N - tm)/N < m/N$.

Lemma 4. *For any finite $S \subseteq \mathcal{U}_k$, and any valid setup public-key $\mathsf{PK} = (\mathbb{G}, g, n)$, the commitment C produced by $\mathsf{commit}(\mathsf{PK}, S)$ is almost uniformly distributed over $\langle g \rangle$.*

Proof. The proof is a simple consequence of Lemma 3.

Note that the commitment C is distributed as g_1^v where v is distributed as $\mathcal{U}_{N|m}$ where $m = \mathrm{ord}(g_1)$. Here $g_1 = g^u$ where $u = \prod_{p_i \in S} p_i$. Since $p_i > n > \mathrm{ord}(g)$ we have $\gcd(u, \mathrm{ord}(g)) = 1$, and therefore $m = \mathrm{ord}(g_1) = \mathrm{ord}(g)$ and $\langle g^u \rangle = \langle g \rangle$. Then applying Lemma 3 with $N = 2^k n$ and $m \leq n$, we get that the statistical distance between the distribution of C and the uniform distribution over $\langle g \rangle$ is at most 2^{-k} which is negligible in the security parameter k.

That means, we can define the distribution $\mathsf{simcommit}$ from the definition of statistical hiding (Section 2) as $\mathsf{simcommit}(\mathbb{G}, g, n) = \{\sigma \leftarrow \mathcal{U}_{\langle g \rangle} : (\sigma, \sigma)\}$, where $\mathcal{U}_{\langle g \rangle}$ is the uniform distribution over the subgroup of $\mathbb{G}$ generated by g. This is for $(\mathbb{G}, g, n)$ being a valid public-key for the setup; otherwise — i.e., if $\mathbb{G}$ is not a valid group with g as an element, or $\mathrm{ord}(g)$ is not upper bounded by n — then commit will abort. Next we specify $\mathsf{simprove}_0$ and $\mathsf{simprove}_1$.

Lemma 5. *For any valid public-key $\mathsf{PK} = (\mathbb{G}, g, n)$, any finite $S \subseteq U_k$, and any commitment (public and private information) $(\sigma, \rho) = (C, uv)$ that can be produced by $\mathsf{commit}(S, \mathsf{PK})$, for each prime $p > \mathrm{ord}(g)$, there are distributions*

$\mathsf{simprove}_0(\mathbb{G}, g, n, C, p)$ *and* $\mathsf{simprove}_1(\mathbb{G}, g, n, C, p)$ *such that for* $p \in S$ *(respectively,* $p \notin S$*), the proof of membership of* p *(respectively of non-membership),* $\mathsf{prove}(p, \rho, \mathsf{PK})$ *is distributed almost as* $\mathsf{simprove}_1(\mathbb{G}, g, n, C, p)$ *(respectively as* $\mathsf{simprove}_0(\mathbb{G}, g, n, C, p)$*).*

Proof. Let $\mathsf{simprove}_1(\mathbb{G}, g, n, C, p)$ be just concentrated on the single unique element (yes, C_0) where $C_0 \in \langle g \rangle$ is such that $C_0^p = C$ (i.e., $C_0 = C^{p^{-1} \bmod \operatorname{ord}(g)}$). Clearly, the proof generated by the honest prover is exactly the same (though the prover does not compute $p^{-1} \bmod \operatorname{ord}(g)$ directly).

$\mathsf{simprove}_0(\mathbb{G}, g, n, C, p)$ is defined as (no, A', B', D') where $A' = g^{a'}, B' = g^{b'}, D' = C^{a'}$, with (a', b') distributed as follows. Let $m = \operatorname{ord}(g)$. Let c be the unique number in $\{0, \dots, m-1\}$ such that $C = g^c$. Then (a', b') is distributed uniformly over the set

$$Z_{c,p,m} = \{(a', b') | a', b' \in \mathbb{Z}_m \text{ and } a'c + b'p \equiv 1 \pmod{m}\}.$$

We argue that the proof of $p \notin S$ produced by an honest prover (conditioned on $\mathsf{PK} = (\mathbb{G}, g, n)$ and commitment C), no matter which set $S \subseteq U_k \backslash \{p\}$ is used, is almost identically distributed as $\mathsf{simprove}_0(\mathbb{G}, g, n, C, p)$. For this, it is enough to show that (a, b) computed by prove is such that $(a \bmod m, b \bmod m)$ is distributed (almost) identically as (a', b') above – i.e., uniformly over $Z_{c,p,m}$.

Firstly, by Lemma 3, we have that for γ as sampled by the prover (i.e., $\gamma \leftarrow \{0, \dots, N-1\}$), $\gamma \bmod m$ is close to being uniform over $\mathbb{Z}_m$ (the statistical difference being negligible in the security parameter). So we can ignore this difference, and assume that in fact the prover does sample γ such that $\gamma \bmod m$ is uniform over $\mathbb{Z}_m$.

Secondly, note that $uv \equiv c \pmod{m}$, since $C = g^{uv} = g^c$. Hence, when the prover uses the extended Euclidean algorithm on the pair of integers (uv, p) to find $(a_0, b_0) \in \mathbb{Z}^2$ we have that

$$a_0 c + b_0 p \equiv 1 \pmod{m}. \tag{5}$$

We remark that the input to the extended Euclidean algorithm uv, and hence its output (a_0, b_0), do depend on u and hence on the set S being committed. In fact, even $(a_0 \bmod m, b_0 \bmod m)$ depends on uv and not just on c.

But recall that the prover sets (a, b) to be $(a_0 + \gamma b, b_0 - \gamma a)$ where γ is sampled such that $\gamma \bmod m$ is (almost) uniform over $\mathbb{Z}_m$. We make the following observations:

1. $((a_0 + \gamma p) \bmod m, (b_0 - \gamma c) \bmod m) \in Z_{c,p,m}$ for every integer γ, by equation (5).

2. Since $\gcd(p, m) = 1$, if $((a_0 + \gamma_1 p) \bmod m, (b_0 - \gamma_1 c) \bmod m) = ((a_0 + \gamma_2 p) \bmod m, (b_0 - \gamma_2 c) \bmod m)$ then $\gamma_1 \equiv \gamma_2 \pmod{m}$. So for each distinct value of $\gamma \bmod m$ there is a unique element in $Z_{c,p,m}$ that $((a_0 + \gamma p) \bmod m, (b_0 - \gamma c) \bmod m)$ evaluates to.

3. Finally, $|Z_{c,p,m}| = m$ because for each value of $a \in \mathbb{Z}_m$ there is a unique $b = (1 - ac) \cdot p^{-1} \in \mathbb{Z}_m$ such that $(a, b) \in Z_{c,p,m}$. So for every $(a', b') \in Z_{c,p,m}$ there is exactly one $\gamma \in \mathbb{Z}_m$ such that $(a', b') = ((a_0 + \gamma p) \bmod m, (b_0 - \gamma c) \bmod m)$.

So we conclude that since the prover samples γ (almost) uniformly from $\mathbb{Z}_m$, (a, b) it computes is such that $(a \bmod m, b \bmod m)$ is indeed (almost) uniform over $Z_{c,p,m}$.

To conclude we derive the following from the above.

Lemma 6. *Construction 1 satisfies the statistical hiding property.*

Proof. This follows from Lemmas 4 and 5. The two lemmas give distributions simcommit(PK), and $\mathsf{simprove}_0(\mathsf{PK}, \rho, x)$ and $\mathsf{simprove}_1(\mathsf{PK}, \rho, x)$, as required in the definition of statistical hiding property. That they indeed statisfy the condition there, is an easy consequence of the fact that the adversary $\mathcal{A}$ is allowed auxiliary information about the set S in the experiments.

More formally, we define $t + 1$ hybrid distributions, by modifying the experiments in the two distributions in the definition of the hiding property: commit is used to generate the commitment, but in the ith hybrid, for the first $i - 1$ iterations of the for loop simprove is used,[5] and in the rest prove is used. The first of these hybrids is identical to the left hand side distribution in the definition, and the last one is statistically close to the right hand side distribution by Lemma 4. Two adjacent hybrids are statistically close by Lemma 5.

Acknowledgment. The second author wishes to thank Manoj Prabhakaran, Mike Rosulek, and colleagues at UIUC for their warm hospitality during his stay there. He also acknowledges Professor N. Li and Dr. J. Li for the many discussions related to this topic.

References

1. Ateniese, G., Camenisch, J., Joye, M., Tsudik, G.: A practical and provably secure coalition-resistant group signature scheme. In: Bellare, M. (ed.) CRYPTO 2000. LNCS, vol. 1880, p. 255. Springer, Heidelberg (2000)
2. Ateniese, G., de Medeiros, B.: Efficient group signatures without trapdoors. In: Laih, C.-S. (ed.) ASIACRYPT 2003. LNCS, vol. 2894, pp. 246–268. Springer, Heidelberg (2003)
3. Barić, N., Pfitzmann, B.: Collision-free accumulators and fail-stop signature schemes without trees. In: Fumy, W. (ed.) EUROCRYPT 1997. LNCS, vol. 1233, pp. 480–494. Springer, Heidelberg (1997)
4. Bellare, M., Palacio, A.: The knowledge-of-exponent assumptions and 3-round zero-knowledge protocols. In: Franklin, M. (ed.) CRYPTO 2004. LNCS, vol. 3152, pp. 273–289. Springer, Heidelberg (2004)
5. Benaloh, J.C., de Mare, M.: One-way accumulators: A decentralized alternative to digital signatures. In: De Santis, A. (ed.) EUROCRYPT 1994. LNCS, vol. 950. Springer, Heidelberg (1995)
6. Camenisch, J., Lysyanskaya, A.: Dynamic accumulators and application to efficient revocation of anonymous credentials. In: Yung, M. (ed.) CRYPTO 2002. LNCS, vol. 2442, p. 61. Springer, Heidelberg (2002)

---

[5] Note that simprove takes C as an input; C produced by commit will be given.

7. Camenisch, J., Michels, M.: Separability and efficiency for generic group signature schemes. In: Wiener, M. (ed.) CRYPTO 1999. LNCS, vol. 1666, p. 413. Springer, Heidelberg (1999)
8. Catalano, D., Dodis, Y., Visconti, I.: Mercurial commitments: Minimal assumptions and efficient constructions. In: Halevi, S., Rabin, T. (eds.) TCC 2006. LNCS, vol. 3876, pp. 120–144. Springer, Heidelberg (2006)
9. Catalano, D., Fiore, D., Messina, M.: Zero-knowledge sets with short proofs. In: Smart, N.P. (ed.) EUROCRYPT 2008. LNCS, vol. 4965, pp. 433–450. Springer, Heidelberg (2008)
10. Chase, M., Healy, A., Lysyanskaya, A., Malkin, T., Reyzin, L.: Mercurial commitments with applications to zero-knowledge sets. In: Cramer, R. (ed.) EUROCRYPT 2005. LNCS, vol. 3494, pp. 422–439. Springer, Heidelberg (2005)
11. Cramer, R., Shoup, V.: Signature schemes based on the strong RSA assumption. In: ACM Conference on Computer and Communications Security (CCS) (1999)
12. Damgard, I.: Towards practical public-key cryptosystems provably-secure against chosen-ciphertext attacks. In: Brickell, E.F. (ed.) CRYPTO 1992. LNCS, vol. 740. Springer, Heidelberg (1993)
13. Dent, A.W., Galbraith, S.D.: Hidden pairings and trapdoor DDH groups. In: Hess, F., Pauli, S., Pohst, M. (eds.) ANTS 2006. LNCS, vol. 4076, pp. 436–451. Springer, Heidelberg (2006)
14. Fujisaki, E., Okamoto, T.: Statistical zero knowledge protocols to prove modular polynomial relations. In: Kaliski Jr., B.S. (ed.) CRYPTO 1997. LNCS, vol. 1294, pp. 16–30. Springer, Heidelberg (1997)
15. Galbraith, S.D., McKee, J.F.: Pairings on elliptic curves over finite commutative rings. In: Smart, N.P. (ed.) Cryptography and Coding 2005. LNCS, vol. 3796, pp. 392–409. Springer, Heidelberg (2005)
16. Gennaro, R., Halevi, S., Rabin, T.: Secure hash-and-sign signatures without the random oracle. In: Stern, J. (ed.) EUROCRYPT 1999. LNCS, vol. 1592, p. 123. Springer, Heidelberg (1999)
17. Gennaro, R., Micali, S.: Independent zero-knowledge sets. In: Bugliesi, M., Preneel, B., Sassone, V., Wegener, I. (eds.) ICALP 2006. LNCS, vol. 4052, pp. 34–45. Springer, Heidelberg (2006)
18. Goodrich, M.T., Tamassia, R., Hasic, J.: An efficient dynamic and distributed cryptographic accumulator. In: Chan, A.H., Gligor, V.D. (eds.) ISC 2002. LNCS, vol. 2433, p. 372. Springer, Heidelberg (2002)
19. Hada, S., Tanaka, T.: On the existence of 3-round zero-knowledge protocols. In: Krawczyk, H. (ed.) CRYPTO 1998. LNCS, vol. 1462, p. 408. Springer, Heidelberg (1998)
20. Li, J., Li, N., Xue, R.: Universal accumulators with efficient nonmembership proofs. In: Katz, J., Yung, M. (eds.) ACNS 2007. LNCS, vol. 4521, pp. 253–269. Springer, Heidelberg (2007)
21. Liskov, M.: Updatable zero-knowledge databases. In: Roy, B. (ed.) ASIACRYPT 2005. LNCS, vol. 3788, pp. 174–198. Springer, Heidelberg (2005)
22. Micali, S., Rabin, M., Kilian, J.: Zero-knowledge sets. In: FOCS 2003 (2003)
23. Mireles, D.: An attack on disguised elliptic curves. Cryptology ePrint Archive, Report 2006/469 (2006), http://eprint.iacr.org/

24. Ostrovsky, R., Rackoff, C., Smith, A.: Efficient consistency proofs for generalized queries on a committed database. In: Díaz, J., Karhumäki, J., Lepistö, A., Sannella, D. (eds.) ICALP 2004. LNCS, vol. 3142, pp. 1041–1053. Springer, Heidelberg (2004)
25. Prabhakaran, M., Xue, R.: Statistical zero-knowledge sets using trapdoor DDH groups. Cryptology ePrint Archive, Report 2007/349 (2007), http://eprint.iacr.org/
26. Rivest, R.L.: On the notion of pseudo-free groups. In: Naor, M. (ed.) TCC 2004. LNCS, vol. 2951, pp. 505–521. Springer, Heidelberg (2004)
27. Xue, R., Li, N., Li, J.: A new construction of zero knowledge sets secure in random oracle model. In: The First International Symposium of Data, Privacy, & E-Commerce. IEEE Press, Los Alamitos (2007)

Adaptively Secure Two-Party Computation
with Erasures

Andrew Y. Lindell

Aladdin Knowledge Systems and Bar-Ilan University, Israel
andrew.lindell@aladdin.com, lindell@cs.biu.ac.il

Abstract. In the setting of multiparty computation a set of parties with private inputs wish to compute some joint function of their inputs, whilst preserving certain security properties (like privacy and correctness). An adaptively secure protocol is one in which the security properties are preserved even if an adversary can adaptively and dynamically corrupt parties during a computation. This provides a high level of security, that is arguably necessary in today's world of active computer break-ins. Until now, the work on adaptively secure multiparty computation has focused almost exclusively on the setting of an honest majority, and very few works have considered the honest minority and two-party cases. In addition, significant computational and communication costs are incurred by most protocols that achieve adaptive security.

In this work, we consider the two-party setting and assume that honest parties may *erase* data. We show that in this model it is possible to securely compute any two-party functionality in the presence of *adaptive semi-honest adversaries*. Furthermore, our protocol remains secure under concurrent general composition (meaning that it remains secure irrespective of the other protocols running together with it). Our protocol is based on Yao's garbled-circuit construction and, importantly, is as efficient as the analogous protocol for static corruptions. We argue that the model of adaptive corruptions with erasures has been unjustifiably neglected and that it deserves much more attention.

1 Introduction

In the setting of multiparty computation, a set of parties with private inputs wish to jointly compute some function of those inputs. Loosely speaking, the security requirements are that even if some of the participants are adversarial, nothing is learned from the protocol other than the output (*privacy*), and the output is distributed according to the prescribed functionality (*correctness*). The definition of security that has become standard today [17,25,1,5] blends these two conditions (and adds more). This setting models essentially any cryptographic task of interest, including problems ranging from key exchange and authentication, to voting, elections and privacy-preserving data mining. Due to its generality, understanding what can and cannot be computed in this model, and at what complexity, has far reaching implications for the field of cryptography.

M. Fischlin (Ed.): CT-RSA 2009, LNCS 5473, pp. 117–132, 2009.
© Springer-Verlag Berlin Heidelberg 2009

One important issue regarding secure computation is the environment in which it takes place. In the basic setting, called the stand-alone model, the secure protocol is run only once and in isolation (or equivalently, the adversary attacks only this execution). A more advanced setting is that of composition, where the secure protocol may be run many times concurrently with itself and arbitrary other protocols. This setting is called universal composability [6], or equivalently security under concurrent general composition [20], and more accurately models the real-world security needs than the stand-alone model.

A central question that needs to be addressed in this setting is the power of the adversary. An adversary can be semi-honest (in which case it follows the protocol specification exactly but attempts to learn more information than it should by analyzing the messages it receives) or malicious (in which case it can take arbitrary actions). In addition, one can consider static corruptions (meaning that the set of corrupted parties that are under the control of the adversary is fixed before the protocol execution begins) or adaptive corruptions (in which case the adversary can choose to corrupt parties during the computation based on its view). There are two main models that have been considered for adaptive corruptions. In both models, upon corrupting an honest party the adversary receives the internal state of that party. The difference lies in the question of what is included in that state. In the non-erasure model, honest parties are not assumed to be able to reliably erase data. Therefore, the internal state of a party includes its input, randomness and *all* of the messages that it received in the past. In contrast, in the model with erasures, honest parties may erase data if so instructed by the protocol and so the state includes all data as above except for data that has been explicitly erased. (Of course, not all intermediate data can be erased because the party needs to be able to run the protocol and compute the output.) In this paper, we consider the problem of achieving security in the presence of *adaptive semi-honest adversaries with erasures*. We remark that adaptive corruptions model the setting where hackers actively break in to computers during secure computations. As such, it more accurately models real-world threats than the model of static corruptions.

To erase or not to erase. In the cryptographic literature, the non-erasure model of adaptive corruptions has received far more attention than the erasure model (see prior work below). The argument has typically been that it is generally hard to ensure that parties fully erase data. After all, this can depend on the operating system, and in real life passwords and other secret data can often be found on swap files way after they were supposedly erased. We counter this argument by commenting that non-swappable memory is provided by all modern operating systems today and it is possible to use this type of memory for the data which is to be erased (as specified by the protocol). Of course, it is more elegant to assume that there are no erasures. However, the price of this assumption has been very great. That is, the complexity and communication of protocols that are secure under adaptive corruptions without erasures are all much higher than the analogous protocols that are secure under static corruptions (for example, we don't even have a constant-round protocol for general two-party computation that is secure under adaptive corruptions). We argue that the result of this is that no protocol designer today would even consider adaptive corruptions if the aim is to construct an efficient protocol that could possibly be used in practice.

Our results. We begin by studying the stand-alone model and note that the current situation is actually very good, as long as erasures are considered. Specifically, by combining results from Beaver and Haber [3], Canetti [5], and Canetti et al. [7], we show the following:

Theorem 1. *Let f be any two-party functionality and let π be a protocol that securely computes f in the presence of static malicious (resp., semi-honest) adversaries, in the stand-alone model. Then, assuming the existence of one-way functions, there exists a highly efficient transformation of π to π' such that π' securely computes f in the presence of adaptive malicious (resp., semi-honest) adversaries with erasures, in the stand-alone model.*

We have no technical contribution in deriving Theorem 1; rather our contribution here is to observe that a combination of known results yields the theorem. To the best of our knowledge, the fact that this theorem holds has previously gone unnoticed.

The only drawback of Theorem 1 is that it holds only for the stand-alone model (see Section 3 for an explanation as to why). As we have mentioned, this is a relatively unrealistic model in today's world where many different protocols are run concurrently. Our main technical contribution is therefore to show that it is possible to construct secure protocols for general two-party computation in the presence of semi-honest adaptive adversaries (and with erasures) that are secure under concurrent general composition [20] (equivalently, universally composable [6]). Importantly, the complexity of our protocol is analogous to the complexity of the most efficient protocol known for the case of semi-honest static adversaries (namely, Yao's protocol [27]). We prove the following theorem:

Theorem 2. *Assume that there exist enhanced[1] trapdoor permutations. Then, for every two-party probabilistic polynomial-time functionality f there exists a constant-round protocol that securely computes f under concurrent general composition, in the presence of adaptive, semi-honest adversaries with erasures.*

The contributions of Theorem 2 are as follows:

1. *Round complexity:* Our protocol for adaptive two-party computation requires a constant number of rounds. The only other protocols for general adaptive two-party computation that are secure under concurrent composition follow the GMW paradigm [16] and the number of rounds is therefore equal to the depth of the circuit that computes the function f; see [10]. We stress that [10] does not assume erasures, whereas we do.
2. *Hardness assumptions:* Our protocol requires the minimal assumption for secure computation in the static model of enhanced trapdoor permutations. In contrast, all known protocols for adaptive oblivious transfer (and thus adaptive

---

[1] See [15, Appendix C.1].

secure computation) without erasures assume seemingly stronger assumptions, like a public-key cryptosystem with the ability to sample a public-key without knowing the corresponding secret key. In fact, in a recent paper, it was shown that adaptively secure computation *cannot* be achieved in a black-box way from enhanced trapdoor permutations alone [24]. Thus, without assuming erasures, it is not possible to construct secure protocols for the adaptive model under this minimal assumption (at least, not in a black-box way).

In addition to the above, our protocol has the same complexity as Yao's protocol for static adversaries in all respects. We view this as highly important and as a sign that it is well worth considering this model when constructing efficient secure protocols. In particular, if it is possible to achieve security in the presence of adaptive corruptions with erasures "for free", then this provides a significant advantage over protocols that are only secure for static corruptions (of course, as long as such erasures can really be carried out). In addition, the typical argument against considering security under composition is that the resulting protocols are far less efficient. Our results demonstrate that this is not the case for the setting of semi-honest adaptive adversaries with erasures. Indeed our protocol is no less efficient than the basic protocol for the semi-honest stand-alone setting.

We remark that our protocol is very similar to the protocol of Yao and we only need to slightly change the order of some operations and include some erase instructions (i.e., in some sense, the original protocol is "almost" adaptively secure). Nevertheless, our *proof of security* is significantly different and requires a completely different simulation strategy to that provided in [22].

Related work. The vast majority of work on adaptive corruptions for secure computation has considered the setting of multiparty computation with an honest majority [3,8,2] and thus is not applicable to the two-party setting. To the best of our knowledge, the only two works that considered the basic question of adaptive corruptions for *general secure computation* in the setting of no honest majority are [10] and [7]. Canetti et al. [7] study the relation between adaptive and static security and present a series of results that greatly clarifies the definitions and their differences. However, this work only relates to the stand-alone setting. In the setting of composition, Canetti et al. [10] presented a protocol that is universally composable (equivalently, secure under concurrent general composition). The construction presented there considers a model with no erasures. As such, it is far less efficient (e.g., it is not constant-round), far more complicated, and relies on seemingly stronger cryptographic hardness assumptions. Regarding secure computation of functions of specific interest, there has also been little work on achieving adaptive security, with the notable exception of threshold cryptosystems [9,13,19] and oblivious transfer [14].

2   Definitions

Due to lack of space in this extended abstract, we refer to [5] for definitions of security in the stand-alone model for both static and adaptive corruptions.

We use the definition of security from [5] with the exception that the post-execution corruption phase and external environment are removed (as stated in [5, Remark 5] and further discussed in Section 3, these properties are not needed in the case of erasures). Our definitions also refer to the stand-alone model only. In order to derive security under concurrent general composition (or equivalently, universal composability [6]), we rely on [18] who proved that any protocol that has been proven secure in the stand-alone model, using a simulator that is *black-box* and *straight-line* and so doesn't rewind the adversary, is secure under concurrent-general composition. In actuality, an additional requirement for this to be true is something called *initial synchronization*. In the two-party setting, this just means that the parties send each other an init message before actually running the protocol. Our proofs of security therefore all refer to the stand-alone model only; security is derived for the setting of concurrent general composition because all of our simulators are black-box and straight-line. See the full version of this paper for more details [21].

3   Stand-Alone Two-Party Computation for Malicious Adversaries

In this section, we observe that in the *stand-alone model*, any two-party protocol that is secure in the presence of static adversaries can be efficiently converted into a protocol that is secure in the presence of adaptive adversaries, as long as erasures are allowed. This powerful result is another good reason why it is worth considering erasures – indeed, adaptive security is obtained almost for free. In order to see why this is true, we combine three different results:

1. First, Beaver and Haber [3] proved that any protocol that is adaptively secure in a model with ideally secure communication channels can be efficiently converted into a protocol that is adaptively secure in a model with regular (authenticated) communication channels, assuming erasures. The transformation of [3] requires public-key encryption and thus assumes the existence of trapdoor permutations. We stress that under specific assumptions, it can be implemented at very low cost.
2. Next, Canetti et al. [7] consider a modification of the standard definition of security for adaptive adversaries. The standard definition includes a *post-execution corruption phase* (known as PEC for short); this phase is necessary for obtaining sequential composition as described in [5].[2] Nevertheless, most of the results in [7] consider a modified definition where there is no PEC phase. Amongst many other results, it is proven in [7] that in a model with ideally secure channels and no PEC, any two-party protocol that is secure in the presence of static malicious adversaries is also secure in the presence of adaptive malicious adversaries. (The same holds also for semi-honest adversaries.)

---

[2] We remark that there is no PEC requirement for the definition of security in the presence of static adversaries.

A combination of the results of [3] and [7] yields the result that any two-party protocol that is secure in the presence of static malicious adversaries can be efficiently transformed into a protocol that is secure in the presence of adaptive malicious adversaries under a definition without PEC. (The requirement of [7] for ideally secure channels is removed by [3].) This result is still somewhat lacking because PEC is in general a necessary definitional requirement.

3. The post-execution corruption phase is not needed in the adaptive model where erasures are allowed [5, Remark 5]. In particular, modular sequential composition holds in this model even without this phase. (There is one requirement: the honest parties must erase the internal data they used at the end of the secure protocol execution, and can store only the input and output. There is no problem doing this because the input and output is all that they need.) Thus, in a model allowing erasures, the results of [7] guarantee adaptive security under a definition of security that is sufficient (and in particular implies sequential composition).

Combining the above three observations, we obtain the following theorem:

Theorem 3. *Consider the stand-alone model of computation. Let f be a two-party functionality and let π be a protocol that securely computes f in the presence of malicious static adversaries. Then, assuming the existence of trapdoor permutations, there exists a protocol π' that securely computes f in the presence of malicious adaptive adversaries, with erasures.*

The theorem statement hides the fact that the transformation of π to π' is highly efficient and thus the boosting of the security guarantee from static to adaptive adversaries is obtained at almost no cost. Before concluding, we stress again that this result *only holds in a model assuming that honest parties can safely erase data*. This is due to the fact that in the non-erasure model the PEC requirement *is* needed, and so the combination of the results of [3] and [7] yields a protocol that is not useful. (In particular, it is not necessarily secure with PEC and so may not be secure under sequential composition.)

Concurrent composition. The above relates to the stand-alone model (and so, of course, also to sequential composition). What happens when considering concurrent composition? An analogous result cannot be achieved because the proof of [7] relies inherently on the fact that the simulator can *rewind* the adversary. Specifically, [7] prove the equivalence of adaptive and static security in the following way. The adaptive simulator begins by running the static simulator for the case that no party is corrupted. Then, if the adversary corrupts a party (say party P_1), the adaptive simulator rewinds the adversary and begins the simulation from scratch running the static simulator for the case that P_1 is corrupted. The adaptive simulator runs the static simulator multiple times until the adversary asks to corrupt P_1 in the same place as the first time. Since the static simulator assumes that P_1 is corrupted, it can complete the simulation. This is the general idea of the simulation strategy; for more details and the actual proof, see [7]. In any case, since rewinding is an inherent part of the strategy, their proof cannot be used in the setting of concurrent composition (where rewinding simulation strategies do not work).

4   Two-Party Computation for Semi-honest Adversaries

4.1   Adaptively-Secure Oblivious Transfer

We start by observing that in the setting of concurrent composition, the oblivious transfer protocol of [12] is *not* adaptively secure (at least, it is not simulatable without rewinding). In order to see this, recall that this protocol works by the sender P_1 choosing an enhanced trapdoor permutation f with its trapdoor t and sending f to the receiver P_2. Upon input σ, party P_2 then sends P_1 values y_0, y_1 so that it knows the preimage of y_σ but not of $y_{1-\sigma}$. Party P_1 then masks its input bit z_0 with the hard-core bit of $f^{-1}(y_0)$ and masks its input bit z_1 with the hard-core bit of $f^{-1}(y_1)$. The protocol concludes by P_2 extracting z_σ; it can do this because it knows the preimage of y_σ and so can compute the hard-core bit used to mask z_σ. Now, consider an adversarial strategy that waits until P_1 sends its second message and then corrupts the receiver. Following this corruption, the adversary should be able to obtain P_2's state and compute z_σ from P_1's message (the adversary must be able to do this because P_2 must be able to do this). However, when the messages from P_1 are generated by the simulator and no party is corrupted, the simulator cannot know what values of z to place in the message. The simulation will therefore often fail. A similar (and in fact worse) problem appears in the known oblivious transfer protocols that rely on homomorphic encryption.

Our approach to solving this problem is novel and has great advantages. We show that *any* oblivious transfer protocol that is secure for static corruptions can be modified so that with a minor addition adaptive security (with erasures) is obtained. The idea is to run *any* secure oblivious transfer upon random inputs and then use the random values obtained to exchange the actual bit. This method was presented in [26] in order to show a reduction from standard OT to OT with random inputs. Here we use it to obtain adaptive security. We remark that our protocol is exactly that of [26]; our contribution is in observing and proving that it is adaptively secure.

Protocol 1 (oblivious transfer):

- **Inputs:** P_1 *has two strings* $x_0, x_1 \in \{0, 1\}^n$, *and* P_2 *has a bit* $\sigma \in \{0, 1\}$.
- **The protocol:**
  1. P_1 *chooses random strings* $r_0, r_1 \in_R \{0, 1\}^n$ *and* P_2 *chooses a random bit* $b \in_R \{0, 1\}$.
     P_1 *and* P_2 *run an oblivious transfer protocol, using the chosen random inputs. (Note that* P_2*'s output is* r_b*.) At the conclusion of the protocol, P_1 and P_2 erase all of the randomness that they used, and remain only with their inputs and outputs from the subprotocol (i.e., P_1 remains with (r_0, r_1) and P_2 remains with (b, r_b)).*

2. *P_2 sends P_1 the bit $\beta = b \oplus \sigma$.*
3. *P_1 sends P_2 the pair $y_0 = x_0 \oplus r_\beta$ and $y_1 = x_1 \oplus r_{1-\beta}$.*
4. *P_2 outputs $y_\sigma \oplus r_b$.*

Before proving security, we first show that the protocol is correct. If $\sigma = 0$ then $\beta = b$ and so $y_0 = x_0 \oplus r_b$, implying that P_2 outputs $y_0 \oplus r_b = x_0$ as required. Likewise, if $\sigma = 1$ then $\beta = b \oplus 1$ and so $y_1 = x_1 \oplus r_b$, implying that P_2 outputs $y_1 \oplus r_b = x_1$ as required. We have the following theorem.

Theorem 4. *If the oblivious transfer used in Step 1 of Protocol 1 is secure in the presence of semi-honest static adversaries in the stand-alone model, then Protocol 1 is secure under concurrent general composition in the presence of semi-honest adaptive adversaries with erasures.*

Proof: Before beginning the proof, we remark that we cannot analyze the security of the protocol in a hybrid model where the oblivious transfer of Step 1 is run by a trusted party. This is because the oblivious transfer protocol of Step 1 is only secure in the presence of static adversaries, and we are working in the adaptive model. We now proceed with the proof. Intuitively, the protocol is adaptively secure because any corruptions that occur before Step 1 are easily dealt with due to the fact that even the honest parties use random inputs in this stage (and thus inputs that are independent of their real input). Furthermore, any corruptions that take place after Step 1 can be dealt with because the oblivious transfer protocol used in Step 1 is statically secure, and so hides the actual inputs used. Given that the parties erase their internal state after this step, the simulator is able to lie about what "random" inputs the parties actually used.

Let $\mathcal{A}$ be a probabilistic polynomial-time real adversary. We construct a simulator $\mathcal{S}$ for Protocol 1, separately describing its actions for every corruption case (of course, $\mathcal{S}$ doesn't know when corruptions occur so its actions are the same until corruptions happen). Upon auxiliary input z, simulator $\mathcal{S}$ invokes $\mathcal{A}$ upon input z and works as follows:

1. *No corruption, or corruption at the end:* $\mathcal{S}$ begins by choosing random r_0, r_1 and b and playing the honest parties in the oblivious transfer protocol of Step 1. Then, $\mathcal{S}$ simulates P_2 sending a random β to P_1, and P_1 replying with two random strings (y_0, y_1).

   If $\mathcal{A}$ carries out corruptions following the execution, then $\mathcal{S}$ acts as follows, according to the case:

   (a) *Corruption of P_1 first:* $\mathcal{S}$ corrupts P_1 and obtains its input pair (x_0, x_1). Then, $\mathcal{S}$ sets $r_\beta = x_0 \oplus y_0$ and $r_{1-\beta} = x_1 \oplus y_1$, where β, y_0, y_1 are as above (and the values r_0, r_1 chosen in the simulation of the oblivious transfer subprotocol are ignored). $\mathcal{S}$ generates the view of P_1 based on this (r_0, r_1).

   If $\mathcal{A}$ corrupts P_2 following this, then $\mathcal{S}$ corrupts P_2 and obtains its input bit σ. Then, $\mathcal{S}$ sets the value b (that P_2 supposedly used in its input to the OT subprotocol) to be $\beta \oplus \sigma$. $\mathcal{S}$ generates the view of P_2 based on it using input b to the OT and receiving output r_b, where the value of r_b is as fixed after the corruption of P_1.

(b) *Corruption of P_2 first:* $\mathcal{S}$ corrupts P_2 and obtains its input bit σ together with its output string x_σ. Then, $\mathcal{S}$ sets $b = \sigma \oplus \beta$ and $r_b = x_\sigma \oplus y_\sigma$, where β is the value set above and likewise y_σ is from the pair (y_0, y_1) above. $\mathcal{S}$ then generates the view of P_2 based on its input to the OT being b and its output being r_b.

   If $\mathcal{A}$ corrupts P_1 follows this, then $\mathcal{S}$ corrupts P_1 and obtains its input pair (x_0, x_1) (note that x_σ was already obtained). Then, $\mathcal{S}$ sets $r_{1-b} = x_{1-\sigma} \oplus y_{1-\sigma}$ and generates the view of P_1 such that its input to the OT subprotocol was (r_0, r_1) as generated upon the corruption of P_2 and the later corruption of P_1.

2. *Corruption of both P_1 and P_2 at any point until Step 1 concludes:* $\mathcal{S}$ begins by emulating the OT subprotocol with random (r_0, r_1) and b as described above. Then, upon corruption of party P_i, simulator $\mathcal{S}$ corrupts P_i and obtains its input. It can then just hand $\mathcal{A}$ the input of P_i together with its view in the emulated subprotocol using the inputs (r_0, r_1) and b.

3. *Corruption of P_1 up until Step 1 concludes and P_2 after it concludes:* The corruption of P_1 is dealt with exactly as in the previous case. We remark that once P_1 is corrupted, $\mathcal{S}$ continues to play P_2 while interacting with $\mathcal{A}$ controlling P_1 as if in a real execution (and using the random input b that was chosen). After the OT subprotocol concludes, $\mathcal{S}$ simulates P_2 sending a random β to P_1, and obtains back a pair (y_0, y_1) from $\mathcal{A}$ who controls P_1.[3] If $\mathcal{A}$ corrupts P_2 at this point, then $\mathcal{S}$ corrupts P_2 and obtains σ. $\mathcal{S}$ sets P_2's input in the OT subprotocol to be $b = \sigma \oplus \beta$ and generates the view accordingly.

4. *Corruption of P_2 up until Step 1 concludes and P_1 after it concludes:* The corruption of P_2 is dealt with exactly as in the corruption case in item 2 above. As previously, once P_2 is corrupted $\mathcal{S}$ continues to play P_1 while interacting with $\mathcal{A}$ controlling P_2 as if in a real execution (and using the random input (r_0, r_1) that was chosen). Let σ be P_1's input and let x_σ be its output. After the OT subprotocol concludes, $\mathcal{S}$ obtains a random bit β from $\mathcal{A}$ controlling P_2 and sets $y_\sigma = x_\sigma \oplus r_b$ where b is the input used by P_2 in the OT subprotocol (whether corrupted or not) and r_b is from above. Furthermore, $\mathcal{S}$ chooses a random $y_{1-\sigma} \in_R \{0,1\}^n$, and simulates P_1 sending (y_0, y_1) to P_2.

   If $\mathcal{A}$ corrupts P_1 at this point (or before (y_0, y_1) were sent – but it makes no difference), then $\mathcal{S}$ corrupts P_1 and obtains (x_0, x_1). It then redefines

---

[3] Note that if $\mathcal{A}$ was malicious and not semi-honest, then the simulation at this point would not work because $\mathcal{S}$ cannot know which inputs $\mathcal{A}$ used in the oblivious transfer (this is due to the fact that the corruption occurred in the middle of the oblivious transfer and the static simulator may not necessarily be able to deal with this). For this reason, we have only been able to prove our transformation for the semi-honest model.

the value of r_{1-b} to be $y_{1-\sigma} \oplus x_{1-\sigma}$, and generates the view based on these values.

This covers all corruption cases. We now proceed to prove that

$$\left\{\text{IDEAL}_{OT,\mathcal{S}(z)}(n, x_0, x_1, \sigma)\right\}_{x_0,x_1,\sigma,z;n\in\mathbb{N}} \overset{c}{\equiv} \left\{\text{REAL}_{\pi,\mathcal{A}(z)}(n, x_0, x_1, \sigma)\right\}_{x_0,x_1,\sigma,z;n\in\mathbb{N}}$$

We present our analysis following the case-by-case description of the simulator:

1. *No corruption, or corruption at the end:* In order to prove this corruption case, we begin by showing that when no parties are corrupted, every oblivious transfer protocol (that is secure for the *static* corruption model) has the following property. Let $\mathcal{A}$ be a probabilistic polynomial-time adversary that corrupts no parties, and only eavesdrops on the communication in the protocol. Then, for all strings $r_0, r_1, r_0', r_1' \in \{0,1\}^n$ and every probabilistic polynomial-time distinguisher D:

$$|\Pr[D(\text{REAL}_{\pi,\mathcal{A}(z)}(n, r_0, r_1, 0)) = 1] - \Pr[D(\text{REAL}_{\pi,\mathcal{A}(z)}(n, r_0', r_1', 1)) = 1]| \leq \mathsf{negl}(n) \tag{1}$$

for some negligible function negl. The above follows from the following three equations (all equations relate to the same quantification as above over all strings and all distinguishers):

$$|\Pr[D(\text{REAL}_{\pi,\mathcal{A}(z)}(n, r_0, r_1, 0)) = 1] - \Pr[D(\text{REAL}_{\pi,\mathcal{A}(z)}(n, r_0, r_1', 0)) = 1]| \leq \mathsf{negl}(n)$$

This holds because otherwise an adversary corrupting the receiver P_2 could learn something about the second string of P_1's input, even though it used input 0 and so received r_0. (This would contradict the security of the protocol in the ideal model; the formal statement of this is straightforward and so omitted.) Next,

$$|\Pr[D(\text{REAL}_{\pi,\mathcal{A}(z)}(n, r_0, r_1', 0)) = 1] - \Pr[D(\text{REAL}_{\pi,\mathcal{A}(z)}(n, r_0, r_1', 1)) = 1]| \leq \mathsf{negl}(n)$$

This second equation holds because otherwise an adversary corrupting the sender P_1 could distinguish the case that the receiver P_2 has input 0 or 1. Finally,

$$|\Pr[D(\text{REAL}_{\pi,\mathcal{A}(z)}(n, r_0, r_1', 1)) = 1] - \Pr[D(\text{REAL}_{\pi,\mathcal{A}(z)}(n, r_0', r_1', 1)) = 1]| \leq \mathsf{negl}(n)$$

As with the first equation, this holds because otherwise an adversary corrupting P_2 can learn something about the first string of P_1's input even though it used input 1. Combining the above three together, Eq. (1) follows. Now, let r_0', r_1', b' be the values used by $\mathcal{S}$ to simulate the oblivious transfer in Step 1 of the protocol, and let y_0, y_1, β be the random values sent by $\mathcal{S}$ in the later steps. In addition, let x_0, x_1, σ be the real parties' inputs that are received by $\mathcal{S}$ upon corruption of both P_1 and P_2. As in the simulation description, we separately analyze the case that no parties are corrupted, the case that P_1 was corrupted first and the case that P_2 was corrupted first.

   (a) *No corruptions:* In the case of no corruptions, all the adversary sees is the oblivious transfer transcript, a random bit β and two random strings y_0, y_1. Let x_0, x_1, σ be the real inputs of the honest parties. Then, the values y_0, y_1, β seen by the adversary are "correct" if the inputs used in the oblivious transfer are $r'_\beta = x_0 \oplus y_0$ and $r'_{1-\beta} = x_1 \oplus y_1$ and $b' = \beta \oplus \sigma$. However, $\mathcal{S}$ used inputs r_0, r_1, b and not these r'_0, r'_1, b'. Nevertheless, Eq. (1) guarantees that the distribution over the transcript generated using r_0, r_1, b (as in the simulation) is computationally indistinguishable from the distribution over the transcript generated using r'_0, r'_1, b' (as in a real execution). Thus, indistinguishability holds for this case.

(b) *Corruption of P_1 first:* As described in the simulation, $\mathcal{S}$ sets $r_\beta = x_0 \oplus y_0$ and $r_{1-\beta} = x_1 \oplus y_1$. Then, $\mathcal{S}$ sets $b = \beta \oplus \sigma$. If $\mathcal{S}$ had used r_0, r_1, b as defined here in the simulation of Step 1 of the protocol, then the simulation would be perfect (because all of the values are constructed exactly as the honest parties would construct them upon inputs x_0, x_1, σ). Thus, using Eq. (1), we have that the distributions are computationally indistinguishable. (Recall that Eq. (1) refers to the transcript generated when no parties are corrupted. However, this is exactly the case here because the corruptions occur *after* the subprotocol has concluded. Furthermore, because the parties erase their internal state, the only information about the subprotocol that $\mathcal{A}$ receives is the transcript of messages sent, as required.)

(c) *Corruption of P_2 first:* The proof of this is almost identical to the case where P_1 is corrupted first.

2. *Corruption of both P_1 and P_2 at any point until Step 1 concludes:* This case is trivial because in Step 1 both $\mathcal{S}$ and the honest parties use random inputs that are independent of the inputs. Thus the distribution generated by $\mathcal{S}$ is identical as in a real execution.

3. *Corruption of P_1 until Step 1 concludes and P_2 after it concludes:* The distribution of the view of P_1 generated by $\mathcal{S}$ is identical to a real execution, because as in the previous case, the inputs used until the end of Step 1 are random and independent of the party's actual input. Regarding P_2's view, indistinguishability follows from the fact that for any oblivious transfer protocol (that is secure for static corruptions), it holds that when $\mathcal{A}$ has corrupted P_1 only, we have that for all $r_0, r_1 \in \{0, 1\}^n$

$$\left\{ \text{REAL}_{\pi, \mathcal{A}(z)}(n, r_0, r_1, 0) \right\}_{n \in \mathbb{N}} \stackrel{c}{\equiv} \left\{ \text{REAL}_{\pi, \mathcal{A}(z)}(n, r_0, r_1, 1) \right\}_{n \in \mathbb{N}}$$

This follows similarly to Eq. (1) because in the ideal model, a corrupted P_1 cannot know if P_2 has input 0 or 1.

4. *Corruption of P_2 until Step 1 concludes and P_1 after it concludes:* The proof of this case is almost identical to the previous case.

This completes the proof of the theorem. ∎

4.2   The Two-Party Protocol for Semi-honest Adversaries

We present a protocol for securely computing any functionality f that maps two n-bit inputs into an n-bit output. It is possible to generalize the construction to functions for which the input and output lengths vary. However, the security of our protocol relies crucially on the fact that the length of the output of f equals the length of the second input y. (This is because our simulation works by generating a garbled circuit computing $f(x, y) = y$ which must be indistinguishable from a garbled circuit computing $C(x, y)$. Such indistinguishability can only hold if $|C(x, y)| = |y|$.) This can be achieved w.l.o.g. by padding the length of P_2's input y with zeroes. Our description assumes familiarity with Yao's garbled circuit construction; see Appendix A for a full description. Observe that we consider only a "same-output functionality", meaning that both P_1 and P_2 receive the same output value $f(x, y)$. In [22], this was shown to be without loss of generality: given any protocol as we describe here it is possible to construct a protocol where the parties have different outputs with very little additional overhead.

Convention. We require that the circuit C used to compute f is of a given structure. Technically, what we need is that the same structure of the circuit (i.e., the positions of the wires connecting the gates) can be used to compute the function $f'(x, y) = y$ (of course, the actual gates would be different, but this is fine). This can be achieved without difficulty, and so we will not elaborate further.

Protocol 2

– **Inputs:** P_1 *has* $x \in \{0, 1\}^n$ *and* P_2 *has* $y \in \{0, 1\}^n$
– **Auxiliary input:** *A boolean circuit* C *such that for every* $x, y \in \{0, 1\}^n$ *it holds that* $C(x, y) = f(x, y)$, *where* $f : \{0, 1\}^n \times \{0, 1\}^n \to \{0, 1\}^n$.[4]
– **The protocol:**
  1. P_1 *constructs the garbled circuit* $G(C)$ *as described in Appendix A, but does not yet send it to* P_2.
  2. *Let* $w_1, \ldots, w_n$ *be the circuit-input wires corresponding to* x, *and let* $w_{n+1}, \ldots, w_{2n}$ *be the circuit-input wires corresponding to* y. *Then,*
     (a) P_1 *sends* P_2 *the strings* $k_1^{x_1}, \ldots, k_n^{x_n}$.
     (b) *For every* i, P_1 *and* P_2 *execute a 1-out-of-2 oblivious transfer protocol that is adaptively secure for semi-honest adversaries, in which* P_1*'s input equals* (k_{n+i}^0, k_{n+i}^1) *and* P_2*'s input equals* y_i.
     The above oblivious transfers can all be run in parallel.
  3. *After the previous step is completed,* P_1 *erases all of the randomness it used to construct the garbled circuit (and in particular, erases all of the secret keys). Following this erasure,* P_1 *sends* P_2 *the garbled circuit* $G(C)$.
  4. *Following the above,* P_2 *has obtained the garbled circuit and* $2n$ *keys corresponding to the* $2n$ *input wires to* C. *Party* P_2 *then computes the circuit, as described in Appendix A, obtaining* $f(x, y)$. P_2 *then sends* $f(x, y)$ *to* P_1 *and they both output this value.*

---

[4] As in [22], we require that C is such that if a circuit-output wire leaves some gate g, then gate g has no other wires leading from it into other gates (i.e., no circuit-output wire is also a gate-input wire). Likewise, a circuit-input wire that is also a circuit-output wire enters no gates.

The only differences between Protocol 2 and Yao's protocol as it appears in [22] are:

1. P_1 does not send $G(C)$ to P_2 until the oblivious transfers have concluded.
2. P_1 erases all of its internal state before it actually sends $G(C)$.
3. The oblivious transfers must be secure in the presence of *adaptive* semi-honest adversaries.

The above differences make no difference whatsoever to the proof of security in the static case. However, the simulator provided in [22] does not work in the case of adaptive corruptions. In order to see this, recall that the simulator there works by constructing a special fake circuit that outputs a predetermined value. In the setting of static security this suffices because the simulator is given the output value before it begins the simulation; it can therefore set the pre-determined output value of the fake circuit to the party's output. However, in the setting of adaptive corruptions, the simulator may need to generate a fake garbled circuit before it knows any of the party's outputs (in particular, this happens if corruptions occur only at the end of the execution). It therefore does not know the circuit's output when it generates it. One way to overcome this problem is to use an equivocal encryption scheme that can be opened to any value. However, this raises a whole other set of problems, and would also be far less efficient. In our proof below we show that, in fact, the construction need not be modified at all. Rather, the simulator can construct a fake garbled circuit that computes the function $f(x, y) = y$ (instead of a fake circuit that outputs a predetermined value). This means that by choosing appropriate keys for the in-put wires associated with y (i.e., those associated with P_2's input), it is possible for the simulator to cause the circuit to output any value that it wishes. More specifically, after constructing such a fake garbled circuit, when the simulator receives an output value $z = f(x, y)$ it can simply choose the keys to the input wires associated with P_2's input to be those that are associated with the *output value* z. This will then result in the circuit being opened to z, as required. Our proof shows that such a circuit is computationally indistinguishable from a real garbled circuit, and so the simulation works.

In the full version of this work [21], we use this simulation strategy to formally prove the following theorem:

Theorem 5. *Let f be a deterministic same-output functionality. Furthermore, assume that the oblivious transfer protocol is secure in the presence of adaptive semi-honest adversaries (with erasures), and that the encryption scheme has indistinguishable encryptions under chosen plaintext attacks, and has an elusive and efficiently verifiable range. Then, Protocol 2 securely computes f in the presence of adaptive semi-honest adversaries (with erasures).*

Since, as we have shown, adaptively secure oblivious transfer (in the erasure model) can be achieved assuming only the existence of enhanced trapdoor per-mutations, Theorem 5 implies Theorem 2 as stated in the introduction.

References

1. Beaver, D.: Foundations of secure interactive computing. In: Feigenbaum, J. (ed.) CRYPTO 1991. LNCS, vol. 576, pp. 377–391. Springer, Heidelberg (1992)
2. Beaver, D.: Plug and play encryption. In: Kaliski Jr., B.S. (ed.) CRYPTO 1997. LNCS, vol. 1294, pp. 75–89. Springer, Heidelberg (1997)
3. Beaver, D., Haber, S.: Cryptographic protocols provably secure against dynamic adversaries. In: Rueppel, R.A. (ed.) EUROCRYPT 1992. LNCS, vol. 658, pp. 307–323. Springer, Heidelberg (1993)
4. Ben-Or, M., Goldwasser, S., Wigderson, A.: Completeness Theorems for Non-Cryptographic Fault-Tolerant Distributed Computation. In: 20th STOC, pp. 1–10 (1988)
5. Canetti, R.: Security and Composition of Multiparty Cryptographic Protocols. Journal of Cryptology 13(1), 143–202 (2000)
6. Canetti, R.: Universally Composable Security: A New Paradigm for Cryptographic Protocols. In: 42nd FOCS, pp. 136–145 (2001), http://eprint.iacr.org/2000/067
7. Canetti, R., Damgård, I., Dziembowski, S., Ishai, Y., Malkin, T.: Adaptive versus Non-Adaptive Security of Multi-Party Protocols. Journal of Cryptology 17(3), 153–207 (2004)
8. Canetti, R., Feige, U., Goldreich, O., Naor, M.: Adaptively Secure Multi-Party Computation. In: 28th STOC, pp. 639–648 (1996)
9. Canetti, R., Gennaro, R., Jarecki, S., Krawczyk, H., Rabin, T.: Adaptive security for threshold cryptosystems. In: Wiener, M. (ed.) CRYPTO 1999. LNCS, vol. 1666, pp. 98–115. Springer, Heidelberg (1999)
10. Canetti, R., Lindell, Y., Ostrovsky, R., Sahai, A.: Universally Composable Two-Party and Multi-Party Computation. In: 34th STOC, pp. 494–503 (2002), http://eprint.iacr.org/2002/140
11. Chaum, D., Crépeau, C., Damgå, I.: rd. Multi-party Unconditionally Secure Protocols. In: 20th STOC, pp. 11–19 (1988)
12. Even, S., Goldreich, O., Lempel, A.: A Randomized Protocol for Signing Contracts. Communications of the ACM 28(6), 637–647 (1985)
13. Frankel, Y., MacKenzie, P.D., Yung, M.: Adaptively-secure optimal-resilience proactive RSA. In: Lam, K.-Y., Okamoto, E., Xing, C. (eds.) ASIACRYPT 1999. LNCS, vol. 1716, pp. 180–195. Springer, Heidelberg (1999)
14. Garay, J.A., MacKenzie, P.D., Yang, K.: Efficient and Universally Composable Committed Oblivious Transfer and Applications. In: Naor, M. (ed.) TCC 2004. LNCS, vol. 2951, pp. 297–316. Springer, Heidelberg (2004)
15. Goldreich, O.: Foundations of Cryptography. Basic Applications, vol. 2. Cambridge University Press, Cambridge (2004)
16. Goldreich, O., Micali, S., Wigderson, A.: How to Play any Mental Game – A Completeness Theorem for Protocols with Honest Majority. In: 19th STOC, pp. 218–229 (1987); for details see [15]
17. Goldwasser, S., Levin, L.A.: Fair computation of general functions in presence of immoral majority. In: Menezes, A., Vanstone, S.A. (eds.) CRYPTO 1990. LNCS, vol. 537, pp. 77–93. Springer, Heidelberg (1991)
18. Kushilevitz, E., Lindell, Y., Rabin, T.: Information-Theoretically Secure Protocols and Security Under Composition. In: The em 38th STOC, pp. 109–118 (2006)
19. Jarecki, S., Lysyanskaya, A.: Adaptively secure threshold cryptography: Introducing concurrency, removing erasures (Extended abstract). In: Preneel, B. (ed.) EUROCRYPT 2000. LNCS, vol. 1807, pp. 221–242. Springer, Heidelberg (2000)

20. Lindell, Y.: General Composition and Universal Composability in Secure Multi-Party Computation. In: 44th FOCS, pp. 394–403 (2003)
21. Lindell, Y.: Adaptively Secure Two-Party Computation with Erasures (full version of this paper). Cryptology ePrint Archive (2009)
22. Lindell, Y., Pinkas, B.: A Proof of Security of Yao's Protocol for Two-Party Computation. the Journal of Cryptology (to appear)
23. Lindell, Y., Pinkas, B.: An efficient protocol for secure two-party computation in the presence of malicious adversaries. In: Naor, M. (ed.) EUROCRYPT 2007. LNCS, vol. 4515, pp. 52–78. Springer, Heidelberg (2007)
24. Lindell, Y., Zarosim, H.: Adaptive Zero-Knowledge Proofs and Adaptively Secure Oblivious Transfer. In: Reingold, O. (ed.) TCC 2009. LNCS, vol. 5444, pp. 183–201. Springer, Heidelberg (2009)
25. Micali, S., Rogaway, P.: Secure computation. In: Feigenbaum, J. (ed.) CRYPTO 1991. LNCS, vol. 576, pp. 392–404. Springer, Heidelberg (1992)
26. Wolf, S., Wullschleger, J.: Oblivious transfer is symmetric. In: Vaudenay, S. (ed.) EUROCRYPT 2006. LNCS, vol. 4004, pp. 222–232. Springer, Heidelberg (2006)
27. Yao, A.: How to Generate and Exchange Secrets. In: 27th FOCS, pp. 162–167 (1986)

A   Yao's Garbled Circuit Construction

In this section, we describe Yao's protocol for secure two-party computation in the presence of semi-honest adversaries [27]. The specific construction described here is from [22], where a full proof of security is also provided.

Let C be a Boolean circuit that receives two inputs $x, y \in \{0, 1\}^n$ and outputs $C(x, y) \in \{0, 1\}^n$ (for simplicity, we assume that the input length, output length and the security parameter are all of the same length n). We also assume that C has the property that if a circuit-output wire comes from a gate g, then gate g has no wires that are input to other gates.[5] (Likewise, if a circuit-input wire is itself also a circuit-output, then it is not input into any gate.)

We begin by describing the construction of a single garbled gate g in C. The circuit C is Boolean, and therefore any gate is represented by a function $g : \{0, 1\} \times \{0, 1\} \to \{0, 1\}$. Now, let the two input wires to g be labelled w_1 and w_2, and let the output wire from g be labelled w_3. Furthermore, let $k_1^0, k_1^1, k_2^0, k_2^1, k_3^0, k_3^1$ be six keys obtained by independently invoking the key-generation algorithm $G(1^n)$; for simplicity, assume that these keys are also of length n. Intuitively, we wish to be able to compute $k_3^{g(\alpha,\beta)}$ from k_1^α and k_2^β, without revealing any of the other three values $k_3^{g(1-\alpha,\beta)}, k_3^{g(\alpha,1-\beta)}, k_3^{g(1-\alpha,1-\beta)}$. The gate g is defined by the following four values

$$c_{0,0} = E_{k_1^0}(E_{k_2^0}(k_3^{g(0,0)}))$$

$$c_{0,1} = E_{k_1^0}(E_{k_2^1}(k_3^{g(0,1)}))$$

---

[5] This requirement is due to our labelling of gates described below, that does not provide a unique label to each wire (see [22] for more discussion). We note that this assumption on C increases the number of gates by at most n.

$$c_{1,0} = E_{k_1^1}(E_{k_2^0}(k_3^{g(1,0)}))$$

$$c_{1,1} = E_{k_1^1}(E_{k_2^1}(k_3^{g(1,1)}))$$

where E is from a private key encryption scheme (G, E, D) that has indistinguishable encryptions for multiple messages, and has an elusive efficiently verifiable range; see [22]. The actual gate is defined by a *random permutation* of the above values, denoted as c_0, c_1, c_2, c_3; from here on we call them the **garbled table** of gate g. Notice that given k_1^α and k_2^β, and the values c_0, c_1, c_2, c_3, it is possible to compute the output of the gate $k_3^{g(\alpha,\beta)}$ as follows. For every i, compute $D_{k_2^\beta}(D_{k_1^\alpha}(c_i))$. If more than one decryption returns a non-$\bot$ value, then output **abort**. Otherwise, define k_3^γ to be the only non-$\bot$ value that is obtained. (Notice that if only a single non-$\bot$ value is obtained, then this will be $k_3^{g(\alpha,\beta)}$ because it is encrypted under the given keys k_1^α and k_2^β. Later we will show that except with negligible probability, only one non-$\bot$ value is indeed obtained.)

We are now ready to show how to construct the entire garbled circuit. Let m be the number of *wires* in the circuit C, and let $w_1, \ldots, w_m$ be labels of these wires. These labels are all chosen uniquely with the following exception: if w_i and w_j are both output wires from the same gate g, then $w_i = w_j$ (this occurs if the fan-out of g is greater than one). Likewise, if an input bit enters more than one gate, then all circuit-input wires associated with this bit will have the same label. Next, for every label w_i, choose two independent keys $k_i^0, k_i^1 \leftarrow G(1^n)$; we stress that all of these keys are chosen independently of the others. Now, given these keys, the four garbled values of each gate are computed as described above and the results are permuted randomly. Finally, the output or decryption tables of the garbled circuit are computed. These tables simply consist of the values $(0, k_i^0)$ and $(1, k_i^1)$ where w_i is a *circuit-output wire*. (Alternatively, output gates can just compute 0 or 1 directly. That is, in an output gate, one can define $c_{\alpha,\beta} = E_{k_1^\alpha}(E_{k_2^\beta}(g(\alpha,\beta)))$ for every $\alpha, \beta \in \{0,1\}$.)

The entire garbled circuit of C, denoted $G(C)$, consists of the garbled table for each gate and the output tables. We note that the structure of C is given, and the garbled version of C is simply defined by specifying the output tables and the garbled table that belongs to each gate. This completes the description of the garbled circuit.

Let $x = x_1 \cdots x_n$ and $y = y_1 \cdots y_n$ be two n-bit inputs for C. Furthermore, let $w_1, \ldots, w_n$ be the input labels corresponding to x, and let $w_{n+1}, \ldots, w_{2n}$ be the input labels corresponding to y. It is shown in [22] that given the garbled circuit $G(C)$ and the strings $k_1^{x_1}, \ldots, k_n^{x_n}, k_{n+1}^{y_1}, \ldots, k_{2n}^{y_n}$, it is possible to compute $C(x, y)$, except with negligible probability.

Short Redactable Signatures Using Random Trees[*]

Ee-Chien Chang, Chee Liang Lim, and Jia Xu

School of Computing
National University of Singapore

Abstract. A redactable signature scheme for a string of objects supports verification even if multiple substrings are removed from the original string. It is important that the redacted string and its signature do not reveal anything about the content of the removed substrings. Existing schemes completely or partially leak a piece of information: the lengths of the removed substrings. Such length information could be crucial in many applications, especially when the removed substring has low entropy. We propose a scheme that can hide the length. Our scheme consists of two components. The first component $\mathcal{H}$, which is a "collision resistant" hash, maps a string to an unordered set, whereby existing schemes on unordered sets can then be applied. However, a sequence of random numbers has to be explicitly stored and thus it produces a large signature of size at least (mk)-bits where m is the number of objects and k is the size of a key sufficiently large for cryptographic operations. The second component uses RGGM tree, a variant of GGM tree, to generate the pseudo random numbers from a short seed, expected to be of size $O(k + tk \log m)$ where t is the number of removed substrings. Unlike GGM tree, the structure of the proposed RGGM tree is random. By an intriguing statistical property of the random tree, the redacted tree does not reveal the lengths of the substrings removed. The hash function $\mathcal{H}$ and the RGGM tree can be of independent interests.

Keyword: Redactable Signature Scheme, Random Tree, Privacy.

1 Introduction

We are interested in a signature scheme for strings of objects whereby their authenticity can be verified even if some substrings have been removed, that is, the strings are redacted. Let $\mathbf{x} = x_1 x_2 \ldots x_m$ be a string, for example a text document where each object can be a character or a word, or an audio file where each object is a sample. The string $\mathbf{x}$ is signed by the authority and both $\mathbf{x}$ and its signature $\mathbf{s}$ are passed to another party, say Alice. Alice wants to show Bob $\mathbf{x}$ but Bob is not authorized to view certain parts of the string, say $x_2 x_3 x_4$ and x_7. Thus, Alice shows Bob $\widetilde{\mathbf{x}} = x_1 \diamond x_5 x_6 \diamond x_8 \ldots x_m$ where each $\diamond$ indicates the location of a removed substring. On the other hand, Bob may want to verify the authenticity of $\widetilde{\mathbf{x}}$. A redactable signature scheme allows Alice to produce a valid

---

[*] Extended version [5] of this paper is available in Cryptology ePrint Archive.

M. Fischlin (Ed.): CT-RSA 2009, LNCS 5473, pp. 133–147, 2009.
© Springer-Verlag Berlin Heidelberg 2009

signature $\widetilde{\mathbf{s}}$ for the redacted string $\widetilde{\mathbf{x}}$, even if Alice does not have the authority's secret key. From the new signature $\widetilde{\mathbf{s}}$, Bob can then verify that $\widetilde{\mathbf{x}}$ is indeed a redacted version of a string signed by the authority.

Unlike the usual signature schemes, redactable signature scheme has additional requirement on privacy: information of the removed strings should be hidden. In this paper, we consider the stringent requirement that, Bob could not obtain any information of any removed substring, except the fact that a non-empty substring has been removed at each location ⋄. This simple requirement turns out to be difficult to achieve. Existing schemes are unable to completely hide a piece of information: the length of each removed substring. Note that information on length could be crucial if the substring has low entropy. For example, if the substring is either "Approved" or "Not Approved", then its length reveals everything. The redactable signature scheme proposed by Johnson et al. [9] employs a Merkle tree [11] and a GGM tree [7] to generate a short signature. However, it is easy to derive the length from the structures of the redacted Merkle and GGM trees. A straightforward modification by introducing randomness into the tree structure also does not hide the length completely. Schemes by Johnson et al. [9] (set-homomorphic signatures) and Miyazaki et al. [13] are designed for unordered sets and are not applicable for a string. A way to extend their schemes to strings is by assigning a sequence of increasing random numbers to the objects [13]. However, this leads to large signatures since the random numbers have to be explicitly stored, and more importantly, it is insecure since the gaps in the sequence reveal some information about the number of removed objects.

Note that the type of information to be removed varies for different applications. There are applications where the lengths of the removed strings should not be hidden. As noted by Johnson et al. [9], semantic attack could be possible in some scenarios if the length information is hidden. On the other hand, there are also applications where not only the substrings have to be completely purged, the fact that a string has been redacted must be hidden. Our scheme can be modified to cater for the above two scenarios.

In this paper, we propose a scheme that can hide the lengths of the removed substrings. Our scheme incorporates two components: a hash, and a random tree with a hiding property. We first give a scheme $\mathcal{RSS}$ using the first component, and then another scheme $\mathcal{SRSS}$ with both components. The first component hashes a string of objects to an unordered set. For the unordered set, existing redactable schemes [13,9] on unordered sets can be applied. The scheme $\mathcal{RSS}$ satisfies the requirements on unforgeability and privacy preserving under reasonable cryptographic assumptions. However, it produces a large signature. Essentially, the main portion of the signature is a sequence of random numbers $\langle r_1, r_2, \ldots, r_m \rangle$, where each r_i is associated with the i-th object in the string.

The goal of the second component is to reduce the signature size by generating the r_i's from a small seed t. If a substring is removed, the corresponding random numbers have to be removed accordingly. Thus, a straightforward method of generating the random numbers iteratively starting from the seed violates privacy, since the seed t reveals all the random numbers.

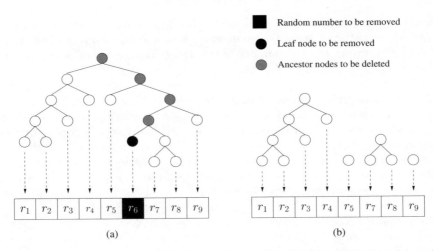

Fig. 1. Redacting the tree in (a) by removing r_6, gives rise to the redacted tree (b)

We employ a variant of GGM binary tree to generate the r_i's in a top-down manner, where the r_i's are at the leaves, and the seed t is at the root. Unlike the GGM tree which is balanced, we use a random binary tree where the *structure* of the binary tree is random. After a substring is removed, the associated leaves and all their ancestors are to be removed, resulting in a collection of subtrees (Figure 1). The roots of the subtrees collectively form the new seed $\tilde{t}$ for the redacted r_i's. Note that from the structures of the subtrees, an adversary might still derive some information of the length of a removed substring. Our main observation is that, by choosing an appropriate tree generation algorithm, the structure of the subtrees reveals nothing about the size of the original tree. Consider a game between- Alice and Bob. Suppose Alice randomly picks a binary tree and it is equal likely that the tree contains 1000 leaves or 9 leaves. Now Alice redacts the tree by removing one substring and only 8 leaves are left. From the structure of the remaining subtrees (for example Figure 1(b)), Bob tries to guess the size of the original tree. Now, if Alice employs a tree generation algorithm with the hiding property, Bob cannot succeed with probability more than 0.5. This hiding property is rather counter-intuitive. Since the size of the tree is involved in the tree generation and thus intuitively the information about the size of the tree is spread throughout the tree. It is quite surprising that the global information on size can be completely removed by deleting some nodes.

Contribution and Organization

1. We propose a "collision resistant" hash $\mathcal{H}$ that maps strings to unordered sets. From $\mathcal{H}$ we obtain $\mathcal{RSS}$, a redactable signature scheme for strings. Unlike previously known methods, $\mathcal{RSS}$ is able to hide the lengths of the removed substrings. We show that $\mathcal{RSS}$ is secure against chosen message attack (Theorem 2) and privacy preserving (Theorem 3) under assumptions weaker than the random oracle assumption. However, the signature size is large. It consists

of $km + kt + \kappa$ bits, where κ is the size of the signature produced by a known redactable signature scheme for unordered sets, m is the number of objects in the redacted string, t is the number of substrings removed, and k is a security parameter (e.g. $k = 1024$).

2. We observe a hiding property of a random tree (Theorem 4). Based on the observation, we propose RGGM, a pseudo random number generator which can be viewed as a randomized version of GGM [7]. If multiple substrings of pseudo random numbers are to be removed, we can efficiently find a new seed that generates the retained numbers, and yet it is computationally difficult to derive the content and length of each removed substring from the new seed, except the locations of the removed substrings.

3. We propose $\mathcal{SRSS}$ by incorporating RGGM into $\mathcal{RSS}$. The expected size of the signature is in $\kappa + O(k + kt \log m)$. $\mathcal{SRSS}$ is secure against chosen message attack (Corollary 5) and privacy preserving (Corollary 6).

2   Related Work

Johnson et al. [9] introduced redactable signature schemes which enable verification of a redacted signed document. Signature scheme with similar property has also been proposed for XML documents [15], where the redaction operation is to remove XML nodes. Redactable signatures are examples of homomorphic signatures which are introduced by Rivest in his talks on "Two New Signature Schemes" [14] and formalized by Johnson et al. [9]. Micali et al. [12] gave a transitive signature scheme as the first construction of homomorphic signatures. They also asked for other possible "signature algebras". The notions on homomorphic signatures can be traced back to incremental cryptography, introduced by Bellare, Goldreich and Goldwasser [3,4]. Recently, Ateniese et al. [2] introduced sanitizable signature scheme [10,8,16,13] allowing a semi-trusted censor modifies the signed documents in a limited and controlled way.

The redactable signature scheme on strings is closely related to directed transitive signature scheme [12,17]. It is possible to convert a directed transitive signature scheme to a redactable signature scheme on strings. However, existing directed transitive signature schemes do not provide privacy in the sense that the resulting signatures reveal some information about the removed substrings.

There are extensive works on random tree. Aldous [1] considered random trees satisfying this consistency property: removing a random leaf from $\mathcal{R}(k)$ gives $\mathcal{R}(k-1)$, where $\mathcal{R}(k)$ is a random tree with k leaves. Thus, given a tree with k leaves, it can be originated from a tree with $k+t$ leaves, and then with t randomly chosen leaves removed, for any t. This consistency property is similar to the hiding property we seek. Unfortunately, it cannot be applied in our problem, since the leaves to be removed are not randomly chosen.

3   Formulation and Background

Johnson et al.[9] gave definitions on homomorphic signature schemes and their security for binary operators. The next two definitions (Definition 1 & 2) are based on the notations by Johnson et al.[9].

A string is a sequence of objects from an object space (or alphabet) $\mathbb{O}$. For example, $\mathbb{O}$ can be the set of ASCII characters, collection of words, or audio samples, etc. We assume that the first and last object in $\mathbf{x}$ can not be removed. This assumption can be easily met by putting a special symbol at the front and back of the string. After a few substrings are removed from $\mathbf{x}$, the string $\mathbf{x}$ may break into substrings, say $\mathbf{x}_1, \mathbf{x}_2, \ldots, \mathbf{x}_u$. The redacted string $(\widetilde{\mathbf{x}}, \mathbf{e})$, which we call *annotated string*[1], is represented by the string $\widetilde{\mathbf{x}} = \mathbf{x}_1 \| \mathbf{x}_2 \| \ldots \| \mathbf{x}_u$ and an *annotation* $\mathbf{e} = \langle m, b_1, b_2, \ldots, b_v \rangle$ where $\|$ denotes concatenation, b_i's is a strictly increasing sequence indicating the locations of the removed substrings, m is the number of objects in $\widetilde{\mathbf{x}}$, and $v \in \{u - 1, u, u + 1\}$. For each i, b_i indicates that a non-empty substring has been removed in between the b_i-th and $(1 + b_i)$-th locations. If $b_1 = 0$ or $b_v = m$, this indicates that a non-empty substring has been removed at the beginning or end of the string respectively. For example, $(abcda, \langle 5, 0, 3 \rangle)$ is a redacted string of the original $xxxabcyyyda$. For convenient, we sometimes write a sequence of objects as $\langle x_1, x_2, x_3, \ldots, x_m \rangle$ or as a string $x_1 x_2 x_3 \ldots x_m$.

Let us define a binary relation $\succ$ between annotated strings. Given two annotated strings $\mathcal{X}_1 = (\mathbf{x}_1, \mathbf{e}_1)$ and $\mathcal{X}_2 = (\mathbf{x}_2, \mathbf{e}_2)$, we say $\mathcal{X}_1 \succ \mathcal{X}_2$, if either $\mathbf{x}_2$ can be obtained from $\mathbf{x}_1$ by removing a non-empty substring in $\mathbf{x}_1$, and the $\mathbf{e}_2$ is updated from $\mathbf{e}_1$ accordingly, or there is a $\mathcal{X}$ s.t. $\mathcal{X}_1 \succ \mathcal{X}$ and $\mathcal{X} \succ \mathcal{X}_2$.

DEFINITION 1 (REDACTABLE SIGNATURE SCHEME [9]). *A redactable signature scheme with respect to binary relation $\vdash$, is a tuple of probabilistic polynomial time algorithms* (KGen, Sign, Verify, Redact), *such that*

1. *for any message x, $\sigma = \mathrm{Sign}_{\mathcal{SK}}(x) \Rightarrow \mathrm{Verify}_{\mathcal{PK}}(x, \sigma) = \mathrm{TRUE}$;*
2. *for any messages x and y, such that $x \vdash y$,*

$$\mathrm{Verify}_{\mathcal{PK}}(x, \sigma) = \mathrm{TRUE} \ \wedge \ \sigma' = \mathrm{Redact}_{\mathcal{PK}}(x, \sigma, y) \Rightarrow \mathrm{Verify}_{\mathcal{PK}}(y, \sigma') = \mathrm{TRUE},$$

where $(\mathcal{PK}, \mathcal{SK}) \leftarrow \mathrm{KGen}(1^k)$ and k is the security parameter.

Both Johnson et al.[9] and Miyazaki et al.[13] presented a redactable signature scheme w.r.t superset relation. Johnson et al.[9] also gave security definition for homomorphic signature schemes. We adapt their definition for redactable signature scheme. Let $\vdash$ denote a binary relation. For any set S, let $span_\vdash(S)$ denote the set $\{x : \exists y \in S, \text{ s.t. } y \vdash x\}$.

DEFINITION 2 (UNFORGEABILITY OF REDACTABLE SIGNATURE SCHEME [9]). *A redactable signature scheme* $\langle \mathrm{KGen}, \mathrm{Sign}, \mathrm{Verify}, \mathrm{Redact} \rangle$ *is (t, q, ϵ)-unforgeable against existential forgeries with respect to $\vdash$ under adaptive chosen message*

---

[1] A string with an annotation which specifies the locations of redactions.

attack, if any adversary $\mathcal{A}$ that makes at most q chosen-message queries adaptively and runs in time at most t, has advantage $Adv\mathcal{A} \leq \epsilon$. The advantage of an adversary $\mathcal{A}$ is defined as the probability that, after queries on ℓ ($\ell \leq q$) messages $x_1, x_2, \ldots, x_\ell$, $\mathcal{A}$ outputs a valid signature σ for some message $x \notin span_\vdash(\{x_1, x_2, \ldots, x_\ell\})$. Formally,

$$Adv\mathcal{A} = \Pr \left[\begin{array}{c} (\mathcal{PK}, \mathcal{SK}) \leftarrow \mathtt{KGen}(1^k); \ \mathcal{A}^{\mathtt{Sign}_{\mathcal{SK}}} = (x, \sigma); \\ \mathtt{Verify}_{\mathcal{PK}}(x, \sigma) = \mathtt{TRUE} \ and \ x \notin span_\vdash(\{x_1, x_2, \ldots, x_\ell\}) \end{array} \right],$$

where the probability is taken over the random coins used by $\mathtt{KGen}, \mathtt{Sign}$ and $\mathcal{A}$.

Redactable signature schemes have an additional security requirement on privacy [2]: the adversary should not be able to derive any information about the removed substrings from a redacted string and its signature.

DEFINITION 3 (PRIVACY PRESERVING). *A redactable signature scheme* $\langle \mathtt{KGen}, \mathtt{Sign}, \mathtt{Verify}, \mathtt{Redact} \rangle$ *is privacy preserving if, given the public key* $\mathcal{PK}$ *and any annotated strings* $\mathcal{X}_1, \mathcal{X}_2, \mathcal{X}$, *such that* $\mathcal{X}_1 \succ \mathcal{X}$ *and* $\mathcal{X}_2 \succ \mathcal{X}$, *the following distributions* $\mathcal{S}_1$ *and* $\mathcal{S}_2$ *are computationally indistinguishable:*

$$\mathcal{S}_1 = \{\sigma : \sigma = \mathtt{Redact}_{\mathcal{PK}}(\mathcal{X}_1, \mathtt{Sign}_{\mathcal{SK}}(\mathcal{X}_1; \mathbf{r}_1), \mathcal{X}; \mathbf{r}_2)\},$$
$$\mathcal{S}_2 = \{\sigma : \sigma = \mathtt{Redact}_{\mathcal{PK}}(\mathcal{X}_2, \mathtt{Sign}_{\mathcal{SK}}(\mathcal{X}_2; \mathbf{r}_1), \mathcal{X}; \mathbf{r}_2)\},$$

where $\mathbf{r}_1$ and $\mathbf{r}_2$ are random bits used by $\mathtt{Sign}$ and $\mathtt{Redact}$ respectively, and public/private key $(\mathcal{PK}, \mathcal{SK})$ is generated by $\mathtt{KGen}$.

4 $\mathcal{RSS}$: Redactable Signature Scheme for Strings

We propose $\mathcal{RSS}$, a redactable signature scheme for strings that is able to hide the lengths of the removed substrings. Our approach is as follows: we first propose a hash function $\mathcal{H}$ that maps an annotated string $\mathcal{X}$ and an auxiliary input $\mathbf{y}$ to an unordered set. This hash is "collision resistant" and satisfies some properties on substring removal. Using $\mathcal{H}$ and some known redactable signature schemes for unordered sets, we have a redactable signature scheme for strings.

4.1 Hashing Strings to Unordered Sets

Let $\mathcal{H}$ be a hash function that maps an annotated string $\mathcal{X}$ and an auxiliary input $\mathbf{y}$ to a (unordered) set of elements from some universe. The auxiliary could be a sequence of numbers from a finite ring, and is not of particular interest right now. In our construction (Table 1), $\mathcal{H}$ maps the input to a set of 3-tuples in $\mathbb{Z}_n \times \mathbb{Z}_n \times \mathbb{Z}_n$, where n is some chosen parameter.

DEFINITION 4 (COLLISION RESISTANT). $\mathcal{H}$ *is* (t, ϵ)-*collision-resistant if, for any algorithm* $\mathcal{A}$ *with running time at most* t,

$$\Pr[\mathcal{X}_1 \not\succ \mathcal{X}_2 \ \wedge \ \mathcal{H}(\mathcal{X}_2, \mathbf{y}_2) \subset \mathcal{H}(\mathcal{X}_1, \mathbf{y}_1)] \leq \epsilon,$$

where $(\mathcal{X}_1, \mathcal{X}_2, \mathbf{y}_2)$ is the output of $\mathcal{A}$ on input $\mathbf{y}_1$, and the probability is taken over uniformly randomly chosen $\mathbf{y}_1$ and random bits used by $\mathcal{A}$.

To be used in constructing a secure scheme, besides collision resistance, the hash function $\mathcal{H}$ is also required to be,

1. redactable, that is, given $\mathcal{X}_1$, $\mathcal{X}_2$ and $\mathbf{y}_1$, such that $\mathcal{X}_1 \succ \mathcal{X}_2$, it is easy to find $\mathbf{y}_2$ such that $\mathcal{H}(\mathcal{X}_1, \mathbf{y}_1) \supset \mathcal{H}(\mathcal{X}_2, \mathbf{y}_2)$; and
2. privacy preserving, that is, $\mathcal{H}(\mathcal{X}_2, \mathbf{y}_2)$ must not reveal any information about the removed substring.

The property on privacy preserving is essential and used in the proof of Theorem 3. However, for simplicity, we will not explicitly formulate the requirement here.

4.2 Construction of $\mathcal{H}$

We present a hash function $\mathcal{H}(\cdot, \cdot)$ in Table 1 based on some hash functions h that output odd numbers. In practice, we may use popular cryptographic hash function like SHA-2 as h, but with the least significant bit always set to 1. For security analysis, we choose functions with certain security requirements as stated in Lemma 1.

Redactable requirement. Note that the hash $\mathcal{H}$ is redactable as mentioned in Section 4.1, that is, given $(\mathbf{x}_1, \mathbf{e}_1)$, $(\mathbf{r}_1, \mathbf{w}_1)$ and $(\mathbf{x}_2, \mathbf{e}_2)$ where $(\mathbf{x}_1, \mathbf{e}_1) \succ (\mathbf{x}_2, \mathbf{e}_2)$, it is easy to find a $(\mathbf{r}_2, \mathbf{w}_2)$ such that

$$\mathbf{H}((\mathbf{x}_1, \mathbf{e}_1), (\mathbf{r}_1, \mathbf{w}_1)) \supset \mathbf{H}((\mathbf{x}_2, \mathbf{e}_2), (\mathbf{r}_2, \mathbf{w}_2)).$$

The design of $\mathcal{H}$ is "inspired" by the following observation. Let us view the sequence $\langle t_1, t_2, \ldots, t_m \rangle$ as the outputs of an iterative hash. We can rewrite t_i's in the form: $t_{i+1} = C(t_i, x_{i+1}, r_{i+1})$, where C is the basic block in the iterative hash. In the event that a substring, say at location $i-1$ and i, is to be removed, both (x_{i-1}, r_{i-1}) and (x_i, r_i) also have to be removed. Yet, we want the iterative hash can still be computed. This can be achieved with the help of the witness w_i's.

Remarks on r_i's. It is crucial that the value of r_i is explicitly represented in t_i for each i (Table 1). If the r_i's are omitted in the design, for instance, by using this alternative definition,

$$\widetilde{\mathcal{H}}((\mathbf{x}, \mathbf{e}), (\mathbf{r}, \mathbf{w})) \triangleq \{\hat{t}_i : \hat{t}_i = (x_i, \ (w_i^{\prod_{j=1}^{i} h(r_j)} \mod n))\},$$

then there would be no linkage between the r_i's and x_i's. Such lack of linkage can be exploited to find collisions.

Table 1. Definition of $\mathcal{H}(\cdot, \cdot)$

Let n be a RSA modulus, and $h : \mathbb{Z}_n \to \mathbb{Z}_n$ be a hash function. Given $\mathbf{x} = x_1 x_2 \ldots x_m$ associated with annotation $\mathbf{e}$, $\mathbf{r} = r_1 r_2 r_3 \ldots r_m$, and $\mathbf{w} = w_1 w_2 w_3 \ldots w_m$, where for each i, $x_i, r_i, w_i \in \mathbb{Z}_n$ (i.e. $\mathbf{x}, \mathbf{r}$ and $\mathbf{w}$ are strings over alphabet $\mathbb{Z}_n$), we define $\mathcal{H}$ as

$$\mathcal{H}((\mathbf{x}, \mathbf{e}), (\mathbf{r}, \mathbf{w})) \triangleq \{t_i : t_i = (x_i, \ r_i, \ (w_i^{\prod_{j=1}^{i} h(r_j)} \mod n)), 1 \le i \le m\}.$$

Table 2. $\mathcal{RSS}$: KGen

KGen.   Given security parameter k.

1. Choose a RSA modulus n, and an element g of large order in $\mathbb{Z}_n$.
2. Run key generating algorithm keygen on input 1^k to get key $(\mathcal{PK}, \mathcal{SK})$.
3. Output $(n, g, \mathcal{PK})$ as public key and $\mathcal{SK}$ as private key.

LEMMA 1. *The hash function $\mathcal{H}$ as defined in Table 1, is $(poly_1(k), \frac{1}{poly_2(k)})$-collision-resistant for any positive polynomials $poly_1(\cdot)$ and $poly_2(\cdot)$, where k is the security parameter, i.e. the bit length of n, assuming that h is division intractable[2] and always outputs odd prime integers, and Strong RSA Problem is hard.*

Essentially, the proof reduces Strong RSA Problem or Division Problem [6] to the problem of finding collisions. Gennaro et al.[6] gave a way to construct a hash function that is division intractable and always outputs odd prime numbers. Thus the conditions of the Lemma 1 can be achieved.

4.3   Construction of $\mathcal{RSS}$

We construct a redactable signature scheme $\mathcal{RSS}$, which consists of four algorithms KGen, Sign, Verify, and Redact, for strings with respect to binary relation $\succ$ based on the hash function $\mathcal{H}$ defined in Table 1 and a redactable signature scheme for (unordered) sets with respect to superset relation $\supseteq$.

The signer chooses a RSA modulus n and an element g of large order in $\mathbb{Z}_n^*$. Both n and g are public. Let the object space be $\mathbb{Z}_n$, that is, a string is a sequence of integers from $\mathbb{Z}_n$. Let $h : \mathbb{Z}_n \to \mathbb{Z}_n$ be a hash which satisfies security requirement stated in Lemma 1. Note that in practice, it may be suffice to employ popular cryptographic hash like SHA-2 (but with the least significant bit of the output always set to 1) as the function h. Let SSS = (keygen, sig, vrf, rec) be a redactable signature scheme for unordered sets w.r.t superset relation $\supseteq$. The signer also needs to choose the public and secret key pair $(\mathcal{PK}, \mathcal{SK})$ of the underlying signature scheme SSS. The details of KGen, Sign, Verify, and Redact are presented in Table 2, Table 3, Table 4 and Table 5 respectively.

The final signature of a string $x_1 x_2 \ldots x_m$ consists of m random numbers $r_1, r_2, \ldots, r_m$, the *witnesses* $w_1, w_2, \ldots, w_m$ where $r_i, w_i \in \mathbb{Z}_n$ for each i, and a signature **s** constructed by SSS.

Initially, the witness is set to be $w_i = g$ for each i (Step 1 in Table 3). The witness will be modified during redactions. By comparing the neighboring value within the witness **w**, we can deduce the locations of the removed substrings. Specifically, for any $1 < i \leq m$, $w_{i-1} \neq w_i$ if and only if a non-empty substring has been removed between x_{i-1} and x_i. Recall that the first and last object in

---

[2] Division intractability [6] implies collision resistance.

Table 3. $\mathcal{RSS}$: Sign

Sign. Given $\mathbf{x} = x_1 x_2 \ldots x_m$ and its associated annotation $\mathbf{e} = \langle m \rangle$.

1. Let $w_i = g$ for each i. Choose m distinct random numbers $r_1, r_2, \ldots, r_m$. Let $\mathbf{r} = r_1 r_2 r_3 \ldots r_m$ and $\mathbf{w} = w_1 w_2 w_3 \ldots w_m$. Compute

$$\mathbf{t} = \mathcal{H}((\mathbf{x}, \mathbf{e}), (\mathbf{r}, \mathbf{w})).$$

2. Sign the set $\mathbf{t}$ using SSS with the secret key $\mathcal{SK}$ to obtain $\mathbf{s}$:

$$\mathbf{s} = \mathrm{sig}_{\mathcal{SK}}(\mathbf{t}).$$

3. The final signature consists of the random numbers r_i's, witnesses w_i's, and the signature $\mathbf{s}$. That is,

$$(\mathbf{r}, \mathbf{w}, \mathbf{s}) \text{ or } (r_1, r_2, \ldots, r_m; \ \ w_1, w_2, \ldots, w_m; \ \ \mathbf{s})$$

Table 4. $\mathcal{RSS}$: Verify

Verify. Given a string $\mathbf{x} = x_1 x_2 \ldots x_m$ associated with annotation $\mathbf{e}$, its signature $(\mathbf{r}, \mathbf{w}, \mathbf{s})$, the public information n, g, and the public key $\mathcal{PK}$ of SSS.

1. If $\mathbf{e}$ and $\mathbf{w}$ are not consistent, output FALSE.
2. Compute $\mathbf{t} = \mathcal{H}(\mathbf{x}, (\mathbf{r}, \mathbf{w}))$.
3. $(\mathbf{r}, \mathbf{w}, \mathbf{s})$ is a valid signature of $\mathbf{x}$ under $\mathcal{RSS}$, if and only if $\mathbf{s}$ is a valid signature of $\mathbf{t}$ under SSS, i.e.

$$\mathrm{vrf}_{\mathcal{PK}}(\mathbf{t}, \mathbf{s}) = \text{TRUE}.$$

---

the string cannot be removed (Section 3) and thus we do not have to consider cases when $i = 1$ and $i - 1 = m$.

Since the witness $\mathbf{w}$ should be consistent with the annotation $\mathbf{b}$, and the $\mathcal{H}$ is collision-resistant, it can be used to verify the integrity of $\mathbf{b}$, as in the Step 1 of Table 4.

THEOREM 2. $\mathcal{RSS}$ *is* $(t, q, \frac{\epsilon_1}{1-\epsilon_2})$-*unforgeable against existential forgeries with respect to relation* $\succ$, *if* SSS *is* $(t + qt_0, q, \epsilon_1)$-*unforgeable against existential forgeries with respect to superset relation* $\supseteq$, *and* $\mathcal{H}$ *is* $(t + qt_1, \epsilon_2)$-*collision-resistant, where* t_0 *is the running time of* $\mathcal{H}$ *and* t_1 *is the time needed by* $\mathcal{RSS}$ *to sign a document.*

Our construction of $\mathcal{H}$ (Table 1) is collision resistant (Lemma 1). Johnson et al.[9] showed their redactable signature scheme Sig (in Section 5 of [9]) is (t, q, ϵ)-unforgeable under reasonable assumptions (see Theorem 1 in [9]), for some proper parameters t, q and ϵ. Miyazaki et al.[13] also showed a similar result on the unforgeability of the redactable signature scheme they proposed. Hence, conditions in Theorem 2 can be satisfied.

Table 5. $\mathcal{RSS}$: Redact

Redact.   Given a string $\mathbf{x} = x_1 x_2 \ldots x_m$ associated with annotation $\mathbf{e}$, and its signature $(\mathbf{r}, \mathbf{w}, \mathbf{s})$, where $\mathbf{r} = r_1 r_2 \ldots r_m$, $\mathbf{w} = w_1 w_2 \ldots w_m$, the public information n, g, public key $\mathcal{PK}$ for SSS, and (i, j) the location of the string to be removed (that is $x_i x_{i+1} \ldots x_j$ is to be removed).

1. Update $\mathbf{e}$ to obtain new annotation $\hat{\mathbf{e}}$. Compute $u = \prod_{k=i}^{j} h(r_k)$, to update the witnesses in the following way: for each $\ell > j$, update w_ℓ

$$\hat{w}_\ell \leftarrow w_\ell^u \mod n.$$

2. Let $\hat{\mathbf{x}} = x_1 x_2 \ldots x_{i-1} x_{j+1} \ldots x_m$, $\hat{\mathbf{r}} = r_1 r_2 \ldots r_{i-1} r_{j+1} \ldots r_m$ and $\hat{\mathbf{w}} = w_1 w_2 \ldots w_{i-1} \hat{w}_{j+1} \hat{w}_{j+2} \ldots \hat{w}_m$. Compute

$$\hat{\mathbf{t}} = \mathcal{H}((\hat{\mathbf{x}}, \hat{\mathbf{e}}), (\hat{\mathbf{r}}, \hat{\mathbf{w}})).$$

3. Compute

$$\hat{\mathbf{s}} = \text{rec}_{\mathcal{PK}}(\mathbf{t}, \mathbf{s}, \hat{\mathbf{t}})$$

where $\mathbf{t} = \mathcal{H}((\mathbf{x}, \mathbf{e}), (\mathbf{r}, \mathbf{w}))$.
4. Output $(\hat{\mathbf{r}}, \hat{\mathbf{w}}, \hat{\mathbf{s}})$ as the signature of $(\hat{\mathbf{x}}, \hat{\mathbf{e}})$.

THEOREM 3. *The redactable signature scheme $\mathcal{RSS}$ is privacy preserving (as defined in Definition 3), assuming that hash function h satisfies the property: the two distributions $X = g^{h(U_1)h(U_2)} \mod n$ and $Y = g^{h(U_1')} \mod n$ are computationally indistinguishable, where n is a RSA modulus, g is an element of large order in $\mathbb{Z}_n^*$ and U_i's and U_j''s are all independent uniform random variables over $\mathbb{Z}_n$.*

Note that the scheme SSS does not need to satisfy requirement on privacy, this is because information is already removed before algorithms of SSS are applied.

4.4   Efficiency

The size of $\mathbf{s}$ depends on SSS, and let us assume it requires κ bits. The number of distinct w_i's is about the same as the number of redactions occurred. So w_i's can be represented in $t(k + \lceil \log m \rceil)$ bits, where t is the number of substrings removed, and k is the bit length of n. Thus the total number of bits required is at most $k(m + t) + t \lceil \log m \rceil + \kappa$. The dominant term is km, which is the total size of the random numbers r_i's.

Disregarding the time taken by the scheme SSS, and the time required to compute the hash $h(\cdot)$, during signing, $O(m)$ of k-bits exponentiation operations are required. During redaction, if ℓ consecutive objects are to be removed between position i and j, and t' number of redactions have been made after position j, then the number of k-bit exponentiation operations is at most $\ell(t' + 1)$, which is in $O(\ell m)$. During verification, $O(tm)$ number of k-bits exponentiation operations are required. Hence, our scheme is suitable for small t, which is reasonable in practice. In sum, the main drawback of $\mathcal{RSS}$ is the size of its signature. In the next section, we will reduce its size using a random tree.

5   RGGM: Random Tree with Hiding Property

We propose RGGM, a variant of GGM tree [7] to generate a sequence of pseudo random numbers, where the structure of the tree is randomized. This generator provides us with the ability to remove multiple substrings of pseudo random numbers, while still being able to generate the retained numbers from a short seed. The expected size of the new seed is in $O(k + tk \log m)$ where t is the number of removed substrings, m is the number of pseudo random numbers, and k is a security parameter. More importantly, the new seed does not reveal any information about the size nor the content of the removed substrings.

Pseudo random number generation.   To generate m pseudo random numbers we employ a method similar to that in the redactable signature scheme proposed by Johnson et al. [9], which is based on the GGM tree [7]. Let $G : \mathcal{K} \to \mathcal{K} \times \mathcal{K}$ be a length-doubling pseudo random number generator. First pick an arbitrary binary tree T with m leaves, where all internal nodes of T have exactly two children, the left and right child. Next, pick a seed $t \in \mathcal{K}$ uniformly at random, and associate it with the root. The pseudo random numbers $r_1, r_2, \ldots, r_m$ are then computed from t in the usual top-down manner along the binary tree.

Hiding random numbers.   If r_i is to be removed, the associated leaf node and all its ancestors will be removed, as illustrated by the example in Figure 1(b). The values associated with the roots of the remaining subtrees, and a description of the structure of the subtrees, form the new seed, whereby the remaining random values r_j's $(j \neq i)$ can be re-computed. By the property of G, it is computationally difficult to guess the removed value r_i from the new seed.

Unlike the method proposed by Johnson et al. [9], our tree T is randomly generated. If the tree is known to be balanced (or known to be of some fixed structure), some information on the number of leaf nodes removed can be derived from the redacted tree. Our random trees are generated by the probabilistic algorithm TreeGen in Table 6. Note that descriptions of the structure of the tree are required for the regeneration of the random values r_i's.

At the moment, for ease of presentation, the descriptions are stored together with the seed. This increases the size of the seed. To reduce the size, we can

Table 6. TreeGen: a random tree generation algorithm

TreeGen: Given m, output a binary tree T with m leaves:

1. Pick a p uniformly at random from $\{1, 2, \ldots, m - 1\}$.
2. Recursively generate a tree T_1 with p leaves.
3. Recursively generate a tree T_2 with $m - p$ leaves.
4. Output a binary tree with T_1 as the left subtree and T_2 as the right subtree.

---

replace the description by another short random seed $\hat{t}$, which is assigned to the root. The random input required in Step 1 of the algorithm can be generated from $\hat{t}$ using G. A difference between the two methods of storing the (redacted) tree structure information is that in the former, we will have an information theoretic security result, whereas in the later, the security depends on G.

Our main observation is as follows: *after a substring of leaves is removed from the random tree, the remaining subtrees do not reveal (information theoretically) anything about the number of leaves removed, except the fact that at least one leaf has been removed at that location.*

Notations.   Given a binary tree T, its leaf nodes can be listed from left to right to obtain a sequence. We call a subsequence of consecutive leaves a substring of leaves. After multiple substrings of leaves and all of their ancestor nodes are deleted, the remaining structures form a *redacted* tree[3] represented by two sequences, $\mathbf{T} = \langle T_1, T_2, \ldots, T_v \rangle$ and $\mathbf{b} = \langle m, b_1, b_2, \ldots, b_u \rangle$, where T_i's are the subtrees retained, and each b_i indicates that a substring was removed between the b_i-th and $(b_i + 1)$-th locations in the remaining sequence of leaf nodes. Let q_i be the number of leaves that were removed in this substring. We call the sequence $\langle m, (b_1, q_1), (b_2, q_2), \ldots, (b_u, q_u) \rangle$ the *original annotation* of $\mathbf{b}$. Thus, the total number of leaf nodes removed is $\sum_{i=1}^{u} q_i$.

Let us consider this process. Given an original annotation $\mathbf{b}_1 = \langle m, (b_1, q_1), (b_2, q_2), \ldots, (b_u, q_u) \rangle$, a random tree T of size $m + \sum_{i=1}^{u} q_i$ is generated using TreeGen, and then redacted according to $\mathbf{b}_1$. Let RED($\mathbf{b}_1$) be the redacted tree.

From an adversary's point of view, he has RED($\mathbf{b}_1$), represented as $(\mathbf{T}, \mathbf{b})$, and wants to guess the q_i's in the original annotation $\mathbf{b}_1$. We can show that the additional knowledge of $\mathbf{T}$ does not improve his chances, compared to another adversary who only has the annotation $\mathbf{b}$ but not the tree $\mathbf{T}$. It is suffice to show that, given any $\mathbf{b}$ and any two possible original annotations $\mathbf{b}_1$ and $\mathbf{b}_2$, the conditional probabilities of obtaining $(\mathbf{T}, \mathbf{b})$ are the same. That is,

THEOREM 4. *For any redacted tree* $(\mathbf{T}, \mathbf{b})$*, any distribution* $\mathcal{B}$ *on the original annotation, and* $\mathbf{b}_1 = \langle m, (b_1, q_1), (b_2, q_2), \ldots, (b_u, q_u) \rangle$, $\mathbf{b}_2 = \langle m, (b_1, q_1'), (b_2, q_2'), \ldots, (b_u, q_u') \rangle$,

$$Prob(\text{RED}(\mathcal{B}) = (\mathbf{T}, \mathbf{b}) \mid \mathcal{B} = \mathbf{b}_1) = Prob(\text{RED}(\mathcal{B}) = (\mathbf{T}, \mathbf{b}) \mid \mathcal{B} = \mathbf{b}_2)$$

6 $\mathcal{SRSS}$: A Short Redactable Signature Scheme for Strings

$\mathcal{RSS}$ produces a large signature, whose main portion is a sequence of true random numbers r_i's. We can combine RGGM with $\mathcal{RSS}$ to produce a short signature by replacing the r_i's with pseudo random numbers generated by RGGM. Let us call this combined scheme $\mathcal{SRSS}$, short redactable signature scheme for

---

[3] Although strictly speaking it is a forest.

strings. It is easy to show that $\mathcal{SRSS}$ is unforgeable and privacy preserving from Lemma 1, Theorem 2, Theorem 3, Theorem 4, and the fact that RGGM is a pseudo random number generator.

Unforgeability.   From the definition of cryptographic secure pseudo random number generator and Theorem 2, we conclude that $\mathcal{SRSS}$ is unforgeable.

COROLLARY 5. *For any positive polynomials (in κ) t and q, $\mathcal{SRSS}$ is $(t, q, \frac{\epsilon_1}{1-\epsilon_2})$-unforgeable against existential forgeries with respect to $\succ$, if SSS is $(t + qt_0, q, \epsilon_1)$-unforgeable against existential forgeries with respect to $\supseteq$, $\mathcal{H}$ is $(t + qt_1, \epsilon_2)$-collision-resistant, and G is a cryptographic secure pseudo random number generator, where t_0 is the running time of $\mathcal{H}$, t_1 is the time needed by $\mathcal{SRSS}$ to sign a document, and κ is the security parameter.*

Privacy.   From the definition of cryptographic secure pseudo random number generator, Theorem 3 and Theorem 4, we conclude that $\mathcal{SRSS}$ is privacy preserving.

COROLLARY 6. *The redactable signature scheme $\mathcal{SRSS}$ is privacy preserving (as defined in Definition 3), assuming that the hash function h satisfies the property: the two distributions $X = g^{h(U_1)h(U_2)} \mod n$ and $Y = g^{h(U_1')} \mod n$ are computationally indistinguishable, and G is a cryptographic secure pseudo random number generator, where n is a RSA modulus, g is an element of large order in $\mathbb{Z}_n^*$ and U_i's and U_j''s are all independent uniform random variables over $\mathbb{Z}_n$, and $h(\cdot)$ is used to define $\mathcal{H}$ in Table 1.*

Efficiency.   The improvement of $\mathcal{SRSS}$ is in signature size. Given the unredacted string, the size of the signature is $\kappa + 2k$, where κ is the signature size of SSS, and k is the length of each seed. Recall that we need two seeds in RGGM, one for the generation of the numbers, and the other for the tree structure. If t substrings are removed, the signature size is $\kappa + tk + O(kt \log m)$, where the term tk is for the witness, and $O(kt \log m)$ is required for the RGGM.

7   Other Variants

7.1   Allowing Removal of Empty Substring

Both $\mathcal{RSS}$ and $\mathcal{SRSS}$ do not allow removal of empty substrings. In fact, it is considered to be a forgery if a censor declares that a substring has been removed but actually the censor does not remove anything. However, some applications may want to allow removal of empty substrings. This can be achieved by slight modifications to our schemes. To sign a string $x_1 x_2 \ldots x_m$, special symbol $\natural$ is inserted to obtain the expanded string $\widetilde{\mathbf{x}} = \natural x_1 \natural x_2 \natural \ldots \natural x_m \natural$ which will be signed directly using $\mathcal{RSS}$ or $\mathcal{SRSS}$. To remove a substring $\mathbf{x}_0$, the expanded substring of $\mathbf{x}_0$ is actually removed. In the case where a substring has already being removed in front or at the end of $\mathbf{x}_0$, the $\natural$ is not included at the front or the end accordingly. To remove an empty substring, simply remove the $\natural$ at intended location.

7.2   Hiding the Fact That the String Is Redacted

There is a question on whether one should hide the location of a removed substring or even the occurrence of redaction. This requirement is also known as *invisibility* or *transparency* [2,13]. For a small object space, if invisibility is satisfied, a censor may take a long signed string, remove some substrings to form an arbitrary "authentic" short string. Nevertheless, some applications may need invisibility.

Here is a simple variation of $\mathcal{RSS}$ that achieves this. To sign a string, simply add a special symbols ♯ in-between any two consecutive objects. Sign the expanded string and then immediately redact it by removing all ♯'s. Redaction and verification is the same as before. However, this variant produces a large signature even if we use $\mathcal{SRSS}$. Furthermore, the computation during verification is high. At least $\Omega(m^2)$ exponentiation operations are required.

To reduce the size of signature, there is an alternative: sign all the pairs of objects. To sign the string $\mathbf{x} = x_1 x_2 x_3 \ldots x_m$, first generate random numbers $r_1, r_2, \ldots, r_m$ such that $r_i \| x_i$'s are distinct. Next, let $\mathbf{t}$ be the set of all pairs $\{(r_i \| x_i, r_j \| x_j)\}_{i<j}$ and employ SSS to sign $\mathbf{t}$. When an object x_i is to be removed, simply remove all the pairs that involve x_i from $\mathbf{t}$. Since the role of r_i is to ensure that all elements are distinct, the size of each r_i can be smaller than the random numbers required by $\mathcal{RSS}$.

8   Discussion and Conclusion

We considered a simple but difficult requirement in redactable signature scheme: hiding the lengths of the removed substrings. We exploited an intriguing statistical property of random trees, and employed a hash from strings to unordered sets to achieve the requirement. Although the signature is short, its size still depends on the number of substrings removed and the length of the string. In contrast, there are known schemes for unordered sets, whose signature size is a constant. Hence, it is interesting to find out whether it is possible to close the gap.

The two main components, the hash $\mathcal{H}$ and the RGGM tree, proposed in this paper, could be of independent interests. The hash function may play a role in the design of transitive signature with additional property on privacy preservation. Many secure outsourced database applications involve Merkel tree or GGM tree. The hiding property of the RGGM tree may be useful in those applications.

References

1. Aldous, D.: The continuum random tree III. The Annals of Probability 21, 248–289 (1993)
2. Ateniese, G., Chou, D.H., de Medeiros, B., Tsudik, G.: Sanitizable signatures. In: de Capitani Vimercati, S., Syverson, P.F., Gollmann, D. (eds.) ESORICS 2005. LNCS, vol. 3679, pp. 159–177. Springer, Heidelberg (2005)

3. Bellare, M., Goldreich, O., Goldwasser, S.: Incremental cryptography: The case of hashing and signing. In: Desmedt, Y.G. (ed.) CRYPTO 1994. LNCS, vol. 839, pp. 216–233. Springer, Heidelberg (1994)
4. Bellare, M., Goldreich, O., Goldwasser, S.: Incremental cryptography and application to virus protection. In: STOC, pp. 45–56 (1995)
5. Chang, E.-C., Lim, C.L., Xu, J.: Short redactable signatures using random trees. Cryptology ePrint Archive, Report 2009/025 (2009), http://eprint.iacr.org/
6. Gennaro, R., Halevi, S., Rabin, T.: Secure hash-and-sign signatures without the random oracle. In: Stern, J. (ed.) EUROCRYPT 1999. LNCS, vol. 1592, p. 123. Springer, Heidelberg (1999)
7. Goldreich, O., Goldwasser, S., Micali, S.: How to construct random functions. J. ACM 33(4), 792–807 (1986)
8. Izu, T., Kanaya, N., Takenaka, M., Yoshioka, T.: Piats: A partially sanitizable signature scheme. In: Qing, S., Mao, W., López, J., Wang, G. (eds.) ICICS 2005. LNCS, vol. 3783, pp. 72–83. Springer, Heidelberg (2005)
9. Johnson, R., Molnar, D., Song, D., Wagner, D.: Homomorphic signature schemes. In: Preneel, B. (ed.) CT-RSA 2002. LNCS, vol. 2271, pp. 244–262. Springer, Heidelberg (2002)
10. Klonowski, M., Lauks, A.: Extended sanitizable signatures. In: Rhee, M.S., Lee, B. (eds.) ICISC 2006. LNCS, vol. 4296, pp. 343–355. Springer, Heidelberg (2006)
11. Merkle, R.: Protocols for public key cryptosystems. In: SP, p. 122 (1980)
12. Micali, S., Rivest, R.L.: Transitive signature schemes. In: Preneel, B. (ed.) CT-RSA 2002. LNCS, vol. 2271, pp. 236–243. Springer, Heidelberg (2002)
13. Miyazaki, K., Hanaoka, G., Imai, H.: Digitally signed document sanitizing scheme based on bilinear maps. In: ASIACCS, pp. 343–354 (2006)
14. Rivest, R.: Two new signature schemes. Presented at Cambridge seminar (2001), http://www.cl.cam.ac.uk/Research/Security/seminars/2000/rivest-tss.pdf
15. Steinfeld, R., Bull, L., Zheng, Y.: Content extraction signatures. In: Kim, K.-c. (ed.) ICISC 2001. LNCS, vol. 2288, pp. 163–205. Springer, Heidelberg (2002)
16. Suzuki, M., Toshiyuki, I., Tanaka, K.: Sanitizable signature with secret information. In: SCIS (2006)
17. Yi, X.: Directed transitive signature scheme. In: Abe, M. (ed.) CT-RSA 2007. LNCS, vol. 4377, pp. 129–144. Springer, Heidelberg (2006)

Divisible On-Line/Off-Line Signatures

Chong-zhi Gao[1], Baodian Wei[2], Dongqing Xie[1], and Chunming Tang[3]

[1] School of Computer Science, Guangzhou University,
Guangzhou 510006, China
czgao@gzhu.edu.cn, dongqing_xie@hotmail.com
[2] Department of Electronics and Communication Engineering,
Sun Yat-sen University, Guangzhou 510275, China
weibd@mail.sysu.edu.cn
[3] Institute of Information Security, Guangzhou University,
Guangzhou 510006, China
tangcm622@hotmail.com

Abstract. On-line/Off-line signatures are used in a particular scenario where the signer must respond quickly once the message to be signed is presented. The idea is to split the signing procedure into two phases: the off-line and on-line phases. The signer can do some pre-computations in off-line phase before he sees the message to be signed.

In most of these schemes, when signing a message m, a partial signature of m is computed in the off-line phase. We call this part of signature the off-line signature token of message m. In some special applications, the off-line signature tokens might be exposed in the off-line phase. For example, some signers might want to transmit off-line signature tokens in the off-line phase in order to save the on-line transmission bandwidth. Another example is in the case of on-line/off-line threshold signature schemes, where off-line signature tokens are unavoidably exposed to all the users in the off-line phase.

This paper discusses this exposure problem and introduces a new notion: divisible on-line/off-line signatures, in which exposure of off-line signature tokens in off-line phase is allowed. An efficient construction of this type of signatures is also proposed. Furthermore, we show an important application of divisible on-line/off-line signatures in the area of on-line/off-line threshold signatures.

Keywords: Signature Schemes, Divisible On-line/Off-line Signatures, On-line/Off-line Threshold Signatures.

1   Introduction

On-line/Off-line signatures are used in a particular scenario where the signer must respond quickly once the message to be signed is presented. This notion was first introduced by Even, Goldreich and Micali in 1990 [10]. The idea of on-line/off-line signatures is to split the signing procedure into two phases. The first phase is off-line: in this phase, the signer does some preparing works before the message to be signed is presented. The second phase is on-line: once the message

M. Fischlin (Ed.): CT-RSA 2009, LNCS 5473, pp. 148–163, 2009.

to be signed is known, the signer utilizes the result of the pre-computation and uses a very short time to accomplish the signing procedure.

Some signature schemes can be naturally viewed as on-line/off-line signature schemes. They include Fiat-Shamir [11], Schnorr [19], El-Gamal [9], DSS [17] and Boneh-Boyen [2] signature schemes.

Up to now, there are two general approaches to convert any signature scheme into an on-line/off-line signature scheme. They are Even et al.'s paradigm [10] based on one time signatures and Shamir and Tauman's paradigm [20] based on trapdoor hash functions. Even et al.'s concrete implementation in [10] has a very long signature length and thus is not practical. Shamir-Tauman paradigm greatly reduces the signature length, whilst the on-line computation is fast. In PKC08, Catalano et al. [4] unified Even et al.'s paradigm and Shamir-Tauman paradigm, in the sense that they both use an ordinary signature scheme and a (weak) one time signature scheme as components[1]. Here the trapdoor hash function in Shamir-Tauman paradigm is viewed as a weak one time signature scheme. However, these two paradigms truly have different security characterizations if we consider the partial signature exposure problem described in the next subsection. See next subsection for more details.

Some recent works in on-line/off-line signatures have also been done in [18, 21, 22, 5, 15, 6, 3]. These schemes aim at some specific goals such as improving the efficiency [18], eliminating the random oracle model assumption [15], constructing ID-based schemes [21], constructing threshold schemes [6, 3], avoiding key exposure [5], or avoiding trapdoor hash primitives [22].

1.1   Divisible On-Line/Off-Line Signatures

In most of the on-line/off-line signature schemes([9, 17, 18, 21, 22, 5, 15, 6, 3, 2] and some variations of [11, 19]), when signing a message m, a partial signature of m is computed in the off-line phase. We call this part of signature the off-line signature token of message m. Although the signature generation is broken into two stages, the transmission of a signature is at one time, i.e., the whole signature of a message is transmitted to the recipient at the end of the on-line phase, while nothing is transmitted in the off-line phase.

A question thus naturally arises: can the off-line signature token be transmitted to the recipient off-line? An equivalent question is: is the signature scheme still secure if the adversary is allowed to query the signing oracle with a message depending on this message's off-line signature token? Addressing this question is meaningful because in some special applications, the off-line signature tokens might be exposed in the off-line phase. For example, some signers might want to transmit off-line signature tokens in the off-line phase in order to save the on-line transmission bandwidth. Another example is in the case of on-line/off-line threshold signature schemes [6, 3], where off-line signature tokens are unavoidably exposed to all the users in the off-line phase.

---

[1] Here the "weak" means the signature scheme is unforgeable only against generic chosen message attack [14].

Unfortunately, most on-line/off-line signature schemes can not be proven to be secure if their off-line signature tokens are exposed in the off-line signing phase. In this paper, we introduce a new notion called *divisible on-line/off-line signatures*, in which exposure of off-line signature tokens in off-line singing phase is allowed. To exemplify this new notion, we give in appendix some on-line/off-line signature schemes extracted from existing literatures, which satisfy the new property of divisibility. This paper also presents an efficient construction satisfying the new requirement, which is based on Boneh and Boyen(BB)'s signature scheme [2].

An informal description. Let $\mathcal{OS}$ be an on-line/off-line signature scheme. When signing a message m submitted by a receiver (or generated randomly), the signer uses the signing algorithm of $\mathcal{OS}$ to obtain a signature, say Σ. Informally, we say scheme $\mathcal{OS}$ is divisible if: i) Σ can be separated into two parts Σ^{off} and Σ^{on}, where Σ^{off} is obtained before the message m is known by the signer. ii) Before the signer knows the message, he can send Σ^{off} to the receiver first. In other word, the message requested to be signed in the attack game can depend on the first part of the signature. A formal definition is presented in Section 3.

An on-line/off-line signature scheme is trivially divisible if its Σ^{off} is *null*. For this reason we restrict to non-trivial divisibility in this paper. In the rest of this paper, the word divisible/divisibility usually means a non-trivial case.

Existing Schemes with divisibility. Some existing on-line/off-line signature schemes are listed in Table 1 to show whether they can be proved divisible. We can see that some schemes are divisible such as Scheme Schnorr-OS and Even et al.'s scheme. However, most schemes like Shamir and Tauman's general paradigm can not be proven to have this property, at least using currently known methods.

It is worthwhile noting that Even et al.'s paradigm, which uses an one time signature scheme as a component, is divisible; whereas Shamir and Tauman's general paradigm cannot be proven divisible because it only uses a weak one time signature scheme.

Remark 1. We argue that El-Gamal signature scheme cannot be proven divisible using the technique in [9]. In short, the simulated hash oracle $H(\cdot)$ should not set the value of $H(m\|\Sigma^{off})$ to a value pre-determined in off-line phase, because that could lead to a hash collision if another $m'\|\Sigma^{off}$ is also requested to the hash oracle before Σ^{off} is used.

Motivations. Considering the exposure problem might be interesting by itself. Besides, there are two main reasons to consider the divisibility of an on-line/off-line signature scheme:

1. *To save the on-line bandwidth.* If an on-line/off-line scheme is divisible, the signer can send the off-line part of the signature in the off-line phase instead of in the on-line phase. This reduces the on-line bandwidth of the communication channel.
2. *To construct on-line/off-line threshold signatures.* An on-line/off-line threshold signature $(\mathcal{OTS})$ scheme [6, 3] is a threshold signature scheme [7, 8] which

Table 1. Some on-line/off-line signature schemes. The second column shows whether they can be proved to be divisible using existing methods.

Schemes	Divisible?	Note
Fiat-Shamir [11]	No	
El-Gamal [9]	No	
DSS [17]	No	
Boneh-Boyen [2]	No	
Shamir and Tauman's paradigm (general) [20]	No	Some specific constructions can be proved divisible. See Appendix A,B.
Xu et al.'s scheme [21]	No	It seems divisible. However a deeper analysis shows it is not.
Chen et al.'s scheme [5]	No	
Even et al.'s scheme [10]	Yes	It has a long signature length.
Schnorr-OS	Yes	It's a variant of Schnorr signature scheme [19]. See Appendix C.
CMTW-OS	Yes	See Appendix A. It is extracted from [6].
BCG-OS	Yes	See Appendix B. It is extracted from [3].

can be partitioned into off-line and on-line phases. In an $\mathcal{OTS}$ scheme, off-line signature tokens will be unavoidably exposed to all the users in the off-line phase.

In the full paper [12] we prove that if an $\mathcal{OTS}$ scheme is simulatable to a divisible on-line/off-line signature scheme $\mathcal{DOS}$, then the unforgeability of $\mathcal{OTS}$ can be reduced to that of $\mathcal{DOS}$. This provides a theoretical basis for securely constructing an $\mathcal{OTS}$ scheme through the simulation approach.

Related work. The notion of divisible on-line/off-line signatures is first explicitly given in this paper, but the original idea goes back to [6, 3]. When proving the unforgeability of an on-line/off-line threshold signature scheme, the authors noticed that the off-line simulation of the scheme should not depend on the message to be signed. From [6, 3], we extract two on-line/off-line signature schemes(CMTW-OS and BCG-OS, see Appendix A,B), which can be proven divisible using the same proof techniques in [6, 3]. Besides, some existing schemes can also be proven divisible. They include Even et al.'s paradigm [10] and Scheme Schnorr-OS (a variant of Schnorr signature scheme, see Appendix C). Even et al.'s work has already contained a proof for their scheme's divisibility. But Schnorr-OS needs a new proof for its divisibility which is given in the full paper [12].

Scheme CMTW-OS and BCG-OS are both based on Shamir and Tauman's hash-sign-switch paradigm [20], which utilizes trapdoor hash functions. But Shamir-Tauman paradigm itself cannot be proven divisible. However, as in CMTW-OS and BCG-OS, if the specific trapdoor hash functions used can be viewed as a fully secure one time signature scheme, Shamir-Tauman paradigm can be unified again into Even et al.'s general paradigm, in the sense that these two paradigms both uses an one time signature scheme as a component and thus can be proven divisible [4].

1.2 Our Contribution

In this paper, we first explicitly give and exemplify the notion of divisible on-line/off-line signatures. Furthermore, without resorting to the random oracle model, we present an efficient divisible scheme, which is based on BB's signature scheme [2]. Compared to divisible schemes extracted from [6, 3], it does not rely on another signature scheme's security and is more efficient. We also show in the full paper that based on a divisible on-line/off-line signature scheme, an on-line/off-line threshold signature scheme can be proven secure if it is simulatable.

1.3 Organization

The rest of this paper is organized as follows. In Section 2, we give some preliminaries. Section 3 gives the security model of divisible on-line/off-line signatures. Section 4 presents an efficient divisible on-line/off-line signature scheme whose security is proven in the standard model. Section 5 concludes the paper with an application to on-line/off-line threshold signatures and some discussions.

2 Preliminaries

2.1 Notations and Definitions

We denote by $\mathbb{N}$ the set of natural numbers, and by $\mathbb{Z}$ the set of integers. If $k \in \mathbb{N}$, we denote by 1^k the concatenation of k ones and by $\{0,1\}^k$ the set of bitstrings of bitlength k. By $\{0,1\}^*$, we denote the set of bitstrings of arbitrary bitlength. "PPT" is an abbreviation for "probabilistic polynomial-time" and "$\|$" represents the concatenation operation.

If S is a set, then the notation $x \xleftarrow{\text{R}} S$ denotes that x is selected randomly from the set S. Similarly, $x \in_R S$ denotes x is a random element of S. If $\mathcal{A}$ is an algorithm, by $\mathcal{A}(\cdot)$ we denote that $\mathcal{A}$ receives only one input. If $\mathcal{A}$ receives two inputs we write $\mathcal{A}(\cdot, \cdot)$ and so on. If $\mathcal{A}(\cdot)$ is a probabilistic algorithm, $y \leftarrow \mathcal{A}^{\mathcal{O}_1, \mathcal{O}_2, \cdots}(x_1, x_2, \ldots)$ means that on input $x_1, x_2, \ldots$ and with access to oracles $\mathcal{O}_1, \mathcal{O}_2, \ldots$, $\mathcal{A}$'s output is y. If $p(\cdot, \cdot, \ldots)$ is a predicate, the notation $\Pr[p(x, y, \ldots) : x \xleftarrow{\text{R}} S; y \xleftarrow{\text{R}} T; \ldots]$ denotes the probability that $p(x, y, \ldots)$ will be true after the ordered execution of the algorithms $x \xleftarrow{\text{R}} S, y \xleftarrow{\text{R}} T, \ldots$, etc.

Definition 1 (Negligible Function). *A function $\epsilon : \mathbb{N} \to \mathbb{R}$ is negligible if for all $c > 0$, $\epsilon(k) < 1/k^c$ for all sufficiently large k.*

Definition 2 (Bilinear Paring). *Let $\mathbb{G}, \mathbb{G}_T$ be two multiplicative cyclic group of prime order p. A bilinear pairing on $(\mathbb{G}, \mathbb{G}_T)$ is a function $e : \mathbb{G} \times \mathbb{G} \to \mathbb{G}_T$ which has the following properties:*

1. *Bilinear: $e(u^a, v^b) = e(u, v)^{ab}$, for all $u, v \in \mathbb{G}$ and $a, b \in \mathbb{Z}$.*
2. *Non-degenerate: $e(u, v) \neq 1$ for some $u, v \in \mathbb{G}$. Here 1 denotes the identity element in $\mathbb{G}_T$.*

3. Computable: paring $e(u, v)$ can be efficiently computed for all $u, v \in \mathbb{G}$.

For generality, one can set $e : \mathbb{G}_1 \times \mathbb{G}_2 \rightarrow \mathbb{G}_T$ where $\mathbb{G}_1 \neq \mathbb{G}_2$. An efficiently computable isomorphism $\psi : \mathbb{G}_2 \rightarrow \mathbb{G}_1$ can convert this general case to the simple case where $\mathbb{G}_1 = \mathbb{G}_2$.

Definition 3 (q-SDH Assumption [2]). *Let $\mathbb{G}$ be a group of prime order p and x be a random element in $\mathbb{Z}_p^*$. Let g be a generator of $\mathbb{G}$. Solving the q-SDH problem in $\mathbb{G}$ is to compute a pair $(c, g^{1/(x+c)})$ where $c \in \mathbb{Z}_p \backslash \{-x\}$, given a $(q + 1)$-tuple $(g, g^x, g^{(x^2)}, \ldots, g^{(x^q)})$. The q-SDH assumption in group $\mathbb{G}$ states that the q-SDH problem in $\mathbb{G}$ is hard to solve, i.e., for any PPT algorithm $\mathcal{A}$, the following probability is negligible in k.*

$$\epsilon(k) = \Pr[\mathcal{A}(g, g^x, g^{(x^2)}, \ldots, g^{(x^q)}) = (c, g^{1/(x+c)}) : x \xleftarrow{\text{R}} \mathbb{Z}_p^*]$$

The following lemma states that given a q-SDH problem instance $(g, g^x, \ldots, g^{(x^q)})$, we can construct a new 1-SDH problem instance (h, h^x) with $q - 1$ known solutions $(c_i, s_i = h^{1/(x+c_i)})$ where any new solution reveals a solution to the original problem instance. Using the same technique in Lemma 9 of [2], this lemma can be easily proved.

Lemma 1. *There exists a PPT algorithm Γ which satisfies:*

- *Its inputs are:*
  1. *$\mathsf{descr}(\mathbb{G})$. A description of a group $\mathbb{G}$ with prime order p.*
  2. *$(g, g^x, \ldots, g^{(x^q)})$. A q-SDH problem instance in Group $\mathbb{G}$ where $q \in \mathbb{N}$.*
  3. *$c_1, \ldots, c_{q-1} \in \mathbb{Z}_p \backslash \{-x\}$.*
- *It outputs a PPT algorithm Δ and a tuple $((h, u), (s_1, \ldots, s_{q-1})) \in \mathbb{G}^2 \times \mathbb{G}^{q-1}$ which satisfy:*
  1. *$u = h^x$.*
  2. *$s_i = h^{1/(x+c_i)}$, i.e., (c_i, s_i) $(1 \le i \le q - 1)$ are solutions of the 1-SDH problem instance (h, h^x).*
  3. *By using Algorithm Δ, any new solution $(c^*, s^*) \neq (c_i, s_i)$ for the 1-SDH problem instance (h, h^x) reveals a solution to the original instance, i.e., on inputs $(c^*, s^*) \in \mathbb{Z}_p \backslash \{-x\} \times \mathbb{G}$ where $(c^*, s^*) \neq (c_i, s_i)$ for all $i \in \{1, \ldots, q-1\}$ and $s^* = h^{1/(x+c^*)}$, Δ can output a pair $(c, g^{1/(x+c)}) \in \mathbb{Z}_p \backslash \{-x\} \times \mathbb{G}$ in polynomial time.*

3 Security Model

We give the security model of divisible online/offline signatures and some security notions.

3.1 Syntax

A divisible online/offline signature scheme ($\mathcal{DOS}$) is a tuple of algorithms (KeyGen, Sign^{off}, Sign^{on}, Ver).

- $(pk, sk) \leftarrow$ KeyGen(1^k). The key generation algorithm, a PPT algorithm which on input a security parameter $k \in \mathbb{N}$, outputs a public/private key pair (pk, sk).
- $(\Sigma_i^{\text{off}}, St_i) \leftarrow$ Sign$^{\text{off}}(sk)$. The i-th $(i \in \mathbb{N})$ executing of the off-line signing algorithm, a PPT algorithm which on input a private key, outputs a (public) off-line signature token Σ_i^{off} and a (secret) state information St_i. The state information is kept secret and will be passed to the i-th executing of the on-line signing algorithm.
- $\Sigma_i^{\text{on}} \leftarrow$ Sign$^{\text{on}}(sk, St_i, m_i)$. The i-th $(i \in \mathbb{N})$ executing of the on-line signing algorithm, a PPT algorithm which on input sk, a state information St_i and a message m_i, outputs an on-line signature token Σ_i^{on}. The signature for m_i is defined as $\Sigma_i = (\Sigma_i^{\text{off}}, \Sigma_i^{\text{on}})$
- $0/1 \leftarrow$ Ver(pk, m, Σ). The verification algorithm, a PPT algorithm which on input the public key pk, a message m and a signature Σ, outputs 0 or 1 for reject or accept respectively.

Completeness: It is required that if $(\Sigma^{\text{off}}, St) \leftarrow$ Sign$^{\text{off}}(sk)$ and $\Sigma^{\text{on}} \leftarrow$ Sign^{on} (sk, St, m), then Ver$(pk, m, \Sigma) = 1$ for all (pk, sk) generated by KeyGen(1^k).

3.2   Security Notion

In the following, we define a security notion for a divisible on-line/off-line signature scheme.

EU-CMA: For a divisible on-line/off-line signature scheme $\mathcal{DOS}$, existential unforgeability against adaptive chosen message attacks (EU-CMA) is defined in the following game. This game is carried out between a challenger $\mathcal{S}$ and an adversary $\mathcal{A}$. The adversary $\mathcal{A}$ is allowed to make queries to an off-line signing oracle Sign$^{\text{off}}(sk)$ and an on-line signing oracle Sign$^{\text{on}}(sk, St, \cdot)$ defined in Section 3.1. We assume that if $\mathcal{A}$ makes the i-th on-line signature query then it has already made the i-th off-line signature query. This requirement is reasonable since the signer always execute his i-th off-line signature signing before his i-th on-line signing. The attack game is as follows:

1. The challenger runs KeyGen on input 1^k to get (pk, sk). pk is sent to $\mathcal{A}$.
2. On input $(1^k, pk)$, $\mathcal{A}$ is allowed to query the oracles Sign$^{\text{off}}(sk)$, Sign$^{\text{on}}(sk, St, \cdot)$ polynomial times. The i-th state information St_i of Sign^{on}, which is kept secret from the adversary, is passed from the i-th executing of Sign$^{\text{off}}(sk)$.
3. $\mathcal{A}$ outputs a pair (m, Σ).

The adversary wins the game if the message m has never been queried to the on-line signing oracle Sign$^{\text{on}}(sk, St, \cdot)$ and Ver$(pk, m, \Sigma) = 1$ holds. Let Adv$_{\mathcal{A}, \mathcal{DOS}}$ be the *advantage* of the adversary $\mathcal{A}$ in breaking the signature scheme, i.e.,

$$\text{Adv}_{\mathcal{A}, \mathcal{DOS}} = \Pr \left[\begin{array}{l} \text{Ver}(pk, m, \Sigma) = 1 : (pk, sk) \leftarrow \text{KeyGen}(1^k); \\ (m, \Sigma) \leftarrow \mathcal{A}^{\text{Sign}^{\text{off}}(sk), \text{Sign}^{\text{on}}(sk, St, \cdot)}. \end{array} \right]$$

where $\mathcal{A}$ has never requested the signature of m from the on-line signing oracle. The probability is taken over the internal coin tosses of the algorithms KeyGen, $\mathsf{Sign^{off}}$, $\mathsf{Sign^{on}}$ and $\mathcal{A}$.

In detail, if $\mathcal{A}$ makes q_{off} off-line signing queries and q_{on} on-line singing queries, $\mathrm{Adv}_{\mathcal{A},\mathcal{DOS}}$ is defined as:

$$\mathrm{Adv}_{\mathcal{A},\mathcal{DOS}} = \Pr \left[\begin{array}{l} \mathsf{Ver}(pk, m, \Sigma) = 1 \text{ and } m \neq m_i \text{ for all } i \in \{1, \ldots, q_{\mathrm{on}}\} : \\[4pt] (pk, sk) \leftarrow \mathsf{KeyGen}(1^k); \\[4pt] (\Sigma_1^{\mathrm{off}}, St_1) \leftarrow \mathsf{Sign^{off}}(sk); \\ \dotfill \\ (\Sigma_{q_{\mathrm{off}}}^{\mathrm{off}}, St_{q_{\mathrm{off}}}) \leftarrow \mathsf{Sign^{off}}(sk); \\[4pt] m_1 \leftarrow \mathcal{A}(pk, \Sigma_1^{\mathrm{off}}, \ldots, \Sigma_{q_{\mathrm{off}}}^{\mathrm{off}}); \\[4pt] \Sigma_1^{\mathrm{on}} \leftarrow \mathsf{Sign^{on}}(sk, St_1, m_1); \\ \dotfill \\ m_{q_{\mathrm{on}}} \leftarrow \mathcal{A}(pk, \Sigma_1^{\mathrm{off}}, \Sigma_1^{\mathrm{on}}, m_1, \ldots, \Sigma_{q_{\mathrm{on}}}^{\mathrm{off}}, \Sigma_{q_{\mathrm{on}}+1}^{\mathrm{off}}, \ldots, \Sigma_{q_{\mathrm{off}}}^{\mathrm{off}}); \\[4pt] \Sigma_{q_{\mathrm{on}}}^{\mathrm{on}} \leftarrow \mathsf{Sign^{on}}(sk, St_{q_{\mathrm{on}}}, m_{q_{\mathrm{on}}}); \\[4pt] (m, \Sigma) \leftarrow \mathcal{A}(pk, \Sigma_1^{\mathrm{off}}, \Sigma_1^{\mathrm{on}}, m_1, \ldots, \Sigma_{q_{\mathrm{on}}}^{\mathrm{off}}, \Sigma_{q_{\mathrm{on}}}^{\mathrm{on}}, m_{q_{\mathrm{on}}}, \ldots, \Sigma_{q_{\mathrm{off}}}^{\mathrm{off}}). \end{array} \right]$$

Definition 4. *An adversary $\mathcal{A}$ $(t, q_{\mathrm{off}}, q_{\mathrm{on}}, \epsilon)$-breaks a divisible online/offline signature scheme $\mathcal{DOS}$ if $\mathcal{A}$ runs in time at most t, makes at most q_{off} queries to the off-line signing oracle, at most q_{on} queries to the on-line signing oracle, and $\mathrm{Adv}_{\mathcal{A},\mathcal{DOS}}$ is at least ϵ.*

A divisible on-line/off-line signature scheme $\mathcal{DOS}$ is EU-CMA if for every PPT adversary $\mathcal{A}$, $\mathrm{Adv}_{\mathcal{A},\mathcal{DOS}}$ is negligible.

Difference to the previous definition. The above definition is different from the attack game for an ordinary on-line/off-line signature scheme where the adversary is only allowed to query the oracle $\mathsf{Sign}(sk, \cdot)$. In other word, in the attack game of an ordinary scheme, the off-line signature token is returned to the adversary only *after* the message to be signed is submitted, whilst in the game for a divisible scheme, the adversary obtains the off-line signature token of a message *before* he submits this message.

Thus, the unforgeability defined above is stronger than the unforgeability defined as usual for ordinary on-line/off-line signatures. Note, however, that the unforgeability defined as usual is enough for the applications where off-line signature tokens are not exposed in the off-line signing phase.

4 A Divisible On-Line/Off-Line Signature Scheme Based on the q-SDH Assumption

In this section, we propose an efficient divisible on-line/off-line signature scheme whose security is proven in the standard model. This scheme is based on Boneh and Boyen's signature scheme [2].

4.1 Construction

Let $\mathbb{G}$ be a bilinear group of prime order p, where p's bit-length depends on the security parameter. Assume the message space is $\mathbb{Z}_p$. Note that using a collision resistant hash function $H : \{0,1\}^* \to \mathbb{Z}_p$, one can extend the message domain to $\{0,1\}^*$. The new divisible on-line/off-line signature scheme is defined as SDH-OS = (KeyGen, Sign^{off}, Sign^{on}, Ver), where

- KeyGen. Pick a random generator $g \in \mathbb{G}$. Choose random $x, y, z \in_R \mathbb{Z}_p^*$, and compute $X = g^x \in \mathbb{G}$, $Y = g^y \in \mathbb{G}$ and $Z = g^z \in \mathbb{G}$. Also compute $v = e(g,g) \in \mathbb{G}_T$. The public key is (g, X, Y, Z, v). The private key is (x, y, z).
- Sign^{off}. (The i-th run). Choose a random $\theta \in \mathbb{Z}_p \backslash \{-x\}$. Compute $\sigma = g^{\frac{1}{(x+\theta)}}$ where $\frac{1}{(x+\theta)}$ is the inverse of $(x + \theta)$ in $\mathbb{Z}_p^*$. Store the state information θ. Output the off-line signature token σ.
- Sign^{on}. (The i-th run, on a message m). Retrieve from the memory the i-th state information θ. Compute $r, w \in \mathbb{Z}_p$ such that:

$$m + yr + zw = \theta.$$

  (This can be done by first selecting a random $r \in \mathbb{Z}_p$, and computing $w = (\theta - m - yr)z^{-1} \mod p$.) Output the on-line signature token (r, w).
- Ver. Given a message $m \in \mathbb{Z}_p$ and a signature (σ, r, w), verify that whether $e(\sigma, X g^m Y^r Z^w) = v$.

Remark 2. To reduce the on-line signing cost, we can move the selection of r and computing $y \cdot r$ to the off-line phase. Thus, the on-line signing requires only 1 modular multiplication in $\mathbb{Z}_p$.

Completeness: Note that

$$
\begin{aligned}
e(\sigma, X g^m Y^r Z^w) &= e(g^{1/(x+\theta)}, g^{x+m+yr+zw}) \\
&= e(g^{1/(x+\theta)}, g^{x+\theta}) \\
&= e(g,g) = v
\end{aligned}
$$

Thus the proposed scheme satisfies the property of completeness.

4.2 Security

Theorem 1. *The divisible on-line/off-line signature scheme* SDH-OS *is EU-CMA, provided that the q-SDH assumption holds in Group $\mathbb{G}$.*

Proof. We prove this theorem by contradiction. Assume there exists an algorithm $\mathcal{A}$ which $(t, q_{\text{off}}, q-1, \epsilon)$ breaks the unforgeability of SDH-OS in the game defined in Section 3. Then we construct an algorithm $\mathcal{B}$ which breaks the q-SDH problem in polynomial time with a non-negligible probability $\epsilon' \geq \frac{\epsilon}{3} - \frac{q-1}{p}$.

Without loss of generality, we assume that $\mathcal{A}$ makes q_{off} off-line signing queries, and makes $q - 1$ on-line signing queries on messages $\{m_i\}_{i \in \{1, \ldots, q-1\}}$ where

$q - 1 \leq q_{\text{off}}$. Let $\{(\sigma_i, r_i, w_i)\}_{i \in \{1,\dots,q-1\}}$ be the $q - 1$ full signatures returned by the signing oracle. At the end of $\mathcal{A}$'s attack game, $\mathcal{A}$ outputs a valid forgery (σ_*, r_*, w_*) on a new message m_* with probability at least ϵ. We can see one of the following cases, which cover all types of successful attacks of $\mathcal{A}$, must hold with probability at least $\epsilon/3$:

Case 1: $h^{m_*} Y^{r_*} Z^{w_*} \neq h^{m_i} Y^{r_i} Z^{w_i}$ for all $i \in \{1,\dots,q-1\}$.
Case 2: $h^{m_*} Y^{r_*} Z^{w_*} = h^{m_i} Y^{r_i} Z^{w_i}$ for some $i \in \{1,\dots,q-1\}$, and $r_* \neq r_i$.
Case 3: $h^{m_*} Y^{r_*} Z^{w_*} = h^{m_i} Y^{r_i} Z^{w_i}$ for some $i \in \{1,\dots,q-1\}$, and $r_* = r_i$, but
$\quad\quad w_* \neq w_i$.

Let $\mathbb{G}$ be a group of prime order p. Let g be a generator of $\mathbb{G}$. Algorithm $\mathcal{B}$ is given a q-SDH problem instance $(g, g^\tau, g^{(\tau^2)}, \dots, g^{(\tau^q)})$. To solve this problem instance, $\mathcal{B}$ selects a list of elements $c_1, \dots, c_{q-1} \in_R \mathbb{Z}_p$. We may assume $c_i + \tau \neq 0$ for all $i \in \{1,\dots,q-1\}$, or else $\mathcal{B}$ has already obtained τ and thus the q-SDH problem is solved. Next, $\mathcal{B}$ feeds Algorithm Γ in Lemma 1 with inputs $\mathsf{descr}(\mathbb{G}), (g, g^\tau, g^{(\tau^2)}, \dots, g^{(\tau^q)}), (c_1, \dots, c_{q-1})$ to get an algorithm Δ and $((h, u), (s_1, \dots, s_{q-1})) \in \mathbb{G}^2 \times \mathbb{G}^{q-1}$. Note that as described in Lemma 1, $u = h^\tau$ and $s_i = h^{1/(\tau + c_i)}$. Algorithm $\mathcal{B}$ computes $v = e(h, h)$ and proceeds for each above case respectively as follows.

[CASE 1.]
Setup: Algorithm $\mathcal{B}$ selects $y, z \in_R \mathbb{Z}_p^*$ and sends to $\mathcal{A}$ a public key (h, X, Y, Z, v)
 where X is set to u, Y is set to h^y, and Z is set to h^z.
Simulating the Signing Oracle (Off-line): Upon the i-th query, if $1 \leq i \leq q-1$, $\mathcal{B}$ returns $\sigma_i = s_i$ as the i-th off-line signature token; else if $q \leq i \leq q_{\text{off}}$, $\mathcal{B}$ just returns a random element in $\mathbb{G} \backslash \{1\}$.
Simulating the Signing Oracle (On-line): Upon the i-th$(1 \leq i \leq q - 1)$ query input m_i, $\mathcal{B}$ selects $r_i \in_R \mathbb{Z}_p$, sets $w_i = (c_i - m_i - yr_i)z^{-1} \mod p$, and outputs (r_i, w_i) as the i-th online signature token. It can be verified that (σ_i, r_i, w_i) is a valid signature on the message m_i.
Output: The simulated off-line/on-line singing oracles are identical to the real ones for $\mathcal{A}$. If Algorithm $\mathcal{A}$ outputs a valid forgery $(m_*, \sigma_*, r_*, w_*)$ satisfying the condition in Case 1, then we get a new solution (c_*, σ_*) for the 1-SDH problem instance (h, h^τ) where $c_* \stackrel{\text{def}}{=} m_* + yr_* + zw_*$. This is because from the verification equation we can get $\sigma_* = h^{1/(\tau + c_*)}$ and from $h^{m_*} Y^{r_*} Z^{w_*} \neq h^{m_i} Y^{r_i} Z^{w_i}$ for all i we can get $c_* \neq c_i$ for all i. From Lemma 1, the original q-SDH problem instance can be solved in polynomial time by Algorithm Δ. Therefore if Case 1 occurs with probability at least $\epsilon/3$, $\mathcal{B}$ can successfully solve the original q-SDH problem instance with the same probability.

[CASE 2.]
Setup: Algorithm $\mathcal{B}$ selects $x, z \in_R \mathbb{Z}_p^*$ and sends to $\mathcal{A}$ a public key (h, X, Y, Z, v)
 where X is set to h^x, Y is set to u, and Z is set to h^z.
Simulating the Signing Oracle (Off-line): Upon the i-th query, if $1 \leq i \leq q-1$, $\mathcal{B}$ selects $r_i \in_R \mathbb{Z}_p^*$, and returns $\sigma_i = (s_i)^{\frac{1}{r_i}}$ as the i-th off-line signature token; else if $q \leq i \leq q_{\text{off}}$, $\mathcal{B}$ just returns a random element in $\mathbb{G} \backslash \{1\}$.

Simulating the Signing Oracle (On-line): Upon the i-th($1 \leq i \leq q-1$) query input m_i, $\mathcal{B}$ sets $w_i = (c_i r_i - x - m_i) z^{-1} \mod p$, and outputs (r_i, w_i) as the i-th online signature token. It can be verified that (σ_i, r_i, w_i) is a valid signature on the message m_i.

Output: From $\mathcal{A}$'s view, the simulated oracles are indistinguishable to the real ones for $\mathcal{A}$. In particular, the only difference is that in the simulation r_i is uniformly distributed in $\mathbb{Z}_p^*$ whereas in the real world r_i is uniformly distributed in $\mathbb{Z}_p$. Thus for one signature the statistical difference is $1/p$ and for the whole game the difference is at most $(q-1)/p$. If Algorithm $\mathcal{A}$ outputs a valid forgery $(m_*, \sigma_*, r_*, w_*)$ satisfying the condition in Case 2, then for some i, $h^{m_*} Y^{r_*} Z^{w_*} = h^{m_i} Y^{r_i} Z^{w_i}$ and $r_* \neq r_i$ hold. Algorithm $\mathcal{B}$ can check to find this i and get $\tau = \mathrm{dl}_h u = \mathrm{dl}_h Y = [(w_i - w_*) z + (m_i - m_*)] \cdot (r_* - r_i)^{-1} \mod p$. Therefore if Case 2 occurs with probability at least $\epsilon/3$ in the real world, $\mathcal{B}$ can successfully solve the original q-SDH problem instance with probability at least $\epsilon/3 - (q-1)/p$.

[CASE 3.]

Setup: Algorithm $\mathcal{B}$ selects $x, y \in_R \mathbb{Z}_p^*$ and sends to $\mathcal{A}$ a public key (h, X, Y, Z, v) where X is set to h^x, Y is set to h^y, and Z is set to u.

Simulating the Signing Oracle (Off-line): Upon the i-th query, if $1 \leq i \leq q-1$, $\mathcal{B}$ selects $w_i \in_R \mathbb{Z}_p^*$, and returns $\sigma_i = (s_i)^{\frac{1}{w_i}}$ as the i-th off-line signature token; else if $q \leq i \leq q_{\mathrm{off}}$, $\mathcal{B}$ just returns a random element in $\mathbb{G} \backslash \{1\}$.

Simulating the Signing Oracle (On-line): Upon the i-th($1 \leq i \leq q-1$) query input m_i, $\mathcal{B}$ sets $r_i = (c_i w_i - x - m_i) y^{-1} \mod p$, and outputs (r_i, w_i) as the i-th online signature token. It can be verified that (σ_i, r_i, w_i) is a valid signature on the message m_i.

Output: The argument is similar to that of Case 2. Finally Algorithm $\mathcal{B}$ can get $\tau = \mathrm{dl}_h u = \mathrm{dl}_h Z = [(r_i - r_*) y + (m_i - m_*)] \cdot (w_* - w_i)^{-1} \mod p$ for some i with probability at least $\epsilon/3 - (q-1)/p$, where $(m_*, \sigma_*, r_*, w_*)$ is a valid forgery output by $\mathcal{A}$ satisfying $h^{m_*} Y^{r_*} Z^{w_*} = h^{m_i} Y^{r_i} Z^{w_i}$ and $w_* \neq w_i$.

To sum up, there exists an algorithm $\mathcal{B}$, which can break the original q-SDH problem instance with probability at least $\epsilon/3 - (q-1)/p$, in polynomial time. This contradicts the q-SDH assumption and thus the theorem is proved. ∎

The purpose of introducing $'z'$. Introducing an additional trapdoor z to the BB's original scheme enables the off-line signing oracle to generate the off-line signature token without knowing the message. Here $f(m, r, w) \stackrel{\text{def}}{=} h^m Y^r Z^w$ plays a role of "double-trapdoor hash function" in the scheme. Boneh and Boyen mentioned in [2] that the exposure of the off-line tokens (and the unused state informations) causes no harm if these tokens will not subsequently used to create signatures. However we note that for a divisible on-line/off-line signature scheme, an exposed (and unused) token also should can be used.

4.3 Comparison

We compare our new scheme **SDH-OS** with some known divisible on-line/off-line signature schemes in Table 2. To achieve the same security level, we assume the

Table 2. Comparisons amongst divisible on-line/off-line signature schemes. The word "stand." refers to operations or signature length of the underlying standard signature scheme. "Sig" in the assumption column means the security also depends on the security of the underlying standard signature scheme. Abbreviations used are: "sq." for squaring, and "mult." for multiplication.

Schemes	Sign^{off}	Sign^{on}	Ver	Signature Size Off-line/ On-line	Assumptions
New Scheme	k sq. in $\mathbb{G}$ $\frac{k}{2}$ mult. in $\mathbb{G}$	1 mult. in $\mathbb{Z}_p$	k sq. in $\mathbb{G}$ $\frac{7}{8}k$ mult. in $\mathbb{G}$ 1 pairing	k bits / $2k$ bits	q-SDH
CMTW-OS (Appendix A)	k sq. in $\mathbb{G}$ $\frac{3}{4}k$ mult. in $\mathbb{G}$ 1 stand. sig	1 mult. in $\mathbb{Z}_p$	k sq. in $\mathbb{G}$ $\frac{3}{4}k$ mult. in $\mathbb{G}$ 1 stand. ver	1 stand.sig / k bits	Sig, one-more-discrete-log
BCG-OS (Appendix B)	k sq. in $\mathbb{G}$ $\frac{7}{8}k$ mult. in $\mathbb{G}$ 1 stand. sig	1 mult. in $\mathbb{Z}_p$	k sq. in $\mathbb{G}$ $\frac{7}{8}k$ mult. in $\mathbb{G}$ 1 stand. ver	1 stand.sig / $2k$ bits	Sig, discrete log
Schnorr-OS (Appendix C)	k sq. in $\mathbb{G}$ $\frac{k}{2}$ mult. in $\mathbb{G}$	1 mult. in $\mathbb{Z}_p$	k sq. in $\mathbb{G}$ $\frac{3}{4}k$ mult. in $\mathbb{G}$	k bits/ k bits	ROM, one-more-discrete-log

parameter p in our new scheme and Schemes CMTW-OS, BCG-OS and Schnorr-OS are all k-bit long. When using an elliptic curves with $k = 160$, our scheme has the same security level with a 1024-bit key RSA signature [2]. In this case, our scheme has a 160-bit off-line signature length and a 320-bit on-line signature length. In comparison, we omit additions in the signing algorithm.

To our knowledge, the most efficient divisible on-line/off-line signature scheme is Scheme Schnorr-OS. However, its security proof is based on the random oracle model(ROM). Our scheme preserves all advantages of BB's original scheme: its security is proven in the standard model; its overall computational cost of signing is only one scalar exponentiation in the group $\mathbb{G}$ (i.e., roughly k squarings and $k/2$ multiplications[2] in $\mathbb{G}$), which is comparable to Scheme Schnorr-OS and is superior to other schemes whose security is proved in the standard model. Our new scheme's on-line signing requires only 1 modular multiplication in $\mathbb{Z}_p$. This is very efficient and comparable to other efficient schemes.

5 Conclusion and Discussions

We propose a new notion called divisible on-line/off-line signatures, in which off-line signature tokens can be sent to others before the messages to be signed are seen. We also propose an efficient construction, and prove its security under

---

[2] Suppose g_i are in some group $\mathbb{G}$, e_i are all k-bit random values and t is small compared to k. By using a variant of the "square-and-multiply" method for exponentiation(Algorithm 14.88, [16]), computing $g_1^{e_1} g_2^{e_2} \ldots g_t^{e_t}$ requires roughly k squarings and $(1 - \frac{1}{2^t})k$ multiplications in $\mathbb{G}$.

the new definition without resorting to the RO model. Below we end with an important application of the new notion and some discussions.

1. *Application to on-line/off-line threshold signature schemes.* Gennaro et al. [13] has proved that if a threshold signature scheme is simulatable, then its unforgeability can be reduced to the unforgeability of its underlying signature scheme. This provides a way to simplify the security proof of a threshold signature scheme. However, this result cannot be applied to on-line/off-line threshold signature schemes due to the partial signature exposure problem. In the full paper [12] we provide a similar result called simulation theorem for on-line/off-line threshold signature schemes. This theorem states that a sufficient condition for the security reduction of an on-line/off-line threshold signature scheme is that it is simulatable to a divisible on-line/off-line signature scheme.
2. *The gap between the security models of an ordinary on-line/off-line signature scheme and a divisible one.* It seems unlikely that Shamir-Tauman's general paradigm or BB's original scheme can be proven divisible, whilst these schemes are secure in a common sense. So Intuition tells us there exists a gap between the ordinary security model and the new one. However we cannot present a substantial attack against these schemes under the new model to illustrate this gap. This leaves us an open problem to find this potential gap.
3. *More shorter on-line signature length.* The main drawback of our divisible scheme is that the on-line signature length is $2 \log_2 p$, which is twice the length of Scheme CMTW-OS or Schnorr-OS. Thus, it remains an unsolved problem to find a divisible scheme whose security is proven in the standard model and whose performance is comparable to Scheme Schnorr-OS.

Acknowledgement

We thank the anonymous reviewers for their valuable comments and suggestions.

References

[1] Bellare, M., Namprempre, C., Pointcheval, D., Semanko, M.: The One-More-RSA-Inversion problems and the security of Chaum's blind signature scheme. Report 2001/002, Cryptology ePrint Archive (May 2002)
[2] Boneh, D., Boyen, X.: Short signatures without random oracles and the sdh assumption in bilinear groups. J. Cryptology 21(2), 149–177 (2008)
[3] Bresson, E., Catalano, D., Gennaro, R.: Improved on-line/Off-line threshold signatures. In: Okamoto, T., Wang, X. (eds.) PKC 2007. LNCS, vol. 4450, pp. 217–232. Springer, Heidelberg (2007)
[4] Catalano, D., Di Raimondo, M., Fiore, D., Gennaro, R.: Off-line/On-line signatures: Theoretical aspects and experimental results. In: Cramer, R. (ed.) PKC 2008. LNCS, vol. 4939, pp. 101–120. Springer, Heidelberg (2008)
[5] Chen, X., Zhang, F., Susilo, W., Mu, Y.: Efficient generic on-line/off-line signatures without key exposure. In: Katz, J., Yung, M. (eds.) ACNS 2007. LNCS, vol. 4521, pp. 18–30. Springer, Heidelberg (2007)

[6] Crutchfield, C., Molnar, D., Turner, D., Wagner, D.: Generic on-line/Off-line threshold signatures. In: Yung, M., Dodis, Y., Kiayias, A., Malkin, T.G. (eds.) PKC 2006. LNCS, vol. 3958, pp. 58–74. Springer, Heidelberg (2006)

[7] Desmedt, Y.: Threshold cryptography. European Transactions on Telecommunications 5(4), 449–457 (1994)

[8] Desmedt, Y.: Some recent research aspects of threshold cryptography. In: Okamoto, E. (ed.) ISW 1997. LNCS, vol. 1396, pp. 158–173. Springer, Heidelberg (1998)

[9] ElGamal, T.: A public key cryptosystem and a signature scheme based on discrete logarithms. IEEE Transactions on Information Theory IT-31(4), 469–472 (1985)

[10] Even, S., Goldreich, O., Micali, S.: On-line/Off-line digital signatures. In: Brassard, G. (ed.) CRYPTO 1989. LNCS, vol. 435, pp. 263–275. Springer, Heidelberg (1990)

[11] Fiat, A., Shamir, A.: How to prove yourself: Practical solutions to identification and signature problems. In: Odlyzko, A.M. (ed.) CRYPTO 1986. LNCS, vol. 263, pp. 186–194. Springer, Heidelberg (1987)

[12] Gao, C., Wei, B., Xie, D., Tang, C.: Divisible on-line/off-line signatures. Cryptology ePrint Archive, Report 2008/447 (2008), http://eprint.iacr.org/

[13] Gennaro, R., Jarecki, S., Krawczyk, H., Rabin, T.: Robust and efficient sharing of RSA functions. In: Koblitz, N. (ed.) CRYPTO 1996. LNCS, vol. 1109, pp. 157–172. Springer, Heidelberg (1996)

[14] Goldwasser, S., Micali, S., Rivest, R.L.: A digital signature scheme secure against adaptive chosen-message attacks. SIAM Journal on Computing 17(2), 281–308 (1988); special issue on cryptography

[15] Kurosawa, K., Schmidt-Samoa, K.: New online/Offline signature schemes without random oracles. In: Yung, M., Dodis, Y., Kiayias, A., Malkin, T.G. (eds.) PKC 2006. LNCS, vol. 3958, pp. 330–346. Springer, Heidelberg (2006)

[16] Menezes, A.J., van Oorschot, P.C., Vanstone, S.A.: Handbook of Applied Cryptography. CRC Press, Boca Raton (1997)

[17] National Institute of Standards and Technology (NIST). The Digital Signature Standard, 186. Federal Information Processing Standards Publication (FIPS PUB) (May 1994)

[18] Schmidt-Samoa, K., Takagi, T.: Paillier's cryptosystem modulo $p^2 q$ and its applications to trapdoor commitment schemes. In: Dawson, E., Vaudenay, S. (eds.) Mycrypt 2005. LNCS, vol. 3715, pp. 296–313. Springer, Heidelberg (2005)

[19] Schnorr, C.P.: Efficient signature generation by smart cards. Journal of Cryptology 4, 161–174 (1991)

[20] Shamir, A., Tauman, Y.: Improved online/Offline signature schemes. In: Kilian, J. (ed.) CRYPTO 2001. LNCS, vol. 2139, pp. 355–367. Springer, Heidelberg (2001)

[21] Xu, S., Mu, Y., Susilo, W.: Online/Offline signatures and multisignatures for AODV and DSR routing security. In: Batten, L.M., Safavi-Naini, R. (eds.) ACISP 2006. LNCS, vol. 4058, pp. 99–110. Springer, Heidelberg (2006)

[22] Yu, P., Tate, S.R.: An online/offline signature scheme based on the strong rsa assumption. In: AINA Workshops (1), pp. 601–606 (2007)

In appendices, we give elliptic curve analogues of existing divisible on-line/off-line signature schemes in order to fairly compare them with our proposed scheme. Suppose E is an elliptic curve over a finite field $\mathbb{F}$. Let g be a point in E with prime order p and let $\mathbb{G}$ be a group generated by g.

A. Crutchfield et al.'s Divisible On-Line/Off-Line Scheme CMTW-OS

The on-line/off-line signature scheme CMTW-OS is extracted from [6]. In [6], the authors construct an on-line/off-line threshold signature scheme which is a threshold version of this basic scheme. Let $\mathcal{S} = (\mathsf{G}, \mathsf{S}, \mathsf{V})$ be an ordinary signature scheme. The on-line/off-line signature scheme CMTW-OS $=$ (KeyGen, Sign^{off}, Sign^{on}, Ver), where

- KeyGen. Choose $x \in_R \mathbb{Z}_p$, and let $h = g^x$. Run the key generation algorithm of $\mathcal{S}$ to obtain (pk, sk) which is public/private key pair for $\mathcal{S}$. The public key of $\mathcal{OS}$ is (E, p, g, h, pk), and the private key is (x, sk).
- Sign^{off}. (The i-th run). Choose $r, m \in_R \mathbb{Z}_p$, and compute $u = g^r h^m$. Use the signing algorithm of $\mathcal{S}$ to obtain $\sigma = \mathsf{S}_{sk}(u)$. Store the state information r, m. The off-line signature token is σ.
- Sign^{on}. (The i-th run, on a message $m' \in \mathbb{Z}_p$). Retrieve m, r from the memory. Compute $r' = r + (m - m')x \mod p$. The on-line signature token is r'.
- Ver. (On a message-signature pair (m', Σ) where $\Sigma = (\sigma, r')$). Verify that whether $\mathsf{V}_{pk}(g^{r'} h^{m'}, \sigma) = 1$.

Theorem 2. *The on-line/off-line signature scheme constructed above is divisible and existentially unforgeable under adaptive chosen message attacks, provided that the underlying signature scheme $\mathcal{S}$ is existentially unforgeable against generic chosen message attacks and the one-more-discrete-log assumption [1] holds in $\mathbb{G}$.*

The proof of this theorem is omitted. Please refer to Theorem 2 of [6] for details.[3]

B. Bresson et al.'s Divisible On-Line/Off-Line Scheme BCG-OS

The on-line/off-line signature scheme BCG-OS is extracted from [3]. In [3], the authors construct an on-line/off-line threshold signature scheme which is a threshold version of this basic scheme. Let $\mathcal{S} = (\mathsf{G}, \mathsf{S}, \mathsf{V})$ be an ordinary signature scheme. The on-line/off-line signature scheme BCG-OS $=$ (KeyGen, Sign^{off}, Sign^{on}, Ver), where

- KeyGen. Choose $x, y \in_R \mathbb{Z}_p$, and let $h_1 = g^x, h_2 = g^y$. Run the key generation algorithm of $\mathcal{S}$ to obtain (pk, sk) which is public/private key pair for $\mathcal{S}$. The public key is (E, p, g, h_1, h_2, pk), and the private key is (x, y, sk).
- Sign^{off}. (The i-th run). Choose $r, s, m \in_R \mathbb{Z}_p$, and compute $u = g^m h_1^r h_2^s$. Use the signing algorithm of $\mathcal{S}$ to obtain $\sigma = \mathsf{S}_{sk}(u)$. Store the state information r, s, m. The off-line signature token is σ.

---

[3] The proof in [6] reduces the security of the on-line/off-line scheme CMTW-OS to the one-more-discrete-log assumption, or the collision resistance of a trapdoor hash function, or the unforgeability of $\mathcal{S}$. A small modification of this proof can simply reduce the security to the one-more-discrete-log assumption or the unforgeability of $\mathcal{S}$.

– Sign^{on}. (The i-th run, on a message m'). Retrieve r, s, m from the memory. Choose $r' \in_R \mathbb{Z}_p$ and compute $s' = s + y^{-1}[(m - m') + (r - r')x] \mod p$. The on-line signature token is (r', s').

– Ver. (On a message-signature pair (m', Σ) where $\Sigma = (\sigma, r', s')$). Verify that whether $V_{pk}(g^{m'} h_1^{r'} h_2^{s'}, \sigma) = 1$.

Remark 3. To reduce the on-line signing cost, we can move the selection of r' and computing $(r - r') \cdot x$ to the off-line phase. Thus, the on-line signing requires only 1 modular multiplication in $\mathbb{Z}_p$.

Theorem 3. *The on-line/off-line signature scheme constructed above is divisible and existentially unforgeable under adaptive chosen message attacks, provided that the underlying signature scheme S is existentially unforgeable against generic chosen message attacks and the discrete logarithm assumption holds in $\mathbb{G}$.*

The proof of this theorem is also omitted. Please refer to Theorem 1 of [3] for details.

C. Proving the Schnorr Signature Scheme [19] is Divisible

A variant of the Schnorr signature scheme can be naturally viewed as a divisible on-line/off-line signature scheme: Schnorr-OS $=$ (KeyGen, Sign^{off}, Sign^{on}, Ver).

– KeyGen. Choose $x \in_R \mathbb{Z}_p$, and let $h = g^x$. Let H be a hash function: $H : \{0,1\}^* \to \mathbb{Z}_p$. The public key is (E, p, g, h, H), and the private key is x.

– Sign^{off}. (The i-th run). Choose $r \in_R \mathbb{Z}_p$, and compute $u = g^r$. Store the state information r. The off-line signature token is u.

– Sign^{on}. (The i-th run, on a message m). Retrieve r from the memory. Set $c = H(m\|u)$ and compute $s = r - cx \mod p$. The on-line signature token is s.

– Ver. (On a message-signature pair (Σ, m) where $\Sigma = (u, s)$). Verify that whether $g^s h^{H(m\|u)} = u$.

Remark 4. The signature token is defined as $(u, s) \in \mathbb{G} \times \mathbb{Z}_p$ instead of $(c, s) \in \mathbb{Z}_p \times \mathbb{Z}_p$. This is to decrease the on-line signature length because the value of c can't be computed in the off-line phase.

Theorem 4. *The divisible on-line/off-line signature scheme Schnorr-OS is existentially unforgeable against adaptive chosen message attacks in the random oracle model, provided that the one-more-discrete-log assumption holds in $\mathbb{G}$.*

Please consult the full paper for the proof.

Speeding up Collision Search for Byte-Oriented Hash Functions

Dmitry Khovratovich, Alex Biryukov, and Ivica Nikolic

University of Luxembourg
{dmitry.khovratovich,alex.biryukov,ivica.nikolic}@uni.lu

Abstract. We describe a new tool for the search of collisions for hash functions. The tool is applicable when an attack is based on a differential trail, whose probability determines the complexity of the attack. Using the linear algebra methods we show how to organize the search so that many (in some cases — all) trail conditions are always satisfied thus significantly reducing the number of trials and the overall complexity.

The method is illustrated with the collision and second preimage attacks on the compression functions based on Rijndael. We show that slow diffusion in the Rijndael (and AES) key schedule allows to run an attack on a version with a 13-round compression function, and the S-boxes do not prevent the attack. We finally propose how to modify the key schedule to resist the attack and provide lower bounds on the complexity of the generic differential attacks for our modification.

1 Introduction

Bit-oriented hash functions like MD5 [17] and SHA [10] have drawn much attention since the early 1990s being the de-facto standard in the industry. Recent cryptanalytic efforts and the appearance of real collisions for underlying compression functions [19,18,4,13] motivated researchers to develop new fast primitives. Several designs were presented last years (LASH [2], Grindahl [12], RadioGatun [3], LAKE [1]) but many have already been broken [14,15,6].

While many hash functions are designed mostly from scratch one can easily obtain a compression function from a block cipher. Several reliable constructions, such as so-called Davies-Meyer and Miyaguchi-Preneel modes, were described in a paper by Preneel et al. [16].

In this paper we present a method for speeding up collision search for byte-oriented hash functions. The method is relatively generic, which is an advantage in the view that most collision attacks exploit specific features of the internal structure of the hash function (see, e.g., the attack on Grindahl [15]) and can be hardly carried on to other primitives.

Most of the collision attacks are differential in nature. They consider a pair of messages with the difference specified by the attacker and study propagation of this difference through the compression function — a *differential trail*. The goal of a cryptanalyst is to find a pair of messages that follows the trail (a conforming pair). Our idea is to deal with fixed values of internal variables as sufficient

M. Fischlin (Ed.): CT-RSA 2009, LNCS 5473, pp. 164–181, 2009.

conditions for the trail to be followed. We express all internal transformations as equations and rearrange them such that one can quickly construct a pair of executions that fits the trail.

We illustrate the method with a cryptanalysis of AES [7] in the Davies-Meyer mode. The AES block cipher attracted much attention and was occasionally considered as a basis for a compression function (mostly unofficially, though some modes were proposed [5] and AES-related hash designs were also investigated [12,11]).

This paper is organized as follows. In the next section we give an idea and formal description of the algorithm. Then in Section 3 we show how to find collisions, semi-free-start collisions and second preimages for the compression functions based on the versions of RIJNDAEL. Section 4 is devoted to the properties of the RIJNDAEL internal transformations, which are weaknesses in the hash function environment. Finally, we propose a modification to the original key schedule, which prevents our attack, and provide some lower bounds for attacks based on the differential techniques. The resulting hash function, which we call Cheetah, is formally introduced in Appendix.

2   Idea in Theory

State of the art. Most of the recent attacks on compression functions deal with a *differential trail*. Informally, a trail is a sequence of pairs of internal states with a restriction on the contents. An adversary looks for the pair of messages that produces such states in each round.

More formally, suppose that the compression function takes the initial value IV and the message M as an input and outputs the hash value H. The whole transformation is usually defined as a sequence of smaller transformations — rounds. Then the execution of a k-round compression function looks as follows:

$$IV \xrightarrow{f(M_1, \cdot)} S_1 \xrightarrow{f(M_2, \cdot)} S_2 \cdots S_{k-1} \xrightarrow{f(M_k, \cdot)} H,$$

where f is a round function of two arguments: the *message block* M_i and the current *internal state* S_{i-1}. The message blocks are a result of the message schedule — a transformation of the original message. A *collision* is a pair of messages (M, M') producing the same hash value H given the initial value IV.

In the collision search the exact contents of internal states are not important; the only conditions are the fixed IV and the coincidence of resulting hash values. Thus an adversary considers a pair of executions where the intermediate values are not specified. However, a naive collision search for a pair of colliding messages would have complexity $2^{n/2}$ queries (with the help of birthday paradox) where n is the bit length of the hash value. Thus the search should be optimized.

In the differential approach an adversary specifies the difference in message blocks M_i and internal states S_i. A pair of executions can be considered as the execution that deals with the differences: it starts and ends with a zero difference, and some internal differences are also specified.

A *differential trail* is a set of conditions on the pair of executions. We assume that this set of conditions is final, i.e. the attacker use them as is and tries message pairs one by one till all the conditions are satisfied. A trail may include non-differential conditions, ex. constraints on specific bits, in order to maximize the chances for the trail to be fulfilled. The complexity of the attack with a trail is then defined by the probability that a message pair produce an execution that fits the trail. This probability is determined by the nonlinear components (e.g. S-boxes) that affect the propagation of the difference.

First idea. As we pointed out before, an adversary may try all the possible pairs and check every condition. He may also strengthen a condition by *fixing* the input value (or the output) of a non-linear function. Then in each trail he checks whether the value is as specified. Thus to find a collision it would be enough to build an execution such that the specified values follow the trail, and the second execution will be derived by adding the differences to the first one. Our algorithm (given below) deals with this type of trails.

Our improvement. Our goal is to carry out the message trials so that many conditions are always satisfied. In such case the complexity of the attack is determined by the conditions that we do not cover. Before we explain how our algorithm works we introduce the notion of free variables.

First, we express all the transformations as equations, which link the internal variables. Variables refer to bits or bytes/words depending on the trail. Secondly, notice that the IV and the message fully define all the other variables and thus the full execution. We call *free variables*[1] a set of variables that completely and computationally fast define the execution. If some variables are pre-fixed the number of free variables decreases.

The idea of our method is to build a set of free variables provided that some variables are already fixed. The size of such a set depends on how many variables are fixed. The latter value also defines the applicability of our method. The heart of our tool is an algorithm for the search of free variables. It may vary and be combined with some heuristics depending on the compression function that it is applied for, but the main idea can be illustrated on the following example.

Example 1. Assume we have 7 byte variables s, t, u, v, x, y, and z which are involved in the following equations:

$$F(x \oplus s) \oplus v = 0;$$
$$G(x \oplus u) \oplus s \oplus L(y \oplus z) = 0;$$
$$v \oplus G(u \oplus s) = 0;$$
$$H(z \oplus s \oplus v) \oplus t = 0;$$
$$u \oplus H(t \oplus x) = 0.$$

---

[1] Recall the Gaussian elimination process. After a linear system has been transformed to the row echelon form all the variables are divided into 2 groups: bound variables and free variables. Free variables are to be assigned with arbitrary values; and bound variables are derived from the values of free variables.

where F, G, H, and L are bijective functions. Note that y is involved in only one equation so it can be assigned after all the other variables have been defined. Thus we temporarily exclude the second equation from the system. Then note that z is involved in only one equation among the remaining ones, so we again exclude the equation from the system. This Gaussian-like elimination process leads to the following system:

$$
\begin{aligned}
F(y \oplus z) \oplus L(\qquad\qquad u\oplus \qquad\qquad x)\oplus\ s\ &= 0;\\
z \oplus H^{-1}(\ t)\oplus \qquad\qquad v\oplus \qquad\quad\ s\ &= 0;\\
t \oplus H^{-1}(\ u)\oplus \qquad\qquad x \qquad\quad &= 0;\\
u \oplus G^{-1}(\ v)\oplus \qquad\qquad s\ &= 0;\\
v \oplus F(\ x \oplus\ s) &= 0.
\end{aligned}
$$

Evidently, x and s can be assigned randomly and fully define the other three variables. Thus x and s are free variables. Varying them we easily get other solutions.

Now assume that the variable u is pre-fixed to a value a. Then the system is transformed in a different way:

$$
\begin{aligned}
F(y \oplus z) \oplus L(\qquad x\oplus \qquad\qquad\ a\ ) \oplus s &= 0;\\
z \oplus H^{-1}(\ t)\oplus \qquad\quad v\oplus \qquad\qquad\ s &= 0;\\
t\oplus\ x\oplus \qquad\quad H^{-1}(a) \qquad\quad &= 0;\\
x\oplus\ F^{-1}(v)\oplus \qquad\qquad s &= 0;\\
G^{-1}(v)\oplus \qquad a\oplus \qquad s &= 0.
\end{aligned}
$$

Here only one variable — s — is free.

Now we provide a more formal description of the algorithm.

1. Build a system of equations based on the compression function. The values defined by the trail are fixed to constants.
2. Mark all the variables and all the equations as non-processed.
3. Find the variable involved in only one non-processed equation. Mark the variable and the equation as processed. If there is no such variable — exit.
4. If there exist non-processed equations go to Step 3.
5. Mark all non-processed variables as free.
6. Assign random values to free variables and derive variables of processed variables.

Depending on the structure of the equations, some heuristics can be applied at step 3. For example, if there are many linear equations, real Gaussian elimination can be applied. If there are terms of degree 2, one variable can be fixed to 0, and so on.

When the algorithm can be applied. If there is no restriction on the internal variables, the algorithm always succeeds: the message variables can be taken as free. As soon as we fix some internal variables we have fewer options for choosing free variables. In terms of block cipher based compression functions, we say that the more active S-boxes we have the fewer free variables exist.

The main property of the compression function that affects the performance of the algorithm is diffusion. The slower diffusion is, the more rounds can be processed by the algorithm. As we show in Section 3, a slow diffusion in the message schedule can be enough to maintain the attack.

This algorithm is not as universal as other algorithms dealing with non-linear equations: SAT-solver based and Gröbner basis based. However, if it works a cryptanalyst can generate a number of solutions in polynomial time while generic algorithms have exponential complexity. This is a real benefit, since we can use the algorithm at the top or at the bottom part of the trail, thus increasing probability of a solution.

Equation properties. There is a desired property of equations: each variable should be uniquely determined by the other ones. If this is not the case (fixing all but one variable may not give a bijection) then the last step of the algorithm becomes probabilistic (some values of free variables do not lead to the solution) or, on the contrary, some variables can be assigned by one of a few values. We can also emphasize not a single variable but a group of variables if it is fully determined by the other variables involved in the equation.

We conclude that under these assumptions the exact functions that link variables do not matter. The only requirement is that they can be easily inverted, which is typically true for the internal functions of a block cipher or a hash function. If no heuristics which mix rows are applied then the algorithm does not need the information about the non-linear functions, only the variables that are involved in. Thus we consider not a system of equations but a *matrix of dependencies* where rows correspond to equations, and columns to variables. The following matrix represents the system from Example 1:

$$
\text{Before triangulation:} \begin{pmatrix} s\ t\ u\ v\ x\ y\ z \\ \hline 1\ 0\ 0\ 1\ 1\ 0\ 0 \\ 1\ 0\ 1\ 0\ 1\ 1\ 1 \\ 1\ 0\ 1\ 1\ 0\ 0\ 0 \\ 1\ 1\ 0\ 1\ 0\ 0\ 1 \\ 0\ 1\ 1\ 0\ 1\ 0\ 0 \end{pmatrix}. \quad \text{After:} \begin{pmatrix} y\ z\ t\ u\ v\ |\ x\quad s \\ \hline 1\ 1\ 0\ 1\ 0\ |\ 1\quad 1 \\ 0\ 1\ 1\ 0\ 1\ |\ 0\quad 1 \\ 0\ 0\ 1\ 1\ 0\ |\ 1\quad 0 \\ 0\ 0\ 0\ 1\ 1\ |\ 0\quad 1 \\ 0\ 0\ 0\ 0\ 1\ |\ 1\quad 1 \\ \quad\quad\quad\quad\quad |\ _{free} \\ \quad\quad\quad\quad\quad |\ _{variables} \end{pmatrix}.
$$

Comparison to other methods. Several tools that reduce the search cost by eliminating some conditions were recently proposed. They are often referred to as *message modification* (a notion introduced by Wang in attacks on SHA) though there is often no direct "modification". The idea is to satisfy conditions by restricting internal variables to pre-fixed values and trying to carry out those restrictions from internal variables to message bits, which are controlled.

Compared to message modification and similar methods, our algorithm may give a solution even if the restrictions can not be carried out to message bits directly. If a system of equations is solved, the algorithm produces one or many solutions that satisfy the restrictions. Even if all the restrictions can not be processed then one may try to solve the most expensive part. Thus we expect our method to work in a more general and automated way compared to the dedicated methods designed before.

3   Idea in Practice

We illustrate our approach with the cryptanalysis of a hash function based on RIJNDAEL [7] in the Davies-Meyer mode. The security of RIJNDAEL as a compression function has been frequently discussed in both official and non-official talks though no clear answer was provided in favour of or against such a construction. Additionally, the exact parameters of RIJNDAEL as a hash function are a subject of a discussion.

The Davies-Meyer mode has been chosen due to its message length/block length ratio, which is crucial for the performance. Assume we want to construct a compression function with performance comparable to SHA-1. The AES block length of 128 bits is too small against birthday attack so 160 bits would be the minimal admissible value. The size of a message block to be hashed should be also increased in order to achieve better performance. However, there should be a tradeoff between the message length and the number of rounds. A simple solution is to take the message block equal to 2 internal blocks (320 bits). The 14-round RIJNDAEL-based construction gives us performance comparable to SHA-1. We will concentrate on this set of parameters though other ones will be also pointed out.

3.1   Properties of Rijndael Transformations. How to Build a Trail

RIJNDAEL is surprisingly suitable for the analysis with our method due to simplicity of its operations and properties of its S-boxes. A differential trail provides a set of active S-boxes. Due to the special differential properties of RIJNDAEL S-boxes (2^{-6} maximal differential probability) the number of possible input/output values is limited to not more than 4 possibilities. We take one of the few values of an S-box input that provides the propagation of differences as a sufficient condition. We found a 12-round trail (Figure 4) which has 50 active S-boxes (44 in the SubBytes transformations and 6 in the KeySchedule). However, most of the active S-boxes are in the upper part of the trail, which allows us to use the algorithm.

The crucial weakness of the RIJNDAEL key schedule, which is exactly the message schedule procedure in the considered compression function, is the XOR operation that produces columns of the next subkey. It provides a good diffusion as a key schedule, which was the goal of the RIJNDAEL design, but is not adapted for the use in a compression function, where all the internal variables are known to an adversary.

The `KeySchedule` transformation for a key of size 256 bits and more is given by the following expressions:

$$
\begin{aligned}
k_{i,0} &\leftarrow S(k_{i+1,NK-1}) \oplus C_r, & 0 \leq i \leq 3; \\
k_{i,j} &\leftarrow k_{i,j-1} \oplus k_{i,j}, & 0 \leq i \leq 3,\ 1 \leq j \leq 3; \\
k_{i,4} &\leftarrow S(k_{i,3}) \oplus k_{i,4}, & 0 \leq i \leq 3; \\
k_{i,j} &\leftarrow k_{i,j-1} \oplus k_{i,j}, & 0 \leq i \leq 3,\ 5 \leq j \leq NK-1,
\end{aligned}
\tag{1}
$$

where $S()$ stands for the `SubBytes` transformation, and C_r — for the round-dependant constant. It is easy to check that the first byte in a row affects all the other bytes in the row, so that any difference will propagate through all the xor operations. NK is a parameter equal to 8 for a 256-bit key. However, the `KeySchedule` transformation is invertible, and its inversion has a slow diffusion. This is the fact that we exploit. More precisely, the formulas for the inversion are as following:

$$
\begin{aligned}
k_{i,j} &\leftarrow k_{i,j-1} \oplus k_{i,j}, & 0 \leq i \leq 3,\ NK-1 \geq j \geq 5, \\
k_{i,4} &\leftarrow S(k_{i,3}) \oplus k_{i,4}, & 0 \leq i \leq 3; \\
k_{i,j} &\leftarrow k_{i,j-1} \oplus k_{i,j}, & 0 \leq i \leq 3,\ 1 \leq j \leq 3; \\
k_{i,0} &\leftarrow S^{-1}(k_{i+1,NK-1} \oplus C_r), & 0 \leq i \leq 3.
\end{aligned}
\tag{2}
$$

We build two trails: for 12 rounds and for 7 rounds. We use a local collision illustrated in Figure 1 as a base for both of them. There, a one byte difference is injected by the `AddRoundKey` transformation and spread to 4-byte difference after `MixColumns`. The 4-byte difference is canceled out by the next `AddRoundKey`. Due to a long message block both differences can be arranged into different columns. The 4-byte difference is fully determined by the contents of the one active S-box. We mark this value by a in Figure 1. It is a *sufficient* condition for the local collision.

If we start with this pattern and go down all the bytes of the message block will likely have the difference. However, the backward propagation is much different. We can build a 7-round trail with only 9 active S-boxes (Figure 5). In order to build a longer trail we swap the left and the right halves of the message block and use some ad-hoc tricks in the first rounds. As a result, we obtain 12-round trail with 50 active S-boxes (Figure 4).

3.2 Collisions, Second Preimages and the Matrix of Dependencies for the Rijndael-Based Hash

Matrix of Dependencies. First we explain in details our usage of variables and equations. We consider *byte* variables: the IV (4*NB variables[2]), the output (4*NB), the message (4*NK per message schedule round), the internal states. We deal with two internal states per round: after the `SubBytes` transformations and after the `MixColumns` transformation. Thus we obtain 8*NB variables per round. The equations are derived from the following transformations:

---

[2] NB is the number of columns in the internal state. NB is equal to 8 in RIJNDAEL-256.

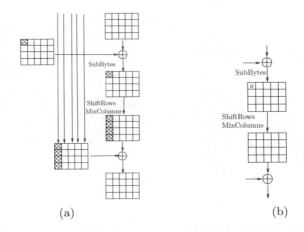

(a)   (b)

Fig. 1. Local collision: non-zero differences (a) and fixed values (b)

   I `SubBytes∘AddRoundKey` transformations: 4*NB equations in each internal
     round;
  II `MixColumns∘ShiftRows` transformations: NB equations in each internal
     round;
 III `KeySchedule` transformations: 4*NK equations in each schedule round.

   The `MixColumns` transformation is actually a set of 4 linear transformations.
In the latter ones any 4 of 5 variables uniquely determine the other one. For the
`MixColumns` transformation as a whole a more complicated property holds: any
4 of 8 variables are determined by the other ones.
   The variables that are predefined by a trail are substituted into the equations
and are not considered in the matrix.

320/160. 5 rounds. The simplest challenge is to build a 5-round collision for
the RIJNDAEL-based hash with 320-bit message block and 160-bit internal state.
The trail is derived by removing the first two rounds from the 7-round trail
(Figure 5). There are 5 active S-boxes, which fix 5 of the 320 internal variables.
There are 225 equations. The resulting matrix of dependencies is presented in
Figure 2. The non-zero elements are color pixels with green ones representing
the `MixColumns` transformation.
   The value of the one-byte difference is chosen randomly as well as that of
the active S-box. Let us denote the one-byte difference by δ and the input to
the active S-box by a. Then the 4-byte difference is the `MixColumns` matrix M
multiplied by $(S(a) + S(a + \delta), 0, 0, 0)$.
   The matrix is easily triangulated (Figure 3). We obtain 55 free variables. Any
assignment of those variables and 5 fixed S-box inputs fully determine the IV
and the message.

320/160. 7 rounds. The trail in Figure 5 is a 7-round collision trail with 9
active S-boxes. Although the triangulation algorithm can not be directly applied

320 variables

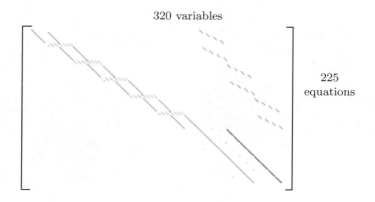

225
equations

Fig. 2. The matrix of dependencies for the 5-round trail before the triangulation

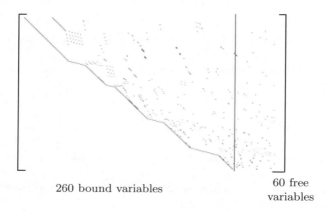

260 bound variables

60 free
variables

Fig. 3. The matrix of dependencies after the triangulation

to the resulting matrix, we can run it for the first 5 rounds, which contain 8 of 9 active S-boxes. Since we have several free variables we are able to generate many colliding pairs. About one of 2^7 pairs satisfy the condition on the one remaining active S-box so we repeat the last step of the algorithm 2^7 times and obtain a 7-round collision. The exact colliding messages are presented in Table 1.

320/160. 12 rounds. A 12-round trail is presented in Figure 4. We swap the left and the right halves of the message block and use some ad-hoc tricks in the first rounds. As a result, we obtain a 12-round trail with 50 active S-boxes with only 6 of them in KeySchedule transformations.[3]

---

[3] The most of non-zero columns in the differences between message blocks are of the form $(a, 0, 0, 0)$ or (b, c, d, e) where a, b, c, d and e are the same in all message blocks. Those values are that are used for obtaining a local collision (Figure 1). They are marked as grey cells in Figures 5 and 4. If they interleave some other values are produced. The latter ones are marked as olive cells.

Table 1. 7-round collision for the RIJNDAEL-based compression function. Message bytes with different values are emphasized.

IV	$b8$ 29 68 $d1$ $f8$ $5b$ $d0$ 01 bd 17 05 83 $8a$ 43 $4b$ 40 40 $9e$ $0c$ $c5$	Message 1	**77 e6 a7 1e** $e3$ **40 e6 ef 56** 26 $7e$ $1b$ aa $2b$ fa **44 70 88 66** $0c$ 04 $2b$ $7b$ $e1$ $d6$ **df 4d 09 52** $5c$ $4a$ 81 31 98 $b6$ **df 67 79 c6** ab
Output	83 $e5$ 06 $a4$ 46 $5f$ $e7$ $7c$ ba 49 $8e$ $7d$ $1e$ bd 96 $b8$ $d4$ $e3$ $e9$ $a0$	Message 2	**76 e7 a6 1f** $e3$ **42 e4 ed 54** 26 $7e$ $1b$ aa $2b$ fa **45 71 89 67** $0c$ 04 $2b$ $7b$ $e1$ $d6$ **de 4c 08 53** $5c$ $4a$ 81 31 98 $b6$ **dc 64 7a c5** ab

The trail is too long to be processed by the triangulation algorithm directly. Instead we fix not all S-box inputs. More precisely, we fix the 6 variables that enter the active S-boxes in the first three `KeySchedule` transformations (actually all active S-boxes in the message scheduling) and the 35 variables that are the outputs of active S-boxes in first 4 rounds. There are 9 active S-boxes left unfixed. We have 11 free variables and generate $2^{7*9} = 2^{63}$ colliding pairs so that one of them pass through those 9 S-boxes and gives the 12-round collision. The resulting complexity is 2^{63} compression function calls.

Fixed IV. So far we considered that the IV is constant but can be freely chosen, mainly because we do not attack an already existing standard or a particular proposal. Nevertheless, compression functions with a similar structure, which may be designed later, would require an attack with the fixed IV.

The algorithm described before may be easily adapted to this case. We just mark all the input variables in the trail as pre-fixed, which is equivalent to just the removal of the corresponding columns from the matrix of dependencies. The number of equations is not changed so the probability of successful triangulation can only decrease, not increase. This is the case: now we are not able to reduce the matrix for the 5-round trail, but for the 3-round one we can still do this. This fact does not imply that the 3-round collisions is the maximum achieved level. Actually we just bypass the next two rounds with some probability. If the number of active S-boxes in the trail after these 3 rounds is not large, this probability may still be reasonable.

For example, we can use the trail for 7-rounds collision and process by the algorithm only first three rounds. Then we have to bypass through 3 active S-boxes, which requires about 2^{21} evaluations of the compression function and can be done in real time.

512/256. If we just increase the hash length keeping the message/hash ratio we actually get a much weaker compression function.

For example, a differential trail for 13 rounds with no active S-boxes in the message scheduling can be easily built from the trail in Figure 5. The matrix triangulation algorithm works for 7 rounds, and a 13-round collision can be found after 2^{35} computations of the compression function, which is substantially faster than the birthday attack.

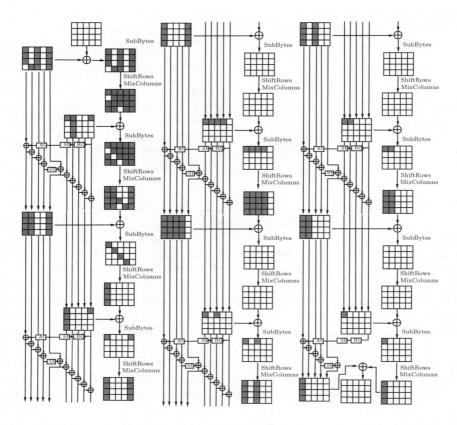

Fig. 4. 12-round differential trail for the RIJNDAEL-based compression function with the 320-bit message block and the 160-bit internal state (RIJNDAEL-hash 320/160)

3.3 Second Preimage Attack

320/160. Here we assume that the message is fixed, the IV is constant but not fixed and we have to find a message such that it produces the same hash value as the first one. Our goal is to obtain a second preimage faster than for 2^{160} calls. We just take any trail such that the conditions on the message variables do not confuse with the pre-fixed values. For example, the 7-round trail (Figure 5) do not impose such restrictions on message variables.

We mark all the message variables as fixed and run the triangulation algorithm on first three rounds. We obtain $60 - 40 - 6 = 14$ variables that can be assigned randomly. We generate 2^{21} pairs (IV, second message) so that one of them passes the three other active S-boxes in rounds 4-7. The resulting complexity of the second-preimage search is about 2^{21} compression function calls.

Although we have a longer collision trail (Figure 4), it can not be used because the number of active S-boxes is bigger than the number of the degrees of freedom.

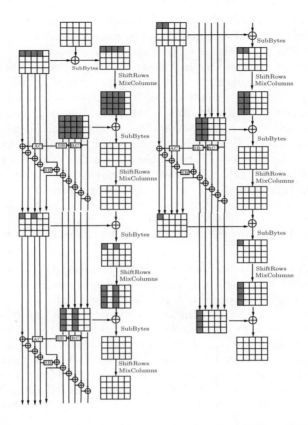

Fig. 5. 7-round differential trail for RIJNDAEL-hash 320/160

4   Rijndael Properties That May Lead to Weaknesses in Compression Functions

Here we summarize the properties of RIJNDAEL that allow us to attack RIJN-DAEL-based compression functions.

Let us first look at the RIJNDAEL key schedule ((1), (2)). The weakness that we exploited is a non-symmetric diffusion. More precisely, one byte in block i affects only two or three bytes in block $i-1$. Furthermore, one can build a trail in a key schedule without active S-boxes for NK−4 schedule rounds. Full diffusion in key schedule may take up to 4*NK schedule rounds if we consider one-byte difference in a corner byte and proceed backwards.

Even active S-boxes in a trail give some additional power to an adversary. Due to the differential properties of the RIJNDAEL S-box non-zero difference Δa can be converted to any of about 127 differences Δb; only half of differences can not be reached. The exact value of the output difference is guaranteed by the value of the S-box input variable.

The derivation of unknown internal variables from the known ones is also easier than that in the bit-oriented hash functions such as the SHA-family. In RIJNDAEL if we know two of three variables in the bitwise addition or four of eight in MixColumns then we can uniquely determine the other variables. We would not be able to do this if we had equations of type $x_1 x_2 = x_3$. So we claim that equations in RIJNDAEL are actually pseudo-linear rather that non-linear. Though we do not know how to exploit this fact in attacks on RIJNDAEL as a block cipher, it is valuable if we consider a RIJNDAEL-based compression function.

Finally we note that our attacks are only weakly dependant on some RIJN-DAEL parameters such as actual S-box tables, the MixColumns matrix coefficients and ShiftRows rotation values. For example, we only need a row that is not rotated by the ShiftRows but it does not matter matter where it is exactly located.

5 Modification to the Message Schedule — Our Proposal

In this section we propose an improvement to the RIJNDAEL message schedule, which prevents the low-weight trails that were shown above.

The key idea is to use primitives providing good diffusion. The RIJNDAEL round function was designed so that no low-weight trails can be built for the full cipher. We propose to use a modified version of this round function in the message schedule for future RIJNDAEL-based hash functions.

First, we significantly extend the size of the message block that is processed by one call of the compression function. Secondly, 256 bits is going to be the main digest size for SHA-3 [9]. A 1024-bit message block combined with a 256-bit internal state give a good security/performance tradeoff, which is justified below.

The message block is treated as a 8×16 byte square and passes through 3 iterations of a round function. Like that of RIJNDAEL, the round function we propose is a composition of the SubBytes, the ShiftRows, and the MixColumns transformations. While S-boxes remain the same, the ShiftRows and MixColumns operations are modified in order to get a maximal possible diffusion. We propose to use the following offset table and the MixColumns matrix:

The matrix A, which is modified version of the matrix used in Grindahl [12], is an MDS-matrix.

Table 2. ShiftRows and MixColumns parameters for a new message schedule proposal

Index	Offset	Index	Offset
0	0	4	5
1	1	5	6
2	2	6	7
3	3	7	8

$$A = \begin{pmatrix} 02 & 0c & 06 & 08 & 01 & 04 & 01 & 01 \\ 01 & 02 & 0c & 06 & 08 & 01 & 04 & 01 \\ 01 & 01 & 02 & 0c & 06 & 08 & 01 & 04 \\ 04 & 01 & 01 & 02 & 0c & 06 & 08 & 01 \\ 01 & 04 & 01 & 01 & 02 & 0c & 06 & 08 \\ 08 & 01 & 04 & 01 & 01 & 02 & 0c & 06 \\ 06 & 08 & 01 & 04 & 01 & 01 & 02 & 0c \\ 0c & 06 & 08 & 01 & 04 & 01 & 01 & 02 \end{pmatrix}$$

This round function provides full diffusion after 3 rounds thus giving 4096 bits to inject to the internal state. Since the internal state is 16 times smaller, we propose to increase the number of rounds to 16. The resulting compression function is more formally introduced in Appendix.

Resistance to attacks. We did not manage to find a good trail for the resulting compression function so we can not apply our attack. Furthermore, now we give arguments supporting that low-weight trails are impossible in the resulting design: we give a lower bound on the number of active S-boxes in such a trail.

We consider a 16-round trail which starts and ends with a zero-difference state. Let us denote the number of non-zero differences in the internal state before the SubBytes transformation by s_i, $1 \leq i \leq 16$. The last s_i is equal to zero. Let us also denote by c_i the number of non-zero differences in the internal after the internal MixColumns transformation. The last c_i is equal to 0 as well. Finally, we denote by m_i the number of non-zero differences in the round message block that is xored to the internal state. These differences either cancel non-zero differences in the internal state or create them. Thus the following condition holds

$$s_i + c_{i-1} \geq m_i \tag{3}$$

Due to the branch number of the internal MixColumns transformation c_i is upper bounded: $c_i \leq 4s_i$. Thus we obtain the following:

$$s_i + 4s_{i-1} \geq m_i \ \Rightarrow \ \sum_i s_i + \sum_i 4s_{i-1} \geq \sum_i m_i \ \Rightarrow \ S \geq \frac{M}{5},$$

where S is the number of active S-boxes in the internal state of the compression function, and M is the number of non-zero byte differences in the expanded message.

Now we estimate the minimum number of non-zero byte differences in the message scheduling only. First we note that this number is equal to the number of the active S-boxes in the message scheduling extended to 4 round. Such a 4-round transformation is actually a RIJNDAEL-like block cipher, which can be investigated using the theory of the wide trail design by Daemen and Rijmen [8].

Daemen and Rijmen estimated the minimum number of active S-boxes in 4 rounds of a RIJNDAEL-like block cipher (Theorem 3, [8]). The sufficient condition to apply their theorem is that the ShiftRows should be diffusion optimal: bytes from a single column should be distributed to different columns, which is the case. Thus the number of active S-boxes can be estimated as the square of the branch number of the 8×8 MixColumns matrix, which is equal to 9. As a result, any pair of different message blocks has difference in at least $M = 81$ bytes of ExpandedBlock. This implies the lower bound 17 for S. Thus we obtain the following proposition.

Proposition 1. *Any collision trail has at least 17 active S-boxes in the internal state.*

Thus any attack using such minimal trail as is would be only slightly faster than the birthday attack. However, we expect that the values of M even close to minimal do not give collision trails due to the following reasons:

- Small number of active S-boxes in the internal state implicitly assumes many local collisions;
- The distribution of non-zero differences in the message scheduling is not suitable for local collisions due to high diffusion;
- The MixColumns matrix in the message scheduling differs from that of the internal transformation so, e.g., 4-byte difference collapse to 1-byte difference via only one of two transformations.

6   Conclusions and Future Directions

We proposed the triangulation algorithm for the efficient search of the message pairs that fit a differential trail with fixed internal variables. We illustrated the work of the algorithm by applying it to RIJNDAEL in the Davies-Meyer mode with different parameters. Although the trails that we built contain many active S-boxes, the task of the search for a message pair becomes much easier with our algorithm. It allows to build message pairs that satisfy subtrails of an original trails. Such subtrail can be chosen in order to minimize the number of active S-boxes in the other part of the trail.

In Table 3 we summarize our efforts on building collisions and preimages for RIJNDAEL-based compression functions.

Table 3. Summary of attacks

Hash length	Message length	Rounds	Compl.	Type of a collision
160	320	7	2^7	Full collision
160	320	12	2^{63}	Full collision
160	320	7	2^{21}	Second preimage
256	512	13	2^{35}	Full collision

We also investigated why RIJNDAEL as a compression function is vulnerable to collision attacks. We showed how the non-symmetric diffusion in the message schedule allows to build long differential trails.

As a countermeasure, we propose a new version of the message schedule for the RIJNDAEL-based compression functions and provide lower bounds for the probability of differential trails for the resulting function.

Acknowledgements

The authors thank anonymous reviewers for their valuable and helpful comments. Dmitry Khovratovich is supported by PRP "Security & Trust" grant of the University of Luxembourg. Ivica Nikolić is supported by the BFR grant of the FNR.

References

1. Aumasson, J.-P., Meier, W., Phan, R.C.-W.: The hash function family LAKE. In: Nyberg, K. (ed.) FSE 2008. LNCS, vol. 5086, pp. 36–53. Springer, Heidelberg (2008)
2. Bentahar, K., Page, D., Saarinen, M.-J.O., Silverman, J.H., Smart, N.: LASH, Tech. report, NIST Cryptographic Hash Workshop (2006)
3. Bertoni, G., Daemen, J., Peeters, M., van Assche, G.: Radiogatun, a belt-and-mill hash function (2006), http://radiogatun.noekeon.org/
4. De Cannière, C., Rechberger, C.: Finding SHA-1 characteristics: General results and applications. In: Lai, X., Chen, K. (eds.) ASIACRYPT 2006. LNCS, vol. 4284, pp. 1–20. Springer, Heidelberg (2006)
5. Cohen, B.: AES-hash, International Organization for Standardization (2001)
6. Contini, S., Matusiewicz, K., Pieprzyk, J., Steinfeld, R., Jian, G., San, L., Wang, H.: Cryptanalysis of LASH. In: Nyberg, K. (ed.) FSE 2008. LNCS, vol. 5086, pp. 207–223. Springer, Heidelberg (2008)
7. Daemen, J., Rijmen, V.: AES proposal: Rijndael, Tech. report (1999), http://csrc.nist.gov/archive/aes/rijndael/Rijndael-ammended.pdf
8. Daemen, J., Rijmen, V.: The wide trail design strategy. In: IMA Int. Conf., pp. 222–238 (2001)
9. Cryptographic hash project, http://csrc.nist.gov/groups/ST/hash/index.html
10. FIPS 180-2. secure hash standard (2002), http://csrc.nist.gov/publications/
11. International Organization for Standardization, The Whirlpool hash function. iso/iec 10118-3:2004 (2004)
12. Knudsen, L.R., Rechberger, C., Thomsen, S.S.: The grindahl hash functions. In: Biryukov, A. (ed.) FSE 2007. LNCS, vol. 4593, pp. 39–57. Springer, Heidelberg (2007)
13. Manuel, S., Peyrin, T.: Collisions on SHA-0 in one hour. In: Nyberg, K. (ed.) FSE 2008. LNCS, vol. 5086, pp. 16–35. Springer, Heidelberg (2008)
14. Matusiewicz, K., Peyrin, T., Billet, O., Contini, S., Pieprzyk, J.: Cryptanalysis of FORK-256. In: Biryukov, A. (ed.) FSE 2007. LNCS, vol. 4593, pp. 19–38. Springer, Heidelberg (2007)
15. Peyrin, T.: Cryptanalysis of Grindahl. In: Kurosawa, K. (ed.) ASIACRYPT 2007. LNCS, vol. 4833, pp. 551–567. Springer, Heidelberg (2007)
16. Preneel, B., Govaerts, R., Vandewalle, J.: Hash functions based on block ciphers: A synthetic approach. In: Stinson, D.R. (ed.) CRYPTO 1993. LNCS, vol. 773, pp. 368–378. Springer, Heidelberg (1994)
17. Rivest, R.L.: The MD5 message-digest algorithm, request for comments (RFC 1320), Internet Activities Board, Internet Privacy Task Force (1992)
18. Wang, X., Yin, Y.L., Yu, H.: Finding collisions in the full SHA-1. In: Shoup, V. (ed.) CRYPTO 2005. LNCS, vol. 3621, pp. 17–36. Springer, Heidelberg (2005)
19. Wang, X., Yu, H.: How to break MD5 and other hash functions. In: Cramer, R. (ed.) EUROCRYPT 2005. LNCS, vol. 3494, pp. 19–35. Springer, Heidelberg (2005)

A   Hash Function Cheetah-256

Now we formally introduce a hash function proposal, which is based on the ideas discussed in Section 5. Due to space restrictions we limit ourselves to the design of the compression function.

The Cheetah compression function is an iterative transformation based on the RIJNDAEL block cipher. The 128-byte message block is expanded to a 512-byte block by the message schedule. The internal state is of size 32 bytes and is iterated for 16 rounds. The output hash value is 32 bytes (256 bits).

The message block is expanded by the means of the message schedule. The resulting block is divided into 16 vectors, which are xored to the internal state before every round. The Cheetah compression function is then defined by the following pseudo-code:

```
CheetahCompression(IntermediateHashValue, MessageBlock) {
    InternalState = IntermediateHashValue;
    ExpandedBlock = MessageExpansion(MessageBlock);
    for(i=1; i<= 16; i++)
    {
        InternalState +=RoundBlock(ExpandedBlock,i);
        InternalState = InternalRound(InternalState);
    }
    return InternalState;
}
```

The procedures *MessageExpansion*, *RoundBlock*, and *InternalRound* are determined below.

Message Schedule. The *MessageExpansion* procedure is a RIJNDAEL-like transformation, which is defined in pseudocode as follows:

```
MessageExpansion(byte MessageBlock[128]) {
    byte ExpandedBlock[512];
    ExpandedBlock[0..127] = MessageBlock;
    for(i=1; i<=3; i++)
    {
        SubBytes(MessageBlock);
        ShiftRows8(MessageBlock);
        MixColumn8(MessageBlock);
        AddRoundConstant(MessageBlock,i);
        ExpandedBlock[128*i..128*i+127] = MessageBlock;
    }
}
```

The *SubBytes* transformation is the byte-wise SubBytes transformation used in RIJNDAEL. The ShiftRows8 and the MixColumns8 operation parameters were given in Table 2.

The *AddRoundConstant* operation adds a 32-bit constant to the message block. The constant is a function of the round index r:

$$m_{i,0} = S[4 * r + i], \ 0 \leq i \leq 3,$$

where S stands for the S-box.

The *RoundBlock* operation selects a 32-byte block from ExpandedBlock $=$ (E_0, E_1, E_2, E_3). Define the round index r as $r = 4l + m$, $0 \leq l, m \leq 3$. Then the selected block is the 4×8 byte array M_r, that is defined as follows:

$$M_r = (m_{i,j})^{4 \times 8}, \ E_l = (e_{i,j})^{8 \times 16};$$

$$m_{i,j} = e_{4*(m\%2)+i, 4*(m/2)+j}.$$

The selected block is bytewise xored to the InternalState: $a_{i,j}^{\text{new}} \leftarrow a_{i,j} + m_{i,j}$.

Internal round. The *InternalRound* transformation is actually the RIJNDAEL round as it would be used with 32-byte block. It consists of three operations: SubBytes, ShiftRows, and MixColumns.

```
InternalRound(byte InternalState[256]) {
    SubBytes(InternalState);
    ShiftRows4(InternalState);
    MixColumn4(InternalState);
}
```

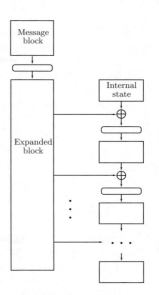

Fig. 6. The outline of the compression function

The SubBytes operation has already been defined above. Both the ShiftRows and MixColumns operations treat the InternalState as a byte array of size 4×8, with 4 rows and 8 columns.

Parameters of the ShiftRows and MixColumns transformations are the same as that of RIJNDAEL-256 (Table 4).

Table 4. ShiftRows and MixColumns parameters for the internal round function

Row index i	Offset c_i
0	0
1	1
2	3
3	4

$$B = \begin{pmatrix} 02 \ 03 \ 01 \ 01 \\ 01 \ 02 \ 03 \ 01 \\ 01 \ 01 \ 02 \ 03 \\ 03 \ 01 \ 01 \ 02 \end{pmatrix}.$$

Hard and Easy Components of Collision Search in the Zémor-Tillich Hash Function: New Attacks and Reduced Variants with Equivalent Security

Christophe Petit[1,*], Jean-Jacques Quisquater[1],
Jean-Pierre Tillich[2], and Gilles Zémor[3]

[1] UCL Crypto Group[**],
Université catholique de Louvain Place du levant 3
1348 Louvain-la-Neuve, Belgium
christophe.petit@uclouvain.be,jjq@uclouvain.be
[2] Equipe SECRET
INRIA Rocquencourt 78153 Le Chesnay, France
jean-pierre.tillich@inria.fr
[3] Institut de Mathématiques de Bordeaux UMR 5251
Université Bordeaux 1
351 Cours de la Libération
33405 Talence, France
Gilles.Zemor@math.u-bordeaux1.fr

Abstract. The Zémor-Tillich hash function has remained unbroken since its introduction at CRYPTO'94. We present the first *generic* collision and preimage attacks against this function, in the sense that the attacks work for any parameters of the function. Their complexity is the *cubic root* of the birthday bound; for the parameters initially suggested by Tillich and Zémor they are very close to being practical. Our attacks exploit a separation of the collision problem into an easy and a hard component. We subsequently present two variants of the Zémor-Tillich hash function with essentially the same collision resistance but reduced outputs of $2n$ and n bits instead of the original $3n$ bits. Our second variant keeps only the hard component of the collision problem; for well-chosen parameters the best collision attack on it is the birthday attack.

1 Introduction

Since its introduction at CRYPTO'94, the Zémor-Tillich hash function has kept on appealing Cryptographers by its originality, its elegance, its simplicity and its security. The function computation can be parallelized and even the serial version is quite efficient as it only requires XOR, SHIFT and TEST operations. Uniform

---

[*] Research Fellow of the Belgian Fund for Scientific Research (F.R.S.-FNRS) at Université catholique de Louvain (UCL).

[**] A member of BCRYPT network.

M. Fischlin (Ed.): CT-RSA 2009, LNCS 5473, pp. 182–194, 2009.
© Springer-Verlag Berlin Heidelberg 2009

distribution of the outputs follows from a graph theoretical interpretation of the hash computation, and collision resistance is strictly equivalent to an interesting group theoretical problem [9].

A few publications have claimed attacks on the Zémor-Tillich hash function. However, a closer look at these papers reveals that the scheme has not been seriously threatened so far. Some of the claimed "attacks" are unpractical, creating very long colliding messages [3]. Others are trapdoor attacks that can be avoided by fixing the parameters in an appropriate way [2,1,8]. A last, important class of attacks are subgroup attacks [8], damaging for particular parameters in a similar way as RSA algorithm can be insecure if the parameters are not correctly generated. For well-chosen parameters, the function has remained unbroken so far.

In this paper, we present new collision and preimage subgroup attacks against the Zémor-Tillich hash function. Unlike previous ones, our attacks are generic in the sense that they work for any parameters of the function. With a time complexity close to $2^{n/2}$, our attacks beat by far the birthday bound and ideal preimage complexities which are $2^{3n/2}$ and 2^{3n} for the Zémor-Tillich hash function. The attacks are practical up to $n \approx 120, 130$ that is very close to the parameter's lower bound $n \geq 130$ initially proposed by Zémor and Tillich. As the attacks include a birthday search in a reduced set of size 2^n they do not invalidate the scheme but rather suggest that the initial parameters were too small.

Our attacks exploit a separation of the collision problem into an easy and a hard component, and suggest that an output of n bits should be extracted from the original $3n$ bits of Zémor-Tillich. We consequently present two reduced versions of Zémor-Tillich, the vectorial and projective versions with output sizes respectively $2n$ and n, and we show that their collision resistance is essentially equivalent to the collision resistance of the original Zémor-Tillich.

This paper is organized as follows: the Zémor-Tillich hash function is recalled in Section 2. In Section 3 we present a general result separating hard and easy components of the collision problem, then we apply this result in Section 4 to obtain a generic collision search algorithm with time complexity close to $2^{n/2}$ (while the birthday bound is $2^{3n/2}$). This collision algorithm is extended in Section 5 to a generic preimage attack with the same complexity (while the ideal bound would be 2^{3n}), and memory free versions of these algorithms are given in Section 6. Finally, we introduce the vectorial and projective versions of Zémor-Tillich in Sections 7 and 8 and conclude the paper in Section 9.

2   The Zémor-Tillich Hash Function

Let $m = m_0 m_1 ... m_k$ be the bit string representation of a message m. Let $P_n(X)$ be an irreducible polynomial of degree n (Tillich and Zémor suggested using $130 \leq n \leq 170$) and let us represent the field $\mathbb{F}_{2^n}$ by $\mathbb{F}_2[X]/(P_n(X))$. Let A_0, A_1 be the matrices of $G := SL(2, \mathbb{F}_{2^n})$ (the group of 2×2 matrices over $\mathbb{F}_{2^n}$ with unitary determinant) defined by

$$A_0 = \begin{pmatrix} X & 1 \\ 1 & 0 \end{pmatrix} \qquad A_1 = \begin{pmatrix} X & X+1 \\ 1 & 1 \end{pmatrix}$$

The Zémor-Tillich hash value of m is defined as the matrix product [9]

$$h_{ZT}(m) := A_{m_0} A_{m_1} ... A_{m_k}.$$

As the group $SL(2, \mathbb{F}_{2^n})$ has size $2^n(2^{2n} - 1)$, the output size is roughly $3n$ bits if the matrices of $SL(2, \mathbb{F}_{2^n})$ are mapped to bitstrings.

3   Hard and Easy Components of Collision Search

The best attack so far against the Zémor-Tillich hash function has been the subgroup attack of Steinwandt et al. [8]. However, as this attack exploits subgroups of $SL(2, \mathbb{F}_{2^n})$ that are specific to composite degrees n and particular polynomials $P_n(X)$, it can be simply prevented by choosing n in an appropriate way.

In this section, we consider the generic subgroups of $SL(2, \mathbb{F}_{2^n})$ (subgroups existing for any parameter n), including the subgroups of diagonal or triangular matrices and the subgroups of matrices with a given left or right eigenvector. We show that finding elements of these subgroups together with their factorization is nearly as hard as finding collisions for the Zémor-Tillich hash function. As our reductions involve solving discrete logarithms in $\mathbb{F}_{2^n}^*$ we do not claim ppt (probabilistic polynomial time) reductions but reductions that are practical for the parameters initially suggested by Zémor and Tillich.

We start with an easy proposition that will simplify our proofs later.

Proposition 1

(a) Let $(\,a\ b\,), (\,a'\ b'\,) \in \mathbb{F}_{2^n}^2 \setminus \{(\,0\ 0\,)\}$ and $M \in SL(2, \mathbb{F}_{2^n})$ such that $(\,a\ b\,) M = (\,a'\ b'\,)$. Then there exists $\epsilon \in \mathbb{F}_{2^n}$ such that $M = \begin{pmatrix} a_0^{-1} & b \\ 0 & a \end{pmatrix} \begin{pmatrix} a' & b' \\ 0 & a'^{-1} \end{pmatrix} + \epsilon \begin{pmatrix} b \\ a \end{pmatrix} (\,a'\ b'\,)$.

(b) If $M_1 = \begin{pmatrix} a_0^{-1} & b_0 \\ 0 & a_0 \end{pmatrix} \begin{pmatrix} a_1 & b_1 \\ 0 & a_1^{-1} \end{pmatrix} + \epsilon_1 \begin{pmatrix} b_0 \\ a_0 \end{pmatrix} (\,a_1\ b_1\,)$ and $M_2 = \begin{pmatrix} a_0^{-1} & b_1 \\ 0 & a_1 \end{pmatrix} \begin{pmatrix} a_2 & b_2 \\ 0 & a_2^{-1} \end{pmatrix} + \epsilon_2 \begin{pmatrix} b_1 \\ a_1 \end{pmatrix} (\,a_2\ b_2\,)$ then $M_1 M_2 = \begin{pmatrix} a_0^{-1} & b_0 \\ 0 & a_0 \end{pmatrix} \begin{pmatrix} a_2 & b_2 \\ 0 & a_2^{-1} \end{pmatrix} + (\epsilon_1 + \epsilon_2) \begin{pmatrix} b_0 \\ a_0 \end{pmatrix} (\,a_2\ b_2\,)$.

PROOF: Part (a) is implied by the two following observations:

- For $\epsilon = 0$ we have $(\,a\ b\,) \begin{pmatrix} a_0^{-1} & b \\ 0 & a \end{pmatrix} \begin{pmatrix} a' & b' \\ 0 & a'^{-1} \end{pmatrix} = (\,a'\ b'\,)$.
- If $M_1, M_2 \in SL(2, \mathbb{F}_{2^n})$ satisfy $(a, b)M_1 = (a, b)M_2 = (a', b')$ then $M_1 + M_2 = \epsilon \begin{pmatrix} b \\ a \end{pmatrix} (\,a'\ b'\,)$. Indeed, let c, d such that $\begin{pmatrix} a & b \\ c & d \end{pmatrix}$ is unitary and let $\begin{pmatrix} a' & b' \\ c_1 & d_1 \end{pmatrix} := \begin{pmatrix} a & b \\ c & d \end{pmatrix} M_1$ and $\begin{pmatrix} a' & b' \\ c_2 & d_2 \end{pmatrix} := \begin{pmatrix} a & b \\ c & d \end{pmatrix} M_2$. As M_1, M_2 and $\begin{pmatrix} a & b \\ c & d \end{pmatrix}$ are in $SL(2, \mathbb{F}_{2^n})$, we have $\det \begin{pmatrix} a' & b' \\ c_1 & d_1 \end{pmatrix} = \det \begin{pmatrix} a' & b' \\ c_2 & d_2 \end{pmatrix} = 1$. We get

$$M_1 + M_2 = \begin{pmatrix} a & b \\ c & d \end{pmatrix}^{-1} \left[\begin{pmatrix} a' & b' \\ c_1 & d_1 \end{pmatrix} + \begin{pmatrix} a' & b' \\ c_2 & d_2 \end{pmatrix} \right] = \begin{pmatrix} d & b \\ c & a \end{pmatrix} \begin{pmatrix} 0 & 0 \\ c_1+c_2 & d_1+d_2 \end{pmatrix}$$
$$= \begin{pmatrix} b \\ a \end{pmatrix} (\,c_1+c_2\ d_1+d_2\,).$$

Moreover, as $(\,c_1+c_2\ d_1+d_2\,) \begin{pmatrix} b \\ a \end{pmatrix} = a(d_1 + d_2) + b(c_1 + c_2) = (ad_2 + bc_2) + (ad_1 + bc_1) = 0$, we get the result.

Part (b) is a straightforward computation. □

We now define the (generalized) representation problem in $\mathbb{F}_{2^n}^*$ and we show how it can be solved for small n (and certainly if $n \leq 170$).

Problem 1 Representation problem in $\mathbb{F}_{2^n}^*$: *Given N (randomly chosen) elements $g_i \in \mathbb{F}_{2^n}^*$, find a factorization $\prod g_i^{e_i} = 1$ such that $\sum |e_i|$ is not too large. Generalized representation problem in $\mathbb{F}_{2^n}^*$: Given N (randomly chosen) elements $g_i \in \mathbb{F}_{2^n}^*$ and a (randomly chosen) element $g_0 \in \mathbb{F}_{2^n}^*$, find a factorization $\prod g_i^{e_i} = g_0$ such that $\sum |e_i|$ is not too large.*

Proposition 2 *The (generalized) representation problem can be solved in groups $\mathbb{F}_{2^n}^*$ where the discrete logarithm problem can be solved.*

PROOF: Let $g_i \in \mathbb{F}_{2^n}^*, i = 0, ...N$. Let g a generator of $\mathbb{F}_{2^n}^*$, and let α_i be the discrete logarithms of g_i with respect to base g. The representation problem amounts to solving the following problem: find $\{e_i\}$ such that $\sum e_i\alpha_i = \alpha_0 \bmod (2^n - 1)$ and $\sum |e_i|$ is *not too large*. A good solution to this problem can be computed with the LLL algorithm [4]. □

If the exponents α_i are random numbers uniformly distributed in $[1, 2^n - 1]$ the smallest solution has expected size $\sum_i |e_i|$ about $N2^{n/N}$ (approximating that there is no collision, the sums $\sum e_i\alpha_i$ for $e_i \leq 2^{n/N}$ produce the $2^n - 1$ possible values). The LLL algorithm actually gives a solution such that $\sum |e_i|^2$ is close to optimal, but this is enough for our purposes. By the LLL approximation bound, the solution provided using LLL has a norm 2 smaller than $\sqrt{N}2^{n/N+N}$ which is subexponential for $N \approx \sqrt{n}$. In practice, LLL performs much better and in the analysis of our algorithms, we will approximate that the isze of the solution given by LLL algorithm is also about $N2^{n/N}$.

With this method, the representation problem in $\mathbb{F}_{2^n}^*$ can be solved if discrete logarithms can be computed, in particular the representation problem can be solved today for $n \leq 170$. The following result follows from Proposition 2.

Proposition 3 *Let n be such that discrete logarithms can be solved in $\mathbb{F}_{2^n}^*$. Let $\mathcal{D}, \mathcal{T}^{up}, \mathcal{T}^{low}, \mathcal{L}^v, \mathcal{R}^v \subset SL(2, \mathbb{F}_{2^n})$ be the subgroups of diagonal, upper and lower triangular matrices and the subgroup of matrices with left or right eigenvector v. If an attacker can compute N random elements M_i of one of these subgroups together with bit sequences m_i of length at most L hashing to these matrices, then he can also find a message m such that $h_{ZT}(m) = I$. The message m has expected size smaller than $NL2^{n/N}$ in the diagonal case and smaller than $NL2^{1+n/N}$ in the other cases.*

PROOF: Clearly any diagonal matrix writes down as $D_i = \begin{pmatrix} a_i & 0 \\ 0 & a_i^{-1} \end{pmatrix}$ for some $a_i \in \mathbb{F}_{2^n}^*$. Let $\{e_i\}$ be a solution to the representation problem with respect to $\{a_i\}$, that is $\prod a_i^{e_i} = 1$. Construct m as the concatenation of e_1 messages m_1, e_2 messages m_2, etc. (in any order). Then $h_{ZT}(m) = \prod D_i^{e_i} = \begin{pmatrix} \prod a_i^{e_i} & 0 \\ 0 & \prod a_i^{-e_i} \end{pmatrix} = I$.

Similarly, an upper triangular matrix T_i writes down as $\begin{pmatrix} a_i & b_i \\ 0 & a_i^{-1} \end{pmatrix}$ for some $a_i \in \mathbb{F}_{2^n}^*, b_i \in \mathbb{F}_{2^n}$. Let $\{e_i\}$ be a solution to the representation problem with respect to $\{a_i\}$, that is $\prod a_i^{e_i} = 1$. Construct m' as the concatenation of e_1 messages m_1, e_2 messages m_2, etc. (in any order) and $m = m'||m'$. Then $h_{ZT}(m') = \begin{pmatrix} 1 & b \\ 0 & 1^{-1} \end{pmatrix}$ for some $b \in \mathbb{F}_{2^n}$ and $h_{ZT}(m) = I$.

By definition each $M_i \in \mathcal{L}^{(a\ b)}$ satisfies $(a\ b)\, M_i = \lambda_i\, (a\ b)$ for some $\lambda_i \in \mathbb{F}_{2^n}^*$. Let $\{e_i\}$ be a solution to the representation problem with respect to $\{\lambda_i\}$, that is $\prod \lambda_i^{e_i} = 1$. Construct m' as the concatenation of e_1 messages m_1, e_2 messages m_2, etc. (in any order) and $m = m'||m'$. Then $(a\ b)\, h_{ZT}(m') = (a\ b)$ which by Proposition 1 implies $h_{ZT}(m') = I + \epsilon \begin{pmatrix} b \\ a \end{pmatrix} (a\ b)$ hence $h_{ZT}(m) = I$.

The proof for $\mathcal{T}^{low}$ and $\mathcal{R}^v$ are similar and the claim on the message lengths follows from our analysis of the representation problem in $\mathbb{F}_{2^n}^*$. □

The part of Proposition 3 concerning $\mathcal{L}^v$ and $\mathcal{R}^v$ has interesting graph interpretations that we give in Appendix A.

4   A New Generic Collision Attack

We now give an algorithm finding N_2 matrices M_i such that $(1\ 0)\, M_i = \lambda_i\, (1\ 0)$ for some $\lambda_i \in \mathbb{F}_{2^n}^*$, and combining them as in Proposition 3 to find collisions for the Zémor-Tillich hash function.

We denote by $\mathbb{P}^1(\mathbb{F}_{2^n})$ the projective space of dimension 1 on $\mathbb{F}_{2^n}$, which is the set of equivalence classes of $\mathbb{F}_{2^n} \times \mathbb{F}_{2^n}$ that results from identifying two vectors $(a_1\ b_1)$ and $(a_2\ b_2)$ if and only if $(a_2\ b_2) = \lambda\, (a_1\ b_1)$ for some $\lambda \in \mathbb{F}_{2^n}^*$. We denote by $[a : b]$ the projective point that is the equivalence class of a vector $(a\ b)$. To any message $m = m_1 m_2 ... m_k$ we associate two projective points $q(m), q_{-1}(m) \in \mathbb{P}^1(\mathbb{F}_{2^n})$ as follows. We define $(a(m)\ b(m)) := (1\ 0) \prod_{i=1}^{k} M_{m_i} = (1\ 0)\, h_{ZT}(m)$ and $(a'(m)\ b'(m)) := (1\ 0) \prod_{i=k}^{1} M_{m_i}^{-1} = (1\ 0)\, h_{ZT}^{-1}(m)$, then $q(m) := [a(m) : b(m)]$ and $q_{-1}(m) := [a'(m) : b'(m)]$.

Our algorithm first performs a birthday attack [11] to find collisions on the q values as follows. Random messages m and m' of size $k > n/2$ are generated and stored together with $q(m)$ and $q(-m')$, until m_1, m_2 are found such that $q(m_1) = q_{-1}(m_2)$ (see Figure 1). As there are $2^n + 1$ points in $\mathbb{P}^1(\mathbb{F}_{2^n})$, the probability that $q(m_1) = q_{-1}(m_2)$ for some m_1, m_2 is $1 - \left(1 - \frac{2^{N_1}}{2^n + 1}\right)^{2^{N_1}}$ after 2^{N_1} steps. In particular, after $2^{N_1} = 2^{n/2}$ steps we have a probability $1 - e^{-1} \approx 0.63$ to know a message $m := m_1 || m_2$ of size $2k$ such that $(1\ 0)\, h_{ZT}(m) = \lambda\, (1\ 0)$ for some $\lambda \in \mathbb{F}_{2^n}^*$.

This collision search is repeated until N_2 distinct messages m_i are found such that $(1\ 0)\, h_{ZT}(m_i) = \lambda_i\, (1\ 0)$ for some $\lambda_i \in \mathbb{F}_{2^n}^*$. To guarantee that the collisions found are all distinct, we may perform each collision search with a different length $k > n/2$, or choose k slightly larger than $n/2 + \log_2(N_2)$, say $k = n/2 + \log_2(N_2) + 10$.

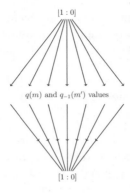

$[1:0]$

$q(m)$ and $q_{-1}(m')$ values

$[1:0]$

Fig. 1. Collision search on q values

The next step of the algorithm combines the messages m_i to get a collision for the Zémor-Tillich hash function. As in the proof of Proposition 3, we compute a solution $\{e_i\}$ to the representation problem in $\mathbb{F}_{2^n}^*$ with respect to the λ_i, that is $\prod \lambda_i^{e_i} = 1$. From this solution, we finally construct a message m' as the concatenation of each message m_i repeated e_i times (in any order), and a message $m = m' \| m'$ that collides with the void message as shown in the proof of Proposition 3.

To analyze this attack, suppose that the N_2 collision searches are done with $k = n/2 + 1, ..., n/2 + N_2$ and that the algorithm described in Section 3 is used to solve the representation problem. The expected size of the collision is then bounded by $(n/2 + N_2)N_2 2^{n/N_2 + 2}$, the memory requirement is $2^{n/2+1}n$ and the time complexity is $N_2 2^{n/2+1}t + t_{REP}$ where t is the time needed to compute one q value and t_{REP} is the time needed to solve the representation problem. In particular for $n = 130$ and $N_2 = 16$, this attack produces a collision to the void message of size about 2^{18} in time $2^{69}t$ and memory requirements 2^{69}. The memory requirements will be removed in Section 6 by using distinguished points techniques [6].

5   A New Generic Preimage Attack

We now extend our ideas to a preimage attack. Interestingly, this attack has essentially the same complexity as the collision attack.

Suppose we want to find a preimage to a matrix $M = \left(\begin{smallmatrix} a & b \\ c & d \end{smallmatrix} \right)$, that is a message $m = m_1 ... m_k$ such that $M = h_{ZT}(m) = \prod M_{m_i}$. As we showed in previous section, random messages m_i of size $L > n$ such that $(\begin{smallmatrix} 1 & 0 \end{smallmatrix}) h_{ZT}(m_i) = \lambda_i (\begin{smallmatrix} 1 & 0 \end{smallmatrix})$ for some $\lambda_i \in \mathbb{F}_{2^n}^*$ can be found with memory $n2^{n/2+1}$ and time $2^{n/2+1}t$. Similarly, random messages $m_i, i = 0, ... N_2$ of size $L > n$ satisfying $(\begin{smallmatrix} 1 & 0 \end{smallmatrix}) h_{ZT}(m_0) = \lambda_0 (\begin{smallmatrix} a & b \end{smallmatrix})$ and $(\begin{smallmatrix} a & b \end{smallmatrix}) h_{ZT}(m_i) = \lambda_i (\begin{smallmatrix} a & b \end{smallmatrix}), i > 0$ for some $\lambda_i \in \mathbb{F}_{2^n}^*$ can also be found with the same time and memory complexities.

Solving a (generalized) representation problem, we can compute $\{e_i\}$ such that $\prod \lambda_i^{e_i} = \lambda_0$, hence we can compute a message m_0' of size $N_2 L 2^{n/N_2}$ and a matrix $M_0 := h_{ZT}(m_0')$ such that $(\begin{smallmatrix} 1 & 0 \end{smallmatrix}) M_0 = (\begin{smallmatrix} a & b \end{smallmatrix})$. Similarly, from N_3 different solutions to the representation problem $\prod \lambda_i^{e_i} = 1$ we get N_3 messages m_i' of size $N_2 L 2^{n/N_2}$ such that $(\begin{smallmatrix} a & b \end{smallmatrix}) h_{ZT}(m_i') = (\begin{smallmatrix} a & b \end{smallmatrix})$. Let $(\begin{smallmatrix} c' & d' \end{smallmatrix}) := (\begin{smallmatrix} 0 & 1 \end{smallmatrix}) h_{ZT}(m_0')$. As $ad' + bc' = ad + bc = 1$, we have $a(d+d') + b(c+c') = 0$, that is $(\begin{smallmatrix} c+c' & d+d' \end{smallmatrix}) = \delta_0 (\begin{smallmatrix} a & b \end{smallmatrix})$ for some $\delta_0 \in \mathbb{F}_{2^n}$.

According to Proposition 1, for all $i > 0$ there exists $\delta_i \in \mathbb{F}_{2^n}$ such that $h_{ZT}(m_i') = (\begin{smallmatrix} 1 & 0 \\ 0 & 1 \end{smallmatrix}) + \delta_i (\begin{smallmatrix} b \\ a \end{smallmatrix}) (\begin{smallmatrix} a & b \end{smallmatrix})$; moreover we have $h_{ZT}(m_{i_1}') h_{ZT}(m_{i_2}') = (\begin{smallmatrix} 1 & 0 \\ 0 & 1 \end{smallmatrix}) + (\delta_{i_1} + \delta_{i_2}) (\begin{smallmatrix} b \\ a \end{smallmatrix}) (\begin{smallmatrix} a & b \end{smallmatrix})$. Suppose the δ_i values generate $\mathbb{F}_{2^n}/\mathbb{F}_2$, which is very likely if N_3 is shortly bigger than n, say $N_3 = n + 10$. Then by solving a binary linear system, we can write $\delta_0 = \sum_{i \in I} \delta_i$ for some $I \subset \{1, ..., N_3\}$ of size $\leq n$ and hence $M_1 := \prod_{i \in I} h_{ZT}(m_i') = (\begin{smallmatrix} 1 & 0 \\ 0 & 1 \end{smallmatrix}) + \delta_0 (\begin{smallmatrix} b \\ a \end{smallmatrix}) (\begin{smallmatrix} a & b \end{smallmatrix})$. Finally, we have $M_0 M_1 = (\begin{smallmatrix} a & b \\ c' & d' \end{smallmatrix}) [(\begin{smallmatrix} 1 & 0 \\ 0 & 1 \end{smallmatrix}) + \delta_0 (\begin{smallmatrix} b \\ a \end{smallmatrix}) (\begin{smallmatrix} a & b \end{smallmatrix})] = (\begin{smallmatrix} a & b \\ c & d \end{smallmatrix})$.

This shows that any message made of m_0' concatenated with any concatenation of the messages $m_i', i \in I$, is a preimage to $(\begin{smallmatrix} a & b \\ c & d \end{smallmatrix})$. The collision size is about bounded by $N_3(n/2 + N_2)N_2 2^{2n/N_2+2}$, that is $12n^2(n + 10)$ if $N_2 = n$ and $N_3 = n + 10$. The memory requirement of this attack is $2^{n/2+1}n$ and the time complexity is $N_2 2^{n/2+1} t + t_{REP}$ where t is the time needed to compute one q value and t_{REP} is the time needed to solve the representation problem (note that finding N_3 solutions to a representation problem essentially requires the same time as finding one solution because both times are essentially determined by the computation of the discrete logarithms). As for our collision attack, the memory requirements can be removed by using distinguished points techniques.

6   Memory-Free Versions of Our Attacks

The attacks of Sections 4 and 5 require storing two databases of about $2^{n/2}$ projective points in $\mathbb{P}^1(\mathbb{F}_{2^n})$ and their corresponding messages. We now remove the memory requirements by using distinguished points techniques [6].

Let $\alpha : \mathbb{P}^1(\mathbb{F}_{2^n}) \to \{0,1\}^k$ and $\beta : \mathbb{P}^1(\mathbb{F}_{2^n}) \to \{0,1\}$ be two "pseudorandom functions" and let $\varphi : \mathbb{P}^1(\mathbb{F}_{2^n}) \to \mathbb{P}^1(\mathbb{F}_{2^n})$ be defined by

$$p \to \varphi(p) = \begin{cases} q(\alpha(p)) & \text{if } \beta(p) = 0 \\ q_{-1}(\alpha(p)) & \text{if } \beta(p) = 1, \end{cases}$$

where $k > n$ is arbitrarily chosen and q and q_{-1} are defined as in Section 4.

The iterates $q_0, \varphi(q_0), \varphi(\varphi(q_0)), ...$ of φ on q_0 all belong to the finite domain $\mathbb{P}^1(\mathbb{F}_{2^n})$ so at some point iterating φ will produce a collision (see Figure 2), that is two points p_1 and p_2 such that $\varphi(p_1) = \varphi(p_2) = c$. If the behavior of φ is sufficiently random then $\beta(p_1) \neq \beta(p_2)$ with a probability $1/2$, in which case $\alpha(p_1)$ and $\alpha(p_2)$ can be combined to produce a message m of size $2k$ such that $(\begin{smallmatrix} 1 & 0 \end{smallmatrix}) h_{ZT}(m) = \lambda (\begin{smallmatrix} 1 & 0 \end{smallmatrix})$ for some $\lambda \in \mathbb{F}_{2^n}^*$.

The functions α and β do not need to be "pseudorandom" in the strong cryptographic meaning, but only "sufficiently pseudorandom" for the above analysis to hold.

Now that the problem of finding a collision on the q values has been translated in the problem of detecting a cycle in the iterates of φ, we can remove the memory requirements by standard techniques. We recall here the method of *distinguished points*; other methods are described in [7]. Let $\mathcal{D}_d := \{q = [a : b] \in \mathbb{P}^1(\mathbb{F}_{2^n}) | b \neq 0, lsb_d(a/b) = 0^d\}$ be sets of 2^{n-d} distinguished q values such that their d last bits are all 0. During the collision search, we only store the q values that belong to $\mathcal{D}$ and only look for collisions on these particular q values. Finding a collision c' on distinguished points requires 2^{d-1} additional steps in average but the memory is reduced to $2^{n/2-d}$; if $d = n/2 - 10$ the time overhead is negligible and the memory requirements are very small (see Figure 3).

From the two distinguished points p'_1 and p'_2 that precede c' in the iterates of φ, we can recover the points p_1 and p_2 that produce the actual collision c as follows. Iterate again φ on p'_1 and p'_2 and store only distinguished points but this time with $d = n/2 - 20$. After about $2^{n/2-10}$ steps on each side (and a small memory of about 2^{11}) a collision c'' and preceding distinguished points p''_1 and p''_2 are found that are closer to the actual collision c, p_1, p_2. Iterating again from p''_1 and p''_2 with a larger distinguished-point set, we finally get the actual collision with small time overhead and small memory.

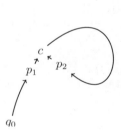

Fig. 2. Iterating φ from some initial point q_0, we eventually get a collision c

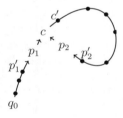

Fig. 3. Collision graph with markers on the distinguished points. The average distance between two distinguished points is 2^d. The average length of the path is $2^{n/2}$. Finding a collision on a distinguished point requires essentially the same time as finding a general collision, as soon as $2^d << 2^{n/2}$.

With this method instead of the trivial collision search steps, our collision and preimage attacks require negligible memory and essentially the same time complexity. As the output of Zémor-Tillich is about $3n$ bits, these attacks are far better than birthday and optimal preimage bounds. In the following sections, we introduce two variants of Zémor-Tillich with reduced output sizes respectively $2n$ and n bits, and we show that these variants are essentially as secure as the original Zémor-Tillich for sufficiently small parameters including the parameters initially suggested in [9].

7   Vectorial Version of Zémor-Tillich

Our first variant h_{ZT}^{vec} is simply the first row of Zémor-Tillich, that is $h_{ZT}^{vec}(m) := (\,a\ b\,)$ if $h_{ZT}(m) = \left(\begin{smallmatrix} a & b \\ c & d \end{smallmatrix}\right)$. This variant was introduced in [5] by Petit et al. but without a proof of its equivalence to the original function. Alternatively, we may parameterize the function h_{ZT}^{vec} by an initial vector $(\,a_0\ b_0\,) \neq (\,0\ 0\,)$ as $h_{ZT}^{vec,(\,a_0\ b_0\,)}(m) := (\,a_0\ b_0\,)\,h_{ZT}(m)$. Clearly, the output has $2n$ bits.

Finding a collision for this variant corresponds to finding two messages m and m' such that $(\,a_0\ b_0\,)\,h_{ZT}(m) = (\,a_0\ b_0\,)\,h_{ZT}(m')$, in particular it is enough to find one message m such that $(\,a_0\ b_0\,)\,h_{ZT}(m) = (\,a_0\ b_0\,)$ (we call such a collision a *cyclic* collision). Finding a preimage to a vector $(\,a\ b\,)$ is finding a message m such that $(\,a_0\ b_0\,)\,h_{ZT}(m) = (\,a\ b\,)$.

The following proposition shows that h_{ZT}^{vec} is collision resistant if and only if the original function h_{ZT} is collision resistant.

Proposition 4 *If there exists a* ppt *(probabilistic polynomial time) algorithm that for randomly chosen starting vectors $(\,a_0\ b_0\,) \neq (\,0\ 0\,)$ finds a collision on $h_{ZT}^{vec,(\,a_0\ b_0\,)}$, then there exists a* ppt *algorithm finding collisions for the original Zémor-Tillich function.*

PROOF: Given a ppt algorithm A^{vec} finding collisions for the vectorial version, we build a ppt algorithm A^{mat} finding collisions for the original matrix version. The algorithm A^{mat} first picks a random matrix $M_0 := \left(\begin{smallmatrix} a_0 & b_0 \\ c_0 & d_0 \end{smallmatrix}\right) \in SL(2, \mathbb{F}_{2^n})$ and runs A^{vec} on (a_0, b_0) to get two messages m_{10} and m_{11} corresponding to matrices M_{10} and M_{11} such that $(a_0, b_0)M_{10} = (a_0, b_0)M_{11} = (a_1, b_1)$. Without loss of generality, we can assume that (a_1, b_1) is randomly uniformly distributed (otherwise we may just append the same randomly chosen sequence of bits to both messages). Algorithm A^{mat} then calls again A^{vec} on (a_1, b_1) to get two matrices M_{20} and M_{21}, etc. It repeats this operation $n + 1$ times.

Let $v_i := (\,a_i\ b_i\,)$ and $\widetilde{v}_i := \left(\begin{smallmatrix} b_i \\ a_i \end{smallmatrix}\right)$. According to Proposition 1(a), the matrices M_{ij} write down as

$$M_{ij} = \begin{pmatrix} a_{i-1}^{-1} & b_{i-1} \\ 0 & a_{i-1} \end{pmatrix} \begin{pmatrix} a_i & b_i \\ 0 & a_i^{-1} \end{pmatrix} + \epsilon_{ij}\widetilde{v_{i-1}}v_i$$

for some $\epsilon_{ij} \in \mathbb{F}_{2^n}$. Applying Proposition 1(b) recursively, for any $e = e_1...e_{n+1} \in \{0, 1\}^{n+1}$, we have

$$\prod_{i=1}^{n+1} M_{ie_i} = \begin{pmatrix} a_0^{-1} & b_0 \\ 0 & a_0 \end{pmatrix} \begin{pmatrix} a_{n+1} & b_{n+1} \\ 0 & a_{n+1}^{-1} \end{pmatrix} + \left(\sum_{i=1}^{n+1} \epsilon_{ie_i} \right) \widetilde{v}_0 v_{n+1}.$$

For $1 \leq i \leq n+1$, let $\epsilon_i := \epsilon_{i0} + \epsilon_{i1}$. Seeing each ϵ_i as a binary vector of length n over $\mathbb{F}_2$, these vectors are linearly dependent. Moreover, finding a subset I of $\{1, ..., n + 1\}$ such that $\sum_{i \in I} \epsilon_i = 0$ simply amounts to invert a binary linear system, which is cubic in $n + 1$.

We now conclude the description of A^{mat}. After computing $I \subset \{1, ..., n+1\}$ such that $\sum_{i \in I} \epsilon_i = 0$, the algorithm A^{mat} returns $m = m_{10}||m_{20}||...||m_{n+1,0}$ and $m' = m_{1e_1}||m_{2e_2}||...||m_{n+1,e_{n+1}}$ where $e_i = 1$ if and only if $i \in I$. By the discussion above, it is clear that

$$h_{ZT}^{mat}(m) = h_{ZT}^{mat}(m') = \begin{pmatrix} a_0^{-1} & b_0 \\ 0 & a_0 \end{pmatrix} \begin{pmatrix} a_{n+1} & b_{n+1} \\ 0 & a_{n+1}^{-1} \end{pmatrix} + \left(\sum_{i=1}^{n+1} \epsilon_{i0} \right) \widetilde{v_0} v_{n+1}.$$

$\square$

The reduction of Proposition 4 is polynomial but not completely tight: the algorithm A^{mat} runs $n+1$ times the algorithm A^{vec}. Note that if instead of A^{vec} we have an algorithm A'^{vec} returning a message m corresponding to a *cycle* for the vectorial version, then the message $m||m$ is a collision for the matrix version. Indeed, if $\begin{pmatrix} a & b \end{pmatrix} M = \begin{pmatrix} a & b \end{pmatrix}$ Proposition 1(a) shows that M writes down as $M = \begin{pmatrix} a^{-1} & b \\ 0 & a \end{pmatrix} \begin{pmatrix} a & b \\ 0 & a^{-1} \end{pmatrix} + \epsilon \begin{pmatrix} b \\ a \end{pmatrix} \begin{pmatrix} a & b \end{pmatrix} = I + \epsilon \begin{pmatrix} b \\ a \end{pmatrix} \begin{pmatrix} a & b \end{pmatrix}$ hence $M^2 = I$.

8   Projective Version of Zémor-Tillich

Our second variant $h_{ZT}^{proj,(a_0 \, b_0)}$ exploits even further Proposition 3. We define

$$h_{ZT}^{proj,(a_0 \, b_0)} := [a : b]$$

where $\begin{pmatrix} a & b \end{pmatrix} := h_{ZT}^{vec,(a_0 \, b_0)}(m)$ and $[a : b] \in \mathbb{P}^1(\mathbb{F}_{2^n})$. Finding a collision for $h_{ZT}^{proj,(a_0 \, b_0)}$ is finding two messages m and m' such that $\begin{pmatrix} a_0 & b_0 \end{pmatrix} h_{ZT}(m) = \lambda \begin{pmatrix} a_0 & b_0 \end{pmatrix} h_{ZT}(m')$ for some λ, in particular it is enough to find a *cyclic* collision which is a message m such that $\begin{pmatrix} a_0 & b_0 \end{pmatrix}$ is a left eigenvector of $h_{ZT}(m)$.

The output of $h_{ZT}^{proj,(a_0 \, b_0)}$ is very close to n bits. For the parameters suggested by Tillich and Zémor, its collision resistance is equivalent to the collision resistance of the original function.

Proposition 5 *If there exists an algorithm that finds collisions on $h_{ZT}^{proj,(a_0 \, b_0)}$, there exists an algorithm that finds collisions on $h_{ZT}^{vec,(a_0 \, b_0)}$, assuming that for some $n' > n$ it is feasible to compute n' discrete logarithms in $\mathbb{F}_{2^n}^*$ and one subset sum problem of size n'.*

If we denote by t^{proj}, t^{DL} and $t^{SS}(n')$ the times needed respectively to find collisions on the projective version, to solve one discrete logarithm problem in $\mathbb{F}_{2^n}^*$ and to solve a subset sum problem of size n', collisions on the vectorial version can be found in time $n'(t^{proj} + t^{DL}) + t^{KN}(n')$.

PROOF: Given an algorithm A^{proj} finding collisions for the projective version, we build an algorithm A^{proj} finding collisions for the vectorial version. Receiving an initial vector $v_0 = (a_0, b_0)$, A^{vec} forwards it to A^{proj} and receives two messages m_{10}, m_{11}. To the two messages correspond two vectors (a_{10}, b_{10}) and $(a_{11}, b_{11}) = \lambda_1(a_{10}, b_{10})$ for some λ_1. The algorithm A^{vec} computes the discrete logarithm

d_1 of λ_1 with respect to some generator g of $\mathbb{F}_{2^n}^*$. The algorithm A^{vec} then runs A^{proj} on the projective point (a_{10}, b_{10}) and computes d_2 similarly, etc.

After n' steps, the algorithm A^{vec} computes a subset $I \subset \{1, ..., n'\}$ such that $\sum_{i \in I} d_i = 0 \mod 2^n - 1$. By concatenating the paths m_{ie_i} where $e_i = 1$ if $i \in I$ and $e_i = 0$ otherwise, algorithm A^{vec} produces a collision with the message $m_{10}||...||m_{n'0}$ for the vectorial version. The output is correct because both messages lead to the vector $\left(\prod_{i \in I} \lambda_i\right)(a_{n'0}, b_{n'0}) = g^{\sum_{i \in I} d_i}(a_{n'0}, b_{n'0}) = (a_{n'0}, b_{n'0})$.

The claim on the running time follows straightforwardly.   □

The best choice for n' depends on the exact values of t^{proj}, t^{DL} and $t^{SS}(n')$. Solving discrete logarithms problems is believed to be hard but is definitely feasible in $\mathbb{F}_{2^n}^*$ if $n < 170$. Computing $I \subset \{1, ..., n'\}$ such that $\sum_{i \in I} d_i = 0 \mod 2^n - 1$ is related to the subset sum problem which is NP-hard but usually easy in average. For the parameters proposed by Zémor-Tillich, lattice reduction algorithms like LLL will probably succeed in performing the reduction. Another method is to use Wagner's "k-lists" algorithm [10] for solving the subset sum problem. This algorithm can solve the subset sum problem in time and space $k2^{n/(1+\log k)}$ which for $k \approx \sqrt{n}$ is roughly $2^{2\sqrt{n}}$ which is about 2^{26} for $n = 170$. The drawback with this method is that n' must also increase to $2^{2\sqrt{n}}$ hence the discrete logarithm costs increase and the quality of the reduction decreases.

Assuming the existence of an algorithm A'^{proj} computing *cyclic* collisions on the projective version (messages m_i such that $(a_0, b_0)h^{mat}(m_i) = \lambda_i(a_0, b_0)$ for some λ_i) the reduction slightly improves. Indeed, A^{vec} must only compute a *small integer* solution $(x_1, ..., x_{n'})$ to $\sum_i x_i d_i = 0 \mod 2^n - 1$ instead of a *binary* solution. The reduction algorithm still has to compute discrete logarithm problems but it must not solve any subset sum problem.

9   Conclusion

We have given new algorithms for computing collisions for the Zémor-Tillich hash function in a time equal to the cubic root of the birthday bound. Our attacks are the first generic ones in the sense that unlike previous attacks they work for any parameters n and $P_n(X)$ of the function. Moreover, they are very close to being practical for the parameters $n \in [130, 170]$ initially suggested in [9].

Interestingly, we could extend our collision attacks to new preimage attacks with the same complexity due to the inherent possibility of "meet-in-the-middle" attacks in Zémor-Tillich and the fact that our collision attacks use a subgroup structure that preserves this possibility.

Our attacks exploit a separation of the collision problem into an easy and a hard component, and suggest that the output of Zémor-Tillich should be of n bits rather than $3n$ bits. We have consequently introduced two variants of this function, the vectorial and the projective versions, with reduced output sizes of respectively $2n$ and n bits. We have proved that the original function is collision resistant if and only if the vectorial variant and ((for small n) if and only if the projective variant are collision resistant.

Acknowledgements. We thank Nicolas Veyrat-Charvillon and Giacomo Demeulenaer for interesting discussions related to this paper. We thank Martijn Stam for a remark on the Zémor-Tillich hash function that motivated the algorithm of Section 5. We thank Phong Nguyen for pointing us out reference [10]. Finally, we would like to thank an anonymous referee of CT-RSA for his very useful comments that considerably improved the paper.

References

1. Abdukhalikov, K.S., Kim, C.: On the security of the hashing scheme based on SL2. In: Vaudenay, S. (ed.) FSE 1998. LNCS, vol. 1372, pp. 93–102. Springer, Heidelberg (1998)
2. Charnes, C., Pieprzyk, J.: Attacking the SL2 hashing scheme. In: Safavi-Naini, R., Pieprzyk, J.P. (eds.) ASIACRYPT 1994. LNCS, vol. 917, pp. 322–330. Springer, Heidelberg (1995)
3. Geiselmann, W.: A note on the hash function of Tillich and Zémor. In: Gollmann, D. (ed.) FSE 1996. LNCS, vol. 1039, pp. 51–52. Springer, Heidelberg (1996)
4. Lenstra, H.W.J.L.L., Lenstra, A.K.: Factoring polynomials with rational coefficients. Mathematische Annalen 261(5), 515–534 (1982)
5. Petit, C., Veyrat-Charvillon, N., Quisquater, J.-J.: Efficiency and Pseudo-Randomness of a Variant of Zémor-Tillich Hash Function. In: IEEE International Conference on Electronics, Circuits, and Systems, ICECS 2008 (2008)
6. Quisquater, J.-J., Delescaille, J.-P.: How easy is collision search? Application to DES. In: Quisquater, J.-J., Vandewalle, J. (eds.) EUROCRYPT 1989. LNCS, vol. 434, pp. 429–434. Springer, Heidelberg (1990)
7. Shamir, A.: Random graphs in cryptography. In: Invited talk at Asiacrypt 2006 (2006)
8. Steinwandt, R., Grassl, M., Geiselmann, W., Beth, T.: Weaknesses in the $SL_2(\mathbb{F}_{2^n})$ hashing scheme. In: Bellare, M. (ed.) CRYPTO 2000. LNCS, vol. 1880, p. 287. Springer, Heidelberg (2000)
9. Tillich, J.-P., Zémor, G.: Hashing with SL_2. In: Desmedt, Y.G. (ed.) CRYPTO 1994. LNCS, vol. 839, pp. 40–49. Springer, Heidelberg (1994)
10. Wagner, D.: A generalized birthday problem. In: Yung, M. (ed.) CRYPTO 2002. LNCS, vol. 2442, pp. 288–303. Springer, Heidelberg (2002)
11. Yuval, G.: How to swindle Rabin. Cryptologia 3, 187–189 (1979)

A   Graphical Interpretation of Proposition 3

The part of Proposition 3 concerning $\mathcal{L}^v$ and $\mathcal{R}^v$ has interesting graph interpretations. To the Zémor-Tillich hash function is associated a Cayley graph $\mathcal{ZT}$, in which each vertex corresponds to a matrix $M \in SL(2, \mathbb{F}_{2^n})$ and each edge to a couple $(M_1, M_2) \in SL(2, \mathbb{F}_{2^n})^2$ such that $M_2 = M_1 A_0$ or $M_2 = M_1 A_1$ [9].

We now construct the graphs $\mathcal{ZT}^{vec}$ and $\mathcal{ZT}^{proj}$ as follows. For $\mathcal{ZT}^{vec}$, associate a vertex to each row vector $(\, a\ b\,) \in \mathbb{F}_{2^n}^{1 \times 2} \backslash \{(\, 0\ 0\,)\}$ and an edge to each couple of such vectors $((\, a_1\ b_1\,), (\, a_2\ b_2\,))$ satisfying $(\, a_2\ b_2\,) = (\, a_1\ b_1\,) A_0$ or $(\, a_2\ b_2\,) = (\, a_1\ b_1\,) A_1$. Alternatively, the graph $\mathcal{ZT}^{vec}$ can be constructed from the graph $\mathcal{ZT}$ by identifying two vertices $M_1 = \left(\begin{smallmatrix} a_1 & b_1 \\ c_1 & d_1 \end{smallmatrix}\right)$ and $M_2 = \left(\begin{smallmatrix} a_2 & b_2 \\ c_2 & d_2 \end{smallmatrix}\right)$ when $(\, a_1\ b_1\,) = (\, a_2\ b_2\,)$. An example of such a graph is shown in Figure 4.

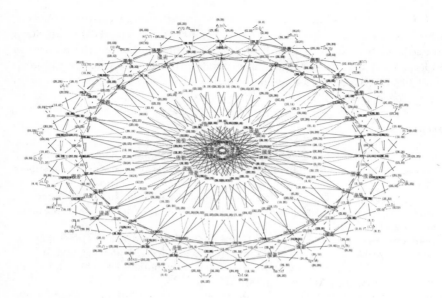

Fig. 4. $\mathcal{ZT}^{vec}$ graph for parameter $P_5(X) = X^5 + X^2 + 1$. The vertices are labeled by matrices. Red dotted (resp. blue solid) arrows correspond to multiplication by matrix A_0 (resp. A_1). Each polynomial $\sum a_i X^i$ is written as $\sum a_i 2^i$.

Similarly, we associate a vertex of $\mathcal{ZT}^{proj}$ to each projective point $q_i = [a_i : b_i] \in \mathbb{P}^1(\mathbb{F}_{2^n})$ and an edge to each couple (q_1, q_2) such that $\lambda (a_2\ b_2) = (a_1\ b_1)\, A_0$ or $\lambda (a_2\ b_2) = (a_1\ b_1)\, A_1$ for some $\lambda \in \mathbb{F}_{2^n}^*$. Alternatively, the graph $\mathcal{ZT}^{proj}$ may be constructed from the graph $\mathcal{ZT}^{vec}$ by identifying two vertices $(a_1\ b_1)$ and $(a_2\ b_2)$ when $(a_1\ b_1) = \lambda (a_2\ b_2)$ for some $\lambda \in \mathbb{F}_{2^n}^*$.

Finding a cycle in $\mathcal{ZT}^{vec}$ is just as hard as finding a cycle in $\mathcal{ZT}$ because if $(a\ b)\, M = (a\ b)$ then $M^2 = I$. The radial symmetry in the graph $\mathcal{ZT}^{vec}$ (Figure 4) is not surprising as it reflects the relation $(a\ b)\, A_i = (a'\ b') \Leftrightarrow [\lambda (a\ b)]\, A_i = [\lambda (a'\ b')]$: multiplying each vertex of $\mathcal{ZT}^{vec}$ by a constant λ is equivalent to a rotation of the graph.

Roughly, a vertex in the graph $\mathcal{ZT}^{vec}$ can be characterized by a radial and an angular position. A cycle in the graph $\mathcal{ZT}^{proj}$ induces a path in the graph $\mathcal{ZT}^{vec}$ from a vertex to another vertex with the same radial coordinate, but not necessarily the same angular coordinate. Clearly, different such paths can be combined to give a cycle in the graph $\mathcal{ZT}^{vec}$. According to Proposition 3 and its proof, this can be done if the discrete logarithm problem, hence the representation problem, can be solved in $\mathbb{F}_{2^n}^*$.

A cycle in $\mathcal{ZT}^{vec}$ induces cycles in both radial and angular coordinates. Proposition 3 means that solving the angular part of the representation problem is easy once the radial part can be solved to produce various points with the same radius.

A Statistical Saturation Attack against the Block Cipher PRESENT

B. Collard* and F.-X. Standaert**

UCL Crypto Group, Microelectronics Laboratory, Université catholique de Louvain,
Place du Levant 3, Louvain-la-Neuve, Belgium
{baudoin.collard,fstandae}@uclouvain.be

Abstract. In this paper, we present a statistical saturation attack that combines previously introduced cryptanalysis techniques against block ciphers. As the name suggests, the attack is statistical and can be seen as a particular example of partitioning cryptanalysis. It extracts information about the key by observing non-uniform distributions in the ciphertexts. It can also be seen as a dual to saturation (*aka* square, integral) attacks in the sense that it exploits the diffusion properties in block ciphers and a combination of active and passive multisets of bits in the plaintexts. The attack is chosen-plaintext in its basic version but can be easily extended to a known-plaintext scenario. As an illustration, it is applied to the block cipher PRESENT proposed by Bogdanov *et al.* at CHES 2007. We provide theoretical arguments to predict the attack efficiency and show that it improves previous (linear, differential) cryptanalysis results. We also provide experimental evidence that we can break up to 15 rounds of PRESENT with $2^{35.6}$ plaintext-ciphertext pairs. Eventually, we discuss the attack specificities and possible countermeasures. Although dedicated to PRESENT, it is an open question to determine if this technique improves the best known cryptanalysis for other ciphers.

Introduction

This paper introduces a statistical attack that is closely related to previous works in *partitioning cryptanalysis* [2,8,9]. Such attacks can be seen as a generalization of the linear cryptanalysis in which one exploits partitions of the plaintexts (*resp.* ciphertexts) leading to significantly non uniform distributions of the ciphertexts (*resp.* plaintexts). While arguably more powerful than linear cryptanalysis, they usually face the question of how to find good partitions for a given cipher. Hence, in practice they generally rely on some specificities that a cryptanalyst may find within the ciphers, *e.g.* in [7,15]. Following these works, our results focus on a (relatively) generic and simple way to find partitions that can, in certain contexts, lead to efficient attacks. For this purpose, we exploit ideas from the *integral cryptanalysis* [13], originally introduced as a specialized attack against the SQUARE block cipher [6] and also known as saturation attacks [11]. Such

---

* Work supported by the Walloon Region under the project Nanotic-Cosmos.
** Associate researcher of the Belgian Fund for Scientific Research (FNRS-F.R.S.).

M. Fischlin (Ed.): CT-RSA 2009, LNCS 5473, pp. 195–210, 2009.
© Springer-Verlag Berlin Heidelberg 2009

attacks are chosen-plaintext and generally study the propagation of well chosen sets of plaintexts through the cipher. In practice, they typically fix a number of plaintext bytes to a constant value and track the evolution of some other bytes having a known distribution. To some extent, the proposed statistical saturation attack can be seen as a dual of the previous saturation attacks. It also takes advantage of several plaintexts with some bits fixed while the others vary randomly. But instead of observing the evolution of the variable bits in the cipher state, we observe the diffusion of the fixed bits during the encryption process. That is, we track the evolution of a non-uniform input plaintext distribution through the cipher. The name statistical saturation attack refers both to the the way the inputs are generated and to the fact that it exploits the diffusion properties (and possibly weaknesses) of the target cipher.

As an illustration, we apply the proposed technique to the block cipher PRE-SENT that was presented at CHES 2007 by Bogdanov *et al.* It is a compact block cipher primarily designed for hardware constrained environments such as RFID tags and sensor networks. The name PRESENT reflects its similarity with the block cipher SERPENT [1], known for its security and hardware performances. In the specifications of the cipher [4], the authors analyze the security of PRESENT against various cryptanalytic attacks. In order to argue about the immunity against linear and differential cryptanalysis, they provide lower bounds for the number of active S-boxes in any linear/differential trail of the cipher. Resistance against structural, algebraic and related-key attacks is also analyzed. The security of PRESENT against differential cryptanalysis was further studied in [17] in which the authors present an attack against 16 rounds that requires the entire codebook and a time complexity of 2^{65} memory accesses.

PRESENT is a good target for the proposed statistical saturation attack because it exhibits a particular weakness in its diffusion layer. As a consequence, our following results improve the complexities of the best reported attacks against this cipher, both in theoretical estimations and in experimental validations. In practice, we broke up to 15 rounds of PRESENT with $2^{35.6}$ plaintext-ciphertext pairs. Additionally to these results, we discuss the specificities of the attack compared to other known cryptanalytic techniques. We show that it depends both on the diffusion and substitution layers in a block cipher. We also show that current criteria for S-box design do not properly capture the non-uniformities that are exploited in our partitions. Due to the generality of its principles, the proposed technique could be applied to other ciphers as well. However, since its effectiveness depends on the diffusion properties of the targets, it is an open question to determine if it can improve other cryptanalytic results.

The rest of this paper is divided in three parts. Section 1 presents the basic principles of our attack with theoretical arguments that support it. A comparison between theoretical predictions and experimental observations is also provided. The second section extends the basic profiling attack of Section 1 to a distinguishing attack that is more efficient, both in terms of computations and data complexity. Section 3 discusses countermeasures and the impact of the S-boxes on the attack performances. We conclude the paper and suggest further research in Section 4. The PRESENT block cipher is additionally described in Appendix.

1   A Basic Profiling Attack

1.1   Principle of the Attack

Our attack is based on a weakness in the diffusion layer of PRESENT. A closer look at the permutation shows that, *e.g.* for the S-boxes 5,6,9 and 10 (counting from S-box 0 at the right), only 8 out of 16 input bits are directed to other S-boxes. Figure 1 illustrates this observation. Note that there exists many other examples of poor diffusion in the permutation (*i.e.* with 8 bits out of 16 remaining in the same 4 S-boxes after permutation). Consequently, if we fix the 16 bits at the input of the S-boxes 5-6-9-10, then 8 bits will be known at the very same input for the next round. We can iteratively repeat this process round by round and observe a non-uniform behavior at the output of the S-boxes 5-6-9-10.

In order to exploit this weakness, we first evaluate theoretically the distribution of the 8 bits in the bold trail of Figure 1 at the output of the S-box layer, for a fixed value of the same 8 bits of plaintext. This requires to guess the 8 subkey bits involved in the trail. One also needs to assume that the bits not situated in the trail are uniformly distributed. This is a reasonable assumption as soon as the 56 remaining bits of plaintext (excluding the 8 bits in the trail) are randomly generated. Then, given the distribution of the 8-bit trail at the input of a round, it is possible to compute the 8-bit distribution at the output of the round with Algorithm 1 (given in Appendix B). By iteratively applying this algorithm, we can compute the distribution for an arbitrary number of rounds. For each key guess, the work needed to compute the theoretical distribution of the target trail after r rounds is equivalent to $r \cdot 2^{16}$ partial encryptions.

Once we have computed the theoretical distributions of the trail for each possible key guess, we can attack the cipher by simply comparing them with a practical distribution obtained by encrypting a large number of plaintexts with

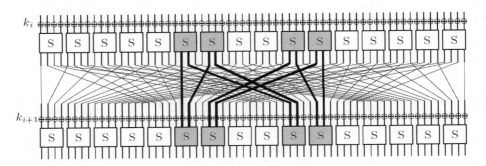

Fig. 1. Permutation layer of PRESENT: bold lines underline the poor diffusion property

a secret key. The key guess minimizing the distance between theoretical and practical distributions is chosen as the correct key. As in [3], we can construct a list of key candidates sorted according to the distance between theory and practice. The better the position of the right key in the list, the better the attack.

1.2   Experimental Results

We evaluated the practicability of our attack a against reduced-round version of PRESENT. In order to reduce the guess work, the key-scheduling of PRESENT was simplified in these experiments and the same subkey was used at each round. With this modification, only 8 bits of the master key have to be estimated and the correct distribution has to be found among 256 possible ones.

Comparison Between Theoretical and Experimental Distributions. Figure 2 depicts the distribution of the 8 bits in the trail after 2,4,6 and 8 rounds for a fixed 8-bit key byte. The theoretical predictions (in black) are compared with experimental results (in grey) generated with 2^{30} plaintexts-ciphertexts pairs. Note that our attack is choosen-plaintext as we have to fix 8 plaintext bits. But it can be turned into a known-plaintext attack by dividing random plaintexts in 256 classes according to the value of the 8 fixed input bits in the trail and observing the output distributions for each of the 256 cases independently.

Both experimental and theoretical distributions present a significant deviation from uniform, even after 8 rounds. The deviation tends to decrease with the number of rounds however. We can observe that predictions match experiments very closely for up to 6 rounds, and then begin to distinguish. This illustrates that the sampling is not sufficient anymore to approximate the distributions.

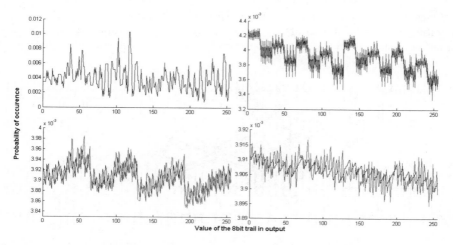

Fig. 2. Comparison between the experimental (in gray) and theoretical (in black) distributions of the target trail output for a given key byte and 2, 4, 6 and 8 rounds

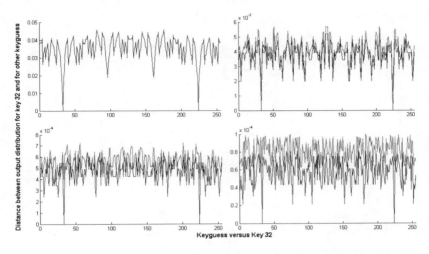

Fig. 3. Distance between the experimental (in gray) and theoretical (in black) distribution of a secret key byte and the theoretical distributions of the 256 possible key guesses. 2^{30} plaintexts were used for the experimental results.

Comparison Between Theoretical and Experimental Distances. Figure 3 shows the evolution of the distance between the distribution corresponding to a secret key byte 32 and the distributions corresponding to the 256 possible key guesses for this secret key byte. The number of rounds in the figure again varies from 2 to 8. The black (*resp.* grey) curves represent the distance between the theoretical (*resp.* experimental) distribution of the correct key byte and the theoretical distributions for each possible key guess. For up to 8 rounds, we observe that position 32 minimizes the distance between theory and practice. Note that we used both an Euclidean and a Kullback-Leibler distance in our experiments: both metrics gave similar results.

In order to confirm the effectiveness of the proposed cryptanalysis in a key recovery context, we also computed the gain of the attack, as defined in [3]:

Gain the Attack. *If an attack is used to recover an n-bit key and is expected to return the correct key after having checked on the average M candidates, then the gain of the attack, expressed in bits, is defined as:*

$$\gamma = -log_2 \frac{2 \cdot M - 1}{2^n} \qquad (1)$$

Intuitively, the gain is a measure of the remaining workload (or number of key candidates to test) after a cryptanalysis has been performed. In the context of our attack, we can produce a list of key candidates sorted according to the distance between their theoretical distribution and the experimental distribution computed with the correct secret key. The gain is simply determined by the position of the secret key in this list. Figure 4 shows the gain of the attack for 1

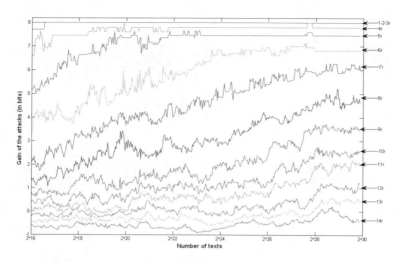

Fig. 4. Gain of the profiling attack against 1- to 14- round PRESENT

to 14 rounds of PRESENT (still with a modified key-scheduling) in function of the data complexity. This experiment used up to 2^{30} plaintext-ciphertext pairs. The gain is bounded by 8, as we guess 8 bits of key material. It increases with the number of texts and decreases with the number of rounds.

Effect of the Key-Scheduling Algorithm. Unlike slide and related key attack, the proposed technique does not use a particular weakness in the key-scheduling. However, the number of subkey bits to guess at each round directly affects the time complexity of the attack, as one must compute the theoretical distribution for each of these key guesses. It may also increase the data complexity as distinguishing more keys generally requires more text pairs. The number of bits to guess according to the number of rounds is given in Table 1 for the complete key-scheduling algorithm. After 12 rounds, we have to guess 63 bits of the key, for a complexity equivalent to $2^{63} \times 2^{16} = 2^{79}$ encryptions. Consequently, this bounds the interest of the profiling attack to 12 rounds of the (simplified) cipher.

Table 1. Number of bits to guess according to the number of rounds in the attack

#rounds	1	2	3	4	5	6	7	8	9	10	11	12	13	14	15	16	17	18	19	20	21
#bits	8	15	24	31	35	39	43	47	51	55	59	63	65	67	69	71	73	75	77	79	80

2   A Distinguishing Attack

In order to get rid of the previous computational limits, we now present a variant of the attack. It has the advantage of not requiring the precomputation of theoretical distributions anymore. The distinguisher is based on the fact that the theoretical distribution at the output of the target trail (as computed with Algorithm 1) is significantly different from uniform, whatever subkey is used.

2.1 Principle of the Attack

The attack is similar to the one presented above, yet it is simpler. We begin by generating a large number of plaintexts with 8 fixed bits. We encrypt those plaintexts using r-rounds PRESENT and record the distribution of the ciphertexts for the 16 bits at the output of the 4 active S-box in the last round. Given this experimental distribution, it is possible to compute the output distribution of the target 8-bit trail one round before by a classical partial decryption process. For one key guess, the evaluation of such an $r-1$-round distribution requires 2^{16} computations. Hence the total time complexity for all the key guesses equals $2^{16} * 2^{16} = 2^{32}$. Additionally using an FFT-based trick similar to the technique presented in [5], this complexity can be decreased to $16 \cdot 2^{16}$. For the correct key guess, the experimental 8-bit distribution in the penultimate round is expected to be more non-uniform than for any other guess. This is because decrypting with a wrong guess is expected to have the same effect as encrypting one more round. We can thus hope to distinguish the correct key from the wrong ones by computing the distance between a partially decrypted distribution and the uniform distribution. If the attack works properly, the distribution with the highest distance should correspond to the correct key.

2.2 Extensions of the Attack

(ext. 1) Increase the fixed part in the plaintext. One can easily gain one round in the attack by simply fixing the 16 bits of plaintext corresponding to the 4 active input S-boxes of the trail. This way, the 8-bit trail in the second round is also fixed and the diffusion is postponed by one round. By fixing 32 bits out of 64 (corresponding to S-boxes 4-5-6-7-8-9-10-11), one can similarly extend the attack by 2 rounds. However, we are then limited in the generation of at most 2^{32} texts. This limitation may be mitigated with the following extension.

(ext. 2) Use multiple fixed plaintext values. The same analysis can be performed multiple times, using different values for the 8-bit (or 16- or 32-bit) fixed part of the plaintexts and then combining the results (*e.g.* taking the sum of the uniform *vs.* measured distances corresponding to the different fixed plaintexts). This allows exploiting more texts and moving to a known-plaintext context. The resulting attack is similar to multiple linear cryptanalysis: each fixed part of the plaintext can be seen as analogous to an additional approximation in [3,12].

(ext. 3) Partial decryption of two rounds instead of one. In this case, 8 S-boxes are active in the last round instead of 4. Therefore, we have to keep a 32-bit distribution table in memory. Additionally, 38 bits of the key must be guessed for the partial decryption (32 bits for the last round + 16 bits for the penultimate round − 10 bits that are redundant). Using this trick, the adversary has to distinguish an $r − 2$-round distribution for the correct key from an $r + 2$-round distribution for the wrong candidates. The time complexity would be $(32 \cdot 2^{32}) \cdot (16 \cdot 2^{16}) = 2^{57}$ using again the results in [5].

2.3   Experimental Results

We have run experiments against reduced-round versions of PRESENT with up to 15 rounds and evaluated the gain of the attack in different contexts. First, Figure 5 represents the mean result of 4 attacks using 2^{34} plaintexts where 16 input bits were fixed (*i.e.* using only **ext. 1**).

To confirm the intuition that non-uniform distributions are observed for the correct key candidate, we represented the distance between the experimental distributions of the trail after partial decryption using a correct key and a uniform distribution. Figure 6 illustrates that this distance decreases with the number of rounds and stabilizes after a sufficient number of plaintexts have been reached.

Figure 7 illustrates the results of a variant of the attack where 32 plaintext bits were fixed and consequently only 2^{32} texts were generated. As expected, the results are slightly better than in the previous experiment.

Figure 8 finally shows an application of **ext. 2**. The graph represents the evolution of the attack gain against 1 to 32-rounds after 2^{32} plaintexts. The top and bottom curves represent the maximum and minimum gains among 12 experiments, while the two other curves represent respectively the average gain and the gain of the attack combining the 12 experiments. We clearly observe that combining the distances corresponding to the 12 experiments and computing a list of key candidates afterwards gives rise to much better result than computing

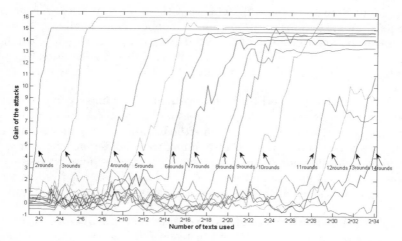

Fig. 5. Average gain of 4 attacks against 2 to 15-round PRESENT (**ext. 1**, 16 fixed bits)

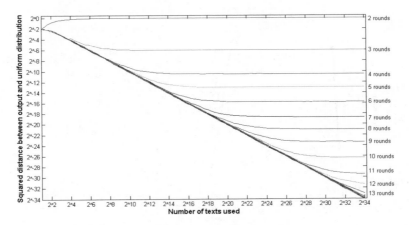

Fig. 6. Distance between uniform and output distributions after 1 to 15 rounds

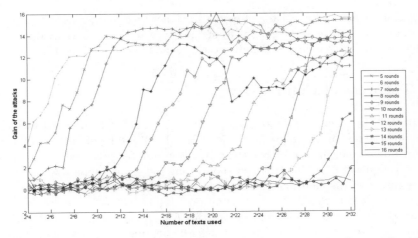

Fig. 7. Avg. gain of 12 attacks against 5 to 16-round PRESENT (**ext. 1**, 32 fixed bits)

a list for each experiment and then taking the average position for each key guess. Using the first method, we reach a significant gain up to 15 rounds.

For discussion purpose, Figure 13 in appendix C represents the gain of a linear cryptanalysis against 6- to 16-rounds PRESENT. The attack is based upon an iterative approximation involving one S-box with bias 2^{-2} in the first round and one S-box with bias 2^{-3} in each other round. It can recover up to 12 bits of the last subkey. Note that this example is not given for comparing the efficiencies of different attacks but to illustrate the big difference between security bounds as provided for the "best possible" linear attacks in [4] and attacks based on approximations that can be found in practice. The one that we proposed in appendix is obviously not optimal, but it is exploits the same iterativeness as our statistical

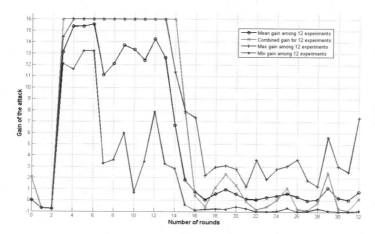

Fig. 8. Gain of attacks against 1- to 32-rounds PRESENT with $12 \cdot 2^{32} = 2^{35.6}$ texts

saturation attack. As a matter of fact, the attacks we discuss in this paper are experimented and therefore cannot be straightforwardly compared with bounds. They more directly relate to actual attacks such as presented in [17].

3   Theoretical Complexity

Intuitively, the efficiency of our distinguisher depends on the extent to which an experimental distribution after r-rounds PRESENT can be distinguished from a uniform distribution. Therefore, it can be nicely related to the theoretical analysis of Baignères et al. in [2] which shows that the data complexity required to distinguish two distributions is proportional to the inverse of the squared Euclidean distance between these distributions. Using Algorithm 1, we can easily compute a theoretical approximation of this Euclidean distance for PRESENT. It directly gives rise to Figure 9 in which the complexity of distinguishing the theoretical distributions at the output of PRESENT from uniform distributions is given for 1 to 16 rounds. We also illustrate the complexity of a linear cryptanalysis using the same approximation as the attack in Appendix C. Again, the difference between the effectiveness of an actual linear attack as in this figure and the security bounds in [4] should be emphasized. But even these security bounds (e.g. 2^{84} plaintext-ciphertext pairs to break 28-rounds PRESENT) suggest that our attack is an improvement compared to the theoretical expectations.

More interesting are the results in Table 2 that summarize the complexity of the attacks against PRESENT known so far (i.e. mainly [17] and the results in this paper). Note that due to the iterative nature of our trail, the time and memory complexities do not vary with the number of rounds in the trail. They only depend on the number of rounds that are partially decrypted. Most importantly, the provided data complexities are based upon the theoretical values given by the graph in Figure 9 and rely on the following assumptions:

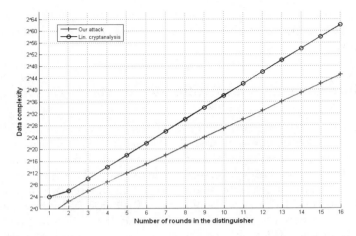

Fig. 9. Theoretical data complexity of the attack against PRESENT

Table 2. Summary of attacks (italic are not experimented and use **ext. 2**)

#rounds	type of attack	data compl.	time compl.	memory compl.	gain	reference
16	Diff. Crypt.	2^{64}	$2^{65}MA$	$6 * 2^{32}bits$	≤ 32	[17]
8	our attack*	$c * 2^{12}$	2^{20} op.*	2^{16} counters	≤ 16	this paper
	our attack**	$c * 2^{9}$	2^{57} op.*	2^{32} counters	≤ 38	this paper
12	our attack*	$c * 2^{24}$	2^{20} op.*	2^{16} counters	≤ 16	this paper
	our attack**	$c * 2^{21}$	2^{57} op.*	2^{32} counters	≤ 38	this paper
16	our attack*	$c * 2^{36}$	2^{20} op.*	2^{16} counters	≤ 16	this paper
	our attack**	$c * 2^{33}$	2^{57} op.*	2^{32} counters	≤ 38	this paper
20	*our attack**	$c * 2^{48}$	2^{20} op.*	2^{16} counters	≤ 16	this paper
	*our attack***	$c * 2^{45}$	2^{57} op.*	2^{32} counters	≤ 38	this paper
24	*our attack**	$c * 2^{60}$	2^{20} op.*	2^{16} counters	≤ 16	this paper
	*our attack***	$c * 2^{57}$	2^{57} op.*	2^{32} counters	≤ 38	this paper

* 1-round decryption, ** 2-round decryption.

- All attacks use **ext. 1** with 32 plaintext bits fixed.
- In 1-round (*resp.* 2-round as suggested in **ext. 3**) decryption attacks, we use a $r - 3$-round (*resp.* $r - 4$-round) distinguisher and have the time and memory complexities discussed in Sections 2.1 and 2.2.
- When the number of plaintexts needed to perform the attack exceeds 2^{32}, we use **ext. 2**. By combining multiple fixed plaintext values, we consider an attack exploiting distributions of larger dimensions, similarly to the multiple linear cryptanalysis in [3]. But it is an open problem to determine exactly the effect of this extension to the attack complexities. At least, our experiments suggest that these estimations are valid up to 16 rounds.

Note that our attack only recovers 16 key bits while [17] recovers the whole key. But as mentioned in Section 1.1, similar trails could be used to recover

32 more key bits with similar complexities. Hence, our results (including the part confirmed experimentally) anyway improve the best reported attack considerably.

4   Countermeasures and Influence of the S-Box

The origin of the proposed statistical saturation attack against PRESENT mainly lies in a weakness of the diffusion layer. A straightforward countermeasure would be to modify the permutation in order to avoid poor diffusion in any subset of S-boxes. But the proposed attack relates to the overall diffusion properties of the cipher. Hence the S-boxes also have an impact with this respect that we shortly study in this section. To do so, we generated 5000 different S-boxes respecting the four conditions imposed in the generation of the PRESENT S-box (see [4], Section 4.3). According to the authors, these constraints ensure that PRESENT is resistant to differential and linear attacks. Figure 14 in Appendix D represents the evolution of the squared distance between the uniform and output distribution of the cipher according to the number of rounds. Each curve represent a different choice for the S-box used in the cipher. It is noticeable that the PRESENT S-box is among the worst possible choices to resist our attack.

To confirm this impact of a weak *vs.* strong S-box in our cryptanalysis, we finally ran new experiments against a tweaked PRESENT where the original S-boxes were replaced by the dashed S-boxes of Figure 14 (*i.e.* those corresponding to the best and worst diffusion properties among our 5000 generated S-boxes). Figure 10 gives the gain of these attacks for different number of rounds (each attack used 2^{30} chosen plaintexts). As expected, the attack against the weak version of the cipher gives the best results. The figure emphasizes that the proposed attack is not directly related to linear or differential cryptanalysis (*i.e.* it is possible to find a cipher that is immune against linear and differential cryptanalysis, but not against the proposed statistical saturation attack).

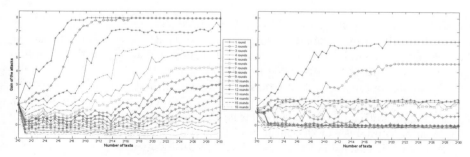

Fig. 10. Comparison between weak (left) and strong (right) S-boxes

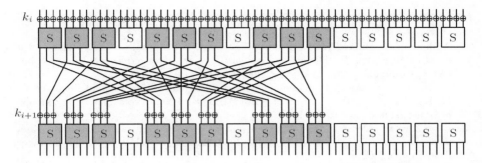

Fig. 11. Another poor diffusion trail in the permutation layer of PRESENT

5   Conclusion and Further Works

In this paper, we presented a new attack against the block cipher PRESENT that improves previously known cryptanalyses against this cipher. Experimentally, it allows us to break 15 rounds with $2^{35.6}$ plaintext-ciphertext pairs. We also present theoretical estimations of attacks than can break more cipher rounds. Additionally, we show that the proposed cryptanalysis is not directly related to linear and differential attacks. In practice, the security of the full cipher does not seem to be compromised by our results although the proposed attack was not discussed in the algorithm specifications. However, it confirms and emphasizes that PRESENT has been designed with little security margins.

Determining if the proposed statistical saturation attacks can improve cryptanalytic results against other ciphers is an interesting open question. Since they only exploit very general principles (namely, uncomplete diffusion after some cipher rounds), they are likely to be applicable to other reduced algorithms. But on the other hand, the proposed attack was particularly efficient against PRESENT due to a weakness in its permutation. Hence, it is not clear if the proposed technique can be as effective against other ciphers, with better diffusion properties. More theoretically, a better theoretical analysis of the attack, in particular the analogy between multiple linear cryptanalysis and the use of multiple fixed plaintext bytes in our context is also worth further investigation. Eventually, it would be interesting to investigate if the multidimensional cryptanalysis presented in [10] could be used to improve our results.

Another research direction would be to use the trail of Figure 11 in which 27 bits out of 36 are redirected to only 9 S-boxes at each round. It means that the lack of diffusion could be worse than in the trail of Figure 1 (we found 4 trails of this kind). However, this weakness is compensated by a larger trail size (36 bits instead of 16) which increases the diffusion inside the trail. Applying the attack presented in this paper to this new trail is also more difficult to experiment because of the 36-bit distributions for which a 1-round decryption would have to be performed. Hence, the efficiency of this attack is an open question.

References

1. Anderson, R., Biham, E., Knudsen, L.: Serpent: A Proposal for the Advanced Encryption Standard. In: The proceedings of the First Advanced Encryption Standard (AES) Conference, Ventura, CA (August 1998)
2. Baignères, T., Junod, P., Vaudenay, S.: How far can we go beyond linear cryptanalysis? In: Lee, P.J. (ed.) ASIACRYPT 2004. LNCS, vol. 3329, pp. 432–450. Springer, Heidelberg (2004)
3. Biryukov, A., De Cannière, C., Quisquater, M.: On multiple linear approximations. In: Franklin, M. (ed.) CRYPTO 2004. LNCS, vol. 3152, pp. 1–22. Springer, Heidelberg (2004)
4. Bogdanov, A., Knudsen, L., Leander, G., Paar, C., Poschmann, A., Robshaw, M., Seurin, Y., Vikkelsoe, C.: PRESENT: An ultra-lightweight block cipher. In: Paillier, P., Verbauwhede, I. (eds.) CHES 2007. LNCS, vol. 4727, pp. 450–466. Springer, Heidelberg (2007)
5. Collard, B., Standaert, F.-X., Quisquater, J.-J.: Improving the time complexity of matsui's linear cryptanalysis. In: Nam, K.-H., Rhee, G. (eds.) ICISC 2007. LNCS, vol. 4817, pp. 77–88. Springer, Heidelberg (2007)
6. Daemen, J., Knudsen, L.R., Rijmen, V.: The Block Cipher Square. In: Biham, E. (ed.) FSE 1997. LNCS, vol. 1267, pp. 149–165. Springer, Heidelberg (1997)
7. Gilbert, H., Handschuh, H., Joux, A., Vaudenay, S.: A Statistical Attack on RC6. In: Schneier, B. (ed.) FSE 2000. LNCS, vol. 1978, pp. 64–74. Springer, Heidelberg (2001)
8. Harpes, C., Kramer, G., Massey, J.: A Generalization of Linear Cryptanalysis and the Applicability of Matsui's Piling-Up Lemma. In: Guillou, L.C., Quisquater, J.-J. (eds.) EUROCRYPT 1995. LNCS, vol. 921, pp. 24–38. Springer, Heidelberg (1995)
9. Harpes, C., Massey, J.: Partitioning Cryptanalysis. In: Biham, E. (ed.) FSE 1997. LNCS, vol. 1267, pp. 13–27. Springer, Heidelberg (1997)
10. Hermelin, M., Cho, J.Y., Nyberg, K.: Multidimensional Linear Cryptanalysis of Reduced Round Serpent. In: Mu, Y., Susilo, W., Seberry, J. (eds.) ACISP 2008. LNCS, vol. 5107, pp. 203–215. Springer, Heidelberg (2008)
11. Hwang, K.: Saturation Attacks on Reduced Round Skipjack. In: Daemen, J., Rijmen, V. (eds.) FSE 2002. LNCS, vol. 2365, pp. 100–111. Springer, Heidelberg (2002)
12. Kaliski, B.S., Robshaw, M.J.B.: Linear Cryptanalysis using Multiple Approximations. In: Desmedt, Y.G. (ed.) CRYPTO 1994. LNCS, vol. 839, pp. 26–39. Springer, Heidelberg (1994)
13. Knudsen, L.R., Wagner, D.: Integral cryptanalysis. In: Daemen, J., Rijmen, V. (eds.) FSE 2002. LNCS, vol. 2365, pp. 112–127. Springer, Heidelberg (2002)
14. Matsui, M.: Linear cryptanalysis method for DES cipher. In: Helleseth, T. (ed.) EUROCRYPT 1993. LNCS, vol. 765, pp. 386–397. Springer, Heidelberg (1994)
15. Minier, M., Gilbert, H.: Stochastic Cryptanalysis of Crypton. In: Schneier, B. (ed.) FSE 2000. LNCS, vol. 1978, pp. 121–133. Springer, Heidelberg (2001)
16. Vaudenay, S.: An experiment on DES - Statistical Cryptanalysis, in the third ACM Conference on Computer Security, New Dehli, India, pp. 139–147 (March 1996)
17. Wang, M.: Differential cryptanalysis of reduced-round PRESENT. In: Vaudenay, S. (ed.) AFRICACRYPT 2008. LNCS, vol. 5023, pp. 40–49. Springer, Heidelberg (2008)

A   The Block Cipher PRESENT

PRESENT is a Substitution-Permutation Network with a block size of 64 bits. The recommended key size is 80 bits, which should be sufficient for the expected applications of the cipher. However a 128-bit key-schedule is also proposed. The encryption is composed of 31 rounds. Each of the 31 rounds consists of a XOR operation to introduce a round key K_i for $1 \leq i \leq 32$, where K_{32} is used for post-whitening, a linear bitwise permutation and a non-linear substitution layer. The non-linear layer uses a single 4-bit S-box S which is applied 16 times in parallel in each round. The cipher is described in pseudo-code in Figure 1.

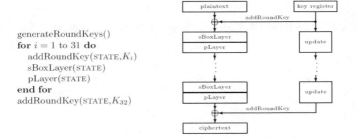

```
generateRoundKeys()
for i = 1 to 31 do
    addRoundKey(STATE,K_i)
    sBoxLayer(STATE)
    pLayer(STATE)
end for
addRoundKey(STATE,K_32)
```

Fig. 12. Top-level algorithmic description of PRESENT according to [4]

The linear permutation is defined by Table 3 where bit i of input is moved to bit position $P(i)$. The 4-bit S-box is defined according to the table 4. The 4-bit nibble i at the input of an S-box is substituted by the 4-bit $\mathsf{S}[i]$ in output.

We don't mention the key-schedule here as we don't make explicit use of it in our attack. We refer to the original paper [4] for the details of the specifications.

Table 3. Permutation layer for PRESENT

i	0	1	2	3	4	5	6	7	8	9	10	11	12	13	14	15
$P(i)$	0	16	32	48	1	17	33	48	2	18	34	50	3	19	35	51
i	16	17	18	19	20	21	22	23	24	25	26	27	28	29	30	31
$P(i)$	4	20	36	52	5	21	37	53	6	22	38	54	7	23	39	55
i	32	33	34	35	36	37	38	39	40	41	42	43	44	45	46	47
$P(i)$	8	24	40	56	9	25	41	57	10	26	42	58	11	27	43	59
i	48	49	50	51	52	53	54	55	56	57	58	59	60	61	62	63
$P(i)$	12	28	44	60	13	29	45	61	14	30	46	62	15	31	47	63

Table 4. S-box table for PRESENT (hexadecimal notation)

i	0	1	2	3	4	5	6	7	8	9	A	B	C	D	E	F
$\mathsf{S}[i]$	C	5	6	B	9	0	A	D	3	E	F	8	4	7	1	2

B   Theoretical Evaluation of the Target Trail Distribution

Algorithme 1

```
1    input: a 8-bit subkey guess sk and the 8-bit input distribution distrib_in[256]
2    output: the 8-bit output distribution distrib_out[256]
3
4    initialize distrib_out[256] to the all-zero state
5    for each 8-bit values text do
6       for each 8-bit values rand do
7          fix the 8-bit trail to text and xor with sk
8          fix the 8-bit non trail to rand
9          apply the sboxes
10         apply the permutation
11         evaluate the value of the 8 bit trail out
12         update distrib_out[out]= distrib_out[out]+ distrib_in[text]/256;
13      end for
14   end for
```

C   Linear Cryptanalysis Using a Single Approximation

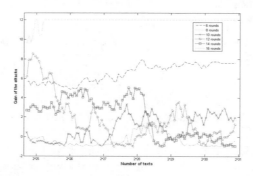

Fig. 13. Gain of a linear cryptanalysis against 6- to 16-rounds PRESENT

D   Influence of the S-Box on the Attack Effectiveness

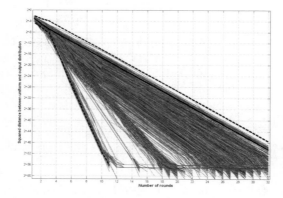

Fig. 14. Evolution of the squared distance between uniform and output distribution for 5000 different S-boxes (the PRESENT S-box is in plain black)

Practical Attacks on Masked Hardware

Thomas Popp[1], Mario Kirschbaum[1], and Stefan Mangard[2]

[1] Institute for Applied Information Processing and Communications (IAIK)
Graz University of Technology
Inffeldgasse 16a, 8010 Graz, Austria
{Thomas.Popp,Mario.Kirschbaum}@iaik.tugraz.at
[2] Infineon Technologies AG
Security Innovation
Am Campeon 1-12, 85579 Neubiberg, Germany
Stefan.Mangard@infineon.com

Abstract. In this paper we analyze recently introduced questions for masked logic styles in general and for one such logic style called MDPL in particular. The DPA resistance of MDPL suffers significantly from a problem called early propagation, which denotes a data-dependent time of evaluation of logic cells depending on input signal-delay differences. Experiments on a prototype chip show that in case of specific MDPL modules like the analyzed AES coprocessor, early propagation does not unconditionally break the DPA resistance of MDPL. Investigations indicate that this might be due to the regular structure of the particular MDPL circuit, which is assumed to cause only relatively "small" signal delay differences. Furthermore, in this article it is shown that the recently proposed, so-called PDF-attack could not be turned into a successful practical attack in our environment. Finally, the recently raised question whether MDPL has special requirements in terms of the generation of random mask bits or not is discussed theoretically.

Keywords: DPA-Resistant Masked Logic Styles, MDPL, Prototype Chip, Hardware AES, PDF-Attack, PRNG.

1 Introduction

In the last years, various masked logic styles have been presented. These logic styles counteract differential power analysis (DPA) attacks at the cell level by randomizing the processed values on a cryptographic device [7].

Many masked logic styles have only been proposed and were never implemented in real silicon. One of the masked logic styles that was also tested in practice is masked dual-rail precharge logic (MDPL) [12]. MDPL received plenty of attention in the last years and there are also rather recent publications about this topic, e.g. [10].

The problems of the first masked logic styles like [4,18] with glitches [8,15] have been avoided for MDPL. Also other masked logic styles like random-switching logic (RSL) [16] have been designed with the glitch-problem in mind. RSL is

M. Fischlin (Ed.): CT-RSA 2009, LNCS 5473, pp. 211–225, 2009.

rather complicated to implement due to special timing constraints between the levels of combinational logic blocks. Dual-rail random-switching logic (DRSL) [1] was later introduced, which simplifies the circuit design of RSL. However, DRSL and especially MDPL suffer from early propagation [14]. Recent proposals like improved MDPL (iMDPL) [11] and precharged masked Reed-Muller logic (PMRML) [5] also take the issue of early propagation into account.

The analysis of an MDPL prototype chip, which implements an 8051-compatible microcontroller [11], clearly showed that for the large signal-delay differences in the range of a nanosecond that occur in this design, no increase in the DPA resistance could be achieved. Still an open question in this context is how this effect scales in a design with smaller delay differences. In our work we analyzed this issue with the help of an AES coprocessor that is also part of the microcontroller on the prototype chip. This AES module is based on a very regular architecture. Thus, the occurring signal-delay differences are much smaller than in the microcontroller module itself. The attack results presented in Section 3 indicate that MDPL might not be broken in all circumstances. So far, we were not able to mount a successful DPA attack on the AES coprocessor.

Also some more general possible vulnerabilities of masked logic styles have been discussed recently. The authors of [13,17] presented an attack which apparently allows to undo the effect of masking in a clock cycle of a masked circuit by analyzing the mean of the power consumption during that clock cycle. With this method, which is called PDF-attack (PDF ... probability density function), it should be possible to remove the effects of the mask completely or at least to introduce a significant bias in the mask. With the help of the prototype chip, we analyze the PDF-attack in practice. It turned out that, at least in our environment, no direct impact on the DPA resistance of the attacked device could be shown. Another recent publication states that MDPL seems to have special requirements for the generation of the mask bits [3]. We elaborate on that issue in a theoretical manner.

This article is structured as follows. In the next section, MDPL and the analyzed prototype chip are shortly introduced. In Section 3, the results of the DPA attacks on the AES hardware module of the MDPL prototype chip are presented in detail. Section 4 describes the analysis results for the PDF-attack and Section 5 goes further into the question of mask generation for MDPL. Conclusions are drawn in Section 6.

2   MDPL and the Prototype Chip

MDPL conceals the characteristic data dependent power consumption of implementations in CMOS logic, which are well known to be vulnerable to DPA attacks. The goal of MDPL is to break the dependency between the processed data and the instantaneous power consumption of a device by applying the same random mask bit to every signal in the device. Every operation within the implementation only processes masked data, and hence, the power consumption of an MDPL circuit only depends on the masked data. Thus, an attacker cannot obtain information about the unmasked data from the power consumption as long as the mask remains secret.

Additionally to masking, MDPL is based on the dual-rail precharge (DRP) principle in order to avoid glitches, which may significantly reduce the DPA resistance of a device. The operating principle of DRP logic and MDPL is the following. Basically, DRP logic tries to break the dependency between the processed data and the power consumption by making the power consumption constant in each clock cycle. For this purpose, every signal in the circuit is represented by two complementary wires d and $\bar{d}$. Depending on the value of the signal that the wires d and $\bar{d}$ belong to, **either** d **or** $\bar{d}$ switches to "1" in each clock cycle, both wires never switch to "1" at the same time. In order to reach a constant power consumption, the wire loads (*i.e.* capacitive and resistive wire loads) need to be perfectly balanced. Unfortunately, this requirement is very hard to achieve in practice, and hence, MDPL uses masking to bypass the need of perfectly balanced wires. One important fact related to MDPL is that DRP logic prevents glitches by adding a precharge phase in each clock cycle. This way it is ensured that every wire in the circuit switches its value at most once per clock cycle.

The analysis of MDPL has been performed by means of a prototype chip that contains several cores implemented in different logic styles using a $0.13\,\mu m$ CMOS process technology. The focus of our work lies on the cores implemented in CMOS and MDPL. Each core contains an 8051-compatible microcontroller with a serial interface as well as an AES coprocessor. A pseudo-random number generator (PRNG) on the prototype chip supplies the MDPL core with a pseudo-random mask bit.

The implementations of the CMOS and the MDPL core on the prototype chip have the following main attributes: the MDPL core ($1\,mm^2$) has about 5 times the size of the CMOS core ($0.2\,mm^2$), reaches approximately half the maximum clock frequency (MDPL runs at $7\,MHz$, CMOS runs at $18\,MHz$), and consumes about 11 times more power than the CMOS core (MDPL approx. $4.5\,mA$, CMOS approx. $0.4\,mA$).

3   DPA Attacks on the MDPL AES Hardware Module

The cryptographic coprocessor attached to every 8051-compatible microcontroller is a hardware implementation of AES-128. Its architecture follows the standard version of the regular and scalable AES hardware architecture presented in [6], which reproduces the logical $4x4$ State layout of AES in hardware. The two main differences are the missing CBC unit and key storage in the key unit. Thus, the AES module is only capable of the ECB mode of operation and the secret key has to be loaded into the AES module before each operation.

The datapath of the AES coprocessor is shown in Figure 1. The third and last main difference to the original design is the single MixColumns module attached to the left column of the AES State. The ShiftRows operation is implemented as a Barrel shifter. The four S-boxes performing the SubBytes operation are combinational and pipelined (one stage) implementations as described in [19].

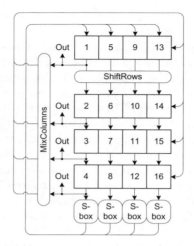

Fig. 1. Architecture of the datapath of the AES hardware module

The initial State values are loaded column-wise from the right (not shown in figure). One round of AES takes 9 clock cycles in this architecture. First, each of the 16 AES State cells applies the AddRoundKey operation to its stored value. Then, the State is shifted row-wise down through the S-boxes and afterwards column-wise left through the MixColumns module. Calculation of the next AES round key happens in parallel to the MixColumns operation. When the AES operation is finished, the final State values are read out column-wise to the left.

3.1   Measurement Setup Description

In this subsection the measurement setup that has been used to investigate the DPA resistance of the MDPL core on the prototype chip is described. The power measurements have been performed using a host PC to obtain and store the power traces, a digital oscilloscope, and a test board holding the prototype chip. The power consumption has been measured via the voltage drop over a $10\,Ohm$ resistor in the VDD line with a differential probe and a digital oscilloscope (both with a bandwidth of $1\,GHz$) to sample the measured data. The sampling rate has been set to $4\,GS/s$ and the prototype chip has been operated with a clock frequency of $3.6864\,MHz$.

3.2   DPA Scenario Description

The target of the DPA attack are the movements of the SubBytes results of the first AES round through the State registers 1 to 16. According to the architecture shown in Figure 1, these values are moved in the AES State from top to bottom during the SubBytes operation and from right to left during the subsequent MixColumns operation. Each of the SubBytes results depends exactly on one byte of the unknown secret key, thus there are 256 possibilities for each of them.

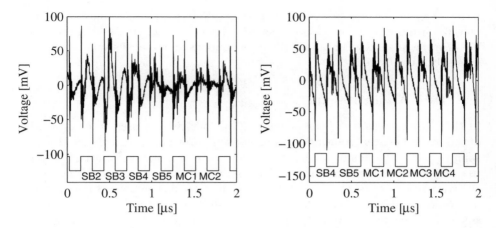

Fig. 2. Power trace and clock signal during the execution of the attacked AES operations for the CMOS core (left) and the MDPL core (right)

As an example, we track the movements of a plaintext byte in the AES State cell 4. In the first clock cycle, the initial AddRoundKey operation with round key byte 4 is performed by the State cell. The result enters the first stage of the S-box until the intermediate result is stored in the pipeline registers. In the next clock cycle, the second stage of the S-box is performed and the substitution result is stored in State cell 1. Since the value we track belongs to the fourth row of the AES State, the byte is rotated three positions to the left in the third clock cycle. Thus, it gets stored in State cell 6. In the following two clock cycles, the byte moves down through cell 7 into cell 8, which is its final position after SubBytes and ShiftRows. In the next clock cycles, the byte is moved to the left because the MixColumns operation is performed.

During its way through the AES State, a SubBytes result overwrites other SubBytes results. In case of the attacks on the CMOS core, the Hamming distance (HD) of such two values is used as the power model in the DPA attack. Note that with our approach, we favor the attacker in the following way. We assume that the secret key byte which is necessary to calculate the overwritten SubBytes result is known. In doing so, we keep the number of possibilities for the secret key in one DPA attack at 256. In the attacks on the MDPL core, we only have to predict the Hamming weight (HW) of the moved SubBytes result. The reason is that before a byte is transferred over a bus, all bus lines get precharged to 0. Furthermore, since MDPL is a dual-rail logic style with more or less randomly imbalanced wire pairs, we can generally only use 1 bit at a time of the 8-bit SubBytes result in a meaningful leakage model. In the following, we call this the HWbit power model.

Figure 2 shows the power trace and the clock signal of the part of the first AES round we have measured for the DPA attacks. An identifier beneath each clock cycle indicates the step of the operation that is performed (SB = SubBytes; MC = MixColumns). A trigger signal was generated by the attacked device to easily locate these important clock cycles. The assumption that one knows when the power consumption needs to be measured also favors the attacker significantly, because the amount of data and the measurement and analysis time are in this case substantially reduced.

3.3   DPA Attack Results

In the DPA attacks, the Pearson correlation coefficient is used to quantify the dependency between measured power consumption and predicted power consumption [2]. The height of the correlation peak in a successful attack is used to calculate the minimum number of needed power traces to get a distinct peak. For correlation peaks $\rho_{ck,ct} \leq 0.2$, the minimum number of needed power traces n is approximated as follows: $n = 28/\rho_{ck,ct}^2$ ($confidence = 99.99\%$, ck ... correct key, ct ... point in time where the correlation peak occurs) [7]. The number n is used to quantify the DPA resistance of the attacked device for the given attack scenario.

DPA Attack on the CMOS Reference Core
The DPA attacks on the hardware AES in the CMOS reference module showed that the leakage (*i.e.* the height of the resulting correlation peak) of the different key bytes significantly depends on the path each key byte takes through the AES State matrix (see Figure 1). It turned out that State cell 4 leaks the most information because its output is connected to an S-box input, a MixColumns input and the AES module output bus. The State cells 1 to 3 leak the second most and cells 8, 12, and 16 leak the third most. The least leakage occurs for all other cells because their output is only connected to their neighboring cells to the left (circuitry for a State shift to the left) and their neighboring cells below (circuitry for a State shift to the bottom).

640 000 traces have been measured to attack the CMOS AES hardware module. As mentioned in Section 3.2, the HD power model has been used to map the byte transitions in the AES State cells to power consumption values. All key bytes have been successfully attacked. The lowest correlation peak value of 0.01 has been achieved for key byte 13 using the byte transition $14 \rightarrow 13$ (*i.e.* data related to key byte 14 is overwritten by data related to key byte 13). This byte transition only occurs in State cell 13, which has a very low leakage. A correlation peak value of 0.01 maps to a number of needed traces n of more than 276 000. This rather high number is caused by the small output load of the involved State cell 13 and the high amount of noise caused by the microcontroller working in parallel. The highest correlation peak value of 0.0382 has been achieved for key byte 8 using the byte transition $4 \rightarrow 8$. This byte transition occurs in State cell 4, which is that one with the highest leakage. A correlation peak value of 0.0382 maps to a number of needed traces n of about 19 200. The transition also occurs one clock cycle earlier in State cell 8, which has a lower leakage (see Figure 3, left). For the transition in State cell 8, a correlation peak value of 0.02 has been achieved.

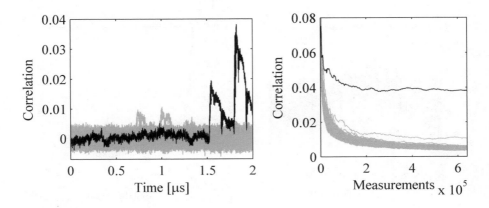

Fig. 3. CMOS AES hardware module: 256 correlation traces for key byte 8 (left, 640 000 measurements, HD power model) and evolution of the maxima of the correlation traces over measurements (right); traces for the correct key are plotted in black

Figure 3 (left) shows the correlation traces for the attack on key byte 8. The correlation trace for the correct key guess is plotted in black. The clock cycles when the attacked byte transition occurs in State cell 8 (lower correlation peak) and one clock cycle later in State cell 4 (higher correlation peak) can clearly be seen. The time axis of the left figure matches that one of Figure 2 (left).

DPA Attack on the MDPL Core with Deactivated PRNG

In order to get some experience with the MDPL AES hardware module, we first attacked it with deactivated PRNG. This gave us the chance to see if our attack scripts are correct and that we used the correct part of the power trace in the attack. We were able to attack more than half of the key bytes with the number of measurements we performed.

The DPA results in case of the MDPL AES module look quite different from those of the CMOS AES module. For MDPL, we could not clearly identify State cells with a higher and State cells with a lower leakage. Furthermore, the findings concerning the amount of leakage of the AES State cells for the CMOS core could not be reproduced for the MDPL core. The reason for it is that in a dual-rail logic style like MDPL, the amount of leakage of a cell does not depend on the absolute load connected the a cell's output but on how well the dual-rail output wires are balanced. Since no explicit balancing was done in the MDPL circuit, the actual balancing situation in the placed and routed circuit is more or less random.

For the DPA attack on the MDPL AES with deactivated PRNG we measured 1 256 000 power traces. With the arguments from the last paragraph and from Section 3.2 in mind, one would typically assume that the HWbit power model is the best choice for this attack. However, our analysis showed that with the HWbit model, we could find only 1 key byte (key byte 12; bit 1 used in attack; resulting correlation peak $\rho_{ck,ct,12} = 0.0056$). Much better results were yielded

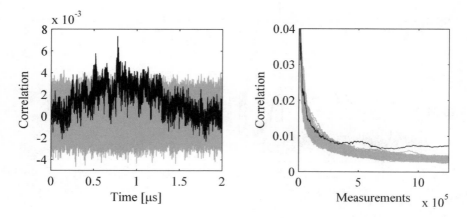

Fig. 4. MDPL AES hardware module with deactivated PRNG: 256 correlation traces for key byte 14 (left, 1 256 000 measurements, HW power model) and evolution of the maxima of the correlation traces over measurements (right); traces for the correct key are plotted in black

by using the HD model and the best results were achieved with the HW model. The reason for this behavior most likely lies in the architecture of the MDPL flip-flops [12]. They store the masked data in a single-rail flip-flop, which is connected to only one part of the complementary network of the input signal. This leads to a more or less uniform direction of the imbalance of the dual-rail signals going into an MDPL flip-flop. Therefore, the HW power model leads to better results than the HWbit model in case such signals are attacked.

When using the HD power model as in the DPA attack on the CMOS AES module, the highest correlation peak we got had a value of 0.0054, which maps to a number of required traces n of around 960 000. We could unambiguously identify 6 key bytes. In case of the HW power model, we got more and higher correlation peaks for the different key bytes (due to the precharging, which also happens in the input stages of MDPL flip-flops). However, we could not find all key bytes with the 1 256 000 measured power traces. The four highest correlation peaks occurred for key bytes 5 ($\rho_{ck,ct,5} = 0.0067$), 13 ($\rho_{ck,ct,13} = 0.0061$), 14 ($\rho_{ck,ct,14} = 0.0074$), and 15 ($\rho_{ck,ct,15} = 0.0061$). This maps to a number of needed power traces of approximately: $n_5 = 624\,000$, $n_{13} = n_{15} = 752\,000$, and $n_{14} = 511\,000$. More than half of the key bytes could be successfully attacked, for the others there were either ambiguous peaks for incorrect key guesses or no peaks at all in the correlation traces. Figure 4 (left) shows the correlation traces for the HW attack on key byte 14. The correlation trace for the correct key guess is plotted in black and the time axis matches the one of Figure 2 (right). In Figure 4 (right), the evolution of the maxima of the correlation traces over the measurements can be seen. It clearly shows that the number of needed power traces $n_{14} = 511\,000$, which has been calculated above, is reasonable.

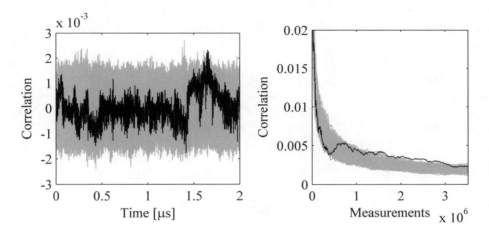

Fig. 5. MDPL AES hardware module with activated PRNG: 256 correlation traces for key byte 14 (left, 3 511 000 measurements, HW power model) and evolution of the maxima of the correlation traces over measurements (right); traces for the correct key are plotted in black

DPA Attack on the MDPL Core with Activated PRNG

Due to the findings described in the last section, we attacked the MDPL core with activated PRNG also by using the power models HWbit, HD, and HW. A similar behavior was observed: No at least "almost" or "weak" correlation peaks could be achieved with the HWbit and the HD power model. With the HW power model, we were also not able to identify a single key byte unambiguously. However, there were 4 key bytes (9, 10, 13, 14) for which one could get the impression that not too many more measurements would be necessary to get a distinct correlation peak. But this is only a speculation.

For the attack on the MDPL core with activated PRNG, we measured three sets of power traces to improve our statistical sampling accuracy. For the biggest set, we measured 3 511 000 power traces, because we expected an increase in the number of needed measurements due to the random mask that is provided to the MDPL AES core.

Figure 5 shows the correlation results for key byte 14 (HW attack), which showed the highest correlation peak in the HW attack on the MDPL core with deactivated PRNG. One can clearly see that the maxima of the correlation trace over measurements for the correct key (right figure, black trace) stay for long periods at the top border of the gray correlation-maxima band. However, its value always declines with more measurements, which is not the case in a successful attack (see Figures 3 and 4). The correlation traces for the other three suspicious key bytes showed a similar behavior. Therefore, we regard the MDPL AES module in the prototype chip (with activated PRNG) as secure for at least 3 511 000 measurements under the given attack scenario.

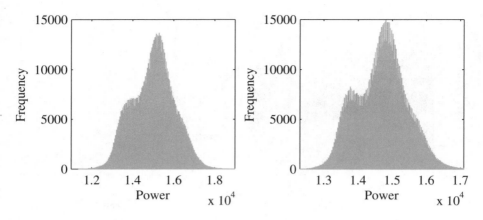

Fig. 6. Power histograms of the PDF-attack on the MDPL core with activated PRNG: clock cycle $MC1$ (left), clock cycle $MC4$ (right). Each power value in the histogram is the sum of the absolute values of the power consumption in the evaluation phase of the respective clock cycle.

4   The PDF-Attack in Practice

In [13] and [17], the authors present a new attack methodology against masked logic styles like MDPL. In an experiment with the MDPL core on our prototype chip, we tested whether we can practically reproduce this PDF-attack. The outcome of the experiment was that we could not mount a successful attack on the MDPL core in our environment.

In Section 3.3, we found out that key byte 14 leaks the most information in an attack on the AES hardware module in the MDPL core when the PRNG is deactivated. In case of an activated PRNG, the correlation result for key byte 14 was ambiguous. In order to test the effectiveness of the PDF-attack, we chose the following approach. If the PDF-attack works (*i.e.* if the mask can be biased), it should be possible to turn the ambiguous result for key byte 14 (DPA attack with activated PRNG) into an unambiguous result.

The first step in the PDF-attack was to build power profiles of the clock cycles where we expect a significant leakage and where we thus want to bias the mask. We selected two clock cycles for this purpose. The first clock cycle was that one where the correlation peak occurred in the attack on key byte 14 on the MDPL core with deactivated PRNG: clock cycle $MC1$ around $0.75\,\mu s$ (see Figure 4 left and Figure 2 right). The other clock cycle was the one where there might be the beginning of a correlation peak in the DPA attack on the MDPL core with activated PRNG: clock cycle $MC4$ around $1.6\,\mu s$ (see Figure 5 left and Figure 2 right).

For the power profiles, we took the power values during the evaluation phase of the clock cycles $MC1$ and $MC4$, respectively, for each measurement sample of the MDPL core with activated PRNG. Then, we calculated the sum of the

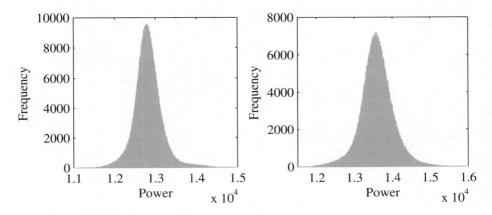

Fig. 7. Power histograms for the MDPL core with deactivated PRNG: clock cycle $MC1$ (left), clock cycle $MC4$ (right). Each power value in the histogram is the sum of the absolute values of the power consumption in the evaluation phase of the respective clock cycle.

absolute values of the selected points of each clock cycle. This is done in accordance to [13] where the analysis is based on toggle counts of clock cycles. As a result, we got two power values for every measurement sample, one for clock cycle $MC1$ and one for clock cycle $MC4$. Figure 6 shows the two histograms of the two groups of power values. The shape of the histograms were rather unexpected, because according to the PDF-attack, there should be two equally high peaks (we verified that the PRNG does not have such a significant bias via simulations, which could also cause such a shape). For comparison, Figure 7 shows the power histograms for the same clock cycles $MC1$ and $MC4$ in case of a deactivated PRNG. Here, the histograms show single peaks as expected.

According to the theory of the PDF-attack, it should now be possible to bias the mask by using just the power traces in a DPA attack that have a power consumption in the profiled clock cycles that is above (below) the mean power consumption in the respective clock cycle. Since we were not sure where the different peaks in the power profile histograms originated from, we selected different values as discerning points: the overall mean of each histogram, the mean of each peak in the histograms, and the lowest value in the valley between two peaks of a histogram (we also tried various other clock cycles and in some of them, we got more separated peaks with a valley between them).

The original number of power traces we had at hand for the PDF-attack were 3 511 000. After the selection process according to the chosen discerning point, we always had around 1.5 million traces left for the PDF-attack. According to the needed power traces to successfully attack the MDPL core with deactivated PRNG, this amount of traces should have been enough. However, in none of the settings for the discerning point, we could turn the DPA attack against key byte

Table 1. Truth table and energy dissipation of an MDPL AND gate

q	a	b	m	bias	q_m	a_m	b_m	E
0	0	0	0	α	0	0	0	δ
0	0	0	1	$1-\alpha$	1	1	1	γ
0	0	1	0	α	0	0	1	δ
0	0	1	1	$1-\alpha$	1	1	0	γ
0	1	0	0	α	0	1	0	δ
0	1	0	1	$1-\alpha$	1	0	1	γ
1	1	1	0	α	1	1	1	γ
1	1	1	1	$1-\alpha$	0	0	0	δ

Table 2. Truth table and energy dissipation of a masked signal q in general

q	m	bias	q_m	E (generic)	E
0	0	α	0	E_1	δ
0	1	$1-\alpha$	1	E_2	γ
1	0	α	1	E_3	γ
1	1	$1-\alpha$	0	E_4	δ

14 in a successful one. Finally, we targeted some other key bytes in PDF-attacks and we used the hypotheses for the HD power model instead of the hypotheses for the HW power model. Also these attacks were not successful.

5   Mask Generation for MDPL

The mask generation is of crucial importance for every countermeasure that is based on masking. Every countermeasure of this type requires that the masks are unknown by the attacker, randomly generated and uniformly distributed. Otherwise masking does not provide protection against power analysis attacks. This property is well known and can be easily shown for every masking scheme. Descriptions of attacks that exploit biased masks can for example be found in [7].

At CHES 2007, Gierlichs presented an article [3] that is devoted to the discussion of the effect of a biased mask on the DPA resistance of MDPL. Unfortunately, the article does not make a clear statement about the fact how much the mask needs to be biased for a successful attack and only talks about slight biases. Hence, one might get the impression that MDPL has different requirements on the mask generation than other masking schemes. In this section we analyze the arguments presented in [3] and we show that these arguments are not specific for MDPL. MDPL does not have other requirements concerning mask generation than any other countermeasure that is based on masking.

The basic idea of the attack presented in [3] can be best illustrated using a truth table of an MDPL AND gate as shown in Table 1. The last column of this table shows the power consumption of the gate. The power consumption δ occurs, if the output q_m is 0 and it is γ, if q_m is 1. Gierlichs proposes to perform a DPA attack by exploiting the difference between the power consumption for the case $(a = 1, b = 1)$ and the case $(a = 0, b = 0)$. The expected value of the difference d between these two cases is given by

$$d_{MDPL} = \alpha\gamma + (1-\alpha)\delta - \alpha\delta - (1-\alpha)\gamma = 2\alpha\gamma - 2\alpha\delta + \delta - \gamma \qquad (1)$$

Clearly, there is only a difference $d_{MDPL} > 0$ between these two cases, if $\alpha \neq 0.5$. We now compare this result with the result of a DPA attack on a masked signal in general. Table 2 shows the energy dissipation that occurs, if a signal q is masked and hence represented by q_m and m in the circuit. This leads to four possible different representations of the signal in the circuit with potentially four different energy dissipations $E_1 \ldots E_4$. In a DPA attack on the value q, the attacker calculates the difference between the cases for $q = 0$ and $q = 1$. The expected value of the difference is hence given by

$$d_{generic} = \alpha E_3 + (1 - \alpha)E_4 - \alpha E_1 - (1 - \alpha)E_2 \tag{2}$$

In this generic case, there can be several reasons why $d_{generic} > 0$. In an implementation where m and q_m are processed independently (*i.e.* the total power consumption is the sum of the power consumption of the mask and the power consumption of the masked value), it holds that $E_1 = E_4 = \delta$ and $E_2 = E_3 = \gamma$. In this case, $d_{generic}$ can only be bigger than zero, if there is a bias on the mask. In fact, in this case it holds that $d_{generic} = d_{MDPL}$. In implementations, where there are glitches or other joint timing properties of the mask and the data (see for example [9], [14]) it can hold that $E_1 \neq E_2 \neq E_3 \neq E_4$. In this case there can be a leakage even if there is no bias on the mask.

In summary, the argumentation for exploiting a bias in a mask of an MDPL circuit that is provided in [3] is not different than it is for any other masked implementation. In fact, the presented DPA attack in [3] is equivalent to a DPA attack on a generic masked signal in a circuit where the mask and masked data are processed independently.

6   Conclusions

In our work, we showed that the masked logic style MDPL is not completely broken because of early propagation. In regular designs where the signal-delay differences are small, MDPL still provides an acceptable level of protection against DPA attacks. No key byte of the attacked MDPL AES coprocessor was revealed with 3 511 000 power measurements. For the recently presented PDF-attack we could show that the attack does not work offhand in practice. This is also clear from the following perspective. With the settings chosen for the PDF-attack in theory (no electronic and other forms of noise, no perfect balancing of dual-rail wire pairs, ...), it can easily be shown that all DPA-resistant logic styles that try to achieve a constant power consumption are also completely broken. However, various publications draw opposite conclusions. Finally, we showed that mask generation is not particularly difficult for MDPL.

Acknowledgements. This work has been supported by the Austrian Government through the research program FIT-IT Trust in IT Systems (Project GRANDESCA, Project Number 813434).

References

1. Chen, Z., Zhou, Y.: Dual-Rail Random Switching Logic: A Countermeasure to Reduce Side Channel Leakage. In: Goubin, L., Matsui, M. (eds.) CHES 2006. LNCS, vol. 4249, pp. 242–254. Springer, Heidelberg (2006)
2. Coron, J.-S., Kocher, P.C., Naccache, D.: Statistics and Secret Leakage. In: Frankel, Y. (ed.) FC 2000. LNCS, vol. 1962, pp. 157–173. Springer, Heidelberg (2001)
3. Gierlichs, B.: DPA-Resistance Without Routing Constraints? In: Paillier, P., Verbauwhede, I. (eds.) CHES 2007. LNCS, vol. 4727, pp. 107–120. Springer, Heidelberg (2007)
4. Golić, J.D., Menicocci, R.: Universal Masking on Logic Gate Level. IEE Electronic Letters 40(9), 526–527 (2004)
5. Lin, K.J., Fang, S.C., Yang, S.H., Lo, C.C.: Overcoming Glitches and Dissipation Timing Skews in Design of DPA-Resistant Cryptographic Hardware. In: Lauwereins, R., Madsen, J. (eds.) 2007 Design, Automation and Test in Europe Conference and Exposition (DATE 2007), Nice, France, April 16-20, 2007, pp. 1265–1270. ACM Press, New York (2007)
6. Mangard, S., Aigner, M., Dominikus, S.: A Highly Regular and Scalable AES Hardware Architecture. IEEE Transactions on Computers 52(4), 483–491 (2003)
7. Mangard, S., Oswald, E., Popp, T.: Power Analysis Attacks – Revealing the Secrets of Smart Cards. Springer, Heidelberg (2007)
8. Mangard, S., Popp, T., Gammel, B.M.: Side-Channel Leakage of Masked CMOS Gates. In: Menezes, A. (ed.) CT-RSA 2005. LNCS, vol. 3376, pp. 351–365. Springer, Heidelberg (2005)
9. Mangard, S., Schramm, K.: Pinpointing the Side-Channel Leakage of Masked AES Hardware Implementations. In: Goubin, L., Matsui, M. (eds.) CHES 2006. LNCS, vol. 4249, pp. 76–90. Springer, Heidelberg (2006)
10. Moradi, A., Salmasizadeh, M., Shalmani, M.T.M.: Power Analysis Attacks on MDPL and DRSL Implementations. In: Nam, K.-H., Rhee, G. (eds.) ICISC 2007. LNCS, vol. 4817, pp. 259–272. Springer, Heidelberg (2007)
11. Popp, T., Kirschbaum, M., Zefferer, T., Mangard, S.: Evaluation of the Masked Logic Style MDPL on a Prototype Chip. In: Paillier, P., Verbauwhede, I. (eds.) CHES 2007. LNCS, vol. 4727, pp. 81–94. Springer, Heidelberg (2007)
12. Popp, T., Mangard, S.: Masked Dual-Rail Pre-Charge Logic: DPA-Resistance without Routing Constraints. In: Rao, J.R., Sunar, B. (eds.) CHES 2005. LNCS, vol. 3659, pp. 172–186. Springer, Heidelberg (2005)
13. Schaumont, P., Tiri, K.: Masking and Dual-Rail Logic Dont Add Up. In: Paillier, P., Verbauwhede, I. (eds.) CHES 2007. LNCS, vol. 4727, pp. 95–106. Springer, Heidelberg (2007)
14. Suzuki, D., Saeki, M.: Security Evaluation of DPA Countermeasures Using Dual-Rail Pre-charge Logic Style. In: Goubin, L., Matsui, M. (eds.) CHES 2006. LNCS, vol. 4249, pp. 255–269. Springer, Heidelberg (2006)
15. Suzuki, D., Saeki, M., Ichikawa, T.: DPA Leakage Models for CMOS Logic Circuits. In: Rao, J.R., Sunar, B. (eds.) CHES 2005. LNCS, vol. 3659, pp. 366–382. Springer, Heidelberg (2005)
16. Suzuki, D., Saeki, M., Ichikawa, T.: Random Switching Logic: A New Countermeasure against DPA and Second-Order DPA at the Logic Level. IEICE Transactions on Fundamentals of Electronics, Communications and Computer Sciences E90-A(1), 160–168 (2007)

17. Tiri, K., Schaumont, P.: Changing the Odds against Masked Logic. In: Biham, E., Youssef, A.M. (eds.) SAC 2006. LNCS, vol. 4356, pp. 134–146. Springer, Heidelberg (2007), http://rijndael.ece.vt.edu/schaum/papers/2006sac.pdf
18. Trichina, E., Korkishko, T., Lee, K.-H.: Small Size, Low Power, Side Channel-Immune AES Coprocessor: Design and Synthesis Results. In: Dobbertin, H., Rijmen, V., Sowa, A. (eds.) AES 2005. LNCS, vol. 3373, pp. 113–127. Springer, Heidelberg (2005)
19. Wolkerstorfer, J., Oswald, E., Lamberger, M.: An ASIC implementation of the AES SBoxes. In: Preneel, B. (ed.) CT-RSA 2002. LNCS, vol. 2271, pp. 67–78. Springer, Heidelberg (2002)

Cryptanalysis of CTC2

Orr Dunkelman[1,*] and Nathan Keller[2,**]

[1] Département d'Informatique,
École normale supérieure
45 rue d'Ulm,
Cedex 05, Paris 75230, France
orr.dunkelman@ens.fr
[2] Einstein Institute of Mathematics,
Hebrew University
Jerusalem 91904, Israel
nkeller@math.huji.ac.il

Abstract. CTC is a toy cipher designed in order to assess the strength of algebraic attacks. While the structure of CTC is deliberately weak with respect to algebraic attacks, it was claimed by the designers that CTC is secure with respect to statistical attacks, such as differential and linear cryptanalysis. After a linear attack on CTC was presented, the cipher's linear transformation was tweaked to offer more diffusion, and specifically to prevent the existence of 1-bit to 1-bit approximations (and differentials) through the linear transformation. The new cipher was named CTC2, and was analyzed by the designers using algebraic techniques.

In this paper we analyze the security of CTC2 with respect to differential and differential-linear attacks. The data complexities of our best attacks on 6-round, 7-round, and 8-round variants of CTC2 are 64, 2^{15}, and 2^{37} chosen plaintexts, respectively, and the time complexities are dominated by the time required to encrypt the data.

Our findings show that the diffusion of CTC2 is relatively low, and hence variants of the cipher with a small number of rounds are relatively weak, which may explain (to some extent) the success of the algebraic attacks on these variants.

1 Introduction

The merits of algebraic attacks [8,13] are still debated in the cryptographic community. In the field of stream ciphers, algebraic attacks succeeded significantly, allowing to break several ciphers much faster than other known techniques (see [7,12]). At the same time, in the field of block ciphers, the situation is more complicated: while there are several instances in which algebraic attacks can be applied (see [5,11,16]), algebraic attacks did not succeed in attacking any

---

* This author is supported by the France Telecom Chaire.
** This author is supported by the Adams Fellowship Program of the Israel Academy of Sciences and Humanities.

M. Fischlin (Ed.): CT-RSA 2009, LNCS 5473, pp. 226–239, 2009.

well known block cipher better than other techniques. Hence, the applicability of algebraic attacks to block ciphers have stirred quite a debate in the cryptographic community, arguing whether this class of attacks can indeed impose a threat to "strong" block ciphers.

In [8], Courtois presented the block cipher CTC, intended to provide an example of a "cryptographically strong" cipher that can be attacked using algebraic techniques. CTC is an SP network, with scalable parameters. The most interesting set of parameters is when the block and key sizes are of 255 bits, and a large number of 3-bit S-boxes (its six-round variant was broken using algebraic approaches). The components of the cipher are intentionally weak with respect to algebraic attacks: the S-boxes are 3-bit to 3-bit, and the linear transformation is extremely simple. It was claimed in [8] that despite these weaknesses, the cipher is strong against statistical attacks, such as differential and linear cryptanalysis.

The resistance of CTC against linear cryptanalysis was challenged in [14]. It was shown that the linear transformation of CTC, along with the particular S-box used in the cipher, allow for an iterative linear approximation with a single active S-box in every round. This approximation has a bias of $1/4$ per round, thus allowing to mount a simple key recovery attack on r-round CTC with data complexity of about 2^{2r+2} known plaintexts. Moreover, it was shown that for a large portion of the 3-bit S-boxes, similar results would hold.

As a response to the suggested attack, CTC's designers changed the linear transformation of the cipher and introduced CTC2 [9,10]. The new linear transformation no longer allows for one bit to depend only on one bit. Despite this fact, the diffusion of the new linear transformation is still very weak as most of the bits after the linear transformation depend on only two bits before the linear transformation (except for one bit which depends on three bits). The weak diffusion can be combined with the 3-bit to 3-bit S-boxes to construct high probability differentials and linear approximations for a small number of rounds.

In this paper we examine the security of CTC2 against statistical attacks. We start by examining 3-round and 4-round variants of the cipher. We present a meet-in-the-middle attack on 4-round CTC2 with data complexity of 4 chosen plaintexts and a negligible time complexity. Furthermore, we present many "structural" differential characteristics (with probability 2^{-14} each) and linear approximations (with bias 2^{-8} each) for 3-round CTC2, and also a 4-round differential with probability 2^{-18} and a 4-round linear approximation with bias 2^{-10}.

We then present truncated differential attacks against 5-round, 6-round, and 7-round variants of CTC2. The data complexities of the attacks are 24, 64, and 2^{15} chosen plaintexts, respectively, and the time complexities are dominated by the time required for encrypting the data.

Finally, we present differential-linear attacks against 7-round and 8-round variants of the cipher. The 8-round attack has data complexity of 2^{37} chosen plaintexts and time complexity of 2^{37} encryptions.

It is important to note that our attacks do not imply that algebraic attacks on block ciphers are unfit to play a role in the cryptanalytic toolbox. However, it seems that our findings show that the diffusion in CTC2 is insufficient to offer security when the number of rounds is relatively small (e.g., 8 or less), which may explain the success of the algebraic attack on CTC2 with a small number of rounds.

The rest of the paper is organized as follows: Section 2 shortly describes the CTC and CTC2 block ciphers. In Section 3 we present an efficient meet-in-the-middle attack on 4-round CTC2. We discuss short differentials and linear approximations of CTC2 in Section 4. In Section 5 these differentials are used to present truncated differential attacks on CTC2, and in Section 6 the differentials and the linear approximations are combined in differential-linear attacks on CTC2. We discuss our results in Section 7.

2   A Description of CTC2

CTC is a toy cipher presented for the sake of cryptanalysis using algebraic attacks [8]. The cipher supports a variable block size, and a variable number of rounds. After an iterative linear approximation with a large bias was presented in [14], the designers of CTC introduced a tweaked version of the cipher called CTC2 [9,10].

The most discussed version of CTC has a block size and a key size of 255 bits. Each round is composed of an XOR with a subkey, parallel application of the same S-box, and a simple linear transformation. After the last round another subkey is XORed to the output.

The 3-bit to 3-bit S-box used in CTC is $S[\cdot] = \{7, 6, 0, 4, 2, 5, 1, 3\}$. The state is initialized to the plaintext XORed with the first subkey, and the bits enter the S-boxes in groups of three consecutive bits, where bit 0 is the least significant bit of the first S-box (which we denote by S-box 0) and bit 254 is the most significant bit of the 85th S-box (which we denote by S-box 84).

The linear transformation of CTC is very simple, and each output bit, denoted by Z_i, depends on one or two of the input bits, denoted by Y_i. We note that in [8] the notations are slightly different (as they include the round number before the number of the bit):

$$\begin{cases} Z_2 = Y_0 \\ Z_{i \cdot 202 + 2 \bmod 255} = Y_i \oplus Y_{i+137 \bmod 255} \text{ for } i = 1, \ldots 254 \end{cases}$$

The key schedule of CTC is also very simple: The ith round subkey is obtained from the secret key by a cyclical rotation to the left by i bits.

The main difference between CTC and CTC2 is a different linear transformation used in CTC2 [9,10]:

$$\begin{cases} Z_{151} = Y_2 \oplus Y_{139} \oplus Y_{21} \\ Z_{i \cdot 202 + 2 \bmod 255} = Y_i \oplus Y_{i+137 \bmod 255} \text{ for } i = 0, 1, 3, 4 \ldots 254 \end{cases}$$

In addition, the key schedule of CTC2 is slightly different: The ith round subkey is obtained from the secret key by a *right* rotation by i bits (instead of a left rotation used in CTC).

3   A Meet-in-the-Middle Approach

The relatively slow diffusion of CTC2 allows to mount simple meet-in-the-middle attacks on small number of rounds of the cipher. As an example, we present a 4-round attack following the methodology presented in [15]. The attack exploits the fact that there are bits in the intermediate state after three full rounds of encryption, that depend on less than 255 plaintext bits. Consider a concrete bit x having this property, and denote the set of plaintext bits x depends on by S_1. It is clear that if two plaintexts have the same value in all the bits of S_1, then the corresponding intermediate values after three rounds agree on the bit x.

To attack 4-round CTC2, we consider several plaintexts having the same value in the bits of S_1. We would like to partially decrypt the ciphertexts through the last round and check whether the intermediate values after three rounds agree on the bit x. Since the fourth round is the last one, we can swap the order of the linear transformation and the key addition in the fourth round, and hence guessing three bits of an equivalent subkey is sufficient to recover the value of the bit x.[1] For each guess of these three bits, we check whether the suggested values of the bit x are the same for all the encryptions, and if not, we discard the key guess. The right key guess is expected to pass this filtering, and the probability that at most one wrong key guess passes the filtering is as high as 78%, if 4 plaintexts are used in the attack (for 6 plaintexts, only the right subkey remains with probability 80%). Hence, the attacker gains two bits of key information. The time complexity of the attack is dominated by the partial decryption of the first two ciphertexts, since after them the number of candidate subkeys is smaller. By using a simple precomputed table, the attacker can retrieve the possible subkey values by a simple table query.

We note that while the attack seems to retrieve only a small number of subkey bits, it is possible to obtain additional key information by repeating the attack with a different bit instead of x. This would result in an increase in the data complexity of the attack, but this increase can be moderated by using structures, allowing to re-use the same data in attacks with different values of the bit x. For example, the sets of unaffected plaintext bits corresponding to the two least significant bits (i.e., $x = e_0$ and $x = e_1$)[2], share seven joint bits, and hence the attacks with $x = e_0$ and $x = e_1$ can be mounted simultaneously using the same data set.

3.1   Attacking More Rounds Using the 4-Round Meet-in-the-Middle Attack

It is possible to use the above attack to speed-up exhaustive key search on more than 4 rounds of CTC2 when only four chosen plaintexts are available

---

[1] This standard technique is used in all the attacks presented in this paper. In the sequel, we do not mention it explicitly, and use the term "equivalent subkey", referring to the subkey resulting from the swap.

[2] Throughout the paper e_i denotes a value of 0's in all positions but position i. Similarly, $e_{i,j} = e_i \oplus e_j$, $e_{i,j,k} = e_{i,j} \oplus e_k$, etc.

to the attacker. The attack is based on guessing the subkey of the last rounds, partially decrypting the ciphertexts, and then checking whether the meet-in-the-middle attack "succeeds" (i.e., retrieves a subkey which agrees with the partial decryption).

To attack r-round CTC2, for each key guess, the attacker has to decrypt two of the ciphertexts (which takes time proportional to $2 \cdot (r - 4)$ rounds in a trivial implementation). Then, the attacker checks whether the subkey the two ciphertexts suggest agrees with the key guess. In case the check succeeds, then the attacker partially decrypts the remaining pairs, and checks them. Only if the 4-round attack succeeds with the four plaintext/ciphertext pairs, then the attacker performs the full trial encryption. As the first step is the most time consuming, the time complexity of this approach is about $r/(2 \cdot (r - 4))$ times faster than exhaustive key search (about 60% speed-up for a 5-round attack, a 33% speed-up for a 6-round attack, and a 14% speed-up for a 7-round attack).

Moreover, the attacker can use a slightly better implementation of the partial decryptions (which are done under the same subkeys). For example, by implementing the partial decryption step in a bit-slice manner (which is expected to be the faster implementation of CTC2's decryption in any case), and using the fact that most modern CPUs can support several operations in parallel (or at least have higher throughput by scheduling the operations correctly), one can reduce the time required for the decryption of two values to almost the time required for decrypting one value. In such a case, the expected speed-up is going to be about $r/(r - 4)$, i.e., offer a 80% speed-up for a 5-round attack and a 66% speed-up for a 6-round attack.

4   Linear Approximations and Differentials of CTC2

4.1   Linear Approximations

Due to the low diffusion of CTC2, there are many linear approximations for a small number of rounds with a relatively high bias. These approximations exploit the following two properties of CTC2:

1. The S-box of CTC2 has several one-bit to one-bit linear approximations. We note that this property is not a weakness of the particular S-box used in CTC2. The majority of the 3-bit to 3-bit S-boxes have such approximations.
2. The diffusion of the linear transformation in the backward direction is extremely low: Each bit after the linear transformation, except for one, is the XOR of only two bits before the linear transformation. The remaining bit is the XOR of three bits.

Using these properties, it is easy to construct many 3-round linear approximations having only seven active S-boxes: four active S-boxes in the first round, two active S-boxes in the second round, and one active S-box in the third round. All these approximations have a bias of $\pm 1/4$ for each active S-box, or a total of $q = \pm 2^6 \cdot (2^{-2})^7 = \pm 2^{-8}$. Most of them can be easily extended by one (or two) more rounds in the backward direction, where in the additional rounds there are eight (or 24, respectively) more active S-boxes. The following is an example of such approximation:

$$\lambda_P = e_{14,104,134,241} \xrightarrow{S} e_{14,104,132,241} \qquad p = \tfrac{1}{2} - 2^{-5}$$
$$\xrightarrow{LT} e_{38,154} \xrightarrow{S} e_{36,154} \qquad p = \tfrac{1}{2} + 2^{-3}$$
$$\xrightarrow{LT} e_0 \xrightarrow{S} e_2 = \lambda_C \qquad p = \tfrac{1}{2} + 2^{-2}$$

We note that there are many ways to change the approximation slightly and get similar approximations with bias 2^{-8} and the same output mask. Hence, attacks which exploit multiple linear approximations [4,6] are expected to succeed against CTC2 significantly better than a standard linear attack.

While the diffusion of the linear transformation in the backward direction is very low, the diffusion in the forward direction is extremely high.[3] Most of the bits before the linear transformation are the XOR of more than 50 bits after the linear transformation, and hence most of the linear approximations presented above cannot be extended in the forward direction. However, there exist several special bits whose linear diffusion in the forward direction is weak. In particular, bit 2 before the linear transformation is the XOR of bits 30 and 151 after the linear transformation. As a result, the specific approximation presented above can be extended by one more round by concatenating the following 1-round approximation:

$$e_2 \xrightarrow{LT} e_{30,151} \xrightarrow{S} e_{32,151} = \lambda_C \qquad p = \tfrac{1}{2} - 2^{-3}$$

The resulting 4-round approximation (with bias 2^{-10}) cannot be extended further due to the high diffusion of bits 32 and 151 in the forward direction.

In total, the best linear approximations we detected for three, four and five rounds of the cipher have biases of $2^{-6}, 2^{-10}$, and 2^{-18}, respectively.

4.2   Differential Characteristics

The analysis of short differentials of CTC2 is similar to the analysis of short linear approximations presented above. The only difference is that for differentials, the diffusion in the *forward* direction is very weak (i.e., a difference in a single bit before the linear transformation influences only two or three bits after the linear transformation), while the diffusion in the *backward* direction is extremely high.[4] As a result, there are many 3-round differentials with only seven active S-boxes, that can be extended in the forward direction by one or two more rounds

---

[3] We note that this situation is not typical in block ciphers. In most of the ciphers, the diffusions in both directions are comparable, and hence usually approximations have a round with a single active S-box in the middle, and then are extended to both directions. In CTC2, the round with a single active S-box can be placed only at the end of the approximation.

[4] This situation is quite general in block ciphers. Differentials in the forward direction usually correspond to linear approximations in the backward direction and vice versa, see [1].

(with 8 or 24 more active S-boxes, respectively). One of these differentials is the following:

$$\Omega_P = e_2 \xrightarrow{S} e_0 \qquad\qquad p = 2^{-2}$$
$$\xrightarrow{LT} e_{2,123} \xrightarrow{S} e_{0,123} \qquad\qquad p = 2^{-4}$$
$$\xrightarrow{LT} e_{2,113,123,234} \xrightarrow{S} e_{0,111,123,234} = \Omega_C \qquad p = 2^{-8}$$

Due to the low diffusion of bit 2 in the backward direction, this specific differential can be extended in the backward direction by adding the following one-round differential:

$$e_{30,150} \xrightarrow{S} e_{30,151} \xrightarrow{LT} e_2 \qquad p = 2^{-4}$$

The resulting four-round differential has probability 2^{-18}. In total, the best differential characteristics we detected for $3, 4$, and 5 rounds of the cipher have probabilities $2^{-14}, 2^{-18}$, and 2^{-34}, respectively.

5   Differential Attacks on CTC2

5.1   A 5-Round Differential Attack

The differentials presented above can be used to construct a high-probability truncated differential for 5 rounds of CTC2. The truncated differential starts with the first two rounds of the above 4-round differential:

$$\Omega_P = e_{30,150} \xrightarrow{S} e_{30,151} \qquad p = 2^{-4}$$
$$\xrightarrow{LT} e_2 \xrightarrow{S} e_0 \qquad p = 2^{-2}$$

From e_0 (before the linear transformation of the second round), the difference propagates unconstrained for 3 more rounds. If the two first rounds of the differential hold, then there is no difference in 35 bits after the fifth round. Denote the set of these 35 bits by S_2.[5]

This truncated differential can be used to mount a simple attack on 5-round CTC2. The attacker considers 64 pairs of plaintexts with difference Ω_P, and checks whether in the corresponding ciphertext pairs, the difference in the bits of S_2 is zero. Since for a random pair, the probability that a difference in 35 bits is zero is 2^{-35}, it is expected that only pairs satisfying the differential (the right pairs) pass this filtering. The probability of the differential is $1/64$, and hence it is expected that only the right pair remains after the filtering. Once identified, this pair can be used to retrieve the subkey used in S-boxes 10 and 50 in the first round.

---

[5] For completeness we give the set S_2:

$S_2 = \{13, 14, 24, 35, 50, 56, 59, 65, 66, 74, 79, 80, 82, 112, 116, 118, 131, 132, 135, 148, 165,$
$\qquad 169, 171, 184, 187, 190, 199, 200, 201, 222, 226, 237, 240, 243, 252\}.$

The data complexity of the attack can be reduced using structures. The attacker considers a structure of 24 plaintexts, in which the value in all the S-boxes except for S-boxes 10 and 50 is constant, and the values in these two S-boxes are distinct. These plaintexts can be combined into $24 \cdot 23/2 = 276$ pairs, and about 4 of them[6] are expected to have difference only in bit 2 after the first round. Since the probability of the second round of the differential is 2^{-2}, it is expected that the data contains one right pair. As in the basic attack, this right pair can be detected immediately (by checking whether the ciphertexts have zero difference in the bits of S_2) and used to retrieve the subkeys used in S-boxes 10 and 50 in the first round. The identification of the right pair is immediate when using a hash table, and the complexity of the attack is dominated by the encryptions of the 24 plaintexts.

5.2   A 6-Round Differential Attack

Using the low diffusion of CTC2, the truncated differential presented above can be used to attack 6-round CTC2.

As in the 5-round attack, we consider plaintext pairs with input difference Ω_P. First, we note that if the differential is satisfied, then the input difference to each of the S-boxes 5, 27, and 67 in round 6 has at most one active bit (since the difference in bits $13, 14, 79, 80, 199$, and 200 in the output of round 5 is zero by the differential). As a result, the output difference in each of these S-boxes can assume only 5 out of the 8 possible values.

Second, we observe that given a right pair, we can easily construct additional right pairs based on it. We note that bit 3 of the plaintext does not influence the bits of S-boxes 10 and 50 in round 1 and S-box 1 in round 2, which are the only S-boxes used in the differential. Hence, if (P_1, P_2) is a right pair, then $(P_1 \oplus e_3, P_2 \oplus e_3)$ is also a right pair.

Using these two observations, we can attack 6-round CTC2 in the following manner. We consider 64 plaintext pairs with difference Ω_P, and for each pair (P_1, P_2), we consider also the pair $(P_1 \oplus e_3, P_2 \oplus e_3)$. For each pair of pairs, we check whether the ciphertext differences satisfy the constraint in S-boxes 5, 27, and 67. For a random pair of pairs, the probability of passing this filtering is $((5/8)^2)^3 = 2^{-4.06}$, and hence four pairs of pairs are expected to remain. These pairs are expected to contain one pair of right pairs (since the probability of the differential is $1/64$).

For each of the remaining pairs of pairs, we guess the 9 bits of the equivalent subkeys used in S-boxes 5, 27, and 67 of round 6, and check whether the difference in bits $13, 14, 79, 80, 199$, and 200 in the input of round 6 is indeed zero. Each pair of pairs is expected to suggest one consistent subkey proposal on average for the nine subkey bits under attack. Thus, out of the four pairs of pairs, we expect four subkey suggestions for 9-bit information about the key. These suggestions contain

---

[6] In a "full" structure of 64 plaintexts, there are $64 \cdot 63/2 = 2016$ possible pairs, of which 32 satisfy the required input difference. Hence, amongst 276 pairs, we expect $32 \cdot 276/2016 = 4.38$ pairs with the required input difference.

the correct value of the subkey bits (the one suggested by the pair of right pairs). More key information can be obtained by attaching to the right pairs another "companion" pair, using other input bits that do not affect the differential (e.g., bit 4). In addition, the attacker can gain information by analyzing the S-boxes in round 6 in which the differential predicts the difference in a single input bit.

Hence, using two structures of 32 plaintexts each, it is possible to deduce the equivalent of seven subkey bits for the last round's subkey quickly and efficiently.

5.3   A Differential Attack on 7-Round CTC2

In order to mount a differential attack on a 7-round variant of the cipher, we have to extend the 5-round truncated differential presented above to 6 rounds. A natural way to extend the differential is to add one round in which the difference is "fully specified", such that the differential will consist of three constrained rounds and three unconstrained rounds. However, we checked all the possible differentials of this class and found that the number of output bits in which the difference is assured to be zero is too small, and hence cannot be used to detect the right pairs. Therefore, we use in the attack a differential consisting of four fully specified rounds and two unconstrained rounds. The first four rounds of the differential are:

$$\Omega_P = e_{30,150} \xrightarrow{S} e_{30,151} \qquad\qquad p = 2^{-4}$$
$$\xrightarrow{LT} e_2 \xrightarrow{S} e_0 \qquad\qquad p = 2^{-2}$$
$$\xrightarrow{LT} e_{2,123} \xrightarrow{S} e_{0,123} \qquad\qquad p = 2^{-4}$$
$$\xrightarrow{LT} e_{2,113,123,234} \xrightarrow{S} e_{0,111,125,236} \qquad p = 2^{-8}$$

These four rounds are followed by two rounds where there is no restriction on the development of the difference. If the differential is satisfied, then in the output of round 6, there is zero difference in S-boxes 22 and 65, and at most one active bit in the differences of 21 more S-boxes (in total, 92 bits are assured to have zero difference).

Like the differential, the attack algorithm should also be modified. Since every input bit affects some of the S-boxes used in the differential, right pairs cannot be used to produce "companion" right pairs anymore. We compensate for that by using the stronger filtering in the last round, along with a filtering in the first round.

In the attack, we examine $2^9 = 512$ structures of plaintexts, where in each structure the input to S-boxes 10 and 50 takes all the 64 possible values, and the rest of the bits are constant. For each structure, we decrypt the ciphertexts through the linear transformation, and detect the pairs whose intermediate values satisfy the restrictions on the difference imposed by the differential (zero difference in S-boxes 22 and 65, and one of the five possible differences in 21 more S-boxes).

In each structure there are $64 \cdot 63/2 \approx 2^{11}$ pairs, 32 of which have difference e_2 after the first round. Hence, in total there are about 2^{20} pairs, 2^{14} of them may satisfy the differential in rounds 2–6. As the probability of rounds 2–6 of the differential is 2^{-14}, the obtained data is expected to contain about one right pair. The probability that a wrong pair satisfies the difference in the intermediate encryption values is

$$(2^{-3})^2 \cdot \left(\frac{5}{8}\right)^{21} = 2^{-6} \cdot 2^{-0.678 \cdot 21} = 2^{-20.2}.$$

Therefore, besides the right pair, about one wrong pair is expected to remain.

At this stage, we consider the remaining pairs and check (for each pair) whether the input difference can lead to the difference e_2 after the first round. Since only 16 of the 64 input differences satisfy this requirement, it is expected that only the right pair remains after this step. As in the 5-round attack, the right pair can be used to suggest two values for the 6 subkey bits used in S-boxes 11 and 51 in round 1. In addition, the pair suggests two values for the 3-bit subkey in each of the S-boxes in round 7, in which the input difference has exactly one active bit.[7] Hence, in total the attack reveals four bits of key information in round 1, and between 0 and 42 bits of key information in round 7.

The data complexity of the attack is 2^{15} chosen plaintexts, and the time complexity is dominated by the time required for encrypting the plaintexts (the filtering condition on the ciphertext pairs can be performed efficiently using a hash table). The memory complexity is also small, as it is sufficient to store one structure at a time.

To extend the attack to 8-round CTC2, one can extend the differential to a 7-round truncated differential, in which the differences are fully specified in the first 6 rounds, and completely unrestricted in the seventh round. This results in an attack whose data and time complexities are about 2^{60}, which is inferior to the differential-linear attack that we present in the next section.

6   Differential-Linear Attacks on CTC2

The truncated differentials and the linear approximations presented in the previous sections can be concatenated to devise differential-linear attacks on a bigger number of rounds. In this section we present differential-linear attacks on 7-round and 8-round CTC2.

6.1   A Differential-Linear Attack on 7-Round CTC2

The differential-linear attack on 7-round CTC2 is based on a 6-round distinguisher, composed of a 5-round truncated differential and a one-round linear approximation. The truncated differential is the same used in the 5-round differential attack (Section 5.1), which assures that the difference in 35 bits after

---

[7] The differential has 23 S-boxes in round 7 for which the input difference has at most one active bit. S-boxes where all the input bits have zero difference do not suggest key information, since in these S-boxes the output difference is zero, regardless of the key.

round 5 is zero. These 35 bits contain bits 50 and 184, and hence the differential can be combined with the following one-round linear approximation:

$$e_{50,184} \xrightarrow{S} e_{48,185} \xrightarrow{LT} e_{131} \qquad p = \tfrac{1}{2} - 2^{-3}$$

We note that this specific linear approximation is chosen in order to have a single active S-box in the round after the distinguisher. Since the probability of the differential is 2^{-6} and the bias of the linear approximation is 2^{-3}, the overall bias of the differential-linear approximation is $2 \cdot 2^{-6} \cdot (2^{-3})^2 = 2^{-11}$ (see [2,17]). Thus, a simple 7-round attack exploiting the distinguisher requires 2^{24} plaintext pairs with input difference Ω_P. After obtaining the ciphertexts, the attacker guesses three bits in the equivalent last round subkey, and checks whether the differential-linear approximation holds or not.

Like in the 5-round differential attack, the data complexity of the attack can be reduced using structures. The attacker considers structures of 64 plaintexts, in which the value in all the S-boxes except for S-boxes 10 and 50 is constant, and the values in these two S-boxes assume all the 64 possible values. For each guess of the six corresponding bits of the first subkey, the attacker can find 32 pairs in each structure satisfying the input difference of the second round of the differential. For these pairs, the attacker can apply a 5-round differential-linear approximation composed of the four last rounds of the differential and the same linear approximation used before. The bias of the reduced approximation is $2 \cdot 2^{-2} \cdot (2^{-3})^2 = 2^{-7}$, and thus 2^{16} pairs are sufficient for the attack. Since each structure contains 32 pairs, the attack requires 2^{11} structures, which are 2^{17} chosen plaintexts.

In order to reduce the time complexity of the attack we use an observation similar to the one presented in [3]. In the partial decryption phase of the attack, the attacker has to decrypt only a single S-box. In this S-box, there are only 64 possible pairs of ciphertext values. Hence, instead of decrypting the same values many times, the attacker can perform the following: First, the attacker counts how many times each of these 64 values occurs amongst the ciphertext pairs. Then, for each guess of the three equivalent last subkey bits, the attacker decrypts the 64 values, and checks for each of them whether the differential-linear approximation holds or not. Finally, the attacker combines the result of the check with the number of occurrences of each of the 64 values to compute the overall probability that the approximation holds, and use this probability to choose the most biased key.

Using this strategy, the time complexity of the decryption phase is negligible compared to the encryption of the plaintexts. The time complexity of the first round subkey guess is $2^6 \cdot 2 \cdot 2^{17}$ S-box computations, which are faster than 2^{17} encryptions. Hence, the overall data complexity of the attack is approximately 2^{17} encryptions.

Following [18], we can calculate the success rate of the attack to be about 81.5% (this is the probability that the right subkey is the most biased).

6.2   A Differential-Linear Attack on 8-Round CTC2

The attack on 8-round CTC2 is based on a 7-round distinguisher, composed of a 6-round truncated differential and a one-round linear approximation. The truncated differential is the same used in the 7-round differential attack (Section 5.3), which assures that the difference in 73 bits after round 6 is zero. These bits contain bit 1, and hence the differential can be combined with the following one-round linear approximation:

$$e_1 \xrightarrow{S} e_2 \xrightarrow{LT} e_{30,151} \qquad p = \tfrac{1}{2} - 2^{-2}$$

The probability of the differential is 2^{-18} and the bias of the linear approximation is 2^{-2}. Thus, the bias of the differential-linear approximation is $2 \cdot 2^{-18} \cdot (2^{-2})^2 = 2^{-21}$. A simple 8-round attack exploiting the 7-round differential-linear approximation requires 2^{45} chosen plaintexts.

This basic attack can be improved using structures exactly in the same manner as in the 7-round attack. This reduces the data complexity to 2^{37} chosen plaintexts, while maintaining a low time complexity dominated by encrypting the plaintexts.

Following [18], we can calculate the success rate of the attack to be about 61.8% (this is the probability that the right subkey is the most biased, with probability 69.5% the right key is amongst the two most biased keys, etc.).

7   Summary and Conclusions

In this paper we have presented several attacks on CTC2. We have showed that due to its slow diffusion, CTC2 is susceptible to standard cryptanalytic attacks. In Table 1 we summarize the various attacks, and their complexities.

We note that our attacks rely very strongly on the very slow diffusion of the cipher. The slow diffusion, as our results exposed, may be the reason for the success of the algebraic attacks on CTC2. While this does not mean that

Table 1. Summary of the Complexities of Our Attacks on CTC2

Attack	Rounds	Complexity	
		Data	Time
Meet in the middle	4	4	4
Differential	5	24	24
Differential	6	64	64
Differential	7	2^{15}	2^{15}
Differential-Linear	7	2^{17}	2^{17}
Differential-Linear	8	2^{37}	2^{37}

Data complexity is given in chosen plaintexts.
Time complexity is given in encryption units.

algebraic attacks are inferior to standard attacks, it does re-open the issue of finding an example where algebraic attacks are better than standard and well understood cryptanalytic techniques.

When we compare our results to the ones in [10], we encounter some similarities. In [10], a 6-round attack with data complexity of 4 chosen plaintexts and time complexity of about 2^{253} encryptions is presented. Without discussing the methodology which was used for the time complexity estimations of [10], we note that by employing the meet-in-the-middle attack presented in Section 3.1, an attacker which is allowed only 4 chosen plaintexts can attack 6-round CTC2 with an expected speed-up of about 66%. In other words, a basic meet-in-the-middle attack, which probably can be further optimized (e.g., there are key considerations which may be useful for improving the attack) achieves a 66% speed-up, to be compared with the 75% speed-up gained in [10].

If only meet-in-the-middle attacks given 4 chosen plaintexts are taken into consideration, it seems that 4-round CTC2 can be easily broken, while 6-round CTC2 requires an exponential time (equivalent to about 2^{253} trial encryptions). Even though time estimates for algebraic attacks on 4-round CTC2 are not given, it would be surprising if such attacks require time complexity which is of the order of magnitude of exhaustive key search. Hence, it seems that the strength of the algebraic technique in attacking 4-round or 6-round CTC2 is approximately equivalent to the strength of the meet-in-the-middle technique.

This leaves the issue of attacking 5-round CTC2 as a challenge for the algebraic technique. We believe that given 4 chosen plaintexts, the best attack on 5-round CTC2 using a simple meet-in-the-middle technique, is the variant of exhaustive search presented in Section 3.1. If algebraic techniques can beat these results by an order of magnitude, this would look as a proof of their merits (assuming of course, that no one finds a better attack using easy to analyze attack methods).

References

1. Biham, E.: On matsui's linear cryptanalysis. In: De Santis, A. (ed.) EUROCRYPT 1994. LNCS, vol. 950, pp. 341–355. Springer, Heidelberg (1995)
2. Biham, E., Dunkelman, O., Keller, N.: Enhancing differential-linear cryptanalysis. In: Zheng, Y. (ed.) ASIACRYPT 2002. LNCS, vol. 2501, pp. 254–266. Springer, Heidelberg (2002)
3. Biham, E., Dunkelman, O., Keller, N.: Differential-linear cryptanalysis of serpent. In: Johansson, T. (ed.) FSE 2003. LNCS, vol. 2887, pp. 9–21. Springer, Heidelberg (2003)
4. Biryukov, A., De Cannière, C., Quisquater, M.: On multiple linear approximations. In: Franklin, M. (ed.) CRYPTO 2004. LNCS, vol. 3152, pp. 1–22. Springer, Heidelberg (2004)
5. Buchmann, J., Pyshkin, A., Weinmann, R.-P.: Block ciphers sensitive to gröbner basis attacks. In: Pointcheval, D. (ed.) CT-RSA 2006. LNCS, vol. 3860, pp. 313–331. Springer, Heidelberg (2006)

6. Collard, B., Standaert, F.-X., Quisquater, J.-J.: Improved and multiple linear cryptanalysis of reduced round serpent. In: Pei, D., Yung, M., Lin, D., Wu, C. (eds.) Inscrypt 2007. LNCS, vol. 4990, pp. 51–65. Springer, Heidelberg (2008)
7. Courtois, N.T.: Fast algebraic attacks on stream ciphers with linear feedback. In: Boneh, D. (ed.) CRYPTO 2003. LNCS, vol. 2729, pp. 176–194. Springer, Heidelberg (2003)
8. Nicolas, T.: Courtois, How Fast can be Algebraic Attacks on Block Ciphers?, IACR ePrint report 2006/168 (2006)
9. Courtois, N.T.: Algebraic Attacks On Block Ciphers. In: Dagstuhl seminar on Symmetric Cryptography (January 2007)
10. Courtois, N.T.: CTC2 and Fast Algebraic Attacks on Block Ciphers Revisited, IACR ePrint report 2007/152 (2007)
11. Courtois, N.T., Bard, G.V., Wagner, D.: Algebraic and slide attacks on keeLoq. In: Nyberg, K. (ed.) FSE 2008. LNCS, vol. 5086, pp. 97–115. Springer, Heidelberg (2008)
12. Courtois, N.T., Meier, W.: Algebraic Attacks on Stream Ciphers with Linear Feedback. In: Biham, E. (ed.) EUROCRYPT 2003. LNCS, vol. 2656, pp. 345–359. Springer, Heidelberg (2003)
13. Courtois, N.T., Pieprzyk, J.: Cryptanalysis of block ciphers with overdefined systems of equations. In: Zheng, Y. (ed.) ASIACRYPT 2002. LNCS, vol. 2501, pp. 267–287. Springer, Heidelberg (2002)
14. Dunkelman, O., Keller, N.: Linear Cryptanalysis of CTC, IACR ePrint report 2006/250 (2006)
15. Dunkelman, O., Sekar, G., Preneel, B.: Improved meet-in-the-middle attacks on reduced-round DES. In: Srinathan, K., Rangan, C.P., Yung, M. (eds.) INDOCRYPT 2007. LNCS, vol. 4859, pp. 86–100. Springer, Heidelberg (2007)
16. Faugére, J.-C., Perret, L.: Algebraic Cryptanalysis of Curry and Flurry using Correlated Messages, IACR ePrint report 2008/402 (2008)
17. Langford, S.K.: Differential-Linear Cryptanalysis and Threshold Signatures, Ph.D. thesis (1995)
18. Selçuk, A.A.: On Probability of Success in Linear and Differential Cryptanalysis. Journal of Cryptology 21(1), 131–147 (2008)

A CCA2 Secure Public Key Encryption Scheme Based on the McEliece Assumptions in the Standard Model

Rafael Dowsley[1], Jörn Müller-Quade[2], and Anderson C.A. Nascimento[1]

[1] Department of Electrical Engineering, University of Brasilia
Campus Universitário Darcy Ribeiro,Brasilia, CEP: 70910-900, Brazil
rafaeldowsley@redes.unb.br, andclay@ene.unb.br
[2] Universität Karlsruhe, Institut für Algorithmen und Kognitive Systeme
Am Fasanengarten 5, 76128 Karlsruhe, Germany
muellerq@ira.uka.de

Abstract. We show that a recently proposed construction by Rosen and Segev can be used for obtaining the first public key encryption scheme based on the McEliece assumptions which is secure against adaptive chosen ciphertext attacks in the standard model.

1 Introduction

Indistinguishability of messages under adaptive chosen ciphertext attacks is the strongest known notion of security for public key encryption schemes (PKE). Many computational assumptions have been used in the literature for obtaining cryptosystems meeting such a strong security requirements. Given one-way trapdoor permutations, we know how to obtain CCA2 security from any semantically secure public key cryptosystem [14,20,12]. Efficient constructions are also known based on number-theoretic assumptions [6] or on identity based encryption schemes [3]. Obtaining a CCA2 secure cryptosystem (even an inefficient one) based on the McEliece assumptions in the standard model has been an open problem in this area for quite a while.

Recently, Rosen and Segev proposed an elegant and simple new computational assumption for obtaining CCA2 secure PKEs: *correlated products* [19]. They provided constructions of correlated products based on the existence of certain *lossy trapdoor functions* [16] which in turn can be based on the decisional Diffie-Hellman problem and on Paillier's decisional residuosity problem [16].

In this paper, we show that the ideas of Rosen and Segev can also be applied for obtaining the first construction of a PKE built upon the McEliece assumptions. Based on the definition of correlated products [19], we define a new kind of PKE called k-repetition CPA secure cryptosystem and show that the construction proposed in [19] directly translates to this new scenario. We then show that a randomized version of the McEliece cryptosystem [15] is k-repetition CPA secure and obtain a CCA2 secure scheme in the standard model. The resulting cryptosystem enciphers many bits as opposed to the single-bit PKE obtained in [19]. We expand the public and private keys and the ciphertext by a factor of k

M. Fischlin (Ed.): CT-RSA 2009, LNCS 5473, pp. 240–251, 2009.
© Springer-Verlag Berlin Heidelberg 2009

when compared to the original McEliece PKE. Additionally, our result implies a new construction of correlated products based on the McEliece assumptions.

In a concurrent and independent work [9], Goldwasser and Vaikuntanathan proposed a new CCA-secure public-key encryption scheme based on lattices using the construction by Rosen and Segev. Their scheme assumed that the problem of learning with errors (LWE) is hard [18].

2 Preliminaries

2.1 Notation

If x is a string, then $|x|$ denotes its length, while $|S|$ represents the cardinality of a set S. If $n \in \mathbb{N}$ then 1^n denotes the string of n ones. $s \leftarrow S$ denotes the operation of choosing an element s of a set S uniformly at random. $w \leftarrow \mathcal{A}(x, y, \ldots)$ represents the act of running the algorithm $\mathcal{A}$ with inputs $x, y, \ldots$ and producing output w. We write $w \leftarrow \mathcal{A}^{\mathcal{O}}(x, y, \ldots)$ for representing an algorithm $\mathcal{A}$ having access to an oracle $\mathcal{O}$. We denote by $\Pr[E]$ the probability that the event E occurs. If a and b are two strings of bits or two matrices, we denote by $a|b$ their concatenation. The transpose of a matrix M is M^T. If a and b are two strings of bits, we denote by $\langle a, b \rangle$ their dot product modulo 2 and by $a \oplus b$ their bitwise XOR. $\mathcal{U}_n$ is an oracle that returns a random element of $\{0,1\}^n$.

2.2 Public-Key Encryption Schemes

A Public Key Encryption Scheme (PKE) is defined as follows:

Definition 1. *(Public-Key Encryption). A public-key encryption scheme is a triplet of algorithms (*Gen, Enc, Dec*) such that:*

- Gen *is a probabilistic polynomial-time key generation algorithm which takes as input a security parameter 1^n and outputs a public key* pk *and a secret key* sk. *The public key specifies the message space $\mathcal{M}$ and the ciphertext space $\mathcal{C}$.*
- Enc *is a (possibly) probabilistic polynomial-time encryption algorithm which receives as input a public key* pk *and a message* m $\in \mathcal{M}$, *and outputs a ciphertext* c $\in \mathcal{C}$.
- Dec *is a deterministic polynomial-time decryption algorithm which takes as input a secret key* sk *and a ciphertext* c, *and outputs either a message* m $\in \mathcal{M}$ *or an error symbol $\bot$.*
- *(Soundness) For any pair of public and private keys generated by* Gen *and any message* m $\in \mathcal{M}$ *it holds that* Dec(sk, Enc(pk, m)) $=$ m *with overwhelming probability over the randomness used by* Gen *and* Enc.

Below we define indistinguishability against chosen-plaintext attacks (IND-CPA) [8] and against adaptive chosen-ciphertext attacks (IND-CCA2) [17]. Our game definition follows the approach of [10].

Definition 2. *(IND-CPA security). To a two-stage adversary $\mathcal{A} = (\mathcal{A}_1, \mathcal{A}_2)$ against* PKE *we associate the following experiment* $\mathsf{Exp}_{\mathsf{PKE},\mathcal{A}}^{cpa}(n)$:

> $(\mathsf{pk}, \mathsf{sk}) \leftarrow \mathsf{Gen}(1^n)$
> $(\mathsf{m}_0, \mathsf{m}_1, state) \leftarrow \mathcal{A}_1(\mathsf{pk})$ *s.t.* $|\mathsf{m}_0| = |\mathsf{m}_1|$
> $b \leftarrow \{0, 1\}$
> $\mathsf{c}^* \leftarrow \mathsf{Enc}(\mathsf{pk}, \mathsf{m}_b)$
> $b' \leftarrow \mathcal{A}_2(\mathsf{c}^*, state)$
> *If* $b = b'$ *return* 1 *else return* 0

We define the advantage of $\mathcal{A}$ in the experiment as

$$\mathsf{Adv}_{\mathsf{PKE},\mathcal{A}}^{cpa}(n) = |Pr[\mathsf{Exp}_{\mathsf{PKE},\mathcal{A}}^{cpa}(n) = 1] - \tfrac{1}{2}|$$

We say that PKE *is indistinguishable against chosen-plaintext attacks (IND-CPA) if for all probabilistic polynomial time (PPT) adversaries $\mathcal{A} = (\mathcal{A}_1, \mathcal{A}_2)$ the advantage of $\mathcal{A}$ in the experiment is a negligible function of n.*

Definition 3. *(IND-CCA2 security). To a two-stage adversary $\mathcal{A} = (\mathcal{A}_1, \mathcal{A}_2)$ against* PKE *we associate the following experiment* $\mathsf{Exp}_{\mathsf{PKE},\mathcal{A}}^{cca2}(n)$:

> $(\mathsf{pk}, \mathsf{sk}) \leftarrow \mathsf{Gen}(1^n)$
> $(\mathsf{m}_0, \mathsf{m}_1, state) \leftarrow \mathcal{A}_1^{\mathsf{Dec}(\mathsf{sk}, \cdot)}(\mathsf{pk})$ *s.t.* $|\mathsf{m}_0| = |\mathsf{m}_1|$
> $b \leftarrow \{0, 1\}$
> $\mathsf{c}^* \leftarrow \mathsf{Enc}(\mathsf{pk}, \mathsf{m}_b)$
> $b' \leftarrow \mathcal{A}_2^{\mathsf{Dec}(\mathsf{sk}, \cdot)}(\mathsf{c}^*, state)$
> *If* $b = b'$ *return* 1 *else return* 0

The adversary $\mathcal{A}_2$ is not allowed to query $\mathsf{Dec}(\mathsf{sk}, \cdot)$ *with* c^*. *We define the advantage of $\mathcal{A}$ in the experiment as*

$$\mathsf{Adv}_{\mathsf{PKE},\mathcal{A}}^{cca2}(n) = |Pr[\mathsf{Exp}_{\mathsf{PKE},\mathcal{A}}^{cca2}(n) = 1] - \tfrac{1}{2}|$$

We say that PKE *is indistinguishable against adaptive chosen-ciphertext attacks (IND-CCA2) if for all probabilistic polynomial time (PPT) adversaries $\mathcal{A} = (\mathcal{A}_1, \mathcal{A}_2)$ that makes a polynomial number of oracle queries the advantage of $\mathcal{A}$ in the experiment is a negligible function of n.*

2.3   McEliece Cryptosystem

In this Section we define the McEliece cryptosystem [13]. We closely follow [15]. The McEliece PKE consists of a triplet of probabilistic algorithms ($\mathsf{Gen}_{\mathrm{McE}}$, $\mathsf{Enc}_{\mathrm{McE}}$, $\mathsf{Dec}_{\mathrm{McE}}$) such that:

- The probabilistic polynomial-time key generation algorithm, $\mathsf{Gen}_{\mathrm{McE}}$, works as follows:

1. Generate a $l \times n$ generator matrix $\mathbf{G}$ of a Goppa code, where we assume that there is an efficient error-correction algorithm Correct which can always correct up to t errors.
2. Generate a $l \times l$ random non-singular matrix $\mathbf{S}$.
3. Generate a $n \times n$ random permutation matrix $\mathbf{T}$.
4. Set $\mathbf{P} = \mathbf{SGT}$, $\mathcal{M} = \{0,1\}^l$, $\mathcal{C} = \{0,1\}^n$.
5. Output $\mathsf{pk} = (\mathbf{P}, t, \mathcal{M}, \mathcal{C})$ and $\mathsf{sk} = (\mathbf{S}, \mathbf{G}, \mathbf{T})$.

– The probabilistic polynomial-time encryption algorithm, $\mathsf{Enc}_{\mathrm{McE}}$, takes the public-key pk and a plaintext $\mathsf{m} \in \{0,1\}^l$ as input and outputs a ciphertext $\mathsf{c} = \mathsf{mP} \oplus \mathsf{e}$, where $\mathsf{e} \in \{0,1\}^n$ is a random vector of Hamming weight t.
– The deterministic polynomial-time decryption algorithm, $\mathsf{Dec}_{\mathrm{McE}}$, works as follows:
  1. Compute $\mathsf{c}\mathbf{T}^{-1} = (\mathsf{m}\mathbf{S})\mathbf{G} \oplus \mathsf{e}\mathbf{T}^{-1}$, where $\mathbf{T}^{-1}$ denotes the inverse matrix of $\mathbf{T}$.
  2. Compute $\mathsf{m}\mathbf{S} = \mathsf{Correct}(\mathsf{c}\mathbf{T}^{-1})$.
  3. Output $\mathsf{m} = (\mathsf{m}\mathbf{S})\mathbf{S}^{-1}$.

In our work we use a slightly modified version of the McEliece PKC. Instead of creating an error vector by choosing it randomly from the set of vectors with Hamming weight t, we generate e by choosing each of its bits according to the Bernoulli distribution $\mathcal{B}_\theta$ with parameter $\theta = \frac{t}{n} - \epsilon$ for some $\epsilon > 0$. Clearly, due to the law of large numbers, the resulting error vector should be within the error capabilities of the code.

2.4   McEliece Assumptions

In this subsection, we briefly introduce and discuss the McEliece assumptions.

We assume that there is no efficient algorithm which can distinguish the scrambled (according to the description in the previous subsection) generating matrix of the Goppa code P and a random matrix of the same size. The best algorithm attacking this assumption is by Courtois et al. [5] and it is based on the *support splitting algorithm* [21].

Assumption 4. *There is no PPT algorithm which can distinguish the public-key matrix P of the McEliece cryptosystem from a random matrix of the same size with non-negligible probability.*

We note that this assumption was utilized in [5] to construct a digital signature scheme.

We also assume that the Syndrome Decoding Problem is hard. This problem is known to be NP-complete [1], and all currently known algorithms to solve this problem are exponential. The best algorithms were presented by Canteaut and Chabaud [4] and recently by Bernstein et al. [2].

Assumption 5. *The Syndrome Decoding Problem problem is hard for every PPT algorithm.*

This problem is equivalent to the problem of learning parity with noise (LPN). Below we give the definition of LPN problem following the description of [15].

Definition 6. *(LPN problem) Let r, a be binary strings of length l. We consider the Bernoulli distribution $\mathcal{B}_\theta$ with parameter $\theta \in (0, \frac{1}{2})$. Let $\mathcal{Q}_{r,\theta}$ be the following distribution:*

$$\{(a, \langle r, a \rangle \oplus v) | a \leftarrow \{0, 1\}^l, v \leftarrow \mathcal{B}_\theta\}$$

For an adversary $\mathcal{A}$ trying to discover the random r, we define its advantage as:

$$\mathsf{Adv}_{\mathsf{LPN}_\theta, \mathcal{A}}(l) = \Pr[\mathcal{A}^{\mathcal{Q}_{r,\theta}} = r | r \leftarrow \{0, 1\}^l]$$

The LPN_θ problem with parameter θ is hard if the advantage of all PPT adversaries $\mathcal{A}$ that makes a polynomial number of oracle queries is negligible.

2.5   Admissible PKE

Below we define admissible PKEs which are known to imply IND-CPA security [15]. In the following, $\mathsf{Enc}(\mathsf{pk}, \mathsf{m}, \mathsf{r})$ denotes a public key encryption scheme enciphering a message m with a public key pk and randomness r.

Definition 7. *(Admissible PKE [15]) A public-key encryption scheme $\mathsf{PKE} = (\mathsf{Gen}, \mathsf{Enc}, \mathsf{Dec})$ with message space $\mathcal{M}$ and random space $\mathcal{R}$ is called admissible if there is a pair of deterministic polynomial-time algorithms Enc_1 and Enc_2 satisfying the following properties:*

- *Dividability: Enc_1 takes as input the public key pk and $\mathsf{r} \in \mathcal{R}$, and outputs a $p(n)$ bit-string. Enc_2 takes as input the public key pk and $\mathsf{m} \in \mathcal{M}$, and outputs a $p(n)$ bit-string. Here p is some polynomial in n. Then for any pk generated by Gen, $\mathsf{r} \in \mathcal{R}$ and $\mathsf{m} \in \mathcal{M}$, $\mathsf{Enc}_1(\mathsf{pk}, \mathsf{r}) \oplus \mathsf{Enc}_2(\mathsf{pk}, \mathsf{m}) = \mathsf{Enc}(\mathsf{pk}, \mathsf{m}, \mathsf{r})$.*
- *Pseudorandomness: Consider a probabilistic polynomial time adversary $\mathcal{A}$ against PKE, we associate with it the following experiment $\mathsf{Exp}_{\mathsf{PKE}, \mathcal{A}}^{ind}(n)$:*

$(\mathsf{pk}, \mathsf{sk}) \leftarrow \mathsf{Gen}(1^n)$
$\mathsf{s}_0 \leftarrow \mathcal{U}_{p(n)}$
$\mathsf{r} \in \mathcal{R}$
$\mathsf{s}_1 \leftarrow \mathsf{Enc}_1(\mathsf{pk}, \mathsf{r})$
$b \leftarrow \{0, 1\}$
$b' \leftarrow \mathcal{A}(\mathsf{pk}, \mathsf{s}_b)$
If $b = b'$ return 1 else return 0

We define the advantage of $\mathcal{A}$ in the experiment as

$$\mathsf{Adv}_{\mathsf{PKE}, \mathcal{A}}^{ind}(n) = |Pr[\mathsf{Exp}_{\mathsf{PKE}, \mathcal{A}}^{ind}(n) = 1] - \tfrac{1}{2}|$$

For all probabilist polynomial time (PPT) adversaries $\mathcal{A}$, the advantage of $\mathcal{A}$ in the experiment must be a negligible function of n.

2.6   Signature Schemes

We explain signature schemes (SS) and define one-time strong unforgeability.

Definition 8. *(Signature Scheme). A signature scheme is a triplet of algorithms* (Gen, Sign, Ver) *such that:*

- Gen *is a probabilistic polynomial-time key generation algorithm which takes as input a security parameter* 1^n *and outputs a verification key* vk *and a signing key* dsk. *The verification key specifies the message space* $\mathcal{M}$ *and the signature space* $\mathcal{S}$.
- Sign *is a (possibly) probabilistic polynomial-time signing algorithm which receives as input a signing key* dsk *and a message* m $\in$ $\mathcal{M}$, *and outputs a signature* $\sigma \in \mathcal{S}$.
- Ver *is a deterministic polynomial-time verification algorithm which takes as input a verification key* vk, *a message* m $\in$ $\mathcal{M}$ *and a signature* $\sigma \in \mathcal{S}$, *and outputs a bit indicating whether* σ *is a valid signature for* m *or not (i.e., the algorithm outputs 1 if it is a valid signature and outputs 0 otherwise).*
- *For any pair of signing and verification keys generated by* Gen *and any message* m $\in$ $\mathcal{M}$ *it holds that* Ver(vk, m, Sign(dsk, m)) $= 1$ *with overwhelming probability over the randomness used by* Gen *and* Sign.

Definition 9. *(One-Time Strong Unforgeability). To a two-stage adversary* $\mathcal{A} = (\mathcal{A}_1, \mathcal{A}_2)$ *against* SS *we associate the following experiment* $\mathsf{Exp}_{\mathsf{SS},\mathcal{A}}^{otsu}(n)$:

$$(\mathsf{vk}, \mathsf{dsk}) \leftarrow \mathsf{Gen}(1^n)$$
$$(\mathsf{m}, state) \leftarrow \mathcal{A}_1(\mathsf{vk})$$
$$\sigma \leftarrow \mathsf{Sign}(\mathsf{dsk}, \mathsf{m})$$
$$(\mathsf{m}^*, \sigma^*) \leftarrow \mathcal{A}_2(\mathsf{m}, \sigma, state)$$
$$If\ \mathsf{Ver}(\mathsf{vk}, \mathsf{m}^*, \sigma^*) = 1\ and\ (\mathsf{m}^*, \sigma^*) \neq (\mathsf{m}, \sigma)\ return\ 1,\ else\ return\ 0$$

We say that a signature scheme SS *is one-time strongly unforgeable if for all probabilist polynomial time (PPT) adversaries* $\mathcal{A} = (\mathcal{A}_1, \mathcal{A}_2)$ *the probability that* $\mathsf{Exp}_{\mathsf{SS},\mathcal{A}}^{otsu}(n)$ *outputs 1 is a negligible function of* n.

3   *k*-Repetition PKE

3.1   Definitions

We define a *k*-repetition Public-Key Encryption.

Definition 10. *(k-repetition Public-Key Encryption). For a* PKE (Gen, Enc, Dec), *we define the k-repetition public-key encryption scheme* (PKE_k) *as the triplet of algorithms* (Gen_k, Enc_k, Dec_k) *such that:*

- Gen_k *is a probabilistic polynomial-time key generation algorithm which takes as input a security parameter* 1^n *and calls the* PKE's *key generation algorithm* k *times obtaining the public keys* $(\mathsf{pk}_1, \ldots, \mathsf{pk}_k)$ *and the secret keys* $(\mathsf{sk}_1, \ldots, \mathsf{sk}_k)$. Gen_k *sets the public key as* $pk = (\mathsf{pk}_1, \ldots, \mathsf{pk}_k)$ *and the secret key as* $sk = (\mathsf{sk}_1, \ldots, \mathsf{sk}_k)$.

- Enc_k *is a (possibly) probabilistic polynomial-time encryption algorithm which receives as input a public key* $pk = (\mathsf{pk}_1, \ldots, \mathsf{pk}_k)$ *and a message* $\mathsf{m} \in \mathcal{M}$, *and outputs a ciphertext* $\mathsf{c} = (\mathsf{c}_1, \ldots, \mathsf{c}_k) = (\mathsf{Enc}(\mathsf{pk}_1, \mathsf{m}), \ldots, \mathsf{Enc}(\mathsf{pk}_k, \mathsf{m}))$.
- Dec_k *is a deterministic polynomial-time decryption algorithm which takes as input a secret key* $sk = (\mathsf{sk}_1, \ldots, \mathsf{sk}_k)$ *and a ciphertext* $\mathsf{c} = (\mathsf{c}_1, \ldots, \mathsf{c}_k)$. *It outputs a message* m *if* $\mathsf{Dec}(\mathsf{sk}_1, \mathsf{c}_1), \ldots, \mathsf{Dec}(\mathsf{sk}_k, \mathsf{c}_k)$ *are all equal to some* $\mathsf{m} \in \mathcal{M}$. *Otherwise, it outputs an error symbol* $\perp$.
- *(Soundness) For any* k *pairs of public and private keys generated by* Gen_k *and any message* $\mathsf{m} \in \mathcal{M}$ *it holds that* $\mathsf{Dec}_k(sk, \mathsf{Enc}_k(pk, \mathsf{m})) = \mathsf{m}$ *with overwhelming probability over the randomness used by* Gen_k *and* Enc_k.

We also define security properties that the k-repetition Public-Key Encryption scheme used in the next sections should meet.

Definition 11. *(Security under uniform k-repetition of IND-CPA schemes).* *We say that* PKE_k *(built from an IND-CPA secure scheme* PKE) *is secure under uniform k-repetition if* PKE_k *is IND-CPA secure.*

Definition 12. *(Verification under uniform k-repetition of IND-CPA schemes).* *We say that* PKE_k *is verifiable under uniform k-repetition if given a ciphertext* $\mathsf{c} \in \mathcal{C}$, *the public key* $pk = (\mathsf{pk}_1, \ldots, \mathsf{pk}_k)$ *and any* sk_i *for* $i \in \{1, \ldots, k\}$, *it is possible to verify if* c *is a valid ciphertext.*

3.2   IND-CCA2 Security from CPA Secure k-Repetition PKE

In this subsection we describe the IND-CCA2 secure public key encryption scheme (PKE_{cca2}) and prove its security. We assume the existence of an one-time strongly unforgeable signature scheme and of a PKE_k that is secure and verifiable under uniform k-repetition.

Key Generation: Gen_{cca2} is a probabilistic polynomial-time key generation algorithm which takes as input a security parameter 1^n. Gen_{cca2} does as follows:

1. Calls the PKE's key generation algorithm $2k$ times obtaining the public keys $(\mathsf{pk}_1^0, \mathsf{pk}_1^1, \ldots, \mathsf{pk}_k^0, \mathsf{pk}_k^1)$ and the secret keys $(\mathsf{sk}_1^0, \mathsf{sk}_1^1, \ldots, \mathsf{sk}_k^0, \mathsf{sk}_k^1)$.
2. Executes the key generation algorithm of the signature scheme obtaining a signing key dsk^* and a verification key vk^*. Denote by vk_i^* the i-bit of vk^*.
3. Sets the public key as $pk = (\mathsf{pk}_1^0, \mathsf{pk}_1^1, \ldots, \mathsf{pk}_k^0, \mathsf{pk}_k^1)$ and the secret key as $sk = (\mathsf{vk}^*, \mathsf{sk}_1^{1-\mathsf{vk}_1^*}, \ldots, \mathsf{sk}_k^{1-\mathsf{vk}_k^*})$.

Encryption: Enc_{cca2} is a (possibly) probabilistic polynomial-time encryption algorithm which receives as input the public key $pk = (\mathsf{pk}_1^0, \mathsf{pk}_1^1, \ldots, \mathsf{pk}_k^0, \mathsf{pk}_k^1)$ and a message $\mathsf{m} \in \mathcal{M}$ and proceeds as follows:

1. Executes the key generation algorithm of the signature scheme obtaining a signing key dsk and a verification key vk. Denote by vk_i the i-bit of vk.
2. Computes $\mathsf{c}_i = \mathsf{Enc}(\mathsf{pk}_i^{\mathsf{vk}_i}, \mathsf{m})$ for $i \in \{1, \ldots, k\}$.

3. Computes the signature $\sigma = \mathsf{Sign}(\mathsf{dsk}, (\mathsf{c}_1, \ldots, \mathsf{c}_k))$.
4. Outputs the ciphertext $\mathsf{c} = (\mathsf{c}_1, \ldots, \mathsf{c}_k, \mathsf{vk}, \sigma)$.

Decryption: Dec_{cca2} is a deterministic polynomial-time decryption algorithm which takes as input a secret key $sk = (\mathsf{vk}^*, \mathsf{sk}_1^{1-\mathsf{vk}_1^*}, \ldots, \mathsf{sk}_k^{1-\mathsf{vk}_k^*})$ and a ciphertext $\mathsf{c} = (\mathsf{c}_1, \ldots, \mathsf{c}_k, \mathsf{vk}, \sigma)$ and proceeds as follows:

1. If $\mathsf{vk} = \mathsf{vk}^*$ or $\mathsf{Ver}(\mathsf{vk}, (\mathsf{c}_1, \ldots, \mathsf{c}_k), \sigma) = 0$, it outputs $\bot$ and halts.
2. For some $i \in \{1, \ldots, k\}$ such that $\mathsf{vk}_i \neq \mathsf{vk}_i^*$, it computes $\mathsf{m} = \mathsf{Dec}(\mathsf{sk}^{\mathsf{vk}_i}, \mathsf{c}_i)$.
3. Verifies if $\mathsf{c}_i = \mathsf{Enc}(\mathsf{pk}_i^{\mathsf{vk}_i}, \mathsf{m})$ for all $i \in \{1, \ldots, k\}$. If the condition is satisfied, it outputs m. Otherwise, it outputs $\bot$.

The probability that $\mathsf{Dec}_{cca2}(sk, \mathsf{Enc}_{cca2}(\mathsf{pk}, \mathsf{m})) \neq \mathsf{m}$ is the same as the probability that $\mathsf{vk} = \mathsf{vk}^*$, but this probability is negligible since the signature scheme is one-time strongly unforgeable.

As in [19], we can apply a universal one-way hash function to the verification keys (as in [7]) and use $k = n^\epsilon$ for a constant $0 < \epsilon < 1$. For ease of presentation, we do not apply this method in our scheme description.

Theorem 1. *Given that* SS *is a one-time strongly unforgeable signature scheme and that* PKE_k *is secure and verifiable under uniform k-repetition, the public key encryption scheme* PKE_{cca2} *is IND-CCA2 secure.*

Proof. In this proof we closely follow [19]. Denote by $\mathcal{A}$ the IND-CCA2 adversary. Let Forge be the event that for some decryption query made by $\mathcal{A}$ we have that $\mathsf{Ver}(\mathsf{vk}, (\mathsf{c}_1, \ldots, \mathsf{c}_k), \sigma) = 1$ and $\mathsf{vk} = \mathsf{vk}^*$. The theorem follow from the two lemmas below.

Lemma 1. $\Pr[\mathsf{Forge}]$ *is negligible.*

Proof. Assume that for a PPT adversary $\mathcal{A}$ against PKE_{cca2} the forge probability $(\Pr[\mathsf{Forge}])$ is non-negligible, then we construct an adversary $\mathcal{A}'$ that forges a signature with the same probability. $\mathcal{A}'$ simulates the IND-CCA2 interaction for $\mathcal{A}$ as follows:

Key Generation: $\mathcal{A}'$ invokes the key generation algorithm of the signature scheme and obtains vk^*. It calls the PKE's key generation algorithm $2k$ times obtaining the public keys $(\mathsf{pk}_1^0, \mathsf{pk}_1^1, \ldots, \mathsf{pk}_k^0, \mathsf{pk}_k^1)$ and the secret keys $(\mathsf{sk}_1^0, \mathsf{sk}_1^1, \ldots, \mathsf{sk}_k^0, \mathsf{sk}_k^1)$ and uses vk^* for forming the secret key of PKE_{cca2}. It sends the public key to $\mathcal{A}$.

Decryption Queries: Whenever $\mathcal{A}$ makes a decryption query, $\mathcal{A}'$ proceeds as follows:

1. If $\mathsf{vk} = \mathsf{vk}^*$ and $\mathsf{Ver}(\mathsf{vk}, (\mathsf{c}_1, \ldots, \mathsf{c}_k), \sigma) = 1$, $\mathcal{A}'$ outputs $((\mathsf{c}_1, \ldots, \mathsf{c}_k), \sigma)$ as the forgery and halts.
2. Otherwise, $\mathcal{A}'$ decrypts the ciphertext using the procedures of PKE_{cca2}.

Challenging Query: Whenever $\mathcal{A}$ makes the challenging query with two messages $\mathsf{m}_0, \mathsf{m}_1 \in \mathcal{M}$ such that $|\mathsf{m}_0| = |\mathsf{m}_1|$, $\mathcal{A}'$ proceeds as follows:

1. Chooses randomly $b \in \{0, 1\}$.

2. Encrypts the message m_b using the procedures of PKE_{cca2}. This is possible because $\mathcal{A}'$ can ask the signature oracle to sign one message, so it asks the oracle to sign the value $(c_1, \ldots, c_k)$ obtained during the encryption process.

As long as the event Forge did not occur, the simulation is perfect, so the probability that $\mathcal{A}'$ breaks the one-time strongly unforgeable signature scheme is exactly $\Pr[\mathsf{Forge}]$. Since the signature scheme is strongly unforgeable by assumption, $\Pr[\mathsf{Forge}]$ is negligible for all PPT adversaries against PKE_{cca2}.

Lemma 2. *Given that* Forge *did not occur, the advantage of a PPT adversary* $\mathcal{A}$ *against* PKE_{cca2},

$$|\Pr[\overline{\mathsf{Forge}} \wedge \mathsf{Exp}^{cca2}_{\mathsf{PKE}_{cca2}, \mathcal{A}}(n) = 1] - \tfrac{1}{2}|,$$

is negligible.

Proof. Assume that for some PPT adversary $\mathcal{A}$ against PKE_{cca2} we have that $|\Pr[\mathsf{Exp}^{cca2}_{\mathsf{PKE}_{cca2}, \mathcal{A}}(n) = 1 \wedge \overline{\mathsf{Forge}}] - \tfrac{1}{2}|$ is non-negligible, then we construct an adversary $\mathcal{A}'$ that breaks the IND-CPA security of PKE_k. $\mathcal{A}'$ simulates the IND-CCA2 interaction for $\mathcal{A}$ as follows:

Key Generation: $\mathcal{A}'$ receives as input the public key $(\mathsf{pk}_1, \ldots, \mathsf{pk}_k)$ of PKE_k. $\mathcal{A}'$ proceeds as follows:
1. Runs the key generation algorithm of the signature scheme and obtain the verification key vk^* and the signing key dsk^*.
2. Sets $\mathsf{pk}_i^{\mathsf{vk}^*_i} = \mathsf{pk}_i$ for $i \in \{1, \ldots, k\}$.
3. Runs PKE's key generation algorithm k times obtaining the public keys $(\mathsf{pk}_1^{1-\mathsf{vk}^*_1}, \ldots, \mathsf{pk}_k^{1-\mathsf{vk}^*_k})$ and the secret keys $(\mathsf{sk}_1^{1-\mathsf{vk}^*_1}, \ldots, \mathsf{sk}_k^{1-\mathsf{vk}^*_k})$.
4. Sets the public key as $pk = (\mathsf{pk}_1^0, \mathsf{pk}_1^1, \ldots, \mathsf{pk}_k^0, \mathsf{pk}_k^1)$ and the secret key as $sk = (\mathsf{vk}^*, \mathsf{sk}_1^{1-\mathsf{vk}^*_1}, \ldots, \mathsf{sk}_k^{1-\mathsf{vk}^*_k})$.
5. Sends the public key to $\mathcal{A}$.

Decryption Queries: Whenever $\mathcal{A}$ makes a decryption query, $\mathcal{A}'$ proceeds as follows:
1. If Forge occurs, then $\mathcal{A}'$ halts.
2. Otherwise, $\mathcal{A}'$ decrypts the ciphertext using the procedures of PKE_{cca2}.

Challenging Query: When $\mathcal{A}$ makes the challenging query with two messages $m_0, m_1 \in \mathcal{M}$ such that $|m_0| = |m_1|$, $\mathcal{A}'$ proceeds as follows:
1. Sends m_0 and m_1 to $\mathcal{A}'$ challenging oracle and obtain as response $(c_1^*, \ldots, c_k^*)$.
2. Signs $(c_1^*, \ldots, c_k^*)$ using dsk^*.
3. Outputs the challenge ciphertext $c^* = (c_1^*, \ldots, c_k^*, \mathsf{vk}^*, \sigma^*)$.

Output: When $\mathcal{A}$ outputs b, $\mathcal{A}'$ also outputs b.

As long as the event Forge does not occur, the advantage of $\mathcal{A}'$ in breaking the IND-CPA-security of PKE_k is the same as the advantage of $\mathcal{A}$ in breaking the IND-CCA2-security of PKE_{cca2}. Since PKE_k is IND-CPA-secure by assumption, we have that PKE_{cca2} is IND-CCA2-secure.

4 The Randomized McEliece Scheme

In [15] it was proved that the cryptosystem obtained by changing the encryption algorithm of the McEliece cryptosystem to encrypt r|m (where r is random padding) instead of just encrypting the message m, the so called Randomized McEliece Cryptosystem, is IND-CPA secure.

We modify the encryption algorithm of the Randomized McEliece Cryptosystem as follows. Instead of choosing the error vector randomly from the bit strings of length n and Hamming weight t, we choose each bit of the error vector according to the Bernoulli distribution $\mathcal{B}_\theta$ with parameter $\theta = \frac{t}{n} - \epsilon$ for some $\epsilon > 0$.

By the law of large numbers, for large enough n the Hamming weight of error vector e generated by this procedure will be between $t - 2n\epsilon$ and t with overwhelming probability. So this cryptosystem meets the soundness condition. The IND-CPA security follows from assumptions 4 and 5, since ϵ can be arbitrarily small (given that n is large enough).

4.1 Security of the k-Repetition Randomized McEliece

We prove that the modified Randomized McEliece is secure and verifiable under k-repetition, i.e., we prove that the cryptosystem formed by encrypting k times r|m with different public and private keys ($\mathsf{PKE}_{k,McE}$) is sound, IND-CPA secure and that it allows the verification of a ciphertext validity given the public keys and one secret key.

By the soundness of each instance, the probability that in one instance $i \in \{1, \ldots, k\}$ a correctly generated ciphertext is incorrectly decoded is negligible. Since k is polynomial, it follows by the union bound that the probability that a correctly generated ciphertext of $\mathsf{PKE}_{k,McE}$ is incorrectly decoded is also negligible. So $\mathsf{PKE}_{k,McE}$ meets the soundness requirement.

In order to prove that the cryptosystem $\mathsf{PKE}_{k,McE}$ is admissible (and so IND-CPA secure [15]), we prove that it meets the pseudorandom property (the dividability follows trivially). Denote by $\mathbf{R}_1, \ldots, \mathbf{R}_k$ random matrices of size $l \times n$, by $\mathbf{P}_1, \ldots, \mathbf{P}_k$ the public key matrices of the McEliece cryptosystem and by $e_1, \ldots, e_k$ the error vectors. Define $l_1 = |r|$ and $l_2 = |m|$. Let $\mathbf{R}_{i,1}$ and $\mathbf{R}_{i,2}$ be the $l_1 \times n$ and $l_2 \times n$ sub-matrices of $\mathbf{R}_i$ such that $\mathbf{R}_i^T = \mathbf{R}_{i,1}^T | \mathbf{R}_{i,2}^T$. Define $\mathbf{P}_{i,1}$ and $\mathbf{P}_{i,2}$ similarly. We need a lemma from [11]:

Lemma 3. *Say there exists an algorithm $\mathcal{A}$ making q oracle queries, running in time t, and such that*

$$|\Pr[\mathcal{A}^{\mathcal{Q}_{r,\theta}} = 1 | r \leftarrow \{0,1\}^{l_1}] - \Pr[\mathcal{A}^{\mathcal{U}_{l_1+1}} = 1]| \geq \delta$$

Then there exists an adversary $\mathcal{A}'$ making $q' = O(q\delta^{-2}\log l_1)$ oracle queries, running in time $t' = O(t l_1 \delta^{-2}\log l_1)$, and such that

$$\mathsf{Adv}_{\mathsf{LPN}_\theta, \mathcal{A}'} \geq \frac{\delta}{4}$$

Setting $q = kn$ in the lemma, we have that $(r\mathbf{R}_{1,1} \oplus e_1)|\ldots|(r\mathbf{R}_{k,1} \oplus e_k)$ is pseudorandom if the LPN_θ is hard.

Now we prove that substituting the random matrices for the public key matrices of the McEliece cryptosystem does not alter the pseudorandomness of the output $(r\mathbf{P}_{1,1} \oplus e_1)|\ldots|(r\mathbf{P}_{k,1} \oplus e_k)$.

Lemma 4. $(r\mathbf{P}_{1,1} \oplus e_1)|\ldots|(r\mathbf{P}_{k,1} \oplus e_k)$ *is pseudorandom.*

Proof. Suppose that some PPT adversary $\mathcal{A}$ has non-negligible advantage in distinguishing $(r\mathbf{R}_{1,1} \oplus e_1)|\ldots|(r\mathbf{R}_{k,1} \oplus e_k)$ from $(r\mathbf{P}_{1,1} \oplus e_1)|\ldots|(r\mathbf{P}_{k,1} \oplus e_k)$. Denote them by H_0 and H_k respectively. For $i \in \{1,\ldots,k-1\}$, let H_i be

$$(r\mathbf{P}_{1,1} \oplus e_1)|\ldots|(r\mathbf{P}_{i,1} \oplus e_i)|(r\mathbf{R}_{i+1,1} \oplus e_{i+1})|\ldots|(r\mathbf{R}_{k,1} \oplus e_k).$$

Since k is polynomial, by the hybrid argument it is possible to build an adversary $\mathcal{A}'$ that uses $\mathcal{A}$ as a black-box and has a non-negligible advantage in distinguishing H_{i-1} from H_i for some $i \in \{1,\ldots,k\}$, but this would imply that $\mathcal{A}'$ has a non-negligible advantage in distinguishing the public-key matrix $\mathbf{P}$ of the McEliece cryptosystem from a random matrix of the same size. By assumption 4, there exists no such $\mathcal{A}'$ and so there cannot exist an adversary $\mathcal{A}$ with non-negligible advantage in distinguishing H_0 from H_k.

Theorem 2. $\mathsf{PKE}_{k,McE}$ *is IND-CPA secure.*

Proof. From the lemmas 3 and 4 we have that $(r\mathbf{P}_{1,1} \oplus e_1)|\ldots|(r\mathbf{P}_{k,1} \oplus e_k)$ is pseudorandom. So the cryptosystem is admissible. The IND-CPA security of the cryptosystem follows from the fact that an admissible cryptosystem is also IND-CPA secure [15].

Theorem 3. $\mathsf{PKE}_{k,McE}$ *is verifiable under k-repetition.*

Proof. To verify if a ciphertext $(c_1,\ldots,c_k)$ is valid given the public keys and any secret key of the McEliece cryptosystem $(\mathbf{S}_j, \mathbf{G}_j, \mathbf{T}_j)$, we simply decrypt c_j obtaining $r|m$ and for all $i \in \{1,\ldots,k\}$ compute $c'_i = (r|m)\mathbf{P}_i$ and verify if the hamming distance between c'_i and c_i is less than or equal to t .

Theorem 4. *It is possible to construct an IND-CCA2 secure public key encryption scheme based on McEliece assumptions.*

Proof. Follows directly from theorems 1, 2 and 3.

References

1. Berlekamp, E.R., McEliece, R.J., van Tilborg, H.C.A.: On the Inherent Intractability of Certain Coding Problems. IEEE Trans. Inf. Theory 24, 384–386 (1978)
2. Bernstein, D.J., Lange, T., Peters, C.: Attacking and defending the McEliece cryptosystem, http://eprint.iacr.org/2008/318
3. Canetti, R., Halevi, S., Katz, J.: Chosen-ciphertext security from identity-based encryption. In: Cachin, C., Camenisch, J.L. (eds.) EUROCRYPT 2004. LNCS, vol. 3027, pp. 207–222. Springer, Heidelberg (2004)

4. Canteaut, A., Chabaud, F.: A new algorithm for finding minimum-weight words in a linear code: application to primitive narrow-sense BCH codes of length 511. IEEE Trans. Inf. Theory 44(1), 367–378 (1998)
5. Courtois, N.T., Finiasz, M., Sendrier, N.: How to achieve a mcEliece-based digital signature scheme. In: Boyd, C. (ed.) ASIACRYPT 2001. LNCS, vol. 2248, pp. 157–174. Springer, Heidelberg (2001)
6. Cramer, R., Shoup, V.: A practical public key cryptosystem provably secure against adaptive chosen ciphertext attack. In: Krawczyk, H. (ed.) CRYPTO 1998. LNCS, vol. 1462, pp. 13–25. Springer, Heidelberg (1998)
7. Dolev, D., Dwork, C., Naor, M.: Non-malleable Cryptography. SIAM J. Comput. 30(2), 391–437 (2000)
8. Goldwasser, S., Micali, S.: Probabilistic Encryption. J. Comput. Syst. Sci. 28(2), 270–299 (1984)
9. Goldwasser, S., Vaikuntanathan, V.: Correlation-secure trapdoor functions from lattices (manuscript) (2008)
10. Hofheinz, D., Kiltz, E.: Secure hybrid encryption from weakened key encapsulation. In: Menezes, A. (ed.) CRYPTO 2007. LNCS, vol. 4622, pp. 553–571. Springer, Heidelberg (2007)
11. Katz, J., Shin, J.S.: Parallel and concurrent security of the HB and HB$^+$ protocols. In: Vaudenay, S. (ed.) EUROCRYPT 2006. LNCS, vol. 4004, pp. 73–87. Springer, Heidelberg (2006)
12. Lindell, Y.: A Simpler Construction of CCA2-Secure Public-Key Encryption under General Assumptions. In: Biham, E. (ed.) EUROCRYPT 2003. LNCS, vol. 2656, pp. 241–254. Springer, Heidelberg (2003)
13. McEliece, R.J.: A Public-Key Cryptosystem Based on Algebraic Coding Theory. In: Deep Space Network progress Report (1978)
14. Naor, M., Yung, M.: Universal One-Way Hash Functions and their Cryptographic Applications. In: 21st STOC, pp. 33–43 (1989)
15. Nojima, R., Imai, H., Kobara, K., Morozov, K.: Semantic Security for the McEliece Cryptosystem without Random Oracles. In: Proceedings of International Workshop on Coding and Cryptography (WCC), INRIA, pp. 257–268 (2007); journal version in Designs. Codes and Cryptography 49(1-3), 289–305 (December 2008)
16. Peikert, C., Waters, B.: Lossy trapdoor functions and their applications. In: STOC 2008. pp. 187–196 (2008)
17. Rackoff, C., Simon, D.R.: Non-Interactive Zero-Knowledge Proof of Knowledge and Chosen Ciphertext Attack. In: Feigenbaum, J. (ed.) CRYPTO 1991. LNCS, vol. 576, pp. 433–444. Springer, Heidelberg (1992)
18. Regev, O.: On lattices, learning with errors, random linear codes, and cryptography. In: STOC, pp. 84–93 (2005)
19. Rosen, A., Segev, G.: Chosen-Ciphertext Security via Correlated Products (2008), http://eprint.iacr.org/2008/116
20. Sahai, A.: Non-Malleable Non-Interactive Zero Knowledge and Adaptive Chosen-Ciphertext Security. In: 40th FOCS, pp. 543–553 (1999)
21. Sendrier, N.: Finding the Permutation Between Equivalent Linear Codes: The Support Splitting Algorithm. IEEE Trans. Inf. Theory 46(4), 1193–1203 (2000)

Square, a New Multivariate Encryption Scheme

Crystal Clough[1], John Baena[1,2], Jintai Ding[1,4],
Bo-Yin Yang[3], and Ming-shing Chen[3]

[1] Department of Mathematical Sciences,
University of Cincinnati,
Cincinnati, OH, 45220, USA
{baenagjb,cloughcl}@email.uc.edu,ding@math.uc.edu
http://math.uc.edu
[2] Universidad Nacional de Colombia
Carrera 65, Medellín, Colombia
http://www.unalmed.edu.co/~dirmate/
[3] Institute of Information Science,
Academia Sinica
Taipei, Taiwan
[4] College of Sciences,
South China University of Technology
Guangzhou, China

Abstract. We propose and analyze a multivariate encryption scheme that uses odd characteristic and an embedding in its construction. This system has a very simple core map $F(X) = X^2$, allowing for efficient decryption. We also discuss ways to make this decryption faster with specific parameter choices. We give heuristic arguments along with experimental data to show that this scheme resists all known attacks.

1 Introduction

Multivariate public-key cryptosystems (MPKCs) are considered viable options for post-quantum cryptography. This is because they are based on the problem of solving a system of multivariate polynomial equations, a problem which seems just as hard for a quantum computer to solve as any other computer [12,20]. There are a few MPKCs that are believed secure and practical. We propose and analyze a new encryption scheme that is both efficient and secure.

One tool used in several systems is the "big field" idea. While the public keys of MPKCs are polynomial maps $k^n \to k^n$ for some finite field k, some are constructed using maps over a "big field" $K \cong k^n$, using vector space isomorphisms to go back and forth between the spaces. This approach is a two-edged sword in the sense that the field structure of K can make decryption easier but can also be utilized by attackers.

Until recently, the systems based on the "big field" idea (such as the original MPKC C^* proposed by Matsumoto and Imai, HFE proposed by Patarin, and their many variants) had other commonalities. All used characteristic 2 fields,

M. Fischlin (Ed.): CT-RSA 2009, LNCS 5473, pp. 252–264, 2009.
© Springer-Verlag Berlin Heidelberg 2009

often $k = \mathbb{F}_2$, and the collection of plaintexts comprised all of K. Both of these conventions have recently been called into question.

Odd-characteristic MPKCs have not been popular, presumably because characteristic 2 is so fast and easy to implement. However, it now appears that even characteristic has a major drawback. The field equations $x_i^2 - x_i = 0$ allow algebraic attacks to be much more successful, as will be discussed below. Recent work shows that odd-characteristic systems can be much simpler than their even-characteristic counterparts while still evading algebraic attacks [2,6].

As for using all of K, the idea of using a "projection" or embedding is not new but has not held significant interest until recently. By using a K larger than k^n, there is hope that the field structure can still be helpful but no longer troublesome. Our data, and that of others, suggests that this is in fact a good idea.

The new system that we propose uses both of these ideas and comes to the surprising conclusion that under specific circumstances, a variant of the original Matsumoto-Imai system (which has been broken for more than 10 years) can be viable. The main idea was also proposed by Patarin [20], but he dismissed it. Also, at the time Patarin's system was published, the powerful algebraic attack tools F_4 and F_5 were not yet invented so he did not have to consider them. The Square system we will describe avoids Patarin's original attack and resists algebraic attacks as well.

This paper is organized as follows. In Section 2, we discuss relevant background material. In Section 3 we describe the new system Square. In Section 4 we analyze the effectiveness of known attacks. In Section 5 we give our parameter suggestions, as well as discuss how the system can be made very fast (see Table 1, [8]). We conclude the paper in Section 6.

Table 1. Speed of Square instances compared to other systems on a Core 2 Duo 2.4GHz

Scheme	q	n	l	PubKey	PrvKey	Encr	Decr
Square-42	31	42	3	28.4 kB	1350 B	9.4 μs	9.6 μs
Square-51	31	51	3	49.6 kB	1944 B	13.6 μs	14.4 μs
NTRU	587	787	n/a	1.5 kB	1854 B	149.2 μs	251.5 μs
McEliece	n/a	n/a	n/a	79.5 kB	137282 B	29.7 μs	444.1 μs

2   MPKCs and Relevant Attacks

2.1 C^* and HFE

Among first MPKCs was an encryption scheme C^* [16]. This system has since been broken [17], but has inspired many new encryption and signature schemes. One of these is HFE (Hidden Field Equations) [18], which can be seen as a generalization of the Matsumoto-Imai idea. Attacks on either of these systems

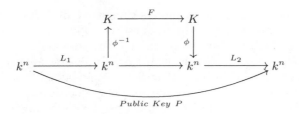

Fig. 1. The MI and HFE systems

or their variants could be relevant to the new system, so we will describe both HFE and C^* here.

Refer to Figure 1. In either system, the plaintext is a vector of length n over k, a field of q elements where q is a power of 2. Since there is a field K of the same size as k^n, we can utilize a nonlinear core map $F \colon K \to K$. In fact, the public key is $P = L_2 \circ \phi \circ F \circ \phi^{-1} \circ L_1$, where L_1 and L_2 are linear maps and ϕ is a vector space isomorphism $K \to k^n$. P is a collection of n polynomials $p_i(x_1, \ldots, x_n)$ in n variables. The decomposition, in particular L_1 and L_2, is the private key.

In the case of C^*, $F(X) = X^{q^\theta + 1}$ for an appropriate θ. In the case of HFE, for some D

$$F(X) = \sum_{\substack{0 \le i < j < n \\ q^i + q^j \le D}} a_{ij} X^{q^i + q^j} + \sum_{\substack{0 \le i < n \\ q^i \le D}} b_i X^{q^i} + c. \tag{1}$$

2.2   Linearization Equations Attack

In the original attack on C^*, Patarin noticed that if $Y = X^{q^\theta + 1}$, then $XY^{q^\theta} - X^{q^{2\theta}} Y = 0$ [17]. This equation forces plaintext-ciphertext pairs from C^* systems to satisfy linearization equations

$$\sum_{i,j=1}^n a_{ij} x_i y_j + \sum_{j=1}^n b_j x_j + \sum_{j=1}^n c_j y_j + d = 0,$$

where $(y_1, \ldots, y_n) = P(x_1, \ldots, x_n)$. Such equations are extremely useful for an attacker because given a ciphertext, the linearization equations yield linear equations satisfied by the plaintext. Also, linearization equations can be found easily from the public key, so an attacker has access to them.

Diene et al showed that the space of linearization equations satisfied by a C^* public key has dimension at least n in most cases [4]. Furthermore Patarin showed that for a given nonzero ciphertext, the space of linear equations satisfied by the correspoding plaintext is at least $n - gcd(n, \theta)$ [17].

Note that in the original C^* construction, $\theta = 0$ cannot be chosen since X^2 is a linear map when $q = 2$.

2.3   Algebraic Attack

Algebraic attacks can be employed against any MPKC. Suppose that some-one, who does not know the private key, wants to recover the plaintext from a ciphertext $(\tilde{y}_1, \ldots, \tilde{y}_n) \in k^n$. This attacker has access only to the public key $(p_1, p_2, \ldots, p_n) : k^n \to k^n$. The most straightforward way to attack is to solve the system of equations

$$
\begin{aligned}
p_1(x_1, \ldots, x_n) - \tilde{y}_1 &= 0 \\
p_2(x_1, \ldots, x_n) - \tilde{y}_2 &= 0 \\
&\vdots \\
p_n(x_1, \ldots, x_n) - \tilde{y}_n &= 0.
\end{aligned}
\tag{2}
$$

Solving these equations directly, without the use of the internal structure of the system, is known as the algebraic attack. Currently the most efficient al-gebraic attacks are the Gröbner basis algorithms F_4 [9] and F_5 [10]. Another algorithm called XL has also been widely discussed but F_4 is seen to be more efficient [1], so we focused our energy on studying algebraic attacks via F_4. Among the best implementations of these algorithms is the F_4 function of MAGMA [15], which represents the state of the art in polynomial solving technology.

Gröbner basis attacks are very fast against the C^* scheme, and also quite effective against HFE as well [11].

2.4   SFlash-Style Attack

SFlash is a C^{*-} signature scheme. The system is constructed in the same way as C^*, except that some number r of the public key polynomials are not published [19]. SFlash was broken using properties of its differential. Recall that for a function f the differential is

$$
Df(a, x) = f(a + x) - f(a) - f(x) + f(0).
$$

In the case of the C^* core map $F(X) = X^{q^\theta + 1}$ for $\xi,\ A,\ X \in K$,

$$
DF(\xi A, X) + DF(A, \xi X) = (\xi + \xi^{q^\theta}) DF(A, X).
\tag{3}
$$

This equation leads to conditions which allow us to identify which matrices in $\mathcal{M}_{n \times n}(k)$ correspond to multiplications in K. These matrices N_ξ have the property that for the C^* public key P,

$$
P(N_\xi(x_1, \ldots, x_n)) = M(P(x_1, \ldots, x_n))
$$

for some linear map M. In other words, the N_ξ mix up the public key polynomi-als, allowing Dubois et al to complete the collection of public key polynomials thus breaking the system [7].

2.5   Kipnis-Shamir Style Attacks

The Kipnis-Shamir attack against HFE exploits the "big field" structure. It uses the following facts, all found in [14]:

– Let $\mathbb{F}_{q^n}$ be the field of q^n elements. If $G \in \mathbb{F}_{q^n}[X]$ such that the q-Haming weight of all monomials is 2 (ie, $G(X) = \sum a_{ij} X^{q^i + q^j}$), then $\exists \mathcal{G} \in \mathcal{M}_{n \times n}(\mathbb{F}_{q^n})$ such that

$$G(X) = \left(X \ X^q \cdots X^{q^{n-1}} \right) \mathcal{G} \begin{pmatrix} X \\ X^q \\ \vdots \\ X^{q^{n-1}} \end{pmatrix}.$$

– If $\mathcal{G}$ is such a matrix for G, S is a linear map $\mathbb{F}_{q^n} \to \mathbb{F}_{q^n}$, and $F = S \circ G$, then

$$\mathcal{F} = \sum_{j=0}^{n} s_j \mathcal{G}^{*j},$$

where the s_j are the coefficients of S and the $\mathcal{G}^{*j}$ are obtained from $\mathcal{G}$ via permutations and Frobenius maps ($x \mapsto x^{q^j}$ for $0 \le j \le n - 1$).
– If $\mathcal{G}$ is such a matrix for G, S is a linear map $\mathbb{F}_{q^n} \to \mathbb{F}_{q^n}$, and $F = G \circ S$, then

$$\mathcal{F} = W \mathcal{G} W^T,$$

where W is obtained from the coefficients of S.

Kipnis and Shamir noted that in the case of HFE, by "lifting" the public key to an extension $L \cong \mathbb{F}_{q^n}$ via some isomorphism $\psi \colon L \to k^n$ and considering the quadratic part, we can find corresponding matrices whose rank must be no more than D. Though L is not necessarily the K used to construct the public key, we still have a decomposition $P = \psi \circ S \circ F \circ T \circ \psi^{-1}$ where S and T are linear and F is some HFE core map.

Using the facts above, Kipnis and Shamir were able to find such a decomposition [14]. The success depends on solving the MinRank problem: Given a collection of matrices, find a linear combination of them that has minimal rank. In general, this problem is NP-complete [3].

3   Design of Square

Having seen what a multivariate encryption scheme is up against, we now describe a new system Square.

See Figure 2. Let k be a field of size q, where $q \equiv 3 \mod 4$. Plaintexts will be vectors in k^n. Let $K \cong k[y]/\langle g(y) \rangle$ be a degree $n + l$ extension of k, where l is such that $n + l$ is odd. The public key will be built up from the following maps:

– $\varphi \colon K \to k^{n+l}$, the vector space isomorphism given by

$$a_{n+l} y^{n+l-1} + \cdots + a_2 y + a_1 \mapsto (a_{n+l}, \ldots, a_1)$$

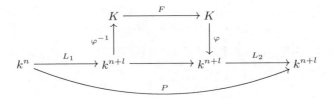

Fig. 2. The Square system

- $F\colon K \to K$, given by $F(X) = X^2$
- $L_1\colon k^n \to k^{n+l}$, an injective affine map
- $L_2\colon k^{n+l} \to k^{n+l}$, an invertible affine map.

From these we construct the public key

$$P = L_2 \circ \varphi \circ F \circ \varphi^{-1} \circ L_1.$$

P will be an $(n+l)$-tuple of quadratic polynomials

$$P(x_1, \ldots, x_n) = \begin{pmatrix} p_1(x_1, \ldots, x_n) \\ p_2(x_1, \ldots, x_n) \\ \vdots \\ p_{n+l}(x_1, \ldots, x_n) \end{pmatrix}.$$

This can be thought of as a C^* system over odd characteristic with $\theta = 0$ and an embedding L_1.

Encryption of a plaintext $(m_1, \ldots, m_n) \in k^n$ is obtained by computing $c_j = p_j(m_1, \ldots, m_n)$ for $j = 1, \ldots, n+l$.

Decryption of a ciphertext $(c_1, \ldots, c_{n+l}) = P(m_1, \ldots, m_n) \in k^{n+l}$ is performed as follows: first, let $Y = \varphi^{-1} \circ L_2^{-1}(c_1, \ldots, c_{n+l})$. Then solve $X^2 = Y$. By choosing $q \equiv 3 \mod 4$ and $n + l$ odd, we ensure that $|K| \equiv 3 \mod 4$. This allows us to use the fact that

$$X = \pm Y^{\frac{q^{n+l}+1}{4}}. \tag{4}$$

This gives two solutions. Since L_1 is affine, in general only one of them will be in the image of $\varphi^{-1} \circ L_1$. The preimage of this solution under $\varphi^{-1} \circ L_1$ will be $(m_1 \ldots, m_n)$.

This simple method to find a preimage under the core map is a major advantage over traditional characteristic-2 HFE systems which require the decryptor to solve a univariate equation of high degree (using Berlekamp's algorithm or its improvements). In fact, encryption and decryption are quite fast. See Table 2 for a summary of times for various choices of q and n. For all experimental data in this paper, we used an Intel(R) Core(TM)2 2.40 GHz processor with 1.99 GB of memory installed.

Table 2. Encryption and decryption times, in seconds, for Square systems. 10 public keys tested and 100 messages encrypted and decrypted per key.

q	n	l	Average Encrypt Time	Average Decrypt Time
31	20	3	0.00022	0.001527
31	32	3	0.00033	0.006781
31	34	3	0.000423	0.003651
43	20	3	0.000234	0.001783
43	34	3	0.000495	0.008135

4   Security Analysis

Let us now present our case for the security of this design. We will explain our motivation and then dig into the specific reasons why each of the aforementioned attacks does not work.

But first, we provide a more thorough comparison to the very similar system proposed by Patarin [20]. His system D^* also uses a square core map, and Patarin broke D^* himself. He did so by finding a way to recover "big field" multiplications without having the big field. As we will see below in Section 4.4, the embedding makes it very hard to recover the multiplicative structure of K. In particular, Patarin's attack on D^* relies on the ability to find pairs of linear maps (C, D) such that for all $\boldsymbol{x_1},\ \boldsymbol{x_2} \in k^n$,

$$C(F(\boldsymbol{x_1} + \boldsymbol{x_2}) - F(\boldsymbol{x_1} - \boldsymbol{x_2})) = F(D(\boldsymbol{x_1}) + \boldsymbol{x_2}) - F(D(\boldsymbol{x_1}) - \boldsymbol{x_2}).$$

When L_1 is invertible, the collection of such pairs forms a vector space of dimension at least n (exactly n according to [20]) which is required for Patarin's attack. In our case, L_1 is a map $k^n \to k^{n+l}$ and thus cannot be invertible.

4.1   Motivation for the Design

All of the ideas used in Square have been seen before; what makes this system novel is that these ideas are combined in such a way that they work.

First, the use of odd characteristic was shown to be a good idea for thwarting algebraic attacks in [6]. This seems to be because the attacker knows that a plaintext $(x_1, \ldots, x_n)$ satisfies not only the public key equations (2) but also the $\mathbb{F}_q$ field equations $x_i^q - x_i = 0$. When q is small, this additional information is very useful, and feeding the field equations into MAGMA along with (2) allows for more efficient solving. However, as discussed in [6], for larger q the field equations do not simplify the algorithm and in fact F_4 runs faster without them.

Secondly, the use of a low-degree core map was inspired by [2], where an odd-characteristic signature scheme with low-degree core map was proposed. It was natural to ask if this idea could be used for encryption as well. However, the signature scheme in [2] uses a vinegar construction, which is not well-suited for encrytion.

This led to the third modification, that of an embedding. Such a tool has been mentioned, but dismissed until recently when a reformulated version of the idea showed promise [5].

Each of these modifications would be weak on their own, but we will make the case below that combined, they are quite strong.

4.2 Linearization Equations Attack

Note that when $\theta = 0$, the equation $XY^{q^\theta} - X^{q^{2\theta}}Y = 0$ that Patarin discovered becomes simply $XY = XY$. So our system should not satisfy any linearization equations other than the trivial one satisfied by any map. In other words, the space of linearization equations should have dimension 0 rather than n. Since the algebraic attack described in 2.3 detects linearization equations, the fact that algebraic attacks are not particularly effective, as descibed below, is an indication that this is in fact the case. To be sure, we did experiments to find the dimension of the space of linearization equations. For each of the 2500 keys we tested, the dimension of this space was 0.

4.3 Algebraic Attack

In order to test the system's resistance to algebraic attacks, we performed the following experiments: We generated a public key and used it to encrypt 50 messages. We then used MAGMA's implementation of F_4 to solve the equations defined by the public key and ciphertext as in 2. We did this for two public keys per choice of parameters q, n, and l. We found that the public key polynomials behave similar to systems of the same size with random polynomials. A sampling of our results are in Table 3.

Plotting this data reveals a clear exponential trend in both time and memory usage as n increases. In fact, linear least-squares approximation on the log of the data has a high correlation. See Figure 3. This trend leads us to believe that $n \geq 33$ is a good choice for a practical system.

Table 3. Average algebraic attack time in seconds and memory usage in MB. $q = 31$

n	l	sec	MB	n	l	sec.	MB	n	l	sec	MB
2	1	0.000	6	3	2	0.000	6	4	3	0.000	6
3	1	0.000	6	4	2	0.001	6	5	3	0.001	6
4	1	0.001	6	5	2	0.001	6	6	3	0.003	6
5	1	0.002	6	6	2	0.003	6	7	3	0.007	6
6	1	0.006	6	7	2	0.008	6	8	3	0.019	6
7	1	0.024	6	8	2	0.021	6	9	3	0.115	7
8	1	0.129	6	9	2	0.165	7	10	3	0.406	8
9	1	0.696	8	10	2	0.818	9	11	3	2.223	12
10	1	4.747	13	11	2	3.137	14	12	3	15.924	26
11	1	27.423	30	12	2	17.294	30	13	3	83.433	68
12	1	215.678	32	13	2	83.516	76	14	3	206.602	137
13	1	1330.657	325	14	2	431.894	262	15	3	2218.500	632

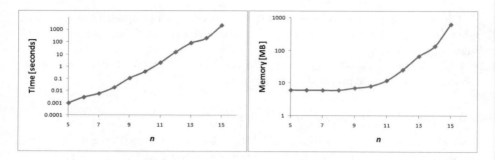

Fig. 3. Algebraic attack for $q=31$, $l=3$, varying n

To inform our choice of q, we tested the effect of changing q while fixing n and l. Our results can be found in Table 4.3 and Figure 4.3. These attacks were done without using field equations $x_i^q - x_i = 0$ for the reasons described in Section 4.1. From this data we see that beyond small values of q, the size of the field does not seem to impact F_4's running time or memory usage. This was expected in light of the results of [2] and [6], and justifies the choice of $q = 31$ for a practical system. Another reason to choose $q = 31$ is that such a choice makes good use of memory, in the sense that the elements of k will require 5 bits to be stored and any larger field will require more bits to store an element.

Table 4. Algebraic attack for $n = 12$, $l = 3$ and varying q

q	Average F_4 Running Time	Memory Usage
3	0.011	6
7	7.082	22
11	9.092	24
19	16.502	26
23	16.308	26
31	15.973	26
43	15.685	26
47	15.618	26

4.4   SFlash-Style Attack

Our public keys are maps $k^n \rightarrow k^{n+l}$. We have constructed the public key P by emebedding the space of plaintexts into a larger space, but one could imagine P as coming from a larger C^* scheme by setting the last l components of the input to 0. Effectively, P is the public key for a (non-embedded) C^* scheme with all monomials involving $x_{n+1}, \ldots,$ or x_{n+l} deleted.

From this point of view, it is important to study the attack on SFlash since its purpose was to recover missing coefficients of the public key (ie, the coefficients

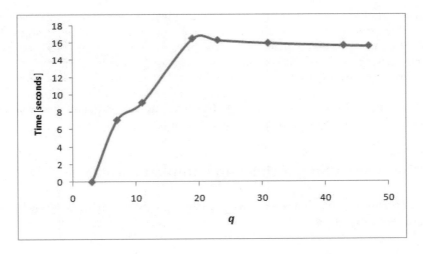

Fig. 4. Algebraic attack for $n = 12$, $l = 3$ and varying q

of the deleted polynomials). Since both SFlash and Square stem from C^*, the differential property (3) exploited by Dubois et al still holds.

While in the case of SFlash this property yields linear equations for the attacker, in our case the property gives us quadratic conditions. In fact, the resulting system of quadratic equations is larger than the system of public key equations. It seemed unlikely that this attack would work. However to be sure, we applied F_4 to these systems of quadratic equations. We quickly realized that the Gröbner basis attack finds no special properties of these systems and takes as long as one might expect. In the "baby" case of $q = 5$, $n = 2$, $l = 1$, this method generates 27 equations in 30 variables which were beyond the abilities of our computer to solve. It stands to reason that with realistic parameters such as $q = 31$, $n = 34$, $l = 3$ these equations will pose no threat.

4.5   Kipnis-Shamir Style Attacks

The attack that Kipnis and Shamir used against HFE depends on finding a combination of matrices derived from the public key which has minimal rank. In our case, we may use the same idea as for the SFlash-style attack and consider the public key as a piece of a C^* public key of $n+l$ variables. In this setting, the rank of the analogous combination of matrices will be 1. As in Kipnis and Shamir's paper [14], we could try to determine the proper combination by finding a basis of the null space of these matrices. The difference between the two systems is that in HFE this yields quadratic equations, while in Square the "missing" coefficients cause the equations to be cubic. One could reduce to quadratic equations, but not without using additional variables.

The HFE attackers claimed that the MinRank problem could be solved in the specific circumstances of HFE [14]. Since that time, doubt has been cast over the original efficiency claims [13]. We tested this attack in the Square case and found that even with the "baby" case of $q = 5$, $n = 2$, $l = 1$, the system that arises from this attack involves 18 cubic equations in 14 variables and solving it exceeded the memory of our computer. We also tried using 2×2 minors as a way to generate equations from the rank condition, but this yields quartic equations with a savings of only 2 variables. This method also exceeded the memory of our computer with $q = 5$, $n = 2$, $l = 1$. Considering this, it is not plausible that such an attack would be dangerous for realistic parameter choices.

5   Parameter Suggestions and Implementations

Based on the security analysis above, an Square system with the following parameter choices will be viable:

Square-34
- $q = 31$
- $n = 34$
- $l = 3$
- Average encryption time: 0.000423 seconds
- Average decryption time: 0.003651 seconds
- Public key size: 15 KB
- Best known attack: $> 2^{80}$ computations.

A system with these parameters will be secure and have relatively fast decryption using the power forumla mentioned in Section 3. Of course, these numbers are very conservative. If we are concerned for speed and not so much for portability, there are ways to get the implementation *much* faster.

Square Roots. Since we are always dealing with a pre-determined field, pre-computation is not a problem. If (field size) $- 1 = 2^a \times$ (odd number o), taking square roots is always possible via raising to the power of $\frac{o-1}{2}$ using a pre-computed table with 2^a elements (the Tonelli-Shanks method [21]).

Consider the case $n = 51$, $l = 3$. Since $n + l = 54$ is even, during decryption we cannot use the formula $X = \pm Y^{\frac{q^{n+l}+1}{4}}$ as in the Square-34 case. However, via raising to a power of $\frac{1}{128}(31^{54} - 65)$, we see that square roots in 1 (mod 4) and 3 (mod 4) field sizes does not differ much, since the number of "total multiplications" does not increase much (at most $1.5\times$ in all our experiments).

Choice of Field. We should choose a field with a "good" irreducible polynomial. For example, for $k = \mathbb{F}_{31}$, only certain $(x^h - \alpha)$ can be irreducible which makes things very fast. Values of h above 34 that is of interest to us here are 45 and 54. and in fact $\mathbb{F}_{31^{45}} \cong k[x]/(x^{45} - 3)$, $\mathbb{F}_{31^{54}} \cong k[x]/(x^{54} - 3)$. Which in and of itself is very simple. But there is a further trick as in the following:

Tower Fields. Tower fields are very common in MPKCs of characteristic two. Here we may use also this trick and use

$$\mathbb{F}_{31^{15}} \cong k[t]/(t^{15} - 3), \quad \mathbb{F}_{31^{45}} \cong \mathbb{F}_{31^{15}}[x]/(x^3 - t); \text{ or}$$
$$\mathbb{F}_{31^{18}} \cong k[t]/(t^{18} - 3), \quad \mathbb{F}_{31^{54}} \cong \mathbb{F}_{31^{18}}[x]/(x^3 - t).$$

Other Techniques. We should delay modulo operations by checking for number sizes appropriately and do not do a modulo-q for as long as we can help it. We should also write the relevant routines in assembly.

Usually the multiplication in an extension of this size would be unwieldy, but due to the tricks mentioned above, "big field" operations have a low computational cost. Our tests show that we can achieve a hundred-fold speed increase.

Hence, we propose alternate parameter choices which can be made even faster than the Square described above as in Tab. 1.

6   Conclusion

In this paper we analyzed a new multivariate encryption scheme that has great promise. In a sense, Square continues the bloodline of the original C^* scheme but our arguments and results above suggest that our system avoids the pitfalls of its predecessors. We showed, via experimental data when possible, that attacks against similar systems are not effective against a reasonably-sized Square system.

We gave parameter choices for a secure system Square-34. We also proposed larger but even more efficient implementations Square-45 and Square-54. Part of our future work will be optimizations of Square systems with other parameters.

References

1. Ars, G., Faugère, J.-C., Imai, H., Kawazoe, M., Sugita, M.: Comparison between XL and gröbner basis algorithms. In: Lee, P.J. (ed.) ASIACRYPT 2004. LNCS, vol. 3329, pp. 338–353. Springer, Heidelberg (2004)
2. Baena, J., Clough, C., Ding, J.: Square-Vinegar signature scheme. In: Buchmann, J., Ding, J. (eds.) PQCrypto 2008. LNCS, vol. 5299. Springer, Heidelberg (2008)
3. Buss, J.F., Frandsen, G.S., Shallit, J.O.: The Computational Complexity of Some Problems of Linear Algebra. Journal of Computer and System Sciences 58(3), 572–596 (1999)
4. Diene, A., Ding, J., Gower, J.E., Hodges, T.J., Yin, Z.: Dimension of the linearization equations of the Matsumoto-Imai cryptosystems. In: Ytrehus, Ø. (ed.) WCC 2005. LNCS, vol. 3969, pp. 242–251. Springer, Heidelberg (2006)
5. Ding, J., Dubois, V., Yang, B.-Y., Chen, O.C.-H., Cheng, C.-M.: Could SFLASH be repaired? In: Aceto, L., Damgård, I., Goldberg, L.A., Halldórsson, M.M., Ingólfsdóttir, A., Walukiewicz, I. (eds.) ICALP 2008, Part II. LNCS, vol. 5126, pp. 691–701. Springer, Heidelberg (2008)
6. Ding, J., Schmidt, D., Werner, F.: Algebraic attack on HFE revisited. In: Wu, T.-C., Lei, C.-L., Rijmen, V., Lee, D.-T. (eds.) ISC 2008. LNCS, vol. 5222, pp. 215–227. Springer, Heidelberg (2008)

7. Dubois, V., Fouque, P.-A., Shamir, A., Stern, J.: Practical cryptanalysis of SFLASH. In: Menezes, A. (ed.) CRYPTO 2007. LNCS, vol. 4622, pp. 1–12. Springer, Heidelberg (2007)
8. eBACS: ECRYPT benchmarking of cryptographic systems, http://bench.cr.yp.to
9. Faugére, J.-C.: A new efficient algorithm for computing Gröbner bases (F_4). J. Pure Appl. Algebra 139(1-3), 61–88 (1999); effective Methods in algebraic geometry (Saint-Malo) (1998)
10. Faugère, J.-C.: A new efficient algorithm for computing Gröbner bases without reduction to zero (F_5). In: Proceedings of the 2002 International Symposium on Symbolic and Algebraic Computation, pp. 75–83. ACM, New York (2002)
11. Faugère, J.-C., Joux, A.: Algebraic cryptanalysis of Hidden Field Equation (HFE) cryptosystems using Gröbner bases. In: Boneh, D. (ed.) CRYPTO 2003. LNCS, vol. 2729, pp. 44–60. Springer, Heidelberg (2003)
12. Garey, M.R., Johnson, D.S., et al.: Computers and Intractability: A Guide to the Theory of NP-completeness. W.H Freeman, San Francisco (1979)
13. Jiang, X., Ding, J., Hu, L.: Public Key Analysis-Kipnis-Shamir Attack on HFE Revisited. In: Pei, D., Yung, M., Lin, D., Wu, C. (eds.) Inscrypt 2007. LNCS, vol. 4990, pp. 399–411. Springer, Heidelberg (2008)
14. Kipnis, A., Shamir, A.: Cryptanalysis of the HFE public key cryptosystem by relinearization. In: Wiener, M. (ed.) CRYPTO 1999. LNCS, vol. 1666, pp. 19–30. Springer, Heidelberg (1999)
15. The MAGMA computational algebra system home page, http://magma.maths.usyd.edu.au/magma
16. Matsumoto, T., Imai, H.: Public quadratic polynomial-tuples for efficient signature-verification and message-encryption. In: Günther, C.G. (ed.) EUROCRYPT 1988. LNCS, vol. 330, pp. 419–453. Springer, Heidelberg (1988)
17. Patarin, J.: Cryptanalysis of the Matsumoto and Imai public key scheme of EUROCRYPT 1988. In: Coppersmith, D. (ed.) CRYPTO 1995. LNCS, vol. 963, pp. 248–261. Springer, Heidelberg (1995)
18. Patarin, J.: Hidden fields equations (HFE) and Isomorphisms of Polynomials (IP): Two new families of asymmetric algorithms. In: Maurer, U.M. (ed.) EUROCRYPT 1996. LNCS, vol. 1070, pp. 33–48. Springer, Heidelberg (1996)
19. Patarin, J., Courtois, N.T., Goubin, L.: FLASH, a fast multivariate signature algorithm. In: Naccache, D. (ed.) CT-RSA 2001. LNCS, vol. 2020, pp. 298–307. Springer, Heidelberg (2001)
20. Patarin, J., Goubin, L.: Trapdoor one-way permutations and multivariate poly-nominals. In: Han, Y., Quing, S. (eds.) ICICS 1997. LNCS, vol. 1334, pp. 356–368. Springer, Heidelberg (1997)
21. Shanks, D.: Five numbertheoretic algorithms. In: Thomas, R.S.D., Williams, H.C. (eds.) Proceedings of the Second Manitoba Conference on Numerical Mathematics, pp. 51–70 (1972)

Communication-Efficient Private Protocols for Longest Common Subsequence

Matthew Franklin, Mark Gondree, and Payman Mohassel

Department of Computer Science
University of California, Davis
{franklin,gondree,mohassel}@cs.ucdavis.edu

Abstract. We design communication efficient two-party and multi-party protocols for the longest common subsequence (LCS) and related problems. Our protocols achieve privacy with respect to passive adversaries, under reasonable cryptographic assumptions. We benefit from the somewhat surprising interplay of an efficient block-retrieval PIR (Gentry-Ramzan, ICALP 2005) with the classic "four Russians" algorithmic design. This result is the first improvement to the communication complexity for this application over generic results (such as Yao's garbled circuit protocol) and, as such, is interesting as a contribution to the theory of communication efficiency for secure two-party and multiparty applications.

1 Introduction

We design communication efficient two-party and multi-party protocols for two variants of the longest common subsequence (LCS) problem, and related problems. The first variant returns only the length of the LCS, while the second outputs the string encoding the subsequence itself. Previous work on this topic [25,34,10] has focused on implementing basic dynamic programming algorithms privately, using techniques that each achieve $O(n^2)$ communication complexity (where n is the length of each input string). Jha, Kruger, and Shmatikov [25] demonstrate that some of these solutions may be practically quite efficient. This previous work, however, does not improve on the asymptotic communication complexity of generic solutions such as Yao's garbled circuit protocol [36]. Thus, our work is both theoretically and practically interesting since it can be interpreted as a new upper bound on the communication complexity of this problem.

Traditionally, the method known as the "four Russians" technique yields only a logarithmic improvement in the running time of the dynamic programming solution for LCS. Somewhat surprisingly, we show how to take advantage of an efficient block-retrieval PIR due to Gentry and Ramzan [21] and use the "four Russians" technique to obtain a communication efficient private protocol for the LCS problem. Specifically, we design protocols that achieve $O(n^2/t)$ total communication cost, for any positive $t < n$.

In our protocol, the computational cost of one party increases (anywhere between linearly and exponentially, depending on the choice of parameter t), while the remaining parties' computation costs reduce to $O(n^2/t)$. In the most

M. Fischlin (Ed.): CT-RSA 2009, LNCS 5473, pp. 265–278, 2009.
© Springer-Verlag Berlin Heidelberg 2009

practical setting, where each party performs only a polynomial amount of computation, we achieve a new, sub-quadratic upper bound for the private protocol's communication complexity.

If some participant can perform more work, our protocol becomes even more communication efficient. This setting, where an asymmetric work load yields communication savings for all participants, may be quite advantageous. This client-server setting may even be realistic, given the large volume of genomic data held by central entities. We also show how to outsource the work needed by our protocol to a set of powerful, dedicated yet untrusted servers.

Motivation: Performing different computational tasks on large biological databases is becoming a more common practice in both public and private institutions. The FBI maintains a database of over four million DNA profiles of criminal offenders, crime scene evidence, and missing persons in its CODIS system [2], and uses the data for forensic studies and DNA-based identification. deCODE Genetics [3], a biopharmaceutical company which studies genomic data for drug discovery and development, has collected the genotypic and medical data of over 50 percent of the population in Iceland. Similar endeavors seek to make these types of databases available for scientific study [5].

The genomic data stored in these databases may be extremely sensitive: an individual's DNA sequence reveals a great deal of information regarding that individual's health, background, and physical appearance [1,4]. It has been shown that a sequence can be linked to the corresponding individual simply by recognizing the presence of certain markers [27]. Protecting a patient's privacy when working with genomic data is recognized as a major challenge for the biomedical research community [8,35]. Furthermore, in the United States, HIPAA's Privacy Rule [32] mandates that a patient's identity must be protected when their data (including genomic data) is shared; failure to assure this may result in legal action, fines, revocation of government funding, and imprisonment.

The pioneering work of Jha, Kruger, and Shamtikov [25] and others [34,10] has recognized the need for private and efficient computation on genomic data in general, and the LCS and edit distance problems in particular. We hope our results continue to motivate improvements to the state-of-the-art in this domain.

2   Related Work

It is possible to use generic solutions or pre-existing protocols to solve the problems considered in this paper. However, we desire protocols that are efficient in terms of communication complexity (the total number of bits sent and received by the participants in the protocol).

While the theoretical lower bound[1] for the running time of an algorithm that solves the LCS problem for two strings of length n is $\Omega(|\Sigma|n)$ for a fixed alphabet Σ [6], the theoretically fastest *known* algorithm solving this problem is that of

---

[1] This lower bound becomes $\Omega(n^2)$ when the alphabet is not fixed, and the basic operations considered are comparisons.

Masek and Patterson [28], achieving $O(n^2/\log n)$. The circuit simulating this algorithm, however, is of size $O(n^2)$. Thus, generic compilers like Yao's garbled circuit protocol [36] would yield communication complexity $O(n^2)$. Naor and Nissim's communication preserving compiler [30] may yield a protocol with $O(n)$ communication complexity, but the protocol would require on-line work that is exponential in n.

We can reduce the LCS problem to the shortest path problem, and then use secure matrix multiplication protocols, e.g. [26], or secure graph protocols [11] to recover a solution. The reduction, however, increases the input size of the problem, so that instead of an input of size $\Theta(n)$, we now must consider a graph or matrix of size $\Theta(n^2)$. Even assuming we have access to private protocols for matrix multiplication or shortest path with optimal communication efficiency, using this reduction will yield a protocol with $\Omega(n^2)$ communication complexity.

We could also re-use secure protocols implementing the Needleman-Wunsch algorithm (or a variation of it, the Smith-Waterman algorithm) [10,34,25]. These algorithms solve a generalization of the edit distance problem, where insertions and deletions have variable costs. When all costs are 1, these algorithms can be used to directly solve the problems we consider here. Szajda, Pohl, Owen, and Lawson [34] provide a heuristic protocol for Smith-Waterman attaining heuristic security, which fails to meet the correctness or privacy needs considered here. Atallah, Kerschbaum, and Du [10] and Jha, Kruger, and Shmatikov [25] provide protocols whose communication complexities are $O(n^2)$, meeting but not improving upon the asymptotic efficiency of the generic solution, using Yao's garbled circuit protocol.

3 Notations and Definitions

Notation $\Omega(f)$ denotes that the asymptotic lower bound f is tight; $\Theta(f)$, means that f is both a lower bound and an upperbound, and $\widetilde{O}(f)$ denotes the asymptotic upper bound $O(f)$, ignoring polylog(f) factors.

Sharing Values. We take advantage of two simple sharing schemes, XOR sharing and additive sharing, in our protocols. Alice and Bob XOR share a value c, if Alice holds the value a and Bob holds the value b, such that $a \oplus b = c$. Similarly, an integer c is additively shared between the parties if $a + b \bmod N = c$, where N is a properly chosen and publicly known integer.

Security. We prove our protocols secure against a passive adversary (also referred to as semi-honest) who follows the steps of the protocol but tries to learn additional information based on the messages he receives throughout the protocol. The security in this model is defined by requiring that any adversary in the real protocol, can be simulated by an adversary in an ideal world where parties send their inputs to a trusted party who computes and sends back their corresponding outputs. For a more formal definition, we refer the reader to [22, Volume 2]. Central to our security claims is the following composition theorem.

Theorem 1 (Composition for Passive Adversaries [22]). *Suppose that g is privately reducible to f and that there exists a protocol for privately computing f. Then, there exists a protocol for privately computing g.*

Using the above theorem, along with simple hybrid arguments, it is straightforward to prove our protocols private against a passive adversary as long as our subprotocols are private.

4   Longest Common Subsequence

Let A and B be two strings over a fixed alphabet Σ of size σ, with lengths $m = |A|$ and $n = |B|$ (without loss of generality, let $m \leq n$). A subsequence of A is a string X such that A can be transformed into X by deleting characters from A. A longest common subsequence (LCS) of A and B is a subsequence of both A and B such that no other common subsequence has greater length.

Algorithms that solve the longest common subsequence problem return one or more of the following outputs:

1. The length of the LCS of A and B.
2. A string which is a LCS of A and B.
3. An embedding α, β of a LCS of A and B

Where an embedding $\alpha \in \{0,1\}^m$ and $\beta \in \{0,1\}^n$ are bit-strings which select an LCS from A and B, respectively.

4.1   LCS Algorithm Using Standard Dynamic Programming Techniques

The following dynamic programming algorithm solving the longest common subsequence problem was independently discovered by many researchers, in both computer science and biology. For a standard presentation of this type of dynamic programming solution, see [14, §16.3] or [23, §11.3]. Let L be the $(m+1) \times (n+1)$ matrix whose entries can be computed (row-by-row or column-by-column) using the following:

$$L[i,j] = \begin{cases} 0 & \text{if } i = 0 \text{ or } j = 0 \\ L[i-1, j-1] + 1 & \text{if } A[i] = B[j] \\ \max(L[i-1,j], L[i, j-1]) & \text{otherwise} \end{cases}$$

Entry $L[m,n]$ holds the length of the LCS for A and B, and simple deterministic backtracking algorithms exist for recovering the value and/or the embedding of an LCS for A and B.

4.2   LCS Algorithm Using the "Four Russians" Technique

Masek and Patterson [28] give a variant on this dynamic programming solution for the LCS problem, using ideas (colloquially known as the "four Russians"

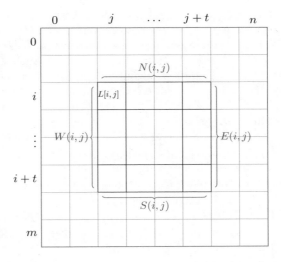

Fig. 1. The $t \times t$ block beginning at position (i, j) in the $(m+1) \times (n+1)$ matrix L

technique) introduced by Arlazarov, Dinic, Kronod and Faradzev [9] for boolean matrix multiplication.

Let t be a positive integer. Then, the boundaries of the $t \times t$ block of the dynamic programming table L starting at position (i, j) can be denoted with the following variables (see Figure 1).

$$N(i, j) = (L[i, j], L[i, j+1], \ldots, L[i, j+t])$$
$$W(i, j) = (L[i, j], L[i+1, j], \ldots, L[i+t, j])$$
$$S(i, j) = (L[i+t, j], L[i+t, j+1], \ldots, L[i+t, j+t])$$
$$E(i, j) = (L[i, j+t], L[i+1, j+t], \ldots, L[i+t, j+t])$$

Let the *offset vectors* $I_1(i, j), I_2(i, j) \in \{0, 1\}^t$ be defined as

$$I_1(i, j)[k] = \begin{cases} 0 & \text{for } k = 1 \\ N(i, j)[k] - N(i, j)[k-1] & \text{for } 1 < k \leq t \end{cases}$$

$$I_2(i, j)[k] = \begin{cases} 0 & \text{for } k = 1 \\ W(i, j)[k] - W(i, j)[k-1] & \text{for } 1 < k \leq t \end{cases}$$

It is simple to check that consecutive values in the L matrix increase by at most one, so the offsets are indeed bit-vectors.

The basic observation underlying the four Russians technique is that the values $E(i, j)$ and $S(i, j)$ are completely determined by $A[i, \ldots, i+t]$, $B[j, \ldots, j+t]$, $I_1(i, j)$, $I_2(i, j)$, and $L[i, j]$. Denote this basic block functionality as bbf($A[i, \ldots, i+t], B[j, \ldots, j+t], I_1(i, j), I_2(i, j), L[i, j]) = (E(i, j), S(i, j))$. Thus, we can compute the entire dynamic programming table in the following manner:

1. **Pre-processing:** pre-compute all possible $t \times t$ blocks, by considering all possible t-length strings and offset vectors, but assume the first value of the block is 0. That is, generate a table summarizing $\mathrm{bbf}(\cdot, \cdot, \cdot, \cdot, 0)$. Note that $\mathrm{bbf}(\cdot, \cdot, \cdot, \cdot, 0) + C = \mathrm{bbf}(\cdot, \cdot, \cdot, \cdot, C)$; *i.e.*, if the first value of the block is C and not zero, the pre-computed outputs differ from the desired outputs in every place by the additive term C. The number of entries in this table is $\sigma^{2t} 2^{2t}$.

2. **Rebuilding the L matrix:** Consider the L matrix to be composed of $t \times t$ blocks that overlap in one row and one column with each other. Retrieve $\mathrm{bbf}(A[i, \ldots, i+t], B[j, \ldots, j+t], I_1(i,j), I_2(i,j), 0) = (\widetilde{E}(i,j), \widetilde{S}(i,j))$ by looking up the appropriate pre-computed block. Any $L[i', j']$ entry situated on the boundary of this block can be calculated by adding $L[i, j]$ to the appropriate value from the retrieved $\widetilde{E}(i,j)$ or $\widetilde{S}(i,j)$ vectors. Iterate in this fashion, using $\widetilde{S}(i,j), \widetilde{E}(i,j)$ to determine $I_1(i+t,j), I_2(i, j+t)$, each time considering the first value of the block to be 0 during look-up and then compensating by adding back the appropriate additive term.

Essentially, we have reduced our $m \times n$ matrix to an $m/t \times n/t$ matrix. In the unit-cost RAM model, partially filling out the L matrix in this fashion takes $O(mn/t^2)$ time.

5   Communication Efficient Protocols for Private LCS

Here, we build a private protocol for determining the length of the LCS. Later, we provide a deterministic backtracking algorithm for privately recovering an actual LCS or LCS embedding.

Definition 1 (Private LCS-length). *A protocol is a private LCS-length protocol between two parties (one holding a private input string A, the other holding a private input string B) if the protocol outputs the length of the LCS of A and B, but reveals no information to a passive adversary other than what she can learn from the output.*

Below we provide details describing a private LCS-length protocol. At a high-level, the protocol executes the algorithm of Masek and Patterson described earlier, but each party holds shares of the L matrix. Basic block function table look-ups are performed using a communication efficient private block retrieval scheme, while the remaining computations are designed to be performed by the parties locally.

Private block retrieval (PBR) was first introduced in the original PIR paper of Chor *et al.* [13]. A PBR scheme is essentially a private information retrieval (PIR) scheme that allows the chooser to retrieve ℓ-bit database entries, as opposed to bit entries. We call such a scheme a symmetric PBR (SPBR) scheme if it also provides privacy for the server (database).

Any secure two-party PBR schemes can be transformed into an SPBR scheme that provides this functionality, via the Naor-Pinkas transform [31], the Aiello-Ishai-Reingold transform [7], or zero-knowledge proofs. We note that, excluding zero-knowledge based techniques, these transforms incur no loss of efficiency for the SPBR.

1. **Pre-computation:** Alice pre-computes the table summarizing the function $\text{bbf}(\cdot, \cdot, \cdot, \cdot, 0)$, in the following way. Let the binary representation of the inputs $A[i, \ldots, i + t]$, $B[j, \ldots, j + t]$, $I_1(i,j)$, $I_2(i,j)$ be x_1, x_2, x_3, x_4. The resulting index $x_1 || x_2 || x_3 || x_4$ is a bit-string of length $2 \log(\sigma) t + 2t$. At this index, Alice stores the following:
   - $\widetilde{E}(i,j), \widetilde{S}(i,j)$.
   - The offset vectors $I_1(i + t, j), I_2(i, j + t)$ associated with $\widetilde{E}(i,j), \widetilde{S}(i,j)$.
   Clearly these two vector pairs are redundant because given either pair we can compute the other, but as each is shared using a different sharing scheme, holding both will make local computations easier, later. Because the first entry of the pre-computed block is set to zero, the values of $\widetilde{E}(i,j), \widetilde{S}(i,j)$ will always be in the interval $[0, t]$. This means each entry of our look-up table will be a bit-string of length $2t + 2t \log t$.

2. **Accessing the basic block function:** From previous database accesses, Alice and Bob will hold XOR shares of $I_1(i + t, j) = x_3$ and $I_2(i + t, j) = x_4$. In all cases, x_1 is known only to Alice and x_2 is known only to Bob. Thus, Alice can simply consider her share of x_2 to be $0^{t \log \sigma}$, and similarly for Bob. This allows Alice and Bob to locally concatenate their values, producing valid XOR shares of the next index $x_1 || x_2 || x_3 || x_4$. The database defining the basic block function is accessed each time in the following way:
   (a) Alice picks some random value and locally blinds each of her database's entries by this value. Particularly, XORing by a random string is used for each offset vector, and adding a random value is used for each $\widetilde{E}(i,j)$ and $\widetilde{S}(i,j)$.
   (b) Alice locally permutes her database by her share of $x_1 || x_2 || x_3 || x_4$.
   (c) Alice and Bob engage in a SPBR protocol using Bob's index share as input, allowing Bob to recover the blinded entries. Alice's shares are the random values (strings) she added (XORed) in the blinding step. Thus, Alice and Bob hold valid shares of the entries of interest.

3. **Reconstructing L:** As in the original Masek and Patterson algorithm, after the appropriate entries of the basic block function are retrieved, Alice and Bob can (non-interactively) use the additive shares of the values to compute shares of entry $L[m, n]$. They then privately exchange these shares to recover $L[m, n]$, the length of the LCS of A and B.

Theorem 2. *The above protocol is a private LCS-length protocol for inputs of length n and m, assuming we have a secure SPBR scheme for a database of N entries each of size ℓ.*

Proof. (sketch) The above security claim follows from the security of the share conversion protocols, our black-box use of SPBR, and from general composition theorems [12,22]. How to specify the parameters N and ℓ to attain the efficiency we desire is explained in the complexity analysis, below.

5.1 Analysis

Theorem 3. *Assume we have an SBR scheme such that: (1) for a database of size N, on query i the scheme returns the i-th ℓ-bit entry of the database for integer parameter ℓ with $2^\ell < N$, and (2) the scheme has $g(N, \ell) = O(\log N)$ communication complexity and requires $f(N, \ell) = O(N)$ work.*

Then the described protocol for LCS-length has the following complexities, as a function of the parameter t.

Communication Complexity	Round Complexity	Alice's Work	Bob's Work
$O(mn/t)$	$O(m/t + n/t)$	$\widetilde{O}(t2^t)$	$O(mn/t)$

Proof. The above protocol fills out the L matrix as in the Masek-Patterson algorithm after mn/t^2 invocations of the SPBR protocol. Naively, the protocol proceeds (row-by-row or column-by-column) in mn/t^2 rounds. In a straightforward manner, we can parallelize many of these steps and improve this to $m/t + n/t$ rounds.

When we set $N = \ell 2^t$, then the query returns the appropriate ℓ-bit entry of the bbf look-up table of size 2^t. When $\ell = t \log t$, the computational complexity of the protocol is $O(2^{2t} * f(\ell N, \ell) * (2t + 2t \log t)/\ell) = \widetilde{O}(t2^t)$. The total communication complexity is $O(mn/t^2 * g(\ell N, \ell) * (2t + 2t \log t)/\ell) = O(mn/t)$.

Claim. Concretely, the assumptions of Theorems 2 and 3 are satisfied by the SPBR scheme attained by transforming the Gentry-Ramzan PBR scheme [21] to an SPBR scheme using either the Aiello-Ishai-Reingold or Naor-Pinkas transforms. The security of this SPBR scheme is based on the hardness of the "extended decision subgroup problem" and the assumptions required for the transform (*e.g.*, secure homomorphic encryption for Aiello-Ishai-Reingold).

The proposed protocol establishes a useful framework for many dynamic programming algorithms, providing a range of efficiencies based on trading computational complexity for communication complexity. At one extreme, when $t = 1$, the communication and computational complexities resemble that of Yao's garbled circuit protocol. At the other extreme, when $t = n$, the costs match those of the communication preserving compiler of Naor and Nissim [30], achieving $O(m)$ communication complexity (matching the simple theoretic lower bound) but requiring work exponential in n. Between these extremes, the protocol yields a smooth trade-off between work and communication, achieving any sub-quadratic communication complexity at the expense of computational complexity.

6 Private Backtracking in Our Protocol

We have provided a protocol for determining the length of an LCS of A and B privately, although a natural remaining question is to ask is if we can recover the value of an LCS string itself (or the bit-string which encodes how an LCS is embedded in A and B) privately. We call this a private LCS-backtracking protocol.

Definition 2 (Private LCS-backtracking). *A private LCS-backtracking protocol between two parties is one in which (1) the parties hold private input strings A and B, (2) the protocol outputs an LCS (or, embedding of the LCS) for A and B, (3) there is some deterministic algorithm that agrees with the protocol's output for any valid input, (4) the protocol reveals no information to a passive adversary other than what she can learn from the output.*

In fact, with some careful planning, we can modify our LCS-length protocol to build a LCS-backtracking protocol. Our LCS-backtracking protocol relies on no stronger assumptions than our LCS-length protocol, and its communication complexity and work are asymptotically no greater than that of our LCS-length protocol. The full details of this protocol are provided in Appendix A of the full version of this paper [18].

7 Related Applications

Our techniques can be viewed as a specialization of Naor-Nissim circuits with look-up tables [30], for a class of dynamic programming problems that have the property that their subproblems can be "efficiently encoded." By this, we mean that the problem can be decomposed into overlapping subproblems of size t, each of which can be encoded using $O(t)$ bits. For the LCS problem, it was essential that adjacent entries in L differed by at most 1, so each $t \times t$ block could be encoded with $O(t)$ bits. If, however, adjacent entries differed by, say, an arbitrary value in $[0, n]$, then any encoding of the subproblem would require $O(t \log n)$ bits; the size of the subproblem's look-up table would be strictly greater than $O(2^t)$, and the cost of iteratively accessing this table would result in no savings when compared to Yao's garbled circuit protocol.

There are natural applications that fall into this general framework of dynamic programming problems with "efficiently encoded" subproblems. We list some below, with remarks on how to solve each with our private protocols for LCS-length and LCS-backtracking, either immediately or with minor modifications.

7.1 Edit Distance

The private LCS-length protocol automatically yields a private algorithm recovering the edit distance (Levenshtein distance), because

$$\text{edit-distance}(A, B) = m + n - 2 * \text{LCS-length}(A, B)$$

This relationship holds because transforming A into B via a series of deletions and insertions is equivalent to deleting $n - \text{LCS-length}$ characters, and inserting $m - \text{LCS-length}$ characters.

Damerau-Levenshtein distance (where the allowable operations are insertions, deletions, substitutions, or transpositions) is not automatically implied by LCS-length, but it is clear that a small modification of our protocol will suffice. In particular, when pre-computing the $t \times t$ block at position (i, j) in L, in order to consider transposition operations we may need to refer back to data in row $i - 1$ and column $j - 1$. That is, the input to the basic block function will require $4t$ offset vectors (instead of $2t$ vectors), and the $t \times t$ blocks will tile the L matrix overlapping with a previous block in two rows or columns (instead of just one). Such a modification does not change the asymptotic complexity of the resultant protocol.

Ulam's metric is a type of similarity measure or distance metric for permutations, similar to Kendall's τ-metric or Spearman's rank correlation metric. For permutations π, π' of order n, the Levenshtein distance of π and π' is equivalent to the Ulam distance. Thus, our protocol gives two parties the ability to privately compute the Ulam similarity of their ordered preference lists of common elements. Note that this is somewhat different from the setting of Freedman *et al.* [20], which performed a privacy-preserving set intersection to measure the similarity between unordered preference lists of different elements.

7.2   Shortest Common Supersequence

The private LCS-length protocol automatically yields a private algorithm recovering the length of the shortest common supersequence (SCS), because

$$\text{SCS-length}(A, B) = m + n - \text{LCS-length}(A, B)$$

This relationship holds because the SCS is the shortest string containing A and B as subsequences, which can always be obtained by adding to an LCS the extra characters from A and B. However, when the number of strings is greater than 2, there is no longer a relationship between the problems [19].

7.3   Other Dynamic Programming Applications

The longest common substring (LCSS) problem and longest increasing subsequence (LIS) problem both have dynamic programming solutions with similar structure. Our protocol can be used directly or with slight modifications to solve these problems. For example, given the string A and an ordering on Σ, $\sigma_1 < \sigma_2 < \ldots < \sigma_{|\Sigma|}$, we have $\text{LIS}(A) = \text{LCS}(A, B)$ where $B = \sigma_1^{|A|} \sigma_2^{|A|} \cdots \sigma_{|\Sigma|}^{|A|}$. The LCSS problem can be solved by modifying the basic block function to utilize the recurrence

$$\text{LCSS}(i, j) = \begin{cases} \text{LCSS}(i - 1, j - 1) + 1 & \text{if } A[i] = B[j] \\ 0 & \text{otherwise} \end{cases}$$

Both the LIS and LCSS problems, however, have more efficient non-dynamic programming solutions: LIS can be solved in $O(n \log n)$ time with arrays and binary search; LCSS can be solved in $O(n + m)$ time with suffix trees. While the circuits implementing these solutions are of size $O(mn)$, it is likely that there are more efficient specialized two-party protocols implementing these solutions.

7.4 Private `diff`

The Unix command "`diff` A B" traditionally returns an annotated file showing the least-costly merge of file A and file B. From the annotations, both A and B can be recovered. Thus, a private implementation of `diff` has little utility, as the output leaks both inputs. It may be natural, however, to consider an asymmetric version of `diff`, where Alice does not learn an annotated file merging A and B but instead learns which parts of A have been removed, which parts have been preserved, and where insertions have been made (analogous to an embedding). Modifying LCS-backtracking to achieve this functionality is straight-forward. Similarly, it may be natural to consider calculating statistics related to the `diff` of two files, and not output the merged file itself. For example, the command

$$\texttt{diff } A\ B \ | \ \texttt{wc -1}$$

returns the edit distance of files A and B and, as we have seen, is easily calculated using our LCS-length protocol.

A private `diff` protocol may be useful in the situation where Alice and Bob want to collaborate to determine if either one has "plagiarized" the other, without leaking their own proprietary data to the other participant who, necessarily, they suspect may be possible of plagiarism. A realistic scenario similar to this one is discovering GPL violations [33]; the protocol may be a useful discovery mechanism, allowing Bob to inspect if Alice has violated the use restrictions of his software licensed under the GPL (by including it in her proprietary code without redistributing it under the GPL) while at the same time respecting the Alice's potentially non-infringing, proprietary code.

7.5 Multiparty Variants of LCS-Length and LCS-Backtracing

Instead of running a protocol between two parties, it may be desirable to outsource the work to some fixed number of parties m, such that no passive adversary corrupting less than m participants can learn any information beyond the output, for privacy reasons. Alternatively, perhaps neither Alice nor Bob want to play the role of the server (who pre-computes the basic block function table) but instead want to leverage the resources of one or more external servers, without having to trust them. We can generalize our LCS-length and LCS-backtracking protocols to this scenario using a multiparty generalization of indirect indexing [17,24]. In particular, the multiparty indirect indexing scheme of Ishai et al. [24] is a general construction using any 2-round OT protocol. To achieve a construction with polylogarithmic communication complexity, Ishai et al. use an OT scheme based on PIR with length-flexible messages (achieved using a length-flexible homomorphic encryption scheme such as Damgård-Jurik [15,16]). To achieve multiparty indirect indexing with a strictly logarithmic communication complexity, we can instead use Gentry-Ramzan. With a database of size N, the m-iterated use of Gentry-Ramzan generates messages of length $O(\log N c^m) = O(\log N)$ for a small constant $c < 8$ [29]. Thus, we can use the construction of Ishai *et al.*, instantiated with an appropriate version of Gentry-Ramzan, to perform multiparty indirect indexing, and generalize our protocols to the multiparty scenario with no asymptotic increase to the costs.

7.6   Fixed m-String LCS

In the multiparty version of LCS-length and LCS-backtracking protocols, it is natural to consider the situation were each party i holds a string A_i as her private input, and the m parties wish to compute the length (or embedding or string) of the LCS of the strings $A_1, \ldots, A_m$. The following is a straight-forward generalization of the standard dynamic programming solution for this problem, using an m-dimensional L matrix.

1: **for** $i_1 \leftarrow 1 \ldots |A_1|$, $i_2 \leftarrow 1 \ldots |A_2|$, $\ldots$, $i_m \leftarrow 1 \ldots |A_m|$ **do**
2:    **if** $(A_1[i_1] = A_2[i_2] = \cdots = A_m[i_m])$ **then**
3:       $L[i_1, i_2, \ldots, i_m] \leftarrow L[i_1 - 1, i_2 - 1, \ldots, i_m - 1] + 1$
4:    **else**
5:       $L[i_1, i_2, \ldots, i_m] \leftarrow \max\{L[i_1 - b_1, i_2 - b_2, \ldots, i_m - b_m] : b_1 b_2 \ldots b_m \in \{0, 1\}^m$ is a bit-string of Hamming weight $m - 1\}$
6:    **end if**
7: **end for**

As before, adjacent rows and columns never differ by more than 1, so the offset vectors can be efficiently encoded using bit-strings. Thus, the four Russians technique can still be used. We can generalize our protocols to simulate this algorithm in a similarly straight-forward fashion.

8   Conclusion

We have presented a private protocol for the longest common subsequence problem, a classic problem for both computer scientists and biologists. While the "four Russians" technique traditionally offers only a logarithmic savings in work for the traditional dynamic programming algorithm for LCS, we use it to achieve a smooth trade-off between communication complexity and work, filling in the theoretical "terrain" between polynomial-work compilers (like Yao) and exponential-work compilers (like Naor-Nissim). Our protocol features an asymmetry in the work required from each participant, making it appealing in a client-server setting where it may be reasonable to believe there is an available server able to perform super-polynomial work to help compute on genomic data. Even when restricting all participants to polynomial resources, the result is a private protocol for the LCS problem achieving a sub-quadratic communication complexity: the first private LCS protocol whose efficiency improves on that of the generic solution using Yao's garbled circuit protocol. We have shown that the protocol and techniques extend (trivially, in some cases) to numerous other string algorithms useful for computing with genomic data. We anticipate this technique to be useful in achieving communication-efficient protocols for many other problems that have similarly-structured dynamic programming algorithms.

References

1. Are guarantees of genome anonymity realistic (2008), http://arep.med.harvard.edu/PGP/Anon.htm
2. CODIS: Combined DNA index system (2008), http://www.fbi.gov/hq/lab/html/codis1.htm
3. deCODE genetics (2008), http://www.decodegenetics.com/
4. The genomic privacy project (2008), http://privacy.cs.cmu.edu/dataprivacy/projects/genetic/
5. HapMap: International HapMap project (2008), http://www.hapmap.org/.
6. Aho, A., Hirschberg, D., Ullman, J.: Bounds on the complexity of the longest common subsequence problem. J. of the ACM 23(1), 1–12 (1976)
7. Aiello, B., Ishai, Y., Reingold, O.: Priced oblivious transfer: How to sell digital goods. In: Pfitzmann, B. (ed.) EUROCRYPT 2001. LNCS, vol. 2045, pp. 119–135. Springer, Heidelberg (2001)
8. Altman, R., Klein, T.: Challenges for biomedical informatics and pharmacogenenomics. Annu. Rev. of Pharmacology and Toxicology 42, 113–133 (2002)
9. Arlazarov, V.L., Dinic, E.A., Kronod, M.A., Faradzev, I.A.: On economic consruction of the transitive closure of a directed graph. Doklady Akademii Nauk SSSR 194, 487–488 (1970)
10. Atallah, M.J., Kerschbaum, F., Du. Secure, W.: private sequence comparisons. In: Proc. of WPES, pp. 39–44 (2003)
11. Brickell, J., Shmatikov, V.: Privacy-preserving graph algorithms in the semi-honest model. In: Roy, B. (ed.) ASIACRYPT 2005. LNCS, vol. 3788, pp. 236–252. Springer, Heidelberg (2005)
12. Canetti, R.: Security and composition of multiparty cryptographic protocols. J. of Cryptology 13, 143–202 (2000)
13. Chor, B., Kushilevitz, E., Goldreich, O., Sudan, M.: Private information retrieval. J. of the ACM 45(6), 965–981 (1998)
14. Cormen, T.H., Leiserson, C.E., Rivest, R.L.: Introduction to Algorithms. MIT Press, Cambridge (2000)
15. Damgård, I., Jurik, M.: A generalisation, a simplification and some applications of Paillier's probabilistic public-key system. In: Kim, K.-c. (ed.) PKC 2001. LNCS, vol. 1992, pp. 119–136. Springer, Heidelberg (2001)
16. Damgård, I., Jurik, M.: A length-flexible threshold cryptosystem with applications. In: Information Security and Privacy, pp. 350–364 (2003)
17. Franklin, M.K., Gondree, M., Mohassel, P.: Multi-party indirect indexing and applications. In: Kurosawa, K. (ed.) ASIACRYPT 2007. LNCS, vol. 4833, pp. 283–297. Springer, Heidelberg (2007)
18. Franklin, M., Gondree, M., Mohassel, P.: Communication-efficient private protocols for longest common subsequence. Cryptology ePrint Archive, Report 2009/019 (2009), http://eprint.iacr.org/
19. Fraser, C.: Subsequences and supersequences of strings. PhD thesis, University of Glasgow (1995)
20. Freedman, M.J., Nissim, K., Pinkas, B.: Efficient private matching and set intersection. In: Cachin, C., Camenisch, J.L. (eds.) EUROCRYPT 2004. LNCS, vol. 3027, pp. 1–19. Springer, Heidelberg (2004)
21. Gentry, C., Ramzan, Z.: Single-database private information retrieval with constant communication rate. In: Caires, L., Italiano, G.F., Monteiro, L., Palamidessi, C., Yung, M. (eds.) ICALP 2005. LNCS, vol. 3580, pp. 803–815. Springer, Heidelberg (2005)

22. Goldreich, O.: Foundations of Cryptography. Cambridge University Press, Cambridge (2001)
23. Gusfield, D.: Algorithms on Strings, Trees, and Sequences. Cambridge University Press, Cambridge (1997)
24. Ishai, Y., Malkin, T.G., Strauss, M.J., Wright, R.N.: Private multiparty sampling and approximation of vector combinations. In: Arge, L., Cachin, C., Jurdziński, T., Tarlecki, A. (eds.) ICALP 2007. LNCS, vol. 4596, pp. 243–254. Springer, Heidelberg (2007)
25. Jha, S., Kruger, L., Shmatikov, V.: Towards practical privacy for genomic computation. In: IEEE Symposium on Security and Privacy (2008)
26. Kiltz, E., Mohassel, P., Weinreb, E., Franklin, M.K.: Secure linear algebra using linearly recurrent sequences. In: Vadhan, S.P. (ed.) TCC 2007. LNCS, vol. 4392, pp. 291–310. Springer, Heidelberg (2007)
27. Malin, B., Sweeney, L.: Re-identification of dna through an automated linkage process. In: AMIA Annual Symposium, pp. 423–427 (2001)
28. Masek, W.J., Paterson, M.S.: A faster algorithm for computing string edit distances. J. of Computer and System Sciences 20, 18–31 (1980)
29. Melchor, C.A., Deswarte, Y.: Single-database private information retrieval schemes : Overview, performance study, and usage with statistical databases. In: Domingo-Ferrer, J., Franconi, L. (eds.) PSD 2006. LNCS, vol. 4302, pp. 257–265. Springer, Heidelberg (2006)
30. Naor, M., Nissim, K.: Communication preserving protocols for secure function evaluation. In: Proc. of STOC, pp. 590–599 (2001)
31. Naor, M., Pinkas, B.: Oblivious transfer and polynomial evaluation. In: Proc. of STOC, pp. 245–254 (1999)
32. Department of Health and Human Services. 45 CFR (Code of Federal Regulations), parts 160–164. Standards for privacy of individually identifiable health information, final rule. Federal Register 67(157), 53182–53273 (August 12, 2002)
33. The GPL Violations project (2008), http://gpl-violations.org/
34. Szajda, D., Pohl, M., Owen, J., Lawson, B.G.: Toward a practical data privacy scheme for a distributed implementation of the Smith-Waterman genome sequence comparison algorithm. In: Proc. of NDSS, pp. 253–265 (2006)
35. Vaszar, L.T., Cho, M.K., Raffin, T.A.: Privacy issues in personalized medicine. Pharmacogenomics 4(2), 107–112 (2003)
36. Yao, A.C.: How to generate and exchange secrets. In: Proc. of FOCS, pp. 162–167 (1986)

Key-Private Proxy Re-encryption

Giuseppe Ateniese, Karyn Benson, and Susan Hohenberger

Johns Hopkins University
ateniese@cs.jhu.edu, kbenson5@jhu.edu, susan@cs.jhu.edu

Abstract. Proxy re-encryption (PRE) allows a proxy to convert a ci-phertext encrypted under one key into an encryption of the same message under another key. The main idea is to place as little trust and reveal as little information to the proxy as necessary to allow it to perform its translations. At the very least, the proxy should not be able to learn the keys of the participants or the content of the messages it re-encrypts. However, in all prior PRE schemes, it is easy for the proxy to determine between which participants a re-encryption key can transform ciphertexts. This can be a problem in practice. For example, in a secure distributed file system, content owners may want to use the proxy to help re-encrypt sensitive information *without* revealing to the proxy the *identity* of the recipients.

In this work, we propose key-private (or anonymous) re-encryption keys as an additional useful property of PRE schemes. We formulate a definition of what it means for a PRE scheme to be secure and key-private. Surprisingly, we show that this property is not captured by prior definitions or achieved by prior schemes, including even the secure ob-fuscation of PRE by Hohenberger et al. (TCC 2007). Finally, we propose the first key-private PRE construction and prove its CPA-security under a simple extension of Decisional Bilinear Diffie Hellman assumption and its key-privacy under the Decision Linear assumption in the standard model.

1 Introduction

In many applications, data protected under one public key pk_1 needs to be distributed to a user with a different public key pk_2. It is not always practical for the owner of sk_1 to be online to decrypt these ciphertexts and then encrypt these contents anew under pk_2. For example, Alice might wish to have her mail server forward her encrypted email to Bob while she is on vacation. However, how can Alice do this without revealing her sk_1 to either her mail server or Bob?

As a solution to this key management problem, the concept of proxy re-encryption (PRE) was introduced [5]. Proxy re-encryption is a cryptosystem with the special property that a proxy, given special information, can efficiently convert a ciphertext for Alice into a ciphertext of the same message for Bob. The proxy should not, however, learn either party's secret key or the contents of the messages it re-encrypts. The main idea is to place as little trust in the proxy as possible. When PRE is used for distributed file systems [2], this absence of trust

M. Fischlin (Ed.): CT-RSA 2009, LNCS 5473, pp. 279–294, 2009.

directly reduces the desirability for an adversary to compromise the distribution server, without compromising functionality.

In addition to hiding the contents of files from the proxy, it is also useful in practice to suppress as much meta-data as possible. For example, we might want the proxy file server to re-encrypt sensitive files for certain recipients *without* revealing to the proxy the recipient's identity. For example, the server might be told to re-encrypt all category one files with key one and category two files with keys two and three, without the proxy being able to deduce the public keys behind these values. This way, if the proxy is compromised, the adversary will not be able to extract a list of "who was speaking privately with whom". This is highly desirable for many encrypted communication scenarios.

This level of privacy for standard encryption schemes was formalized as *key-private* (or anonymous) encryption in 2001 by Bellare, Boldyreva, Desai and Pointcheval (BBDP) [4]. Intuitively, they studied encryption schemes where it is impossible to derive the recipient of a message from the ciphertext and the set of public keys. Consequently, the ciphertext is anonymous; that is, it cannot be linked to a particular public key and its owner. Fortunately, most public key encryption schemes already satisfy this property, such as Elgamal, Cramer-Shoup, and RSA-OAEP.

In this work, we introduce the strictly stronger notion of *key-private* (or anonymous) PRE. Intuitively, it should be impossible for the proxy and a set of colluding users to derive either the sender or receiver's identities from a re-encryption *key* even when given the public keys and flexible interaction ability within the system. As we formalize in Section 2.1, achieving key-private PRE is only possible when the underlying encryption scheme is key-private.

Unfortunately, this condition is far from sufficient. Finding a key-private PRE was a surprisingly difficult task. Whereas most standard encryption schemes are already key-private under the BBDP definition, *none* of the half-dozen existing PRE schemes are key-private under our natural definition in Section 2. This includes even the recent PRE construction of Hohenberger et al. [10], which was proven secure under a very strong *obfuscation* definition. In the next section, we discuss the problems with each existing scheme and the necessary conditions for realizing key-private PRE.

The main contribution of this work, in addition to our formal definition in Section 2, is the first realization of a key-private PRE scheme. Our construction is efficient, reasonably simple, and secure under basic assumptions about bilinear groups in the standard model. Formally, it is a unidirectional, single-hop, CPA-secure PRE with key-privacy. Thus, we show, for the first time, that this natural extension of anonymous encryption is practical and available for many existing PRE applications, as discussed in Section 1.2.

1.1   The Notion of Key-Private PRE and Prior Constructions

In this section, we examine the half-dozen existing proxy re-encryption schemes and discuss why they do not satisfy the notion of key-privacy. Let us first sketch the privacy notion wanted. Intuitively, we want to capture the strong guarantee that even an active proxy colluding with a set of malicious users in the system cannot learn from the re-encryption key the identity of the involved participants nor the contents of their encrypted messages.

Informally, the key-privacy game works as follows. First, the adversary is given the public keys of all honest users and the keypairs of all corrupt users. Next, the adversary is allowed to query two oracles an arbitrary number of times. The adversary may either request: (1) to have a chosen ciphertext under any user i re-encrypted to any user j or (2) to obtain the re-encryption key that translates ciphertexts from any user i to any user j. These oracles will operate regardless of the corruption status of i or j. Finally, the adversary must output a challenge pair of honest users (i^*, j^*), with the restriction that the adversary cannot have asked for this key before. The challenger will then return *either* the re-encryption key from i^* to j^* or a random key in the key space. The adversary wins if he can distinguish these cases with non-negligible probability.

Before discussing the problems with specific PRE constructions, let's get a better sense of what *cannot* possibly work. In Section 2.1, we point out that no *deterministic* re-encryption algorithm can satisfy the key-privacy definition. To see this, consider the generic attack where an adversary asks for a re-encryption of ciphertext C under user i to user j to obtain output C'. The adversary can then challenge on (i, j) and apply the returned re-encryption key to C. Since the re-encryption algorithm is deterministic, this should result in output C' if this is a proper key from i to j and is unlikely to do so for a random key. Unfortunately, the first four (out of six) prior PRE schemes have deterministic re-encryption algorithms, and thus cannot be key-private.

Similarly, in Section 2.1, we show that for a PRE scheme to be key-private (that is, one cannot distinguish the participants from seeing the *key*), the underlying encryption scheme must also be key-private in the sense of Bellare, Boldyreva, Desai and Pointcheval [4] (that is, one cannot distinguish the recipient from seeing the *ciphertext*). Some of the schemes also fail to have this property; mainly because they are in a bilinear setting, where the map can be used for this test.

Let us now discuss some specifics of prior schemes.

BBS PRE. Proxy re-encryption was first proposed by Blaze, Bleumer, and Strauss (BBS) [5] in Eurocrypt 1998. Their scheme, based on Elgamal, works in a group G of prime order p. Anyone can send a message $m \in G$ to user A with public key g^a (with $g \in G$) by computing $(mg^k, (g^a)^k)$, for a random $k \in \mathbb{Z}_p$. A can delegate to B (with public key g^b) her decryption rights by giving the proxy the value $b/a \bmod p$. All ciphertexts for A can be converted to ciphertexts for B by computing $(g^{ak})^{b/a} = g^{bk}$ and then releasing the ciphertext (mg^k, g^{bk}). Unfortunately, this scheme is trivially not key-private, because its re-encryption algorithm is deterministic. But there is an even easier attack: the adversary challenges on (A, B) to obtain challenge key r, this key is correct iff $r = b/a$. Using the public keys (g^a, g^b), the adversary can test this as $(g^a)^r = g^b$.

AFGH PRE. Ateniese, Fu, Green and Hohenberger [2] proposed new PRE schemes that employ bilinear pairings. Their protocols are unidirectional (a re-encryption key from A to B does not imply a key from B to A), an improvement over BBS where the keys are bidirectional. Their schemes require a bilinear map $\mathbf{e} : \mathbb{G} \times \mathbb{G} \to \mathbb{G}_T$, where $g \in \mathbb{G}$ and $Z = \mathbf{e}(g, g) \in \mathbb{G}_T$. In their first scheme, public key for A is g^a and similarly B's public key is g^b. The re-encryption key $rk_{A \to B}$ is $g^{b/a}$. However, this scheme is not key-private, because the adversary can challenge on (A, B) to obtain key r and then test if $r = g^{b/a}$ as $\mathbf{e}(g^a, r) = \mathbf{e}(g^b, g)$. A similar attack also works for their second scheme.[1] But since both schemes are deterministic, the generic attack also applies here.

CH PRE. Canetti and Hohenberger [8] proposed the first CCA-secure bidirectional PRE scheme in the standard model. However, even CCA-security does not ensure key-privacy, because the public keys (e.g., g^a, g^b) and re-encryption keys (e.g., b/a) are the same as in the BBS PRE, so the proxy can attack key-privacy here using the same algorithm from BBS. Part of the re-encryption algorithm of this scheme is also deterministic, and therefore, the generic attack again applies.

LV PRE. Libert and Vergnaud [12] proposed the first CCA-secure unidirectional PRE scheme in the standard model. To achieve CCA-security, they employ a quite interesting technique whereby the encryption of the scheme in [2] is randomized by the proxy via a blinding factor that effectively hides the re-encryption key within the re-encryption (which is also followed by a proof of consistency). Interestingly, their scheme is not key-private even though the re-encryption algorithm is probabilistic. Indeed, A and B have respectively public keys g^a and g^b, and the proxy key is $rk_{A \to B} = g^{b/a}$, just as in AFGH. Thus, as in AFGH [2], the adversary can challenge on (A, B) to obtain key r and then test if $r = g^{b/a}$ as $\mathbf{e}(g^a, r) = \mathbf{e}(g^b, g)$.

HRSV PRE. Recently, Hohenberger, Rothblum, shelat, and Vaikuntanathan [10] presented a CPA-secure unidirectional PRE in the standard model, with probabilistic algorithms for performing encryption and generating re-encryption keys. Moreover, HRSV satisfied a very strong security notion, treating the re-encryption key together with the re-encryption algorithm as an *obfuscated* re-encryption program. That is, a program whose code is scrambled in such a way that: (1) it still produces the correct outputs, and yet (2) it is not possible to "reverse engineer" the program to learn its secrets (i.e., anything that cannot be learned from black-box access to the program.) Interestingly, even their strong obfuscation definition does not imply key privacy and their construction does not satisfy our definition. To see this, recall that their construction is set in a bilinear group, where Alice's public key is of the form (g, g^{a_1}, g^{a_2}) and Bob's public key is of the form (h, h^{b_1}, h^{b_2}) for random $g, h \in \mathbb{G}$ and random exponents a_1, a_2, b_1, b_2. Given these public keys, the adversary can ask to see the re-encryption key

---

[1] To see why the second AFGH scheme is not key-private, consider the following attack. The adversary can ask for the re-encryption key from (C, A) to obtain $r_1 = g^{a_2 c_1}$. The adversary can next challenge on (C, B) to obtain r_2. Then the adversary can test if $r_2 = g^{b_2 c_1}$, making it a valid re-encryption key from C to B, via $\mathbf{e}(r_1, g^{b_2}) = \mathbf{e}(r_2, g^{a_2})$ with public key values g^{a_2} and g^{b_2}. This test determines if two keys have the same delegator, which is not possible under our key-privacy definition.

for (A, B) which will be $(y, y^{b_1/a_1}, y^{b_2/a_2})$, where $y \in \mathbb{G}$ is chosen randomly. The adversary can then challenge on (B, A) to obtain a key (r, r_1, r_2), which if correct, is of the form $(r, r^{a_1/b_1}, r^{a_2/b_2})$ for a random $r \in \mathbb{G}$. The adversary can then test for correctness as $\mathbf{e}(y, r) = \mathbf{e}(y^{b_1/a_1}, r_1) = \mathbf{e}(y^{b_2/a_2}, r_2)$. Thus, even this obfuscation is not key-private. Indeed, our notion seems difficult to satisfy because, unlike the obfuscation definition, we will allow the adversary broad query powers and the ability to collude with system users, whereas the current definition of obfuscation considers the release of only one re-encryption program.

1.2 Applications

PRE has been proposed for use in email-forwarding [5], secure file systems [2], DRM [14], and secure mailing lists [11]. All these applications can benefit from the key-privacy property in some way. In email-forwarding, Alice may not want the mail server to know to whom she is delegating her decryption rights. This is similar in the real world to a P.O. Box address where mail can be sent to a physical location but neither the sender nor the carrier may know who the actual recipient is. Alice can hide the fact that Bob is a delegatee by instructing the server to convert her encrypted emails via a key-private PRE scheme and to forward the results to an anonymous (or group) email address (i.e., an address reachable by Bob but that does not contain any identifiable information on Bob, like a P.O. Box address indeed).

In a distributed file system, PRE schemes can be used as an access control mechanism to specify who can access and read encrypted files [2]. Alice may want Bob to read some of her encrypted files, thus she instructs the file system to convert those files using a proxy re-encryption key from Alice to Bob. In a distributed file system, anyone can access those files but only Bob can read them. If the PRE scheme employed is key-private, nobody can even tell who can access and read any file in the system.

In [14], Taban, Cárdenas, and Gligor describe a secure and interoperable digital rights management (DRM) system based on proxy re-encryption and proxy re-signatures [3]. They specify, implement, and analyze a framework within which different DRM systems can interoperate. Proxy re-encryption is used by a Domain Interoperability Manager (DIM) that translates DRM packaged digital content between devices with distinct DRM systems. The DIM is a semi-trusted entity that is susceptible to compromise, thus encryption is used to ensure privacy of the content and licenses associated with each DRM system. A key-private PRE scheme would also hide the associations between the various devices and their respective DRM systems in case of compromise.

In [11], Khurana, Heo, and Pant propose to use proxy re-encryption for SELS (Secure Email List Services), a system that provides private email discussion lists via encryption. A list is composed of several members that exchange messages internally or with other members outside the list. To send a private message to a list (and to its members), it is enough to encrypt the message under a public-key associated with the list. A List Server (LS) uses a PRE scheme to translate that encryption into encryptions under the public keys of each member of the list, respectively. If the LS server is ever compromised, the secret keys of the list and its members would remain protected as well as the content of any messages exchanged within the list. However, the identities of the members in a list would be exposed by just looking at the re-encryption keys. This may not be desirable in many contexts and thus a key-private PRE scheme would be preferable whenever the privacy of list members must be guaranteed.

In [13], Suriadi, Foo and Smith use proxy re-encryption to develop a credential system with conditional privacy. Their system has many proxies providing keys to parities who wish to remain *anonymous*. They use multiple-hops in their key distribution to help maintain anonymity; it would be possible to instead use a key-private PRE scheme.

2   Key-Private PRE Definitions

We build upon the re-encryption definitions of [2] and [8] to introduce the concept of key privacy. We will only consider a definition for unidirectional, single-hop PREs. By single-hop, we mean that only original ciphertexts (and not re-encrypted ciphertexts) can be re-encrypted.

Definition 1. *(Unidirectional, Single-Hop PRE) A unidirectional, single-hop, proxy re-encryption scheme is a tuple of algorithms $\Pi = ($Setup, KeyGen, ReKeyGen, Enc, ReEnc, Dec$)$ for message space M:*

- Setup$(1^k) \rightarrow PP$. *On input security parameter 1^k, the setup algorithm outputs the public parameters PP.*
- KeyGen$(PP) \rightarrow (pk, sk)$. *On input public parameters, the key generation algorithm* KeyGen *outputs a public key pk and a secret key sk.*
- ReKeyGen$(PP, sk_i, pk_j) \rightarrow rk_{i \rightarrow j}$. *Given a secret key sk_i and a public key pk_j, where $i \neq j$, this algorithm outputs a unidirectional re-encryption key $rk_{i \rightarrow j}$. The restriction that $i \neq j$ is provided as re-encrypting a message to the original recipient is impractical.*
- Enc$(PP, pk_i, m) \rightarrow C_i$. *On input a public key pk_i and a message $m \in M$, the encryption algorithm outputs an original ciphertext C_i.*
- ReEnc$(PP, rk_{i \rightarrow j}, C_i) \rightarrow C_j$. *Given a re-encryption key from i to j and an original ciphertext for i, the re-encryption algorithm outputs a ciphertext for j or the error symbol $\perp$.*
- Dec$(PP, sk_i, C_i) \rightarrow m$. *Given a secret key for user i and a ciphertext for i, the decryption algorithm* Dec *outputs a message $m \in M$ or error symbol $\perp$.*

A PRE scheme Π is correct with respect to domain M if:

- For all $(pk, sk) \in$ KeyGen(PP) and all $m \in M$, it holds that

$$\mathsf{Dec}(PP, sk, \mathsf{Enc}(PP, pk, m)) = m.$$

- For all pairs $(pk_i, sk_i), (pk_j, sk_j) \in$ KeyGen(PP) and re-encryption keys $rk_{i \rightarrow j} \in$ ReKeyGen(PP, sk_i, pk_j), and $m \in M$, it holds that

$$\mathsf{Dec}(PP, sk_j, \mathsf{ReEnc}(PP, rk_{i \rightarrow j}, \mathsf{Enc}(PP, pk_i, m))) = m.$$

Definition 2. (Unidirectional, Single-Hop PRE CPA-Security Game)
Let 1^k be the security parameter. Let $\mathcal{A}$ be an oracle TM, representing the adversary. The PRE-CPA game consists of an execution of $\mathcal{A}$ with the following oracles. The game consists of three phases, which are executed in order. Within each phase, each oracle can be executed in any order, $\text{poly}(k)$ times, unless otherwise specified.
Phase 1:

- **Public Parameter Generation:** *The public parameters are generated and given to $\mathcal{A}$. This oracle is executed first and only once.*
- **Uncorrupted Key Generation:** *Obtain a new key pair (pk, sk) by running* KeyGen(PP). *$\mathcal{A}$ is given pk. Let Γ_H be the set of honest user indices.*
- **Corrupted Key Generation:** *Obtain a new key pair (pk, sk) by running* KeyGen(PP). *$\mathcal{A}$ is given (pk, sk). Let Γ_C be the set of corrupt user indices.*

Phase 2:

- **Re-encryption Key Generation $\mathcal{O}_{rkey}$:** *On input (i, j) by the adversary, where the key pairs for i and j were generated in Phase 1, return the key $rk_{i \to j} = $ ReKeyGen(PP, sk_i, pk_j). All requests where $i = j$ or where $i \in \Gamma_H$ and $j \in \Gamma_C$ are ignored (an output of $\perp$).*
- **Re-encryption $\mathcal{O}_{renc}$:** *On input (i, j, C_i), where the keys for i and j were generated in Phase 1, return $C_j = $ ReEnc$(PP, $ReKeyGen$(PP, sk_i, pk_j), C_i)$. For requests where $i = j$ or where $i \in \Gamma_H$ and $j \in \Gamma_C$, output $\perp$.*
- **Challenge Oracle:** *On input (i, m_0, m_1), the oracle chooses a random $b \leftarrow \{0, 1\}$ and returns the challenge ciphertext $C_i = $ Enc(PP, pk_i, m_b). This oracle can only be queried once, and it is required that $i \in \Gamma_H$.*

Phase 3:

- **Decision:** *Eventually, $\mathcal{A}$ outputs decision $b' \in \{0, 1\}$. $\mathcal{A}$ wins the game if and only if $b = b'$.*

Definition 3. (Unidirectional, Single-Hop PRE CPA Security) *Given security parameter 1^k, a PRE scheme is Unidirectional PRE CPA secure for domain M of messages if is it correct for M and $\forall$ p.p.t. adversaries $\mathcal{A}$, $\exists$ a negligible function ε such that $\mathcal{A}$ wins the unidirectional PRE-CPA game with probability at most $\frac{1}{2} + \varepsilon(k)$.*

Remark 1 (Corruptions). As in many prior re-encryption papers [2,8,12], we work in a static corruption model, where the adversary must chose to either corrupt a party or not at the time the party's keypair is generated. Indeed, the problem of handling dynamic corruptions for any encryption scheme is a classically difficult problem. This rules out allowing the adversary to query $\mathcal{O}_{rkey}$ from an honest to a corrupt user, since this action would corrupt the honest user. Moreover, we also disallow adversarial queries to $\mathcal{O}_{renc}$ from honest to corrupt users, as in [2], since such access could simulate a decryption oracle which we do not consider in CPA-secure constructions.

Next, we turn to what it means for a re-encryption key to be key-private. Informally, we want a proxy to be unable to identify either the delegator i or the delegatee j when given the re-encryption key $rk_{i \to j}$ and flexible interaction with the system (e.g., other re-encryption keys, access to re-encryption oracles, etc.) To capture this idea, we say that the proxy is allowed to choose (i, j) and then cannot distinguish the valid key $rk_{i \to j}$ from a random value in the key space.

Our definition for key privacy is more challenging to realize than CPA-security, because most of the restrictions on how the adversary can call $\mathcal{O}_{rkey}$ and $\mathcal{O}_{renc}$ are now removed. Indeed, we allow a form of dynamic corruption here. For example, it is now legal for the adversary to ask for re-encryption keys and re-encryptions from honest to corrupt parties, and then later to challenge on these honest parties. In other words, key-privacy is maintained even when honest parties unwisely delegate decryption capabilities to corrupt parties.

Definition 4. *(Unidirectional, Single-Hop PRE Key-Privacy Game)* *Let k be the security parameter. Let $\mathcal{A}$ be an oracle TM, representing the adversary. The PRE Key-Privacy Game consists of an execution of $\mathcal{A}$ with the same oracles as before unless specified below. There are three phases.* *Phase 1:*

- *The adversary is given the public parameters, and then may request uncorrupted or corrupted key pairs to be created, as before.*

Phase 2:

- **Re-encryption Key Generation** $\mathcal{O}_{rkey}$: *On input (i, j) by the adversary, where the key pairs for i and j were generated in Phase 1, return the key in table T corresponding to (i, j). If there is no such entry in the table, compute it as $\mathsf{ReKeyGen}(PP, sk_i, pk_j)$, add the key to the table T in cell (i, j), and output this key. The oracle will only produce a single re-encryption key for (i, j). If $i = j$ then the error symbol $\bot$ is returned. Note that there is no longer the restriction that $i \notin \Gamma_H$ or $j \notin \Gamma_C$.*
- **Re-encryption** $\mathcal{O}_{renc}$: *On input (i, j, C_i) where the key pairs for i and j were generated by KeyGen, obtain the re-encryption key s corresponding to (i, j) in table T. If no such key exists, create it as $s \leftarrow \mathsf{ReKeyGen}(PP, sk_i, pk_j)$ and save it in table T. Return either $C_j = \mathsf{ReEnc}(PP, s, C_i)$ or $\bot$ if $i = j$.*
- **Challenge Oracle:** *This oracle can only be challenged once. On input (i, j), the oracle sets s to be the key corresponding to (i, j) in table T. If no such key exists, it creates it as $s \leftarrow \mathsf{ReKeyGen}(PP, sk_i, pk_j)$. The oracle then chooses a bit $b \leftarrow \{0, 1\}$ and then returns the value s if $b = 1$ and a random key in the key space otherwise. The constraints are that $\mathcal{O}_{rkey}$ must not have been queried for (i, j) before, $i \neq j$ and $i, j \in \Gamma_H$.*

Phase 3:

- **Decision:** *Eventually, $\mathcal{A}$ outputs decision $b' \in \{0, 1\}$. We say that $\mathcal{A}$ wins the game if and only if $b = b'$.*

Definition 5. *(Unidirectional, Single-Hop PRE Key Privacy) For security parameter* 1^k, *a PRE scheme is key-private if* $\forall$ *p.p.t. adversaries* $\mathcal{A}$, $\exists$ *a negligible function* ε *such that* $\mathcal{A}$ *wins the unidirectional PRE Key-Privacy Game with probability at most* $\frac{1}{2} + \varepsilon(k)$.

2.1 Impossibility Results for Key-Private Re-encryption

Before seeing the construction, we lay out some necessary, but not sufficient, conditions for satisfying the above definition in two simple lemmas.

Lemma 1. *Any bidirectional or unidirectional re-encryption scheme* (Setup, KeyGen, ReKeyGen, Enc, ReEnc, Dec), *where the* ReEnc *algorithm is deterministic cannot satisfy key-privacy (Definition 5).*

Proof. Suppose ReEnc is deterministic. An adversary $\mathcal{A}$ wins the key-privacy game as follows:

1. Ask for a set of uncorrupted parties to be generated; obtain the public parameters and keys.
2. Choose a random m in the message space and compute $c = \mathsf{Enc}(PP, pk_1, m)$.
3. Query the re-encryption oracle $\mathcal{O}_{renc}(1, 2, c)$ to obtain the response c'.
4. Challenge on identities $(1, 2)$ and obtain the challenge key s. (Recall that, by definition, the key-privacy challenger will return the same key here that it used in the previous step.)
5. Using s, run the deterministic algorithm $\mathsf{ReEnc}(PP, s, c) \rightarrow c''$.
6. If $c' = c''$, output 1, else output 0.

It is easy to see that $\mathcal{A}$ succeeds with overwhelming probability.

This lemma immediately rules out almost all prior PRE constructions [5,2,8] as candidates for key privacy. Nor is it obvious how to transform these constructions into key-private schemes. The schemes by Libert and Vergnaud [12] and Hohenberger et al. [10] employ probabilistic re-encryption algorithms, but they still admit key-privacy attacks. Thus, a probabilistic re-encryption algorithm is a necessary, but not sufficient condition.

Lemma 2. *Any bidirectional or unidirectional re-encryption scheme* (Setup, KeyGen, ReKeyGen, Enc, ReEnc, Dec) *satisfying the key-privacy (Definition 5) implies that* (Setup, KeyGen, Enc, Dec) *admits a key-private encryption scheme according to the standard definition [4].*

In other words, it is not possible for a PRE scheme to be key-private, unless the underlying encryption resulting from a re-encryption is key-private. The point here is that any key-private PRE must admit some key-private encryption scheme. Bellare, Boldyreva, Desai and Pointcheval [4] introduced key-private encryption, where an adversary cannot distinguish the intended recipient from the ciphertext. More formally, the adversary is given two public keys pk_0, pk_1, chooses a message m, is given an encryption of m under one of the two keys

$b \in \{0, 1\}$ chosen at random, and finally issues a guess $b' \in \{0, 1\}$. The security notion requires that all efficient adversaries cannot achieve $b = b'$ with probability non-negligibly better than random guessing.

For a PRE scheme to be key-private, the proxy cannot distinguish the intended recipient from the ciphertext *even when given access to re-encryption keys and re-encryption oracles*. To see this, consider that otherwise an adversary $\mathcal{A}$ can win the key-privacy game as follows:

1. Ask for n uncorrupted parties to be generated; obtain the public parameters and keys.
2. Choose a random m in the message space and compute $c = \mathsf{Enc}(PP, pk_1, m)$.
3. Challenge on identities $(1, 2)$ and obtain the challenge key s.
4. Using s, run the possibly probabilistic algorithm $\mathsf{ReEnc}(PP, s, c) \rightarrow c'$.
5. If c' is a ciphertext under public key pk_2, output 1, else output 0.

The BBS PRE [5] uses Elgamal (in a non-bilinear setting) as its encryption base and thus satisfies anonymous encryption via Bellare et al. [4], although it is not a key-private PRE. Thus, this condition is also necessary, but not sufficient.

3   A Key-Private PRE Scheme

3.1   Algebraic Setting

Bilinear Groups. We write $\mathbb{G} = \langle g \rangle$ to denote that g generates the group $\mathbb{G}$. Let BSetup be an algorithm that, on input the security parameter 1^k, outputs the parameters for a bilinear map as $(q, g, \mathbb{G}, \mathbb{G}_T, \mathbf{e})$, where $\mathbb{G}, \mathbb{G}_T$ are of prime order $q \in \Theta(2^k)$ and $\langle g \rangle = \mathbb{G}$. The efficient mapping $\mathbf{e} : \mathbb{G} \times \mathbb{G} \rightarrow \mathbb{G}_T$ is both: (*Bilinear*) for all $g \in \mathbb{G}$ and $a, b \in \mathbb{Z}_q$, $\mathbf{e}(g^a, g^b) = \mathbf{e}(g, g)^{ab}$; and (*Non-degenerate*) if g generates $\mathbb{G}$, then $\mathbf{e}(g, g) \neq 1$. We consider the following assumptions.

Decisional Bilinear Diffie-Hellman (DBDH) [7]: Let $\mathsf{BSetup}(1^k) \rightarrow (q, g, \mathbb{G}, \mathbb{G}_T, \mathbf{e})$, where $\langle g \rangle = \mathbb{G}$. For all p.p.t. adversaries $\mathcal{A}$, there exists a negligible function ε such that the following probability is less than or equal to $1/2 + \varepsilon(k)$:

$$\Pr[a, b, c, d \leftarrow \mathbb{Z}_q; \ x_1 \leftarrow \mathbf{e}(g, g)^{abc}; \ x_0 \leftarrow \mathbf{e}(g, g)^d; \ z \leftarrow \{0, 1\};$$
$$z' \leftarrow \mathcal{A}(g, g^a, g^b, g^c, x_z) : z = z'].$$

Extended DBDH (EDBDH)   [2]: Let $\mathsf{BSetup}(1^k) \rightarrow (q, g, \mathbb{G}, \mathbb{G}_T, \mathbf{e})$, where $\langle g \rangle = \mathbb{G}$. For all p.p.t. adversaries $\mathcal{A}$, there exists a negligible function ε such that the following probability is less than or equal to $1/2 + \varepsilon(k)$:

$$\Pr[a, b, c, d \leftarrow \mathbb{Z}_q; \ x_1 \leftarrow \mathbf{e}(g, g)^{abc}; \ x_0 \leftarrow \mathbf{e}(g, g)^d; \ z \leftarrow \{0, 1\};$$
$$z' \leftarrow \mathcal{A}(g, g^a, g^b, g^c, \mathbf{e}(g, g)^{bc^2}, x_z) : z = z'].$$

Decision Linear [6]: Let $\mathsf{BSetup}(1^k) \rightarrow (q, g, \mathbb{G}, \mathbb{G}_T, \mathbf{e})$, where $\langle g \rangle = \mathbb{G}$. Let h, f be random generators in $\mathbb{G}$. For all p.p.t. adversaries $\mathcal{A}$, there exists a negligible function ε such that the following probability is less than or equal to $1/2 + \varepsilon(k)$:

$$\Pr[x, y, r \leftarrow \mathbb{Z}_q; \ q_1 \leftarrow f^{x+y}; \ q_0 \leftarrow f^r; \ z \leftarrow \{0, 1\};$$
$$z' \leftarrow \mathcal{A}(g, h, f, g^x, h^y, q_z) : z = z'].$$

3.2 The Construction

Scheme $\Pi = (\mathsf{Setup}, \mathsf{KeyGen}, \mathsf{ReKeyGen}, \mathsf{Enc}, \mathsf{ReEnc}, \mathsf{Dec})$ is described as follows:

Setup (Setup): Run $\mathsf{BSetup}(1^k) \rightarrow (q, g, \mathbb{G}, \mathbb{G}_T, \mathbf{e})$, where $\langle g \rangle = \mathbb{G}$. Choose a random generator $h \in \mathbb{G}$. Compute $Z = \mathbf{e}(g, h)$, and set the public parameters $PP = (g, h, Z)$. In the following, we assume that all parties have PP.

Key Generation (KeyGen): Choose random values $a_1, a_2 \in \mathbb{Z}_q$ and set the public key as $pk = (Z^{a_1}, g^{a_2})$ with secret key $sk = (a_1, a_2)$.

Re-Encryption Key Generation (ReKeyGen): User A with secret key (a_1, a_2) can delegate to user B with public key (Z^{b_1}, g^{b_2}) by selecting random values $r, w \in \mathbb{Z}_q$ and then computing $rk_{A \rightarrow B}$ as:

$$((g^{b_2})^{a_1+r}, h^r, \mathbf{e}(g^{b_2}, h)^w, \mathbf{e}(g, h)^w) = (g^{b_2(a_1+r)}, h^r, Z^{b_2 w}, Z^w).$$

Encryption (Enc): To encrypt a message $m \in \mathbb{G}_T$ under public key $pk_A = (Z^{a_1}, g^{a_2})$, select random value $k \in \mathbb{Z}_q$ and compute the ciphertext as:

$$(g^k, h^k, m \cdot Z^{a_1 k}).$$

We note that, as in prior schemes [2], it is possible to use this same public key to produce a ciphertext that *cannot* be re-encrypted, and thus only opened by the holder of sk_A by selecting a random $k \in \mathbb{Z}_q$ and outputting the Elgamal ciphertext $(\mathbf{e}(g^{a_2}, h)^k, m \cdot Z^k) = (Z^{a_2 k}, m \cdot Z^k)$. We refer to this as a *first-level* ciphertext, and re-encryptable ones as *second-level* ciphertexts.

Re-Encryption (ReEnc): Given a re-encryption key $rk_{A \rightarrow B} = (R_1, R_2, R_3, R_4)$ $= (g^{b_2(a_1+r)}, h^r, Z^{b_2 w}, Z^w)$, it is possible to convert a second-level ciphertext $C_A = (\alpha, \beta, \gamma)$ for A into a first-level ciphertext for B as follows:

1. Verify that the ciphertext is well-formed, by checking that it uses consistent randomness in its first two parts as: $\mathbf{e}(\alpha, h) = \mathbf{e}(g, \beta)$. If this does not hold, output $\perp$ and abort.
2. Otherwise, there exists some $k \in \mathbb{Z}_q$ and $m \in \mathbb{G}_T$ such that $\alpha = g^k$, $\beta = h^k$ and $\gamma = m \cdot Z^{a_1 k}$, and thus, (α, β, γ) is a valid encryption of this m under $pk_A = (Z^{a_1}, g^{a_2})$.
3. Compute $t_1 = \mathbf{e}(R_1, \beta) = \mathbf{e}(g^{b_2(a_1+r)}, h^k) = Z^{b_2 k(a_1+r)}$.
4. Compute $t_2 = \gamma \cdot \mathbf{e}(\alpha, R_2) = m \cdot Z^{a_1 k} \cdot Z^{rk} = m \cdot Z^{k(a_1+r)}$.
5. Select a random $w' \in \mathbb{Z}_q$.
6. Re-randomize t_1 by setting $t_1' = t_1 \cdot R_3^{w'} = Z^{b_2(k(a_1+r)+ww')}$.
7. Re-randomize t_2 by setting $t_2' = t_2 \cdot R_4^{w'} = m \cdot Z^{k(a_1+r)+ww'}$.
8. Publish $C_B = (t_1', t_2') = (Z^{b_2 y}, m \cdot Z^y)$, where $y = k(a_1 + r) + ww'$.

Decryption (Dec): Given secret key (a_1, a_2), to decrypt a first-level ciphertext (α, β), compute $m = \beta/\alpha^{1/a_2}$; and to decrypt a second-level ciphertext (α, β, γ), output $\perp$ if $\mathbf{e}(\alpha, h) \neq \mathbf{e}(g, \beta)$, otherwise output $m = \gamma/\mathbf{e}(\alpha, h)^{a_1}$.

Fortunately, this scheme is practical and multi-purpose. Public keys can be used either for re-encryption purposes or for regular Elgamal encryptions. For completeness, we show in the full version [1] that the CPA-security of the first-level ciphertexts holds under DBDH.

3.3   Security Analysis

We first argue that the above scheme is secure and then that it is key-private.

Theorem 1 (CPA Security). *Under the EDBDH assumption in $\mathbb{G}$, scheme Π is a unidirectional, single-hop, CPA-secure PRE scheme for message domain $\mathbb{G}_T$ according to Definition 3.*

The main difficulty in the proof of Theorem 1 is ensuring that the reduction can properly answer *all* the re-encryption key and re-encryption queries asked by $\mathcal{A}$. It will be easier to work with the following assumption *implied* by EDBDH:

Definition 6. Modified Extended Decisional Bilinear Diffie-Hellman (mEDBDH): *Let* $\mathsf{BSetup}(1^k) \to (q, g, \mathbb{G}, \mathbb{G}_T, \mathbf{e})$, *where* $\langle g \rangle = \mathbb{G}$. *For all p.p.t. adversaries* $\mathcal{A}$, *there exists a negligible function* ε *such that the following probability is less than or equal to* $1/2 + \varepsilon(k)$:

$$\Pr[s, t, u, v \leftarrow \mathbb{Z}_q;\ x_1 \leftarrow \mathbf{e}(g, g)^{st/u};\ x_0 \leftarrow \mathbf{e}(g, g)^v;\ z \leftarrow \{0, 1\};$$
$$z' \leftarrow \mathcal{A}(g, g^s, g^t, g^u, \mathbf{e}(g, g)^{t/u}, x_z) : z = z'].$$

Lemma 3. *If the EDBDH assumption holds in $\mathbb{G}$, then so does the mEDBDH assumption.* (Proof of this Lemma appears in the full version [1].)

We now proceed with the proof of Theorem 1.

Proof. Suppose $\mathcal{A}$ breaks the CPA-security of our PRE construction with probability $1/2 + \mu$, then we create an adversary $\mathcal{B}$ who breaks the mEDBDH assumption with probability $1/2 + \mu/2$. Recall that mEDBDH asks when given $(g, g^s, g^t, g^u, \mathbf{e}(g, g)^{t/u}, Q)$ is $Q = \mathbf{e}(g, g)^{st/u}$. Given a mEDBDH instance Δ, $\mathcal{B}$ handles oracle queries from $\mathcal{A}$ as:

- **Public Parameter Generation** $\mathcal{B}$ sets up the global parameters for $\mathcal{A}$ by selecting a random $n \in \mathbb{Z}_q$ and setting $(g, h, Z) = (g, g^n, \mathbf{e}(g, g)^n)$.
- **Uncorrupted Key Generation** $\mathcal{B}$ chooses random $x, y \in \mathbb{Z}_q$ and outputs the public key $pk = ((\mathbf{e}(g, g)^{t/u})^{nx}, (g^u)^y)$, where the secret key is implicitly defined as $sk = (tx/u, uy)$.
- **Corrupted Key Generation** $\mathcal{B}$ choose random $x_i, y_i \in \mathbb{Z}_q$, and outputs $pk_i = (Z^{x_i}, g^{y_i})$ and $sk_i = (x_i, y_i)$.

- **Re-encryption Key Generation** On input (i, j) to $\mathcal{O}_{rkey}$, do:
  - If (1) i is uncorrupted and j is corrupted or (2) $i = j$, output $\perp$.
  - If i and j are corrupted, pick random $r, w \in \mathbb{Z}_q$ and output $(g^{y_j(x_i+r)}, h^r,$ $\mathbf{e}(g^{y_j}, h)^w, Z^w)$.
  - If i and j are both uncorrupted, select a random $r, w \in \mathbb{Z}_q$ and output $((g^t)^{y_j x_i} \cdot (g^u)^{y_j r}, h^r, \mathbf{e}((g^u)^{y_j}, h)^w, Z^w)$.
  - If i is corrupted and j is uncorrupted, select a random $r, w \in \mathbb{Z}_q$ and output the key $((g^u)^{y_j(x_i+r)}, h^r, \mathbf{e}((g^u)^{y_j}, h)^w, Z^w)$.
- **Re-encryption** On input $(i, j, C_i = (\alpha, \beta, \gamma))$ to $\mathcal{O}_{renc}$, do:
  - If (1) i is uncorrupted and j is corrupted, (2) $i = j$ or (3) $\mathbf{e}(\alpha, h) \neq \mathbf{e}(g, \beta)$, output $\perp$.
  - Otherwise, there exists some $k \in \mathbb{Z}_q$ and $m \in \mathbb{G}_T$ such that $\alpha = g^k$, $\beta = h^k$ and $\gamma = m \cdot Z^{tx_i k/u}$ if i is honest or $\gamma = m \cdot Z^{kx_i}$ if i is corrupted.
  - If i and j are both corrupted, recover $m = \gamma/\mathbf{e}(g, \beta)^{x_i}$, then select a random $r \in \mathbb{Z}_q$ and output $(Z^{y_j r}, m \cdot Z^r)$.
  - If i and j are both uncorrupted, select random $r, w \in \mathbb{Z}_q$ and output $(\mathbf{e}(g^{ty_j x_i} \cdot g^{uy_j r}, \beta) \cdot \mathbf{e}(g^u, h)^{y_j w}, \gamma \cdot \mathbf{e}(\alpha, h^r) \cdot Z^w) = (\mathbf{e}(g^{ty_j x_i} \cdot g^{uy_j r}, h^k) \cdot \mathbf{e}(g^u, h)^{y_j w}, m \cdot Z^{kx_i t/u} \cdot \mathbf{e}(g^k, h^r) \cdot Z^w)$.
  - If i is corrupted and j is uncorrupted, recover $m = \gamma/\mathbf{e}(g, \beta)^{x_i}$, then select a random $r \in \mathbb{Z}_q$ and output $(\mathbf{e}(g^u, h)^{y_j r}, m \cdot Z^r)$.
- **Challenge Oracle** Challenges are of the form (i, m_0, m_1) where i is the index of an uncorrupted user. $\mathcal{B}$ responds by choosing random $d \in \{0, 1\}$ and outputting the ciphertext: $(g^s, (g^s)^n, m_d \cdot Q^{nx_i})$.
- **Decision** $\mathcal{A}$ will submit a guess of $d' \in \{0, 1\}$. If $d = d'$ then $\mathcal{B}$ outputs 1 (is a mEDBDH instance) otherwise it outputs 0 (not a mEDBDH instance).

By construction, the public parameters and all uncorrupted keys, corrupted keys, re-encryption keys, and re-encryptions are correct and distributed properly. As to the challenge ciphertext, we have two cases. In the case that $Q = \mathbf{e}(g, g)^{st/u}$, then the challenge ciphertext is a proper encryption of m_d. $\mathcal{A}$ will output d' such that $d = d'$ with probability $\frac{1}{2} + \mu$. Consequently, $\mathcal{B}$ will determine that Δ was a mEDBDH instance and answer 1 with the same probability. When Q is random, independent of s, t and u then the challenge ciphertext reveals no information about m_d. $\mathcal{A}$ will guess that $d = d'$ with probability of exactly $\frac{1}{2}$, and $\mathcal{B}$ will correctly output 0 (not a mEDBDH) with the same probability. The probability that Δ is a valid mEDBDH instance is $\frac{1}{2}$, and $\mathcal{B}$ will output the correct answer with probability: $(\frac{1}{2})(\frac{1}{2} + \mu) + (\frac{1}{2})(\frac{1}{2}) = \frac{1}{2} + \frac{\mu}{2}$. We apply Lemma 3 to establish the result.

Theorem 2 (Key Privacy). *Under the Decision Linear assumption in $\mathbb{G}$, scheme Π is a unidirectional, single-hop, key-private PRE scheme according to Definition 5.*

The key-privacy proof is more difficult than that of CPA security. In particular, here we must be able to correctly re-encrypt ciphertexts for a special pair of users (I, J) even though we may not be able to compute a valid re-encryption key from I to J. To do this, we designed our encryption scheme in such a way that there is a "back door" for decryption, which in some cases, allows us to decrypt and then encrypt (thus simulating re-encryption) even when we cannot directly compute the re-encryption key needed to run the real re-encryption algorithm.

Proof. We show that if an adversary $\mathcal{A}$ can break the key-privacy game with probability $1/2 + \mu$, then we can construct an adversary $\mathcal{B}$ who can break the Decision Linear assumption with probability roughly $\frac{1}{2} + \frac{\mu}{4n^2}$, where n is the number of honest users. (This loose bound comes from letting the adversary dynamically pick its pair of honest users to challenge on. In prior key-privacy definitions [4], the adversary was restricted to a single pair.)

Given a Decision Linear input $\Delta' = (g, h, f, g^x, h^y, Q \stackrel{?}{=} f^{x+y})$, $\mathcal{B}$ handles oracle queries from $\mathcal{A}$ as follows. Let n be the bound on the number of uncorrupted users which $\mathcal{A}$ will ask to be created. $\mathcal{B}$ randomly chooses two as special users I and J, out of these n, and predicts that $\mathcal{A}$ will challenge on identities (I, J). $\mathcal{B}$ will proceed to set up things, so that first two elements of the valid re-encryption key from I to J will be (f^{x+y}, h^y). At a high-level, if $\mathcal{A}$ challenged on (I, J) then his response will be used to help $\mathcal{B}$, and if $\mathcal{A}$ challenges on anything else $\mathcal{B}$ will abort. Fortunately, we will see that $\mathcal{B}$ does not abort with probability $\geq 1/2n^2$.

Assuming (I, J) are chosen, let's see how $\mathcal{B}$ proceeds:

- **Public Parameter Generation** $\mathcal{B}$ sets up the parameters of $\mathcal{A}$ as $(g, h, Z) = (g, h, \mathbf{e}(g, h))$.
- **Uncorrupted Key Generation**
  - If this is the key for special user I, then select random $a \in \mathbb{Z}_q$ and output $(\mathbf{e}(g^x, h), g^a)$.
  - If this is the key for special user J, then select random $b \in \mathbb{Z}_q$ output (Z^b, f). Denote $f := g^s$, for some $s \in \mathbb{Z}_q$.
  - Otherwise select random $m_i, n_i \in \mathbb{Z}_q$ and output (Z^{m_i}, g^{n_i}).
- **Corrupted Key Generation** Select random $m_i, n_i \in \mathbb{Z}_q$ and output the public key as (Z^{m_i}, g^{n_i}), as well as the private key pair (m_i, n_i).
- **Re-encryption Key Generation** Given a request to encrypt from i to j, $\mathcal{B}$ selects a random $r \in \mathbb{Z}_q$ and proceeds as follows. If i is corrupted, this computation can be done by $\mathcal{A}$.
  - If $\mathcal{B}$ produced a re-encryption key from i to j before or $i = j$, output $\perp$.
  - If i is I and j is J, then abort. (I.e., (I, J) will not be the challenge pair.)
  - If i is I and j is other, then output $((g^x)^{n_j} \cdot g^{n_j r}, h^r, Z^{n_j w}, Z^w)$.
  - If i is J and j is I, then output $(g^{a(b+r)}, h^r, Z^{aw}, Z^w)$.
  - If i is J and j is other, then output $((g^{n_j})^{b+r}, h^r, Z^{n_j w}, Z^w)$.
  - If i is other and j is I, then output $(g^{a(m_i+r)}, h^r, Z^{aw}, Z^w)$.
  - If i is other and j is J, then output $(f^{m_i+r}, h^r, \mathbf{e}(f, h)^w, Z^w)$.
  - If i is other and j is other, then output $(g^{n_j(m_i+r)}, h^r, Z^{n_j w}, Z^w)$.
- **Re-encryption** On input (i, j, C_i),
  - Check that $C_i = (\alpha, \beta, \gamma)$ is properly formed by testing if $\mathbf{e}(\alpha, h) = \mathbf{e}(g, \beta)$. If this check fails or $i = j$, output $\perp$.
  - If (i, j) is (I, J), then decrypt C_i using g^x as $m = \gamma / \mathbf{e}(g^x, \beta)$. Choose a random $r \in \mathbb{Z}_q$ and output $(\mathbf{e}(f, h)^r, m \cdot Z^r) = (Z^{sr}, m \cdot Z^r)$. This is a key step in the proof.

- Otherwise obtain a re-encryption key $(\zeta, \eta, \theta, \lambda)$ of the same form as the re-encryption oracle. (It will not matter here if the same key is used or a new key generated each time, since the next step re-randomizes the output to hide which key was used.) The ciphertext is then re-encrypted and re-randomized by selecting a random $w' \in \mathbb{Z}_q$ and the output is $(\mathbf{e}(\zeta, \beta) \cdot \theta^{w'}, \gamma \cdot \mathbf{e}(\alpha, \eta) \cdot \lambda^{w'})$.

- **Challenge Oracle** Challenges are of the form (i, j) where i and j are indices of uncorrupted users which have not been queried before. If (i, j) is (I, J), $\mathcal{B}$ outputs $(Q, h^y, e(f, h)^w, Z^w)$. Else, $\mathcal{B}$ aborts and makes a random guess.

- **Decision** $\mathcal{A}$ will submit a guess of $d \in \{0, 1\}$. If $d = 1$, then $\mathcal{B}$ outputs 1 (is a Decision Linear instance), else it outputs 0 (not a Decision Linear instance).

The public parameters, key generation algorithm, and all responses of $\mathcal{O}_{rkey}$ and $\mathcal{O}_{renc}$ are correct and properly distributed. When $\mathcal{B}$ does not abort on the challenge and $\mathcal{A}$ does not detect improper queries, we have two cases. If $Q = f^{x+y}$, then the challenge is a valid re-encryption key for (I, J) and $\mathcal{A}$ will output 1 (a good re-encryption key) with probability $\frac{1}{2} + \mu$. $\mathcal{B}$ will output the correct answer (is Decision Linear instance) with the same probability. If Q is random, then $\mathcal{A}$ will output the correct answer of 0 (not a valid re-encryption key) with probability $\frac{1}{2}$. $\mathcal{B}$ outputs the same answer, so it will correctly determine that Δ' is not a Decision Linear instance with the same probability. Given that each case occurs with probability $1/2$ when $\mathcal{B}$ does not abort and $\mathcal{B}$ does not abort with probability $\geq \frac{1}{2n^2}$, the total probability of $\mathcal{B}$'s success is $\geq \frac{1}{2} + \frac{1}{2n^2} \cdot \frac{\mu}{2}$.

4   Conclusions and Open Problems

We formalized the notion of *key-privacy* for proxy re-encryption schemes. We discussed why none of the six or more existing PRE schemes satisfy this simple privacy notion. We then presented the first construction. It is secure under standard assumptions in the standard model.

Our construction realizes CPA-security. It would be interesting to realize key-private CCA-secure PRE. However, some basic approaches, such as applying the CPA-to-CCA transformation of Fujisaki and Okamoto [9] do not appear to maintain the key-privacy properties. It was also surprising that the definition of obfuscation, as in [10], does not capture key-privacy. It would be very interesting to know if a secure obfuscation of PRE could be realized when allowing the proxy and users to collude and allowing all the re-encryption and re-encryption key queries admitted here, as they would be in a real system.

Finally, we suspect that simpler key-private PRE schemes can be devised, although at the cost of stronger assumptions. The extended version of DBDH and the Decision Linear assumptions used here are actually quite mild. Nevertheless, finding more efficient schemes, even under stronger assumptions or in the random oracle model, would be quite useful for several applications.

Acknowledgments

The authors thank Jun Shao and the anonymous reviewers for helpful comments. The authors gratefully acknowledge the support of NSF grant CNS-0716142. Susan Hohenberger is also supported by a Microsoft New Faculty Fellowship.

References

1. Ateniese, G., Benson, K., Hohenberger, S.: Key-private proxy re-encryption (2008),
   http://eprint.iacr.org/2008/463
2. Ateniese, G., Fu, K., Green, M., Hohenberger, S.: Improved Proxy Re-encryption
   Schemes with Applications to Secure Distributed Storage. In: NDSS, pp. 29–43
   (2005)
3. Ateniese, G., Hohenberger, S.: Proxy re-signatures: new definitions, algorithms,
   and applications. In: ACM CCS, pp. 310–319. ACM, New York (2005)
4. Bellare, M., Boldyreva, A., Desai, A., Pointcheval, D.: Key-privacy in public-key
   encryption. In: Boyd, C. (ed.) ASIACRYPT 2001. LNCS, vol. 2248, pp. 566–582.
   Springer, Heidelberg (2001)
5. Blaze, M., Bleumer, G., Strauss, M.J.: Divertible protocols and atomic proxy cryptography. In: Nyberg, K. (ed.) EUROCRYPT 1998. LNCS, vol. 1403, pp. 127–144.
   Springer, Heidelberg (1998)
6. Boneh, D., Boyen, X., Shacham, H.: Short group signatures. In: Franklin, M. (ed.)
   CRYPTO 2004. LNCS, vol. 3152, pp. 41–55. Springer, Heidelberg (2004)
7. Boneh, D., Franklin, M.: Identity-Based Encryption from the Weil Pairing. In:
   Kilian, J. (ed.) CRYPTO 2001. LNCS, vol. 2139, pp. 213–229. Springer, Heidelberg
   (2001)
8. Canetti, R., Hohenberger, S.: Chosen-ciphertext secure proxy re-encryption. In:
   ACM CCS, pp. 185–194. ACM, New York (2007)
9. Fujisaki, E., Okamoto, T.: Secure integration of asymmetric and symmetric encryption schemes. In: Wiener, M. (ed.) CRYPTO 1999. LNCS, vol. 1666, pp. 537–554.
   Springer, Heidelberg (1999)
10. Hohenberger, S., Rothblum, G.N., Shelat, A., Vaikuntanathan, V.: Securely ob-
    fuscating re-encryption. In: Vadhan, S.P. (ed.) TCC 2007. LNCS, vol. 4392, pp.
    233–252. Springer, Heidelberg (2007)
11. Khurana, H., Heo, J., Pant, M.: From proxy encryption primitives to a deployable
    secure-mailing-list solution. In: Ning, P., Qing, S., Li, N. (eds.) ICICS 2006. LNCS,
    vol. 4307, pp. 260–281. Springer, Heidelberg (2006)
12. Libert, B., Vergnaud, D.: Unidirectional chosen-ciphertext secure proxy re-
    encryption. In: Cramer, R. (ed.) PKC 2008. LNCS, vol. 4939, pp. 360–379. Springer,
    Heidelberg (2008)
13. Suriadi, S., Foo, E., Smith, J.: Conditional privacy using re-encryption. In: IFIP
    International Workshop on Network and System Security (2008)
14. Taban, G., Cárdenas, A.A., Gligor, V.D.: Towards a secure and interoperable DRM
    architecture. In: DRM 2006, pp. 69–78. ACM, New York (2006)

Dynamic Universal Accumulators for DDH Groups and Their Application to Attribute-Based Anonymous Credential Systems

Man Ho Au[1], Patrick P. Tsang[2,*], Willy Susilo[1], and Yi Mu[1]

[1] Centre for Computer and Information Security Research
School of Computer Science and Software Engineering
University of Wollongong, Australia
{mhaa456,wsusilo,ymu}@uow.edu.au

[2] Department of Computer Science
Dartmouth College, USA
patrick@cs.dartmouth.edu

Abstract. We present the first dynamic universal accumulator that allows (1) the accumulation of elements in a DDH-hard group $\mathbb{G}$ and (2) one who knows x such that $y = g^x$ has — or has *not* — been accumulated, where g generates $\mathbb{G}$, to efficiently prove her knowledge of such x in zero knowledge, and hence without revealing, e.g., x or y.

We introduce the *Attribute-Based Anonymous Credential System*, which allows the verifier to authenticate anonymous users according to any access control policy expressible as a formula of *possibly negated* boolean user attributes. We construct the system from our accumulator.

1 Introduction

1.1 Background

Accumulators. Introduced by Benaloh and Mare [5], *accumulators* allow the representation of a set of *elements* $Y = \{y_1, y_2, \ldots, y_n\}$ by a single *value* v of size independent of Y's cardinality; using an initial value u, one can accumulate Y into v by invoking the *accumulating function* f as $v := \mathsf{f}(u, Y)$. Accumulators should be *collision-resistant* [3]:

for any element y and any value v, there exists an efficiently computable *witness* w for y w.r.t. v *if and only if* y has been accumulated into v (often abbreviated as "y is *in* v"). To prove that y is in v, one can thus demonstrate the existence of a corresponding w by proving, potentially in zero-knowledge, the knowledge of w.

Several uses of accumulators, e.g., in anonymous credential systems [10], require them to be *dynamic* [12]: one can efficiently update an accumulator value

---

* Supported in part by the Institute for Security, Technology, and Society, under grant 2005-DD-BX-1091, and the National Science Foundation, under grant CNS-0524695. The views in this paper do not necessarily reflect those of the sponsors.

M. Fischlin (Ed.): CT-RSA 2009, LNCS 5473, pp. 295–308, 2009.

by adding elements to — and possibly later deleting them from — the value. Furthermore, when a value is updated, e.g., from v to v', the witness w for some element y w.r.t. v can also be efficiently updated to the witness w' for the same element y w.r.t. the new value v'. Such accumulators are called *dynamic accumulators* (DA's).

Dynamic universal accumulators (DUA's) [20], on the other hand, are DA's with the additional property of *universality*: for any element set Y and any element $\bar{y}$, there exists an efficiently computable *non-membership witness* $\bar{w}$ for $\bar{y}$ w.r.t. value $v = \mathsf{f}(u, Y)$ *if and only if* $\bar{y} \notin Y$. By demonstrating the existence of $\bar{w}$, one can prove that $\bar{y}$ is *not* in v. Non-membership witnesses should allow efficient update.

Several existing DA/DUA constructions have $\mathsf{f} : (u, \{y_1, y_2, \ldots, y_n\}) \mapsto u^{y_1 y_2 \cdots y_n} \bmod N$ as their accumulating function [3,12,20], where N is a safe-prime product and $u \in QR(N)^1$. They permit only primes (up to a certain size) to be accumulated. Their security relies on the Strong RSA (SRSA) assumption [3].

Nguyen [21] constructed a DA from bilinear pairings (to be defined later). It has $\mathsf{f} : (u, \{y_1, y_2, \ldots, y_n\}) \mapsto u^{(s+y_1)(s+y_2)\cdots(s+y_n)}$ as the accumulating function, where s is the master secret of the accumulator instance and u is in some group equipped with a bilinear pairing. The construction allows elements in $\mathbb{Z}_p \setminus \{-s\}$ for some prime p to be accumulated. Its security relies on the q-Strong Diffie-Hellman (q-SDH) assumption [7]. Unlike the above "SRSA-based" constructions, dynamically adding an element to a value in Nguyen's construction requires the knowledge of s.

An accumulator would not be too useful (at least for building anonymous credential systems) without a suite of efficient zero-knowledge protocols for proving various facts about the accumulator values and elements. For instance, all the aforementioned constructions are equipped with a protocol for in zero-knowledge that a commitment c opens to some element in an accumulator value v.

Anonymous Credential Systems. In an anonymous credential system (ACS) [10], those and only those users who have registered to an organization O can authenticate their membership in O to any verifier (e.g., a server, another organization, etc.) anonymously and unlinkably among the set of all members in O. Camenisch and Lysyanskaya [10] constructed the first ACS using *a signature scheme with efficient protocols* [11] (commonly referred to as CL-signatures or P-signatures [4]) as a key building block. Many subsequent works have taken the same approach [11,12,13,4].

In this approach, to join an organization O, a user U first registers her pseudonym, which is simply a commitment of her pre-established private key x_U, e.g., in her PKI credential. Pseudonyms (even those of the same user) are hence unlinkable. O then issues a CL-signature σ_U on x_U according to the issuing protocol for CL-signatures, during which O learns nothing about x_U. U uses σ_U as her anonymous credential.

---

[1] $QR(N)$ denotes the group of quadratic residues modulo N.

To be able to revoke membership efficiently, O can maintain a DA as a "white-list" of users whose membership has *not* yet been revoked [12], by adding each user U's credential σ_U (or its identifier) to its DA when U registers and, when desired, deleting σ_U from DA to revoke U's membership. Therefore, to demonstrate her non-revoked membership in O to a verifier V, U conducts a zero-knowledge proof that (1) she has O's signature on her private key, and that (2) the signature is a credential in O's current DA. Alternatively, O can maintain a DUA as a "blacklist" of users whose membership has been revoked [20]. In this case, to demonstrate her non-revoked membership in O, U instead proves in zero-knowledge that (1) she has O's signature on her private key, and that (2) the signature is *not* a credential in O's current DUA.

1.2   Attribute-Based Anonymous Credential Systems

As a major contribution of this paper, we present the *Attribute-Based Anonymous Credential System* (ABACS), which generalizes the conventional notion of anonymous credential system (ACS) [10], in a fashion analogous to how Ciphertext-Policy Attribute-Based Encryption (CP-ABE) [6] generalizes public-key encryption, and how attribute certificates generalize identity certificates in X.509 PKIs [18].

Credentials in ABACS can be more precisely referred to as *anonymous attribute credentials* — they are issued to users to certify their possession of an attribute, allowing the users to prove various facts to any verifier about their credential ownership and hence attribute possession in some anonymous fashion. ABACS thus enables *privacy-preserving attribute-based access control*, in which a server is willing to grant a user access to an object such as a file or a service so long as the attributes possessed and/or lacked by the user satisfy the server's access control policy on the object, while privacy-concerned users desire to access the object by revealing merely the fact that they satisfy the policy, and can thus conceal, e.g., their identity, how they satisfy the policy, and etc.

In this paper, we confine ourselves to *boolean attributes* only. (Some attributes such as age and weight may take a value from a wider range such as non-negative integers and real numbers, and are hence non-boolean.) Boolean attributes provide rich semantics for labeling objects for access control. For example, they can represent group membership, or "roles" in Role-Based Access Control (RBAC).

Features. ABACS is a credential system with the following features.

- Flexible attribute-based access control. The verifier can choose to enforce *any* access control policy expressible as a boolean attribute formula in disjunctive normal form (DNF), i.e., a disjunction of terms, where each term is a conjunction of *possibly negated* boolean attributes, e.g., "(Student $\wedge$ Bio) $\vee$ ($\neg$Bio)".
- Multiple ACAs. To support an attribute, a corresponding *Attribute Certification Authority* (ACA) is created (during setup or dynamically when needed) to issue credentials to users to certify their possession of that attribute. These

ACAs are mutually independent; an ACA can only certify the possession of attributes for which it was created. This allows them to have different certification procedures with different trust levels, and confines the damages of their compromises.

– **Robust accountability.** The verifier accepts in the authentication only if the authenticating user satisfies the access control policy being enforced, i.e., the corresponding boolean formula evaluates to `true` on input the set of attribute for which the user has acquired a credential[2].
  Hence, a user who has acquired a credential for an attribute can't pretend that she hasn't, and colluding users, none of which alone satisfy the policy, can't satisfy it by pooling together their credentials.

– **Anonymous authentication.** The verifier knows only whether an authenticating user satisfies the access control policy he is enforcing. More precisely, authentication attempts by honest users who (resp. do not) satisfy the verifier's policy are anonymous and unlinkable among the set of all users who also (resp. do not) satisfy the policy.

– **Anonymous certification.** While ACAs must make public some data related to the certification status of users' attribute possession for authentication to be possible, some applications may require that such data reveals no (computational) information about the identity of the certified users, or more generally, no one can tell if two ACAs have issued a credential a common user.

– **Efficiency and practical negation support.** The authentication can be done in $O(|P|)$ time, where $|P|$ is the size of the verifier's policy measured in the number of (negated) attributes in it, and hence regardless of, e.g., the number of users, verifiers, ACAs, or attributes that the authenticating user possesses/lacks.
  Also, a user who lacks an attribute never has to contact anyone (e.g., the corresponding ACA) before she can prove her lack of the attribute.

Applications. The two scenarios below can benefit from ABACS.

The Biology department provides free parking to its students and any visitor from outside the department. The parking lot entrance hence enforces an access control policy of "(Bio ∧ Student) ∨ (¬Bio)". Identifiable authentication solutions[3] would violate the privacy desired by some users. A solution should allow different departments to locally manage their own "membership". Also, a visitor shouldn't have to show up at the Biology Department to get a "¬Bio" credential before he or she can park.

A pharmacist must check that "Fever ∧ ¬Asthma" holds for a patient before dispensing Aspirin (as many asthma sufferers are allergic to Aspirin), while the patient may not want to disclose her entire medical record, e.g., when she has an unrelated genetic disorder. Also, a fever patient with asthma with the "help" from someone without fever or asthma must still be unable to obtain Aspirin.

---

[2] A (resp. negated) attribute in a formula evaluates to `true` *if and only if* it is (resp. not) contained in the user's attribute set.

[3] E.g., waving an RFID card, or an e-token installed with X.509 attribute certificates.

2   Solution Overview

We start with some preliminaries. We then briefly describe how we construct a DUA that allows the accumulation of elements in a DDH-hard group, which we call DUA-DDH. Finally, we highlight how we build ABACS from it.

2.1   Preliminaries

Bilinear pairings. A bilinear pairing is a mapping from a pair of group elements to a group element. Specifically, let $\mathbb{G}_1$ and $\mathbb{G}_2$ be some cyclic groups of prime order p. Let g be a generator of $\mathbb{G}_1$. A function $\hat{e} : \mathbb{G}_1 \times \mathbb{G}_1 \rightarrow \mathbb{G}_2$ is a bilinear pairing if the following holds:

- *Unique Representation.* Each element in $\mathbb{G}_1$, $\mathbb{G}_2$ has unique binary representation.
- *Bilinearity.* $e(A^x, g^y) = e(A, B)^{xy}$ for all $A, B \in \mathbb{G}_1$ and $x, y \in \mathbb{Z}_p$.
- *Non-degeneracy.* $e(g, g) \neq 1$, where 1 is the identity element in $\mathbb{G}_2$.
- *Efficient Computability.* $e(A, B)$ can be computed efficiently (i.e. in polynomial time) for all $A, B \in \mathbb{G}_1$.

Complexity assumptions. The *Decisional Diffie-Hellman* (DDH) problem in $\mathbb{G}$ is defined as follows: On input a quadruple $(h_0, h_1, h_0^x, y^*) \in \mathbb{G}^4$, output 1 if $y^* = h_1^x$ and 0 otherwise. We say that the DDH assumption holds in $\mathbb{G}$ if no PPT algorithm has non-negligible advantage over random guessing in solving the DDH problem in $\mathbb{G}$. We call a group *DDH-hard* if the DDH assumption holds in the group.

The *q-Strong Diffie-Hellman* (q-SDH) problem in $\mathbb{G} = \langle g_0 \rangle$ is defined as follows: On input a $(q + 1)$-tuple $(g_0, g_0^\alpha, g_0^{\alpha^2}, \ldots, g_0^{\alpha^q}) \in \mathbb{G}^{q+1}$, output a pair $(w, y) \in \mathbb{G} \times \mathbb{Z}_p^*$, where p is the order of $\mathbb{G}$, such that $w^{(\alpha+y)} = g_0$. We say that the q-SDH assumption holds in $\mathbb{G}$ if no PPT algorithm has non-negligible advantage in solving the q-SDH problem in $\mathbb{G}$.

Zero-knowledge proof-of-knowledge. In a zero-knowledge proof of knowledge protocol [19], a prover convinces a verifier that some statement is true without the verifier learning anything except the validity of the statement. We use Camenisch and Stadler's notation [14]. For example, $PK\{(x) : y = g^x\}$ denotes a zero-knowledge proof-of-knowledge protocol that proves the knowledge of the discrete logarithm of y to the base g.

2.2   Our Dynamic Universal Accumulators for DDH Groups

To construct DUA-DDH, we take Nguyen's DA construction as the point of departure; we augment *universality* to it. Li et al. [20] presented a technique to augment *universality* to Camenisch and Lysyanskaya's DA construction [12]. The technique, however, requires the unique factorization of integers and relies on the SRSA assumption, and hence is not immediately applicable to Nguyen's

DA. Fortunately, we make the observation that the technique works as long as the domain of accumulatable elements is (a subset of) a Euclidean domain. (In the case of Li et al.'s, the domain is the ring of integers.) Consequently, to augment *universality* to Nguyen's construction, we adapt the technique to work on a different Euclidean domain, namely the ring of polynomials over a finite field.

We also equip our accumulator construction with a few useful zero-knowledge protocols. Of particular importance is the following pair:

$$\begin{cases} PK\,\{(x,y):\ C = \mathsf{Com}_1(x)\ \wedge\ D = \mathsf{Com}_2(y)\ \wedge\ y = g^x\ \wedge\ y \text{ is in } v)\} \\[2mm] PK\,\{(x,y):\ C = \mathsf{Com}_1(x)\ \wedge\ D = \mathsf{Com}_2(y)\ \wedge\ y = g^x\ \wedge\ y \text{ is not in } v)\} \end{cases}$$

where Com_1 and Com_2 are commitment schemes and g generates a DDH group, the elements in which can be accumulated in our accumulator. We construct the protocol using Pedersen's commitment scheme [22] and Camenisch's technique for proving double discrete logarithms [14]. The construction has a complexity of $O(\lambda)$ for a cheating probability of $2^{-\lambda}$.

This protocol is the cornerstone of our ABACS construction.

2.3   Our Attribute-Based Anonymous Credential System

Let $\mathbb{G}$ be a DDH group. Let ACA i be the ACA that certifies users' possession of attribute i. Each ACA i instantiates and maintains a DUA-DDH A_i of its own, but for the same $\mathbb{G}$, and independently picks a generator g_i of $\mathbb{G}$ at random.

Let U be a user with a pre-established private key x. For each attribute i she possesses, she can get certified by ACA i by providing her pseudonym $y_i = g_i^x$ w.r.t. ACA i. ACA i then adds y_i to its A_i. To later revoke the certification, ACA i can simply delete y_i from A_i. Finally, for each attribute j U lacks, she need not do anything (such as contacting ACA j); her pseudonym w.r.t. ACA j is by default *not* in ACA j's A_j.

Each ACA i publishes A_i, g_i (with a proof of their correct generation) and the list of pseudonyms that have been added in A_i. Thanks to the DDH assumption, no one — not even to the ACAs — can tell which user a pseudonym belongs to, or whether two ACAs' pseudonym lists contain a common user (non-negligibly better than random guessing).

From the published information, a user can compute a (resp. non-) membership witness for each attribute i she has (resp. not) been certified. The first-time computation takes $O(|L_i|)$ time when ACA i has certified $|L_i|$ users. This computation can be further reduced to $O(1)$ by moving the computation to ACA which is in possession of the auxiliary information of the accumulator. Updating the witness in the future take constant time per each change in the list of certified users.

User U who possesses attribute i and has been certified by ACA i can prove such fact to any verifier during authentication by proving that she has the knowledge of some x such that $y_i = g_i^x$ is in A_i. Similarly, if U lacks attribute j, she can prove the fact by proving that $y_j = g_j^x$ is not in A_j. These proofs can be accomplished in constant time. Generalizing the proof using a standard technique [16], a user can prove the validity of any DNF boolean attribute formula in time linear in the size of the formula.

Due to page limitation, the presentation of our ABACS ends here. More details about its syntax, construction, security formalism and proofs can be found in the full version of this paper [2].

3 Our Dynamic Universal Accumulators for DDH Groups

3.1 Definitions

We incrementally define *Dynamic Universal Accumulators for DDH Groups* (DUA-DDH's). We start by adapting Li et al.'s definition of *universality* to pairing-based accumulators.

Definition 1 (Universal Accumulators (UAs)). A universal accumulator is a scheme with the following properties:

- **Efficient generation** There exists a *Probabilistic Polynomial-Time (PPT)* algorithm Gen that, on input security parameter 1^λ, outputs a tuple $(\mathsf{f}, \mathsf{g}, \mathcal{Y}_\mathsf{f}, u, t_\mathsf{f})$, where f is a function $\mathcal{U}_\mathsf{f} \times \mathcal{Y}'_\mathsf{f} \to \mathcal{U}_\mathsf{f}$ and g is another function $\mathcal{U}_\mathsf{f} \to \mathcal{U}_\mathsf{g}$ for some domains $\mathcal{Y}_{\mathsf{f}'}, \mathcal{U}_\mathsf{f}, \mathcal{U}_\mathsf{g}$; $\mathcal{Y}_\mathsf{f} \subseteq \mathcal{Y}'_\mathsf{f}$ is the domain for accumulatable elements; t_f is some optional auxiliary information about f; and u is a element in $\mathcal{U}_\mathsf{f}$. We assume the tuple (f, g) is drawn uniformly at random from its domain.
- **Quasi-commutativity** For all $(\mathsf{f}, \mathsf{g}, \mathcal{Y}_\mathsf{f}, \cdot) \leftarrow \mathsf{Gen}(1^\lambda)$, $v \in \mathcal{U}_\mathsf{f}$ and $y_1, y_2 \in \mathcal{Y}'_\mathsf{f}$, we have $\mathsf{f}(\,\mathsf{f}(v, y_1), y_2) = \mathsf{f}(\,\mathsf{f}(v, y_2), y_1)$. Hence, if $Y = \{y_1, \ldots, y_k\} \subset \mathcal{Y}'_\mathsf{f}$, then we can denote $\mathsf{f}(\,\cdots\,\mathsf{f}(\,\mathsf{f}(v, y_1), y_2)\,\cdots\,, y_k)$ by $\mathsf{f}(v, Y)$ unambiguously.
- **Efficient evaluation** For all $(\mathsf{f}, \mathsf{g}, \mathcal{Y}_\mathsf{f}, t_\mathsf{f}, u) \leftarrow \mathsf{Gen}(1^\lambda)$, $v \in \mathcal{U}_\mathsf{f}$, and $Y \subset \mathcal{Y}_\mathsf{f}$ so that $|Y|$ is polynomial in λ, the function $\mathsf{g} \circ \mathsf{f}(v, Y)$ is computable in time polynomial in λ. $v = \mathsf{g} \circ \mathsf{f}(u, Y)$ represents the set Y. We call v the accumulator value for Y and say that y has been accumulated into v (or y is "in" v), for all $y \in Y$.
- **Membership (resp. non-membership) witnesses** For all $(\mathsf{f}, \mathsf{g}, \mathcal{Y}_\mathsf{f}, \cdot) \leftarrow \mathsf{Gen}(1^\lambda)$, there exists a relation Ω (resp. $\overline{\Omega}$) that defines membership (resp. non-membership) witnesses: w (resp. $\overline{w}$) is a valid membership (resp. non-membership) witness for element $y \in \mathcal{Y}_\mathsf{f}$ w.r.t. accumulator value $v \in \mathcal{U}_\mathsf{f}$ if and only if $\Omega(w, y, v) = 1$ (resp. $\overline{\Omega}(\overline{w}, y, v) = 1$). Membership witness (resp. non-membership witness) should be efficiently computable (in polynomial-time in λ) with t_f. □

The security of universal accumulators requires that it is hard to find a valid membership (resp. non-membership) witness for an element that is *not* in (resp. is *indeed* in) an accumulator value w.r.t. that accumulator value. We employ a strong definition in which the adversary is considered successful even if he present an element that is outside the intended domain of the accumulator ($\mathcal{Y}'_\mathsf{f}$ instead of $\mathcal{Y}_f$). Accumulators with this stronger sense of security improves efficiency of systems on which it is based because users within this system needs not conduct proof to demonstrate the elements presented is inside the intended domain of the accumulator. Below we give a precise definition.

Definition 2 (Security of Universal Accumulators (UAs)). A universal accumulator is secure if, for any PPT algorithm $\mathcal{A}$, both P_1 and P_2 are negligible in λ, where:

$$
\begin{cases}
P_1 = \Pr\left[\begin{array}{l}(f, g, \mathcal{Y}_f, u, \cdot) \leftarrow \mathsf{Gen}(1^\lambda); \ (y, w, Y) \leftarrow \mathcal{A}(g \circ f, g, \mathcal{Y}_f, u): \\ Y \subset \mathcal{Y}_f' \ \wedge \ y \in \mathcal{Y}_f' \backslash Y \ \wedge \ \Omega(w, y, g \circ f(u, Y)) = 1\end{array}\right], \\[2ex]
P_2 = \Pr\left[\begin{array}{l}(f, g, \mathcal{Y}_f, u) \leftarrow \mathsf{Gen}(1^\lambda); (y, \overline{w}, Y) \leftarrow \mathcal{A}(g \circ f, g, \mathcal{Y}_f, u): \\ Y \subset \mathcal{Y}_f' \ \wedge \ y \in Y \ \wedge \overline{\Omega}(\overline{w}, y, g \circ f(u, Y)) = 1\end{array}\right].
\end{cases}
$$

$\square$

Definition 3 (Dynamic Universal Accumulators (DUAs)). A DUA is an UA with the following additional properties:

- **Efficient update of accumulator.** There exists an efficient algorithm D_1 such that for all $v = g \circ f(u, Y)$, $y \notin Y$ and $\hat{v} \leftarrow D_1(t_f, v, y)$, we have $\hat{v} = g \circ f(u, Y \cup \{y\})$. If $y \in Y$ instead, then we have $\hat{v} = g \circ f(1, Y \setminus \{y\})$ instead.
- **Efficient update of membership witnesses.** Let v and $\hat{v}$ be the original and updated accumulator values respectively and $\hat{y}$ be the newly added (or deleted) element. There exists an efficient algorithm D_2 that, on input $y, w, v, \hat{v}$ with $y \neq \hat{y}$ and $\Omega(w, y, v) = 1$, outputs $\hat{w}$ such that $\Omega(\hat{w}, y, \hat{v}) = 1$.
- **Efficient update of non-membership witnesses.** Let v and $\hat{v}$ be the original and updated accumulator values respectively and $\hat{y}$ be the newly added (or deleted) element. There exists an efficient algorithm D_3 that, on input $y, \overline{w}, v, \hat{v}$ with $y \neq \hat{y}$ and $\overline{\Omega}(\overline{w}, y, v) = 1$, outputs $\hat{\overline{w}}$ such that $\overline{\Omega}(\hat{\overline{w}}, y, \hat{v}) = 1$.

$\square$

In the above, we call an algorithm "efficient" if its time complexity is independent of the cardinality of the accumulated element set Y. Security of DUA is defined as follows. Capabilities of an adversary is defined through queries to oracle O_D which models a working DUA. O_D is initialized with the tuple $(f, g, \mathcal{Y}_f, u, t_f)$ and maintains a list of elements Y, which is initially empty. O_D responds to two types of queries, namely "add y" and "delete y." It responds to an "add y" query by adding y to the set Y, modifying the accumulator value v using algorithm D_1 and sending back the updated accumulator value $\hat{v}$. It responds to a "delete y" query by deleting it from set Y, modifying the accumulator value v using algorithm D_1 and sending back the updated accumulator value $\hat{v}$. In the end, O_D outputs the current set Y and accumulator value v. The following is the definition of secure DUA.

Definition 4 (Security of Dynamic Universal Accumulators (DUAs)). An universal accumulator is secure if, for any PPT algorithm $\mathcal{A}$, P_3 and P_4 are negligible in λ, where:

$$\begin{cases} P_3 = \Pr \begin{bmatrix} (\mathsf{f}, \mathsf{g}, \mathcal{Y}_\mathsf{f}, u, t_\mathsf{f}) \leftarrow \mathsf{Gen}(1^\lambda); \ (y, w, Y) \leftarrow \mathcal{A}^{O_D(\mathsf{f}, \mathsf{g}, \mathcal{Y}_\mathsf{f}, t_\mathsf{f})}(\mathsf{g} \circ \mathsf{f}, \mathsf{g}, \mathcal{Y}_\mathsf{f}) : \\ Y \subset \mathcal{Y}'_\mathsf{f} \ \wedge \ y \in \mathcal{Y}'_\mathsf{f} \backslash Y \ \wedge \ v = \mathsf{g} \circ \mathsf{f}(u, Y) \ \wedge \ \Omega(w, y, v) = 1 \end{bmatrix} \\[2em] P_4 = \Pr \begin{bmatrix} (\mathsf{f}, \mathsf{g}, \mathcal{Y}_\mathsf{f}, u, t_\mathsf{f}) \leftarrow \mathsf{Gen}(1^\lambda); \ (y, \overline{w}, Y) \leftarrow \mathcal{A}^{O_D(\mathsf{f}, \mathsf{g}, \mathcal{Y}_\mathsf{f}, t_\mathsf{f})}(\mathsf{g} \circ \mathsf{f}, \mathsf{g}, \mathcal{Y}_\mathsf{f}) : \\ Y \subset \mathcal{Y}'_\mathsf{f} \ \wedge \ y \in Y \ \wedge \ v = \mathsf{g} \circ \mathsf{f}(u, Y) \ \wedge \ \overline{\Omega}(\overline{w}, y, v) = 1 \end{bmatrix} \end{cases}$$

$\square$

We state the following theorem. Its proof can be found in the full version of this paper [2].

Theorem 1. A DUA is secure if the underlying UA is secure. $\square$

Finally, a DUA-DDH is a DUA such that there exists a cyclic group $\mathbb{G} \subset \mathcal{Y}_\mathsf{f}$ in which the DDH assumption holds.

3.2   Constructions

We construct our DUA-DDH in stages. We first give a construction of UA for DDH groups. We then adds the necessary algorithms for enabling dynamism.

Our UA construction. This construction can be thought as the extension of Nguyen's accumulator to support *universality*. Our computation of non-membership witnesses involves operations on polynomials over finite fields.

- **Generation.** Let λ be a security parameter. Let $\hat{e} : \mathbb{G}_1 \times \mathbb{G}_1 \to \mathbb{G}_2$ be a bilinear pairing such that $|\mathbb{G}_1| = |\mathbb{G}_2| = p$ for some λ-bit prime p. Let g_0 be a generator of $\mathbb{G}_1$ and $\mathbb{G}_q = \langle h \rangle$ be a cyclic group of prime order q such that $\mathbb{G}_q \subset \mathbb{Z}_p^*$.[4] The generation algorithm Gen randomly chooses $\alpha \in_R \mathbb{Z}_p^*$. For simplicity, we always take the initial element $u = 1$, the identity element in $\mathbb{Z}_p^*$. The function f is defined as $\mathsf{f} : \mathbb{Z}_p^* \times \mathbb{Z}_p^* \to \mathbb{Z}_p^*$ such that $\mathsf{f} : u, y \mapsto u(y + \alpha)$. The function g is defined as $\mathsf{g} : \mathbb{Z}_p^* \times \mathbb{G}_1$ such that $\mathsf{g} : y \mapsto g_0^y$. The domain $\mathcal{Y}_\mathsf{f}$ of accumulatable elements is $\mathbb{G}_q$.[5] The auxiliary information t_f is α.
- **Evaluation.** Computing $\mathsf{g} \circ \mathsf{f}(1, Y)$ efficiently is straightforward with the auxiliary information α. In case one wishes to allow computation of $\mathsf{g} \circ \mathsf{f}$ without α, one can publish $g_0^{\alpha^i}$ for $i = 0$ to k, where k is the maximum number of elements to be accumulated. If we denote the polynomial $\prod_{y \in Y}(y + \alpha) = \sum_{i=0}^{i=k}(u_i \alpha^i)$ of maximum degree k as $v(\alpha)$, one can efficiently compute $\mathsf{g} \circ \mathsf{f}(1, Y)$ as $\mathsf{g} \circ \mathsf{f}(1, Y) = g_0^{v(\alpha)} = \prod_{i=0}^{i=k} g_i^{u_i} \in \mathbb{G}_1$, without the knowledge of α.
- **Membership witnesses.** The relation Ω is defined as $\Omega(w, y, v) = 1$ *if and only if* $\hat{e}(w, g_0^y g_0^\alpha) = \hat{e}(v, g_0)$. For a set of elements $Y := \{y_1, \ldots, y_k\} \in \mathbb{G}_q$, a membership witness for the element $y \in Y$ can be computed in either one of the following ways, depending on whether one knows the auxiliary information.

---

[4] If $p = 2q + 1$, one can choose a random element in $h \in_R \mathbb{Z}_p^*$ with order q and set $\mathbb{G}_q = \langle h \rangle$.

[5] Formally, it is $\mathbb{G}_q \setminus \{-\alpha\}$.

- *(With auxiliary information.)* Compute the witness as $w = [g_0^{\prod_{i=1}^{k}(y_i+\alpha)}]^{\frac{1}{\alpha+y}}$.
- *(Without auxiliary information.)* Let $\mathrm{w}(\alpha)$ be the polynomial $\prod_{i=1,i\neq j}^{k}(y_i+\alpha)$. Expand w and write it as $\mathrm{w}(\alpha) = \sum_{i=0}^{i=k-1}(u_i\alpha^i)$. Compute the witness as $w = g_0^{\mathrm{w}(\alpha)} = \prod_{i=0}^{i=k-1} g_i^{u_i} \in \mathbb{G}_1$.

- **Non-membership witnesses.** The relation $\overline{\Omega}$ for non-membership witnesses is defined as $\overline{\Omega}(\overline{w}, y, v) = 1$ *if and only if* $\overline{w} = (c, d)$ and $\hat{e}(c, g_0^y g_0^\alpha)\hat{e}(g_0, g_0)^d = \hat{e}(v, g_0)$. For a set of elements $Y := \{y_1, \ldots, y_k\} \in \mathbb{G}_q$, a non-membership witness for $\tilde{y} \notin Y$ can be computed in either one of the following ways, depending on whether one knows the auxiliary information:
  - *(With auxiliary information.)* Compute $\overline{w} = (c, d)$ according to $d = \prod_{i=1}^{k}(y_i + \alpha) \bmod (\alpha + \tilde{y}) \in \mathbb{Z}_p$ and $c = g_0^{\frac{\prod_{i=1}^{k}(y_i+\alpha)-d}{\tilde{y}+\alpha}} \in \mathbb{G}_1$.
  - *(Without auxiliary information.)* Denote the polynomial $v(\alpha)$ as $\prod_{i=1}^{k}(y_i + \alpha)$. Compute a polynomial division of $v(\alpha)$ by $(\alpha + \tilde{y})$. Since $(\alpha + \tilde{y})$ is a degree one polynomial and $\tilde{y} \neq y_i$ for all i, there exists a degree $k - 1$ polynomial $\mathrm{c}(\alpha)$ and a constant d such that $v(\alpha) = \mathrm{c}(\alpha)(\alpha + \tilde{y}) + d$. Expand c and write it as $\mathrm{c}(\alpha) = \sum_{i=0}^{i=k-1}(u_i\alpha^i)$. Compute $c = g_0^{\mathrm{c}(\alpha)} = \prod_{i=0}^{i=k-1} g_i^{u_i} \in \mathbb{G}_1$. The non-membership witness of $\tilde{y}$ is $\overline{w} = (c, d)$.

The theorem below states the security of our UA. Its proof can be found in the full version of this paper [2].

Theorem 2 (Security of our UA construction). Under the k-SDH assumption in $\mathbb{G}_1$, the above construction is a secure universal accumulator. $\square$

Our DUA-DDH construction. We present our construction of DUA-DDH by adding the various dynamism algorithms $\mathsf{D}_1, \mathsf{D}_2, \mathsf{D}_3$ to our UAconstruction above. Due to Theorem 1 and 2, our construction is secure under the k-SDH assumption.

- **Update of accumulator (algorithm D_1).** Adding an element $\hat{y}$ to the accumulator value v can be done by computing $\hat{v} = v^{\hat{y}+\alpha}$. Similarly, deleting an element $\hat{y}$ in the accumulator v can be done by computing $\hat{v} = v^{\frac{1}{\hat{y}+\alpha}}$. Both cases require the auxiliary information α.
- **Update of membership witnesses (algorithm D_2).** Let w be the original membership witness of y w.r.t the accumulator value v. Let $\hat{v}$ and $\hat{y}$ be the new accumulator value and the element added (resp. deleted) respectively. Suppose $\hat{y}$ has been added, the new membership witness $\hat{w}$ for y can be computed as $vw^{\hat{y}-y}$. Suppose $\hat{y} \neq y$ has been deleted, the new non-membership witness $\hat{w}$ for y can be computed as $w^{\frac{1}{\hat{y}-y}}\hat{v}^{\frac{1}{y-\hat{y}}}$.
- **Update of non-membership witnesses (algorithm D_3).** Let c, d be the original non-membership witness of y w.r.t. accumulator value v. Let $\hat{v}$ and $\hat{y}$ be the new accumulator value and the element added (resp. deleted) respectively.

- *(Addition.)* Suppose $\hat{y} \neq y$ has been added, the new non-membership witness $\hat{c}, \hat{d}$ of y can be computed as $\hat{c} = vc^{\hat{y}-y} \in \mathbb{G}_1$ and $\hat{d} = d(\hat{y}-y) \in \mathbb{Z}_p^*$. This can be verified as follows:

$$\hat{v} = v^{\alpha+\hat{y}} = v^{(\alpha+y)+(\hat{y}-y)} = v^{\alpha+y}v^{\hat{y}-y} = v^{\alpha+y}(c^{\alpha+y}g_0^d)^{\hat{y}-y}$$

$$= [vc^{\hat{y}-y}]^{\alpha+y}g_0^{d(\hat{y}-y)} = \hat{c}^{\alpha+y}g_0^{\hat{d}}$$

- *(Deletion.)* Suppose $\hat{y}$ has been deleted, the new non-membership witness $\hat{c}, \hat{d}$ of y can be computed as $\hat{c} = (c\hat{v}^{-1})^{\frac{1}{\hat{y}-y}} \in \mathbb{G}_1$ and $\hat{d} = \frac{d}{\hat{y}-y} \in \mathbb{Z}_p^*$. Indeed,

$$\hat{v} = \hat{v}^{\frac{(\alpha+\hat{y})-(\alpha+y)}{\hat{y}-y}} = v^{\frac{1}{\hat{y}-y}}\hat{v}^{\frac{\alpha+y}{y-\hat{y}}} = [c^{\alpha+y}g_0^d]^{\frac{1}{\hat{y}-y}}\hat{v}^{\frac{\alpha+y}{y-\hat{y}}}$$

$$= [(c\hat{v}^{-1})^{\alpha+y}g_0^d]^{\frac{1}{\hat{y}-y}} = [(c\hat{v}^{-1})^{\frac{1}{\hat{y}-y}}]^{\alpha+y}g_0^{\frac{d}{\hat{y}-y}} = \hat{c}^{\alpha+y}g_0^{\hat{d}}$$

4 Zero-Knowledge Protocols for Our **DUA**-DDH

We present several efficient zero-knowledge protocols for our DUA-DDH construction. In the presentation, we give priority to clarity over efficiency; the protocols may be optimized for better performance.

Let $\mathbb{G}_1 = \langle \mathfrak{g} \rangle$ and $\mathbb{G}_q = \langle \mathfrak{h} \rangle$ be cyclic groups of prime order p and q respectively, such that $\mathbb{G}_q \subset \mathbb{Z}_p^*$ is the domain of our DUA-DDH construction. Let $\mathfrak{g}_0, \mathfrak{g}_1$ and $\mathfrak{h}_0, \mathfrak{h}_1, \mathfrak{h}_2$ be independent generators of $\mathbb{G}_1$ and $\mathbb{G}_q$ respectively. Let $y = \mathfrak{h}_0^x \in \mathbb{G}_q$ and let $\mathfrak{C} = \mathfrak{g}_0^y\mathfrak{g}_1^r \in \mathbb{G}_1$ be the commitment of y using random number r. Let v be an accumulator value.

4.1 Proof of Knowledge of the Discrete Logarithm of a Committed Element

This protocol is the main building block of the protocols used in our DUA-DDH construction. We call it PK_1. Let $\mathfrak{D} = \mathfrak{h}_1^x\mathfrak{h}_2^s \in \mathbb{G}_q$ be the commitment of x using some random number s. The goal of PK_1 is to prove the knowledge of x and y such that $y = \mathfrak{h}_0^x$ in zero-knowledge, thus without revealing, e.g., x or y. In other words, we have:

$$\mathrm{PK}_1\left\{ (y, r, x, s) : \mathfrak{C} = \mathfrak{g}_0^y\mathfrak{g}_1^r \ \wedge \ \mathfrak{D} = \mathfrak{h}_1^x\mathfrak{h}_2^s \ \wedge \ y = \mathfrak{h}_0^x \right\}$$

The protocol can be used with the common discrete logarithm relationship proofs [9] to demonstrate relationships of discrete logarithms in $\mathbb{G}_1$ or $\mathbb{G}_q$. Instantiation of PK_1 makes use of the zero-knowledge proof-of-knowledge of double discrete logarithms [14], as we now describe. Let λ_k be a security parameter that determines the cheating probability of the protocol. (The cheating probability is $2^{-\lambda_k}$, we hence suggest $\lambda_k = 80$.) PK_1 consists of PK_{1A} and PK_{1B} as follows.

$$PK_1 \begin{cases} PK_{1A}\left\{(y,r): \mathfrak{C} = \mathfrak{g}_0^y \mathfrak{g}_1^r\right\} \\\\ PK_{1B}\left\{(x,r,s): \mathfrak{C} = \mathfrak{g}_0^{\mathfrak{h}_0^x} \mathfrak{g}_1^r \ \wedge \ \mathfrak{D} = \mathfrak{h}_1^x \mathfrak{h}_2^s\right\} \end{cases}$$

Instantiating PK_{1A} is straightforward. Below we only show how to instantiate PK_{1B}.

(Commitment.) For $i = 1$ to λ_k, the prover randomly generates $\rho_{x,i}, \rho_{s,i} \in_R \mathbb{Z}_q$ and $\rho_{r,i} \in_R \mathbb{Z}_p$, computes $T_{1,i} = \mathfrak{g}_0^{\mathfrak{h}_0^{\rho_{x,i}}} \mathfrak{g}_1^{\rho_{r,i}} \in \mathbb{G}_1$ and $T_{2,i} = \mathfrak{h}_1^{\rho_{x,i}} \mathfrak{h}_2^{\rho_{s,i}} \in \mathbb{G}_q$, and sends $T_{1,i}, T_{2,i}$ to the verifier.

(Challenge.) The verifier randomly generates a λ_k-bit challenge m and sends it to the prover.

(Response.) Denote by $m[i]$ the i-th bit of m, starting from $i = 1$. For $i = 1$ to λ_k, the prover computes $z_{x,i} = \rho_{x,i} - m[i]x \in \mathbb{Z}_q$, $z_{s,i} = \rho_{s,i} - m[i]s \in \mathbb{Z}_q$ and $z_{r_i} = \rho_{r,i} - m[i]\mathfrak{h}_0^x r \in \mathbb{Z}_p$. She sends $\left(z_{x,i}, z_{s,i}, z_{r,i}\right)_{i=1}^{\lambda_k}$ to the verifier.

(Verify.) The verifier outputs 1 if the following holds for all $i = 1$ to λ_k. He outputs 0 otherwise.

$$T_{2,i} \stackrel{?}{=} \mathfrak{D}^{m[i]} \mathfrak{h}_1^{z_{x,i}} \mathfrak{h}_2^{z_{s,i}} \quad \text{and} \quad T_{1,i} \stackrel{?}{=} \begin{cases} \mathfrak{g}_0^{\mathfrak{h}_0^{z_{x,i}}} \mathfrak{g}_1^{z_{r,i}}, & \text{if } m[i] = 0, \\\\ \mathfrak{C}^{\mathfrak{h}_0^{z_{x,i}}} \mathfrak{g}_1^{z_{r,i}}, & \text{otherwise.} \end{cases}$$

It is straightforward to show that PK_1 is Honest-Verifier Zero-Knowledge. It can be converted into a 4-round perfect zero-knowledge protocol using the technique due to Cramer et al. [15] or 3-move concurrent zero-knowledge protocol in the auxiliary string model based on trapdoor commitment schemes [17]. Note that the prover does not need to explicitly prove that the r in PK_{1A} and PK_{1B} are the same; they are bounded to be the same under the discrete logarithm assumption.

4.2 Proof of Knowledge of a Committed Element in an Accumulator Value

Suppose y is in the accumulator value v. That is, there exists witness w such that $\Omega(w, y, v) = 1$. The following protocol demonstrates that the element y, committed as $\mathfrak{C}$, is in the accumulator value v.

$$PK_2\left\{(w,y,r): \hat{e}(w, g_0^y g_0^\alpha) = \hat{e}(v, g_0) \ \wedge \ \mathfrak{C} = \mathfrak{g}_0^y \mathfrak{g}_1^r\right\}$$

PK_2 can be instantiated using the standard proof-of-knowledge of an SDH-tuple [8,1].

Combining PK_1 and PK_2, we have a protocol, denoted as PK_3, that proves the knowledge of the discrete logarithm of an element in an accumulator value:

$$PK_3\left\{(w,y,x): \hat{e}(w, g_0^y g_0^\alpha) = \hat{e}(v, g_0) \ \wedge \ y = \mathfrak{h}_0^x\right\}$$

4.3 Proof of Knowledge of a Committed Element Not in an Accumulator Value

Suppose y is *not* in the accumulator value v. Then there exists witness $\overline{w} = (c, d)$ such that $d \neq 0$ and $\overline{\Omega}(\overline{w}, y, v) = 1$. The following protocol demonstrates that the element y, committed as $\mathfrak{C}$, is *not* in the accumulator value v.

$$\mathrm{PK}_4 \left\{ (c, d, y, r) : \hat{e}(c, g_0^y g_0^\alpha) = \hat{e}(v, g_0) \hat{e}(g_0, g_0)^d \ \wedge \ d \neq 0 \ \wedge \ \mathfrak{C} = \mathfrak{g}_0^y \mathfrak{g}_1^r \right\}$$

PK_4 can be instantiated using standard techniques, which we describe in the full version of this paper [2].

Combining PK_1 and PK_4, we have a protocol, denoted as PK_5, that proves the knowledge of the discrete logarithm of an element not in an accumulator value:

$$\mathrm{PK}_5 \left\{ (c, d, y, x) : \hat{e}(c, g_0^y g_0^\alpha) \hat{e}(g_0, g_0)^d = \hat{e}(v, g_0) \ \wedge \ d \neq 0 \ \wedge \ y = \mathfrak{h}_0^x \right\}$$

5 Concluding Remarks

We have presented the first dynamic universal accumulator construction for accumulating elements in DDH-hard groups and a number of useful zero-knowledge protocols for it. Using this accumulator, we have built an Attribute-Based Anonymous Credential System, which allows the verifier to authenticate anonymous users according to any access control policy expressible as formula of boolean user attributes in the DNF form. Our system features many practicality and scalability properties for a large-scale deployment of privacy-preserving access control in a heterogeneous and decentralized environment.

We end the paper with two research questions that we believe to be worth exploring in the future. The first one is how one can construct ABACS that also efficiently supports numeric attributes. (While one could certainly encode a numerical attribute by a bunch of boolean attributes, that wouldn't be very efficient.) The second question is how one can construct ABACS that avoids the need to prove double discrete logarithms, and hence achieves better efficiency.

References

1. Au, M.H., Susilo, W., Mu, Y.: Constant-Size Dynamic k-TAA. In: De Prisco, R., Yung, M. (eds.) SCN 2006. LNCS, vol. 4116, pp. 111–125. Springer, Heidelberg (2006)
2. Au, M.H., Tsang, P.P., Susilo, W., Mu, Y.: Dynamic Universal Accumulators for DDH Groups and Their Application to Attribute-Based Anonymous Credential Systems. Cryptology ePrint Archive, Report 2009/044 (2009)
3. Barić, N., Pfitzmann, B.: Collision-Free Accumulators and Fail-Stop Signature Schemes without Trees. In: Fumy, W. (ed.) EUROCRYPT 1997. LNCS, vol. 1233, pp. 480–494. Springer, Heidelberg (1997)

4. Belenkiy, M., Chase, M., Kohlweiss, M., Lysyanskaya, A.: P-signatures and Nonin-teractive Anonymous Credentials. In: Canetti, R. (ed.) TCC 2008. LNCS, vol. 4948, pp. 356–374. Springer, Heidelberg (2008)
5. Benaloh, J.C., de Mare, M.: One-Way Accumulators: A Decentralized Alternative to Digital Signatures. In: Helleseth, T. (ed.) EUROCRYPT 1993. LNCS, vol. 765, pp. 274–285. Springer, Heidelberg (1994)
6. Bethencourt, J., Sahai, A., Waters, B.: Ciphertext-policy attribute-based encryp-tion. In: IEEE Symposium on Security and Privacy, pp. 321–334. IEEE Computer Society, Los Alamitos (2007)
7. Boneh, D., Boyen, X.: Short Signatures Without Random Oracles. In: Cachin, C., Camenisch, J.L. (eds.) EUROCRYPT 2004. LNCS, vol. 3027, pp. 56–73. Springer, Heidelberg (2004)
8. Boneh, D., Boyen, X., Shacham, H.: Short Group Signatures. In: Franklin, M. (ed.) CRYPTO 2004. LNCS, vol. 3152, pp. 41–55. Springer, Heidelberg (2004)
9. Camenisch, J.: Group Signature Schemes and Payment Systems Based on the Dis-crete Logarithm Problem. PhD Thesis, ETH Zürich, 1998. Diss. ETH No. 12520, Hartung Gorre Verlag, Konstanz (1998)
10. Camenisch, J., Lysyanskaya, A.: An Efficient System for Non-transferable Anony-mous Credentials with Optional Anonymity Revocation. In: Pfitzmann, B. (ed.) EUROCRYPT 2001. LNCS, vol. 2045, pp. 93–118. Springer, Heidelberg (2001)
11. Camenisch, J.L., Lysyanskaya, A.: A Signature Scheme with Efficient Protocols. In: Cimato, S., Galdi, C., Persiano, G. (eds.) SCN 2002. LNCS, vol. 2576, pp. 268–289. Springer, Heidelberg (2003)
12. Camenisch, J., Lysyanskaya, A.: Dynamic Accumulators and Application to Effi-cient Revocation of Anonymous Credentials. In: Yung, M. (ed.) CRYPTO 2002. LNCS, vol. 2442, pp. 61–76. Springer, Heidelberg (2002)
13. Camenisch, J., Lysyanskaya, A.: Signature Schemes and Anonymous Credentials from Bilinear Maps. In: Franklin, M. (ed.) CRYPTO 2004. LNCS, vol. 3152, pp. 56–72. Springer, Heidelberg (2004)
14. Camenisch, J., Stadler, M.: Efficient Group Signature Schemes for Large Groups. In: Kaliski Jr., B.S. (ed.) CRYPTO 1997. LNCS, vol. 1294, pp. 410–424. Springer, Heidelberg (1997)
15. Cramer, R., Damgård, I.B., MacKenzie, P.D.: Efficient Zero-Knowledge Proofs of Knowledge without Intractability Assumptions. In: Imai, H., Zheng, Y. (eds.) PKC 2000. LNCS, vol. 1751, pp. 354–373. Springer, Heidelberg (2000)
16. Cramer, R., Damgård, I.B., Schoenmakers, B.: Proof of Partial Knowledge and Simplified Design of Witness Hiding Protocols. In: Desmedt, Y.G. (ed.) CRYPTO 1994. LNCS, vol. 839, pp. 174–187. Springer, Heidelberg (1994)
17. Damgård, I.: Efficient Concurrent Zero-Knowledge in the Auxiliary String Model. In: Preneel, B. (ed.) EUROCRYPT 2000. LNCS, vol. 1807, pp. 418–430. Springer, Heidelberg (2000)
18. Farrell, S., Housley, R.: An Internet Attribute Certificate Profile for Authorization (2002)
19. Goldwasser, S., Micali, S., Rackoff, C.: The Knowledge Complexity of Interactive Proof Systems.. SIAM J. Comput. 18(1), 186–208 (1989)
20. Li, J., Li, N., Xue, R.: Universal Accumulators with Efficient Nonmembership Proofs. In: Katz, J., Yung, M. (eds.) ACNS 2007. LNCS, vol. 4521, pp. 253–269. Springer, Heidelberg (2007)
21. Nguyen, L.: Accumulators from Bilinear Pairings and Applications. In: Menezes, A. (ed.) CT-RSA 2005. LNCS, vol. 3376, pp. 275–292. Springer, Heidelberg (2005)
22. Pedersen, T.P.: Non-interactive and Information-Theoretic Secure Verifiable Secret Sharing. In: Feigenbaum, J. (ed.) CRYPTO 1991. LNCS, vol. 576, pp. 129–140. Springer, Heidelberg (1992)

Practical Short Signature Batch Verification

Anna Lisa Ferrara[1], Matthew Green[2], Susan Hohenberger[3],
and Michael Østergaard Pedersen[4]

[1] University of Illinois at Urbana-Champaign
[2] Independent Security Evaluators
[3] Johns Hopkins University
[4] Lenio A/S

Abstract. In many applications, it is desirable to work with signatures that are short, and yet where *many* messages from *different* signers be verified very quickly. RSA signatures satisfy the latter condition, but are generally thousands of bits in length. Recent developments in pairing-based cryptography produced a number of "short" signatures which provide equivalent security in a fraction of the space. Unfortunately, verifying these signatures is computationally intensive due to the expensive pairing operation. Toward achieving "short and fast" signatures, Camenisch, Hohenberger and Pedersen (Eurocrypt 2007) showed how to *batch verify* two pairing-based schemes so that the total number of pairings was independent of the number of signatures to verify.

In this work, we present both theoretical and practical contributions. On the theoretical side, we introduce new batch verifiers for a wide variety of regular, identity-based, group, ring and aggregate signature schemes. These are the first constructions for batching group signatures, which answers an open problem of Camenisch et al. On the practical side, we implement each of these algorithms and compare each batching algorithm to doing individual verifications. Our goal is to test whether batching is practical; that is, whether the benefits of removing pairings significantly outweigh the cost of the additional operations required for batching, such as group membership testing, randomness generation, and additional modular exponentiations and multiplications. We experimentally verify that the theoretical results of Camenisch et al. and this work, indeed, provide an efficient, effective approach to verifying multiple signatures from (possibly) different signers.

1 Introduction

As we move into the era of pervasive computing, where computers are everywhere as an integrated part of our surroundings, there are going to be a host of devices exchanging messages with each other, e.g., sensor networks, vehicle-2-vehicle communications [1,2]. For these systems to work properly, messages must carry some form of authentication, but the system requirements on the authentication are particularly demanding. Any cryptographic solution must simultaneously be:

M. Fischlin (Ed.): CT-RSA 2009, LNCS 5473, pp. 309–324, 2009.
© Springer-Verlag Berlin Heidelberg 2009

1. *Short*: Bandwidth is an issue. Raya and Hubaux argue that due to the limited spectrum available for vehicular communication, something shorter than RSA signatures is needed [3].
2. *Quick to verify large numbers of messages from different sources*: Raya and Hubaux also suggest that vehicles will transmit safety messages every 300ms to all other vehicles within a minimum range of 110 meters [3], which in turn may retransmit these messages. Thus, it is much more critical that authentications be quick to verify rather than to generate.
3. *Privacy-friendly*: Users should be held accountable, but not become publicly identifiable.

Due to the high overhead of using digital signatures, researchers have developed a number of alternative protocols designed to amortize signatures over many packets [4,5], or to replace them with symmetric MACs [6]. Each approach has significant drawbacks; e.g., the MAC-based protocols use time-delayed delivery so that the necessary verification keys are delivered *after* the authenticated messages arrive. This approach can be highly efficient within a restricted setting where synchronized clocks are available, but it does not provide non-repudiability of messages (to hold malicious users accountable) or privacy. Signature amortization requires verifiers to obtain many packets before verifying, and is vulnerable to denial of service. Other approaches, such as the short, undeniable signatures of Monnerat and Vaudenay [7,8] are inappropriate for the pervasive settings we consider, since verification requires interaction with the signer.

In 2001, Boneh, Lynn and Shacham developed a pairing-based signature that provides security equivalent to 1024-bit RSA at a cost of only 170 bits [9] (slightly larger than HMAC-SHA1). This was followed by many signature variants, some of them privacy-friendly, which were also relatively short, e.g., [10,11,12,13]. Unfortunately, the focus was on reducing the signature size, but less attention was paid to the verification cost which require expensive pairing operations.

Recently, Camenisch, Hohenberger and Pedersen [14] took a step toward speeding up the verification of short signatures, by showing how to *batch verify* two short pairing-based signatures so that the total number of dominant (pairing) operations was independent of the number of signatures to verify. However, their solution left open several questions which this work addresses.

First, their work was purely theoretical. To our knowledge, we are the first to provide a detailed *empirical analysis* of batch verification of short signatures. This is interesting, because our theoretical results and those of Camenisch et al. [14] reduce the total number of pairings by *adding* in other operations, such as random number generation and small modular exponentiations, so it was unclear how well these algorithms would perform in practice. Fortunately, in section 5, we verify that these algorithms do work well.

Second, Camenisch et al. [14] dealt only with batching regular and identity-based signatures. They specifically mentioned batching group signatures as an interesting open problem. Here, we present the *first* batch verifier for a group signature scheme, as well as new verifiers for many other types of regular, identity-based, ring and aggregate signatures.

Finally, Camenisch et al. [14] did not address the practical issue of what to do if the batch verification fails. How does one detect *which* signatures in the batch are invalid? Does this detection process eliminate all of the efficiency gains of batch verification? Fortunately, our empirical studies reveal good news: invalid signatures can be detected via a recursive divide-and-conquer approach, and if < 15% of the signatures are invalid, then batch verification is still more efficient than individual verification. At the time we conducted these experiments, the divide-and-conquer approach was the best method known to us. Recently, Law and Matt [15] proposed three new techniques for finding invalid signatures in a batch. One of their techniques allows to save approximately half the time needed by the simple divide-and-conquer approach, for large batch sizes. Thus, while our numbers seem good, they can be further improved.

Overall, we conclude that many interesting short signatures can be batch verified, and that batch verification is an extremely valuable tool for system implementors. As an example of our results in section 5, for the short group signatures of Boneh, Boyen and Shacham [10], we see that when batching 200 group signatures (in a 160-bit MNT curve) individual verification takes 139ms whereas batch verification reduces the cost to 25ms per signature (see Figure 3).

2   Algebraic Setting: Pairings

Let PSetup be an algorithm that, on input the security parameter 1^τ, outputs the parameters for a bilinear pairing as $(q, g_1, g_2, \mathbb{G}_1, \mathbb{G}_2, \mathbb{G}_T, \mathbf{e})$, where $\mathbb{G}_1 = \langle g_1 \rangle$, $\mathbb{G}_2 = \langle g_2 \rangle$ and $\mathbb{G}_T$ are of prime order $q \in \Theta(2^\tau)$. The efficient mapping $\mathbf{e} : \mathbb{G}_1 \times \mathbb{G}_2 \to \mathbb{G}_T$ is both: (*bilinear*) for all $g \in \mathbb{G}_1$, $h \in \mathbb{G}_2$ and $a, b \leftarrow \mathbb{Z}_q$, $\mathbf{e}(g^a, h^b) = \mathbf{e}(g, h)^{ab}$; and (*non-degenerate*) if g generates $\mathbb{G}_1$ and h generates $\mathbb{G}_2$, then $\mathbf{e}(g, h) \neq 1$. This is called the *asymmetric* setting; in the *symmetric* setting, $\mathbb{G}_1 = \mathbb{G}_2$.

In the asymmetric setting, the best we can hope for are group elements in $\mathbb{G}_1, \mathbb{G}_2$ and $\mathbb{G}_T$ of size 160, 512 and 1024 bits respectively. In the symmetric setting, it seems the best curve is a supersingular curve (with $k = 2$), where $\mathbb{G}_1 = \mathbb{G}_2$ and $\mathbb{G}_T$ will be of size 512 and 1024 bits respectively. Most of the signature schemes we discuss can be implemented in the asymmetric setting to take advantage of the smaller group sizes. We discuss this more and the case of batching composite order groups in the full version of this paper [16].

Testing Membership. Our proofs will require that elements of purported signatures are members of $\mathbb{G}_1$, but how efficiently can this fact be verified? Determining whether some data represents a point on a curve is easy. The question is whether it is in the correct subgroup. Assume that the subgroup has order q. The easy way to verify if $y \in \mathbb{G}_1$ is simply to test $y^q = 1$. Since q might be quite large this test is inefficient, but as we will see later the time required to test membership of group elements are insignificant compared to the time required to do the pairings in the applications we have in mind. Yet, in some cases, there are more efficient ways to test group membership [17].

3   Basic Tools for Pairing-Based Batch Verification

Let us begin with a formal definition of a *pairing based* batch verifier. Recall that PSetup is an algorithm that, on input the security parameter 1^τ, outputs the parameters $(q, g_1, g_2, \mathbb{G}_1, \mathbb{G}_2, \mathbb{G}_T, \mathbf{e})$, where $\mathbb{G}_1, \mathbb{G}_2, \mathbb{G}_T$ are of prime order $q \in \Theta(2^\tau)$. Pairing-based verification equation are represented by a *generic pairing based claim* X corresponding to a boolean relation of the following form: $\prod_{i=1}^{k} \mathbf{e}(f_i, h_i)^{c_i} \overset{?}{=} A$, for $k \in \text{poly}(\tau)$ and $f_i \in \mathbb{G}_1, h_i \in \mathbb{G}_2$ and $c_i \in \mathbb{Z}_q^*$, for each $i = 1, \dots, k$. A pairing-based verifier Verify for a generic pairing-based claim is a probabilistic $\text{poly}(\tau)$-time algorithm which on input the representation $\langle A, f_1, \dots, f_k, h_1, \dots, h_k, c_1, \dots, c_k \rangle$ of a claim X, outputs *accept* if X holds and *reject* otherwise. We define a batch verifier for pairing-based claims.

Definition 1 (Pairing-based Batch Verifier). *Let* PSetup$(1^\tau) \to (q, g_1, g_2,$ $\mathbb{G}_1, \mathbb{G}_2, \mathbb{G}_T, \mathbf{e})$. *For each* $j \in [1, \eta]$, *where* $\eta \in \text{poly}(\tau)$, *let* $X^{(j)}$ *be a generic pairing-based claim and let* Verify *be a pairing based verifier. We define a* pairing-based batch verifier *for* Verify *as a probabilistic poly(τ)-time algorithm which outputs:*

- accept *if* $X^{(j)}$ *holds for all* $j \in [1, \eta]$;
- reject *if* $X^{(j)}$ *does not hold for any* $j \in [1, \eta]$ *except with negligible probability.*

3.1   Small Exponents Test Applied to Pairings

Bellare, Garay and Rabin proposed methods for verifying multiple equations of the form $y_i = g^{x_i}$ for $i = 1$ to n, where g is a generator for a group of prime order [18]. One might be tempted to just multiply these equations together and check if $\prod_{i=1}^{n} y_i = g^{\sum_{i=1}^{n} x_i}$. However, it would be easy to produce two pairs (x_1, y_1) and (x_2, y_2) such that the product of them verifies correctly, but each individual verification does not, e.g. by submitting the pairs $(x_1 - \alpha, y_1)$ and $(x_2 + \alpha, y_2)$ for any α. Instead, Bellare et al. proposed the following method, which we will later apply to pairings.

Small Exponents Test: Choose exponents δ_i of (a small number of) ℓ_b bits and compute $\prod_{i=1}^{n} y_i^{\delta_i} = g^{\sum_{i=1}^{n} x_i \delta_i}$. Then the probability of accepting a bad pair is $2^{-\ell_b}$. The size of ℓ_b is a tradeoff between efficiency and security. (In Section 5, we set $\ell_b = 80$ bits.)

Theorem 1. *Let* PSetup$(1^\tau) \to (q, g_1, g_2, \mathbb{G}_1, \mathbb{G}_2, \mathbb{G}_T, \mathbf{e})$ *where* q *is prime. For each* $j \in [1, \eta]$, *where* $\eta \in \text{poly}(\tau)$, *let* $X^{(j)}$ *corresponds to a generic claim as in Definition 1. For simplicity, assume that* $X^{(j)}$ *is of the form* $A \overset{?}{=} Y^{(j)}$ *where* A *is fixed for all* j *and all the input values to the claim* $X^{(j)}$ *are in the correct groups. For any random vector* $\Delta = (\delta_1, \dots, \delta_\eta)$ *of* ℓ_b *bit elements from* $\mathbb{Z}_q$, *an algorithm* Batch *which tests the following equation* $\prod_{j=1}^{\eta} A^{\delta_j} \overset{?}{=} \prod_{j=1}^{\eta} Y^{(j)^{\delta_j}}$ *is a pairing-based batch verifier that accepts an invalid batch with probability at most* $2^{-\ell_b}$.

The proof closely follows the proof of the small exponents test by Bellare et al. [18], we include a full proof of this theorem in the full version of this paper [16]. Thus, Theorem 1 provides a *single* verification equation, which we then want to optimize.

3.2 Basic Batching Techniques

Armed with Theorem 1, let's back up for a moment to get a complete picture of how to develop an efficient batch verifier. This summarizes the ideas we used to obtain the results in Figure 1, which we believe will be useful elsewhere. Immediately after the summary, we'll explain the details.

Summary: Suppose you have η bilinear equations. Batch verify them as follows:

1. Apply Technique 1 to the individual verification equation, if applicable.
2. Apply Theorem 1 to the equations. This *combines* all equations into a single equation after checking membership in the expected algebraic groups and using the small exponents test.
3. Optimize the resulting single equation using Techniques 2, 3 and 4.
4. If batch verification fails, use the divide-and-conquer approach to identify the bad signatures.

Technique 1 *Change the verification equation.* Recall that a Σ-protocol is a three step protocol (commit, challenge, response) allowing a prover to prove various statements to a verifier. Using the Fiat-Shamir heuristic a Σ-protocol can be turned into a signature scheme, by forming the challenge as the hash of the commitment and the message to be signed. The signature is then either (commit, response) or (challenge, response). The latter is often preferred, since the challenge is usually smaller than the commitment, which results in a smaller signature. However, we observed that this often causes batch verification to become very inefficient, whereas using (commit, response) results in a much more suitable verification equation.

We use this technique to help batch the Hess IBS [19] and the group signatures of Boneh, Boyen and Shacham [10] and Boyen and Shacham [11]. Indeed, we believe that prior attempts to batch verify group signatures overlooked this idea and thus came up without efficient solutions.

Combination Step: Given η pairing-based claims, apply Theorem 1 to obtain a single equation. The combination step actually consist of two substeps:

1. *Check Membership*: Check that all elements are in the correct subgroup. Only elements that could be generated by an adversary needs to be checked (e.g., elements of a signature one wants to verify). Public parameters need not be checked, or could be checked only once.
2. *Small Exponents Test*: Combine all equations into one and apply the small exponents test.

Next, optimize this *single* equation using any of the following techniques in any order.

Technique 2 *Move the exponent into the pairing.* When a pairing of the form $\mathbf{e}(g_i, h_i)^{\delta_i}$ appears, move the exponent δ_i into $\mathbf{e}()$. Since elements of $\mathbb{G}$ are usually smaller than elements of $\mathbb{G}_T$, this gives a small speedup when computing the exponentiation.

$$\text{Replace } \mathbf{e}(g_i, h_i)^{\delta_i} \text{ with } \mathbf{e}(g_i^{\delta_i}, h_i)$$

Technique 3. *When two pairings with a common first or second element appear,*

they can be combined. This can reduce η pairings to one. It will work like this:

$$\text{Replace } \prod_{i=1}^{\eta} \mathbf{e}(g_i^{\delta_i}, h) \text{ with } \mathbf{e}(\prod_{i=1}^{\eta} g_i^{\delta_i}, h)$$

In rare cases, it might be useful to apply this technique "in reverse", e.g., splitting a single pairing into two or more pairings to allow for the application of other techniques. For example, we do this when batching Boyen's ring signatures [13], so that we can apply Technique 4 below.

Technique 4 *Waters hash.* In his IBE, Waters described how hash identities to values in $\mathbb{G}_1$ [20], using a technique that was subsequently employed in several signature schemes. Assume the identity is a bit string $V = v_1 v_2 \ldots v_m$, then given public parameters $u_1, \ldots, u_m, u' \in \mathbb{G}_1$, the hash is $u' \prod_{i=1}^{m} u_i^{v_i}$. Following works by Naccache [21] and Chatterjee and Sarkar [22,23] documented the generalization where instead of evaluating the identity bit by bit, divide the k bit identity bit string into z blocks, and then hash. (In Section 5, we SHA1 hash our messages to a 160-bit string, and use $z = 5$ as proposed in [21].) Recently, Camenisch et al. [14] pointed out the following method:

$$\text{Replace } \prod_{j=1}^{\eta} \mathbf{e}(g_j, \prod_{i=1}^{m} u_i^{v_{ij}}) \text{ with } \prod_{i=1}^{m} \mathbf{e}(\prod_{j=1}^{\eta} g_j^{v_{ij}}, u_i)$$

In the full version of this paper [16], we apply this technique to schemes with structures related to the Waters hash; namely, the ring signatures of Boyen [13] and the aggregate signatures of Lu et al. [24].

3.3   Handling Invalid Signatures

If there is even a single invalid signature in the batch, then the batch verifier will reject the entire batch with high probability. In many real-world situations, a signature collection may contain invalid signatures caused by accidental data corruption, or possibly malicious activity by an adversary seeking to degrade service. In some cases, this may not be a serious concern. E.g., sensor networks with a high level of redundancy may choose to simply drop messages that cannot be efficiently verified. Alternatively, systems may be able to cache and/or individually verify important messages when batch verification fails. Yet, in some applications, it might be critical to tolerate some percentage of invalid signatures without losing the performance advantage of batch verification.

In Section 5.2, we employ a recursive *divide-and-conquer* approach, similar to that of Pastuszak, Pieprzyk, Michalek and Seberry [25], as: First, shuffle the incoming batch of signatures, and if batch verification fails, simply divide the collection into two halves, and recurse on the halves. When this process terminates, the batch verifier outputs the index of each invalid signature. Through careful implementation and caching of intermediate results, much of the work of the batch verification (i.e., computing the product of many signature elements) can be performed once over the full signature collection, and need not be repeated when verifying each sub-collection. Thus, the cost of each recursion is dominated by the number of pairings used in the batch verification algorithm. In Section 5.2, we show that even if up to 15% of the signatures are invalid, this technique still performs faster than individual verification.

Recently, Law and Matt [15] proposed three new techniques for finding invalid signatures in a batch. One of their techniques, which is the most efficient for large batch sizes, allows to save approximately half the time needed by the simple divide-and-conquer approach. Thus, it is possible to do even better than the performance numbers we present.

4   Batch Verifiers for Short Signatures

Given the basic batching tools in the last section, it still requires creativity to figure out how best to apply them to batch any given scheme. In this section, we present new results for batch verifying a selection of existing regular, identity-based, group, ring, and aggregate signature schemes. To our knowledge, these are the first such verifiers for group, ring and aggregate signatures. After a search through the existing literature, we present the schemes with the best results.

Figure 1 shows a summary of our theoretical results, together with an indication of which batching techniques were used. Due to space limitations, we cannot describe the details of each scheme. Instead, we demonstrate one example in the Boneh, Boyen and Shacham [10] group signatures and then describe all remaining signatures and their batch verifiers in the full version of this paper [16].

4.1   Batching the Boneh-Boyen-Shacham (BBS) Group Signatures

This scheme does not appear to batch well *without* making some alterations, which increase the signature size by one group element, but where only 2 pairings are sufficient to batch an arbitrary number of signatures. A group signature scheme allows any member to sign on behalf of the group in such a way that anyone can verify a signature using the group public key while nobody, but the group manager, can identify the actual signer. A scheme consists of four algorithms: KeyGen, Sign, Verify and Open, that, respectively generate public

Scheme	Model	Individual-Verify	Batch-Verify	Reference	Techniques
Group Signatures					
BBS [10]	RO	5η	2	§4.1	1,2,3
BS [11]	RO	5η	2	[16]	1,2,3
ID-based Ring Signatures					
CYH [12]	RO	2η	2	[16]	2,3
Ring Signatures					
Boyen [13] (same ring)	plain	$\ell \cdot (\eta + 1)$	$\min\{\eta \cdot \ell + 1, 3 \cdot \ell + 1\}$	[16]	2,3,4
Signatures					
BLS [26]	RO	2η	$s + 1$	[26]	2,3
CHP [14] (time restrictions)	RO	3η	3	[14]	2,3
ID-based Signatures					
Hess [19]	RO	2η	2	[16]	1,2,3
ChCh [27]	RO	2η	2	[15]	2,3
Waters [20,21,28,23]	plain	3η	$\min\{(2\eta + 3), (z + 3)\}$	[14]	2,3,4
Aggregate Signatures					
BGLS [29] (same users)	RO	$\eta(\ell + 1)$	$\ell + 1$	[16]	2,3
Sh [30] (same users)	RO	$\eta(\ell + 2)$	$\ell + 2$	[16]	2,3
LOSSW [24] (same sequence)	plain	$\eta(\ell + 1)$	$\min\{(\eta + 2), (\ell \cdot k + 3)\}$	[16]	2,3,4

Fig. 1. Signatures with Efficient Batch Verifiers. Let η be the number of signatures to verify, s be the number of distinct signers involved and ℓ be either the size of a ring or the size of an aggregate. Boyen batch verifier requires each signature to be issued according to the same ring. Aggregate verifiers work for signatures related to the same set of users. In CHP, only signatures from the same time period can be batched and z is a (small) parameter (e.g., 8). In LOSSW, k is the message bit-length. RO stands for random oracle. The details of each scheme and its batch verifier are provided in the full version of this paper [16].

and private keys for users and the group manager, sign a message on behalf of a group, verify the signature on a message according to the group and trace a signature to a signer. For our purposes, we focus on the verification algorithm.

The Boneh-Boyen-Shacham (BBS) Group Signatures. Let $\mathsf{PSetup}(1^\tau) \rightarrow (q, g_1, g_2, \mathbb{G}_1, \mathbb{G}_2, \mathbb{G}_T, \mathbf{e})$, where $H : \{0, 1\}^* \rightarrow \mathbb{Z}_q$ is a hash function and there exists an efficiently-computable isomorphism $\psi : \mathbb{G}_2 \rightarrow \mathbb{G}_1$. Let ℓ be the number of users in a group.

Key Gen. Select a random $g_2 \in \mathbb{G}_2$ and sets $g_1 \leftarrow \psi(g_2)$. Select $h \xleftarrow{\$} \mathbb{G}_1 \setminus \{1_{\mathbb{G}_1}\}$, $r_1, r_2 \xleftarrow{\$} \mathbb{Z}_q^*$, and set u, v such that $u^{r_1} = v^{r_2} = h$. Select $\gamma \xleftarrow{\$} \mathbb{Z}_q^*$, and set $w = g_2^\gamma$. For $i = 1$ to n, select $x_i \xleftarrow{\$} \mathbb{Z}_q^*$, and set $f_i = g_1^{1/(\gamma + x_i)}$. The public key is $\mathsf{gpk} = (g_1, g_2, h, u, v, w)$, the group manager's secret key is $\mathsf{gmsk} = (r_1, r_2)$ and the secret key of the i'th user is $\mathsf{gsk}[\mathsf{i}] = (f_i, x_i)$.

Sign. Given a group public key $\mathsf{gpk} = (g_1, g_2, h, u, v, w)$, a user private key (f, x) and a message $M \in \{0, 1\}^*$, compute the signature σ as follows: Select $\alpha, \beta, r_\alpha, r_\beta, r_x, r_{\gamma_1}, r_{\gamma_2} \xleftarrow{\$} \mathbb{Z}_q$. Compute $T_1 = u^\alpha$; $T_2 = v^\beta$; $T_3 = f \cdot h^{\alpha+\beta}$, $\gamma_1 = x \cdot \alpha$ and $\gamma_2 = x \cdot \beta$, $R_1 = u^{r_\alpha}$; $R_2 = v^{r_\beta}$; $R_3 = \mathbf{e}(T_3, g_2)^{r_x} \cdot \mathbf{e}(h, w)^{-r_\alpha - r_\beta} \cdot \mathbf{e}(h, g_2)^{-r_{\gamma_1} - r_{\gamma_2}}$; $R_4 = T_1^{r_x} \cdot u^{-r_{\gamma_1}}$; $R_5 = T_2^{r_x} \cdot v^{-r_{\gamma_2}}$. Compute $c = H(M, T_1, T_2, T_3, R_1, R_2, R_3, R_4, R_5)$. Compute $s_\alpha = r_\alpha + c \cdot \alpha$; $s_\beta = r_\beta + c \cdot \beta$;

$s_x = r_x + c \cdot x$; $s_{\gamma_1} = r_{\gamma_1} + c \cdot \gamma_1$; $s_{\gamma_2} = r_{\gamma_2} + c \cdot \gamma_2$. The signature is $\sigma = (T_1, T_2, T_3, c, s_\alpha, s_\beta, s_x, s_{\gamma_1}, s_{\gamma_2})$.

Verify. Given a group public key $\mathsf{gpk} = (g_1, g_2, h, u, v, w)$, a message M and a group signature $\sigma = (T_1, T_2, T_3, c, s_\alpha, s_\beta, s_x, s_{\gamma_1}, s_{\gamma_2})$, compute the values $R_1 = u^{s_\alpha} \cdot T_1^{-c}$, $R_2 = v^{s_\beta} \cdot T_2^{-c}$, $R_3 = \mathbf{e}(T_3, g_2)^{s_x} \cdot \mathbf{e}(h, w)^{-s_\alpha - s_\beta} \cdot \mathbf{e}(h, g_2)^{-s_{\delta_1} - s_{\delta_2}} \cdot \left(\mathbf{e}(T_3, w) \cdot \mathbf{e}(g_1, g_2)^{-1}\right)^c$, and $R_4 = T_1^{s_x} \cdot u^{-s_{\delta_1}}$; $R_5 = T_2^{s_x} \cdot v^{-s_{\delta_2}}$. Accept iff $c \stackrel{?}{=} H(M, T_1, T_2, T_3, R_1, R_2, R_3, R_4, R_5)$.

An Efficient Batch Verifier for BBS Group Signatures. Computing R_3 is the most expensive part of the verification above, but at first glance it is not clear that this can be batched, because each R_3 is hashed in the verification equation. However, as described by Technique 1, the signature and the verification algorithm can be modified at the expense of increasing the signature size by one element. Let $\sigma = (T_1, T_2, T_3, R_3, c, s_\alpha, s_\beta, s_x, s_{\gamma_1}, s_{\gamma_2})$ be the new signature, together with:

New Individual Verify. Given a group public key $\mathsf{gpk} = (g_1, g_2, h, u, v, w)$, a message M and a group signature $\sigma = (T_1, T_2, T_3, R_3, c, s_\alpha, s_\beta, s_x, s_{\gamma_1}, s_{\gamma_2})$, compute the values $R_1 \leftarrow u^{s_\alpha} \cdot T_1^{-c}$; $R_2 \leftarrow v^{s_\beta} \cdot T_2^{-c}$; $R_4 \leftarrow T_1^{s_x} \cdot u^{-s_{\gamma_1}}$; $R_5 \leftarrow T_2^{s_x} \cdot v^{-s_{\gamma_2}}$, then check the following equation

$$\mathbf{e}(T_3, g_2)^{s_x} \cdot \mathbf{e}(h, w)^{-s_\alpha - s_\beta} \cdot \mathbf{e}(h, g_2)^{-s_{\gamma_1} - s_{\gamma_2}} \cdot \left(\mathbf{e}(T_3, w) \cdot \mathbf{e}(g_1, g_2)^{-1}\right)^c \stackrel{?}{=} R_3.$$

Finally check if $c \stackrel{?}{=} H(M, T_1, T_2, T_3, R_1, R_2, R_3, R_4, R_5)$. Accept if all checks succeed, else reject.

Now we define a batch verifier, where the main objective is to use a *constant* number of pairings.

Batch Verify. Let $\mathsf{gpk} = (g_1, g_2, h, u, v, w)$ be the group public key, and let $\sigma_j = (T_{j,1}, T_{j,2}, T_{j,3}, R_{j,3}, c_j, s_{j,\alpha}, s_{j,\beta}, s_{j,x}, s_{j,\gamma_1}, s_{j,\gamma_2})$ be the j'th signature on the message M_j, for each $j = 1, \ldots, \eta$. For each $j = 1, \ldots, \eta$, compute the following values:

$$R_{j,1} \leftarrow u^{s_{j,\alpha}} \cdot T_{j,1}^{-c_j} \qquad R_{j,2} \leftarrow v^{s_{j,\beta}} \cdot T_{j,2}^{-c_j}$$
$$R_{j,4} \leftarrow T_{j,1}^{s_{j,x}} \cdot u^{-s_{j,\gamma_1}} \qquad R_{j,5} \leftarrow T_{j,2}^{s_{j,x}} \cdot v^{-s_{j,\gamma_2}}$$

Now for each $j = 1, \ldots, \eta$, check that $c_j \stackrel{?}{=} H(M_j, T_{j,1}, T_{j,2}, T_{j,3}, R_{j,1}, R_{j,2}, R_{j,3}, R_{j,4}, R_{j,5})$. Then check the following *single* pairing based equation

$$\mathbf{e}\left(\prod_{j=1}^{\eta}(T_{j,3}^{s_{j,x}} \cdot h^{-s_{j,\gamma_1} - s_{j,\gamma_2}} \cdot g_1^{-c_j})^{\delta_j}, g_2\right) \cdot \mathbf{e}\left(\prod_{j=1}^{\eta}(h^{-s_{j,\alpha} - s_{j,\beta}} \cdot T_3^{c})^{\delta_j}, w\right) \stackrel{?}{=} \prod_{j=1}^{\eta} R_{j,3}^{\delta_j}.$$

where $(\delta_1, \ldots, \delta_\eta)$ is a random vector of ℓ_b bit elements from $\mathbb{Z}_q$. Accept iff all checks succeed.

Theorem 2. *For security level ℓ_b, the above algorithm is a batch verifier for the BBS group signature scheme, where the probability of accepting an invalid signature is $2^{-\ell_b}$.* (Proof of this theorem appears in the full version of this paper [16].)

5   Implementation and Performance Analysis

The previous work on batching short signatures [14] considers only asymptotic performance. Unfortunately, this "paper analysis" conceals many details that are revealed only through empirical evaluation. Additionally, the existing work does not address how to handle invalid signatures.

We seek to answer these questions by conducting the first empirical investigation into the feasibility of short signature batching. To conduct our experiments, we built concrete implementations of seven signature schemes described in this work, including two public key signature schemes (BLS, CHP), three Identity-Based Signature schemes (ChCh, Hess, Waters), a ring signature (CYH), and a short group signature scheme (BBS). *For each scheme, we measured the performance of the individual verification algorithm against that of the corresponding batch verifier.* We then turned our attention to the problem of efficiently sorting out invalid signatures.

Experimental Setup. To evaluate our batch verifiers, we implemented each signature scheme in C++ using the MIRACL library for elliptic curve operations [31]. Our timed experiments were conducted on a 3.0Ghz Pentium D 930 with 4GB of RAM running Linux Kernel 2.6. All hashing was implemented using SHA1,[1] and small exponents were of size 80 bits. For each scheme, our basic experiment followed the same outline: (1) generate a collection of η distinct signatures on 100-byte random message strings. (2) Conduct a timed verification of this collection using the batch verifier. (3) Repeat steps (1, 2) four times, averaging to obtain a mean timing. To obtain a view of batching efficiency on collections of increasing size, we conducted the preceding test for values of η ranging from 1 to approximately 400 signatures in intervals of 20. Finally, to provide a baseline, we separately measured the performance of the corresponding *non-batched* verification, by verifying 1000 signatures and dividing to obtain the average verification time per signature. A high-level summary of our results is presented in Figure 3.

Curve	k	$\mathcal{R}(\mathbb{G}_1)$	$\mathcal{R}(\mathbb{G}_T)$	$\mathcal{S}_{RSA}$	Pairing Time
MNT160	6	160 bits	960 bits	960 bits	23.3 ms
MNT192	6	192 bits	1152 bits	1152 bits	33.2 ms
SS512	2	512 bits	1024 bits	957 bits	16.7 ms

Fig. 2. Description of the elliptic curve parameters used in our experiments. $\mathcal{R}(\cdot)$ describes the approximate number of bits to optimally represent a group element. $\mathcal{S}_{RSA}$ is an estimate of "RSA-equivalent" security derived via the approach of Page et al. [32].

---

[1] We selected SHA1 because the digest size closely matches the order of $\mathbb{G}_1$. One could use other hash functions with a similar digest size, *e.g.*, RIPEMD-160, or truncate the output of a hash function such as SHA-256 or Whirlpool. Because the hashing time is negligible in our experiments, this should not greatly impact our results.

Scheme	Signature Size (bits)			Individual Verification			Batched Verification*		
	MNT160	MNT192	SS512	MNT160	MNT192	SS512	MNT160	MNT192	SS512
Signatures									
BLS (single signer)	160	192	512	47.6 ms	77.8 ms	52.3 ms	2.28 ms	2.93 ms	32.42 ms
CHP	160	192	512	73.6 ms	119.0 ms	93.0 ms	26.16 ms	34.66 ms	34.50 ms
BLS cert + CHP sig	1280	1536	1536	121.2 ms†	196.8 ms†	145.3 ms†	28.44 ms†	37.59 ms†	66.92 ms†
Identity-Based Signatures									
ChCh	320	384	1024	49.1 ms	79.7 ms	73.3 ms	3.93 ms	5.24 ms	59.45 ms
Waters	480	576	1536	91.2 ms	138.64 ms	61.1 ms	9.44 ms	11.49 ms	59.32 ms
Hess	1120	1344	1536	49.1 ms	79.0 ms	73.1 ms	6.70 ms	8.72 ms	55.94 ms
Anonymous Signatures									
BBS (modified per §[16])	2400	2880	3008	139.0 ms	218.3 ms	193.0 ms	24.80 ms	34.18 ms	198.03 ms
CYH, 2-member ring	480	576	1536	52.0 ms	77.0 ms	113.0 ms	6.03 ms	8.30 ms	105.69 ms
CYH, 20-member ring	3360	4032	10752	86.5 ms	126.8 ms	829.3 ms	43.93 ms	61.47 ms	932.66 ms

*Average time per verification when batching 200 signatures.
†Values were derived by manually combining data from BLS and CHP tests.

Fig. 3. Summary of experimental results. Timing results indicate verification time *per signature*. With the exception of BLS, our experiments considered signatures generated by distinct signers.

Curve Parameters. The selection of elliptic curve parameters impacts both signature size and verification time. The two most important choices are the size of the underlying finite field $\mathbb{F}_p$, and the curve's embedding degree k. Due to the MOV attack, security is bounded by the size of the associated finite field $\mathbb{F}_{p^k}$. Simultaneously, the representation of elements $\mathbb{G}_1$ requires approximately $|p|$ bits. Thus, most of the literature on short signatures recommends choosing a relatively small p, and a curve with a high value of k. (For example, an MNT curve with $|p| = 192$ bits and $k = 6$ is thought to offer approximately the same level of security as 1152-bit RSA [32].) The literature on short signatures focuses mainly on signature size rather than verification time, so it is easy to miss the fact that using such high-degree curves *substantially* increases the cost of a pairing operation, and thus verification time. To incorporate these effects into our results, we implemented our schemes using two high-degree ($k = 6$) MNT curves with $|p|$ equal to 160 bits and 192 bits. For completeness, we also considered a $|p|=512$ bit supersingular curve with embeddeing degree $k = 2$, and a subgroup $\mathbb{G}_1$ of size 2^{160}. Figure 2 details the curve choices along with relevant details such as pairing time and "RSA-equivalent" security determined using the approach of Page et al. [32].

5.1   Performance Results

Public-Key signatures. Figure 4 presents the results of our timing experiments for the public-key BLS and CHP verifiers. Because the BLS signature does not batch efficiently for messages created by *distinct* signers, we studied the combination suggested in [14], where BLS is used for certificates which are created by a single master authority, and CHP is used to sign the actual messages under users' individual signing keys. Unfortunately, the CHP batch verifier appears to be quite costly in the recommended MNT curve setting. This outcome stems from the requirement that user public keys be in the $\mathbb{G}_2$ subgroup. This

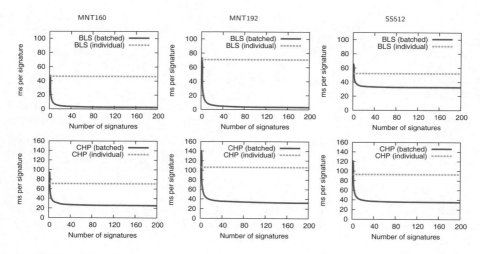

Fig. 4. Public-Key Signature Schemes. Per-signature times were computed by dividing total batch verification time by the number of signatures verified. Note that in the BLS case, all signatures are formulated by the same signer (as for certificate generation), while for CHP each signature was produced by a different signer. Individual verification times are included for comparison.

necessitates expensive point operations in the curve defined over the extension field, which undoes *some* of the advantage gained by batching. However, batching still reduces the per-signature verification cost to as little as 1/3 to 1/4 that of individual verification.

Identity-Based signatures. Figure 5 gives our measurements for three IBS schemes: ChCh, Waters and Hess. (For comparison, we also present CHP signatures with BLS-signed public-key certificates.) In all experiments, we consider signatures generated by different signers. In contrast with regular signatures, the IBSes batch quite efficiently, at least when implemented in MNT curves. The Waters scheme offers strong performance for a scheme not dependent on random oracles.[2] In our implementation of Waters, we first apply a SHA1 to the message, and use the Waters hash parameter $z = 5$ which divides the resulting 160-bit digest into blocks of 32 bits (as in [21]).

Anonymous signatures. Figure 6 gives our results for two privacy-preserving signatures: the CYH ring signature and the modified BBS group signature. As is common with ring signatures, in CYH both the signature size and verification time grow linearly with the number of members in the ring. For our experiments we arbitrarily selected two cases: (1) where all signatures are formed under a 2-member ring (useful for applications such as lightweight email signing [33]), and

---

[2] However, it should be noted that Waters has a somewhat loose security reduction, and may therefore require larger parameters in order to achieve security comparable to alternative schemes.

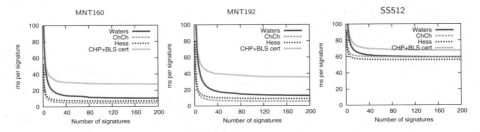

Fig. 5. Identity-Based Signature Schemes. Times represent total batch verification time divided by the number of signatures verified. "CHP+BLS cert" represents the batched public-key alternative using certificates, and is included for comparison.

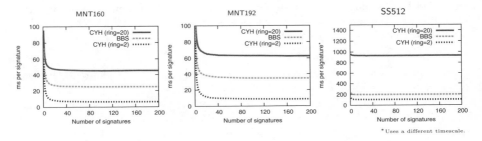

Fig. 6. Anonymous Signature Schemes. Times represent total batch verification time divided by the number of signatures verified. For the CYH ring signature, we consider two distinct signature collections, one consisting of 2-member rings, and another with 20-member rings. The BBS signature verification is independent of the group size.

(2) where all signatures are formed using a 20-member ring.[3] In contrast, both the signature size and verification time of the BBS group signature are independent of the size of the group. This makes group signatures like BBS significantly more practical for applications such as vehicle communication networks, where the number of signers might be quite large.

5.2 Batch Verification and Invalid Signatures

In Section 3.3, we discuss techniques for dealing with invalid signatures. When batch verification fails, this *divide-and-conquer* approach recursively applies the batch verifier to individual halves of the batch, until all invalid invalid signatures have been located. To save time when recursing, we compute products of the form $\prod_{i=1}^{\eta} x_i^{\delta_i}$ so that partial products will be in place for each subset on which me might recurse. We accomplish this by placing each $x_i^{\delta_i}$ at the leaf of a binary tree and caching intermediate products at each level. This requires no additional

---

[3] Although the CYH batch verifier can easily batch signatures formed over differently-sized rings, our experiments use a constant ring size. Our results are representative of any signature collection where the *mean* ring size is 20.

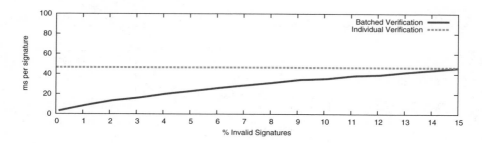

Fig. 7. BLS batch verification in the presence of invalid signatures (160-bit MNT curve). A "resilient" BLS batch verifier was applied to a collection of 1024 purported BLS signatures, where some percentage were randomly corrupted. Per-signature times were computed by dividing the total verification time (including identification of invalid signatures) by the total number of signatures (1024), and averaging over multiple experimental runs.

computation, and total storage of approximately 2η group elements for each product to be computed.

To evaluate the feasibility of this technique, we used it to implement a "resilient" batch verifier for the BLS signature scheme. This verifier accepts as input a collection of signatures where some may be invalid, and outputs the index of each invalid signature found. To evaluate batching performance, we first generated a collection of 1024 valid signatures, and then randomly corrupted an r-fraction by replacing them with random group elements. We repeated this experiment for values of r ranging from 0 to 15% of the collection, collecting multiple timings at each point, and averaging to obtain a mean verification time. The results are presented in Figure 7.

Batched verification of BLS signatures is preferable to the naïve individual verification algorithm even as the number of invalid signatures exceeds 10% of the total batch size. The random distribution of invalid signatures within the collection is nearly the worst-case for resilient verification. In practice, invalid signatures might be grouped together within the batch (e.g., if corruption is due to a burst of EM interference). In this case, the verifier might achieve better results by omitting the random shuffle step or using another re-ordering technique.

6   Conclusion and Open Problems

Our experiments provide strong evidence that batching short signatures is practical, even in a setting where an adversary can inject invalid signatures. We present new algorithms for batching a host of short signature schemes, including the first such verifiers for group, ring and aggregate signatures. At a deeper level, our results indicate that efficient batching depends heavily on the underlying design of a signature scheme, particularly on the placement of elements within the elliptic curve subgroups. For example, the CHP signature and the ChCh IBS have comparable size and security, yet the latter scheme can batch more than

250 signatures per second (each from a different signer), while our CHP implementation clocks in at fewer than 40. Designers should take these considerations into account when proposing new pairing-based signature schemes.

It remains open to batch verify a group signature scheme without random oracles. While many candidate schemes exist, it is not clear how to batch verify them. It also remains open to verify a batch of very short signatures (one group element) in *constant* pairings without the time-period restriction used by Camenisch et al. [14], even with random oracles.

Acknowledgments

Anna Lisa Ferrara and Matthew Green performed part of this work while at the Johns Hopkins University. Matthew Green and Susan Hohenberger were supported by the NSF under grant CNS-0716142 and a Microsoft New Faculty Fellowship. Michael Østergaard Pedersen performed part of this research while at the University of Aarhus.

References

1. Car 2 Car: Communication consortium, http://car-to-car.org
2. SeVeCom: Security on the road, http://www.sevecom.org
3. Raya, M., Hubaux, J.-P.: Securing vehicular ad hoc networks. J. of Computer Security 15, 39–68 (2007)
4. Gennaro, R., Rohatgi, P.: How to sign digital streams. Inf. Comput. 165(1), 100–116 (2001)
5. Lysyanskaya, A., Tamassia, R., Triandopoulos, N.: Multicast authentication in fully adversarial networks. In: IEEE Security and Privacy, pp. 241–253 (2004)
6. Perrig, A., Canetti, R., Song, D.X., Tygar, J.D.: Efficient and secure source authentication for multicast. In: NDSS 2001, The Internet Society (2001)
7. Monnerat, J., Vaudenay, S.: Undeniable signatures based on characters: How to sign with one bit. In: Bao, F., Deng, R., Zhou, J. (eds.) PKC 2004. LNCS, vol. 2947, pp. 69–85. Springer, Heidelberg (2004)
8. Monnerat, J., Vaudenay, S.: Short 2-move undeniable signatures. In: Nguyên, P.Q. (ed.) VIETCRYPT 2006. LNCS, vol. 4341, pp. 19–36. Springer, Heidelberg (2006)
9. Boneh, D., Lynn, B., Shacham, H.: Short signatures from the Weil pairing. In: Boyd, C. (ed.) ASIACRYPT 2001. LNCS, vol. 2248, pp. 514–532. Springer, Heidelberg (2001)
10. Boneh, D., Boyen, X., Shacham, H.: Short group signatures. In: Franklin, M. (ed.) CRYPTO 2004. LNCS, vol. 3152, pp. 41–55. Springer, Heidelberg (2004)
11. Boneh, D., Shacham, H.: Group signatures with verifier-local revocation. In: CCS, pp. 168–177 (2004)
12. Chow, S.S.M., Yiu, S.-M., Hui, L.C.K.: Efficient identity based ring signature. In: Ioannidis, J., Keromytis, A.D., Yung, M. (eds.) ACNS 2005. LNCS, vol. 3531, pp. 499–512. Springer, Heidelberg (2005)
13. Boyen, X.: Mesh signatures. In: Naor, M. (ed.) EUROCRYPT 2007. LNCS, vol. 4515, pp. 210–227. Springer, Heidelberg (2007)
14. Camenisch, J.L., Hohenberger, S., Pedersen, M.Ø.: Batch verification of short signatures. In: Naor, M. (ed.) EUROCRYPT 2007. LNCS, vol. 4515, pp. 246–263. Springer, Heidelberg (2007), http://eprint.iacr.org/2007/172

15. Law, L., Matt, B.J.: Finding invalid signatures in pairing-based batches. In: Galbraith, S.D. (ed.) Cryptography and Coding 2007. LNCS, vol. 4887, pp. 34–53. Springer, Heidelberg (2007)
16. Ferrara, A.L., Green, M., Hohenberger, S., Pedersen, M.Ø.: Practical short signature batch verification, Cryptology ePrint Archive: Report 2008/015 (2008)
17. Chen, L., Cheng, Z., Smart, N.: Identity-based key agreement protocols from pairings, Cryptology ePrint Archive: Report 2006/199 (2006)
18. Bellare, M., Garay, J.A., Rabin, T.: Fast batch verification for modular exponentiation and digital signatures. In: Nyberg, K. (ed.) EUROCRYPT 1998. LNCS, vol. 1403, pp. 236–250. Springer, Heidelberg (1998)
19. Hess, F.: Efficient identity based signature schemes based on pairings. In: Nyberg, K., Heys, H.M. (eds.) SAC 2002. LNCS, vol. 2595, pp. 310–324. Springer, Heidelberg (2003)
20. Waters, B.: Efficient identity-based encryption without random oracles. In: Cramer, R. (ed.) EUROCRYPT 2005. LNCS, vol. 3494, pp. 114–127. Springer, Heidelberg (2005)
21. Naccache, D.: Secure and practical identity-based encryption, Cryptology ePrint Archive: Report 2005/369 (2005)
22. Chatterjee, S., Sarkar, P.: Trading time for space: Towards an efficient IBE scheme with short(er) public parameters in the standard model. In: Won, D.H., Kim, S. (eds.) ICISC 2005. LNCS, vol. 3935, pp. 424–440. Springer, Heidelberg (2006)
23. Chatterjee, S., Sarkar, P.: HIBE with short public parameters without random oracle. In: Lai, X., Chen, K. (eds.) ASIACRYPT 2006. LNCS, vol. 4284, pp. 145–160. Springer, Heidelberg (2006)
24. Lu, S., Ostrovsky, R., Sahai, A., Shacham, H., Waters, B.: Sequential aggregate signatures and multisignatures without random oracles. In: Vaudenay, S. (ed.) EUROCRYPT 2006. LNCS, vol. 4004, pp. 465–485. Springer, Heidelberg (2006)
25. Pastuszak, J., Michałek, D., Pieprzyk, J., Seberry, J.: Identification of bad signatures in batches. In: Imai, H., Zheng, Y. (eds.) PKC 2000. LNCS, vol. 1751, pp. 28–45. Springer, Heidelberg (2000)
26. Boneh, D., Lynn, B., Shacham, H.: Short signatures from the Weil pairing. Journal of Cryptology 17(4), 297–319 (2004)
27. Cha, J.C., Cheon, J.H.: An identity-based signature from gap Diffie-Hellman groups. In: Desmedt, Y.G. (ed.) PKC 2003. LNCS, vol. 2567, pp. 18–30. Springer, Heidelberg (2002)
28. Boyen, X., Waters, B.: Compact group signatures without random oracles. In: Vaudenay, S. (ed.) EUROCRYPT 2006. LNCS, vol. 4004, pp. 427–444. Springer, Heidelberg (2006)
29. Boneh, D., Gentry, C., Lynn, B., Shacham, H.: Aggregate and verifiably encrypted signatures from bilinear maps. In: Biham, E. (ed.) EUROCRYPT 2003. LNCS, vol. 2656, pp. 416–432. Springer, Heidelberg (2003)
30. Shao, Z.: Enhanced aggregate signatures from pairings. In: Feng, D., Lin, D., Yung, M. (eds.) CISC 2005. LNCS, vol. 3822, pp. 140–149. Springer, Heidelberg (2005)
31. Scott, M.: Multiprecision Integer and Rational Arithmetic C/C++ Library (MIRACL). Published by Shamus Software Ltd. (October 2007), http://www.shamus.ie/
32. Page, D., Smart, N., Vercauteren, F.: A comparison of MNT curves and supersingular curves. Applicable Algebra in Eng. Com. and Comp. 17(5), 379–392 (2006)
33. Adida, B., Chau, D., Hohenberger, S., Rivest, R.L.: Lightweight email signatures (Extended abstract). In: De Prisco, R., Yung, M. (eds.) SCN 2006. LNCS, vol. 4116, pp. 288–302. Springer, Heidelberg (2006)

Single-Layer Fractal Hash Chain Traversal with Almost Optimal Complexity

Dae Hyun Yum, Jae Woo Seo, Sungwook Eom, and Pil Joong Lee

Information Security Lab., EEE, POSTECH,
Pohang, Kyungbuk, 790-784, Republic of Korea
{dhyum,jwseo,sweom,pjl}@postech.ac.kr

Abstract. We study the problem of traversing a hash chain with dynamic helper points (called pebbles). Basically, two kinds of algorithms for this problem are known to date. Jakobsson algorithm is a single-layer fractal algorithm with the computational cost of $\lceil \log n \rceil$ (hash evaluations per chain link) and $\lceil \log n \rceil$ pebbles. Coppersmith-Jakobsson algorithm is a complicated double-layer fractal algorithm that improves efficiency at the expense of simplicity; with a complex movement pattern and some extra pebbles, it reduces the computational cost by half. Specifically, Coppersmith-Jakobsson algorithm requires $\lfloor \frac{1}{2} \log n \rfloor$ hash evaluations per chain link and $\lceil \log n \rceil + \lceil \log(\log n + 1) \rceil$ pebbles, which attains an almost optimal complexity. We introduce a new hash chain traversal algorithm that achieves both simplicity and efficiency. While our algorithm is based on the simple single-layer fractal structure of the Jakobsson algorithm, it reduces the computational cost by half without using extra pebbles; specifically, $\lceil \frac{1}{2} \log n \rceil$ hash evaluations per chain link and $\lceil \log n \rceil$ pebbles are needed.

1 Introduction

Hash chains have been used as an important cryptographic tool for various applications including payment systems [1,14], one-time password systems [5], multicast authentication [12,13], secure routing [7], and on-line auctions [16]. The popularity of hash chains mainly stems from the low computational cost for each output link. While public-key cryptographic primitives are still too heavy for many small devices, hash functions can be computed even by RFID tags. However, the situation becomes problematic when the length of hash chains increases. For example, computing links of hash chains with length 2^{32} is overwhelming for most mobile devices. To deal with long chains, hash chain traversal algorithms, the subject of this paper, study time-space tradeoffs and try to reduce the worst case computational cost per chain link with minimal storage.

A hash chain H for a cryptographic hash function $h : \{0,1\}^* \rightarrow \{0,1\}^l$ is a sequence of values $\langle v_0, v_1, \ldots, v_i, \ldots, v_n \rangle$, where v_n is a value chosen uniformly at random from $\{0,1\}^l$ and $v_i = h(v_{i+1})$. The beginning link v_0 and the end link v_n are the public key and the secret key, respectively. The budget b is the worst case computational cost per chain link. In other words, b hash function

M. Fischlin (Ed.): CT-RSA 2009, LNCS 5473, pp. 325–339, 2009.
© Springer-Verlag Berlin Heidelberg 2009

evaluations are allowed per chain link that is output. Hash chain traversal algorithms compute and output the hash chain H, (except the already published link v_0) starting from v_1 and ending with v_n. A trivial hash chain traversal algorithm is to recompute each output link from the secret key v_n. That is, one simply calculates $v_i = h^{n-i}(v_n)$ in round i, where $1 \leq i \leq n$ and $h^0(v_n) = v_n$. The budget of this solution is $n - 1$ and the storage requirement is 1. Another trivial algorithm is to precompute and store all links in memory and output the link v_i in round i by executing lookup operations. However, this solution requires too much memory to be used in small devices. One could easily obtain memory-computation tradeoffs in these trivial algorithms by storing some links and computing each output link from a near stored link. Such variations have a memory-times-computational complexity of $O(n)$.

Jakobsson algorithm. The first breakthrough in the hash chain traversal was made by Jakobsson [9]. Let us assume that there is a dynamic helper point (called a pebble) that stores a hash chain link for the position it is associated with. If the pebble stores a link v_d at the outset, the first output link v_1 can be computed with $d - 1$ hash function evaluations and the second output link v_2 with $d - 2$ hash function evaluations. Once the link v_d stored in the pebble is output, the computational cost of the next output link v_{d+1} will be that of $n - (d + 1)$ hash function evaluations or that of reaching the secret key (or the end link) v_n from the current output position. One can see that the budget is minimized if $d = \frac{n}{2}$, i.e., the pebble is located on the mid-point of the hash chain.

Assume that three links are stored; the secret key and two additional pebbles p_1 and p_2. Instead of spacing the three links $\frac{n}{3}$ apart (which would require a budget $\frac{n}{3} - 1$), p_2 is placed in the middle of the chain and p_1 in the middle of the first half; p_1 at position $\frac{n}{4}$ and p_2 at position $\frac{n}{2}$. One can see that this requires a budget $\frac{n}{4} - 1$ during the first $\frac{n}{2}$ rounds. After the first pebble p_1 at position $\frac{n}{4}$ has been reached in round $\frac{n}{4}$, it is relocated to the position n (of the secret key) and then gradually moved to the middle of the second half (i.e., position $\frac{3n}{4}$). Each such step costs one hash function evaluation, but the cost of all the steps is amortized from round $\frac{n}{4}$ to round $\frac{n}{2}$; the budget is first spent on the output link and then any "leftover computation" is applied to moving p_1. If the pebble p_1 reaches its destination of position $\frac{3n}{4}$ by round $\frac{n}{2}$, the budget $\frac{n}{4} - 1$ will also be enough for the remaining rounds (from round $\frac{n}{2}$ to round n). On the whole, p_1 recursively divides the chain with segments of length $\frac{n}{4}$ (and p_2 with segments of length $\frac{n}{2}$) by utilizing computational leftovers.

Assume that there are $\log n$ pebbles, where $\log \doteq \log_2$. Instead of splitting the chain with equally long segments, these pebbles (including the secret key) are located with exponentially increasing distances, i.e., at positions $2, 2^2, 2^3, ..., \frac{n}{2^2}, \frac{n}{2}, n$. For $1 \leq i \leq \log n$, each pebble p_i is placed at position 2^i in the beginning and recursively divides the chain with segments of length 2^i. Like fractals, where images can be found within images with the same shapes (but with difference sizes), the pebbles move within segments and sub-segments according to a highly symmetric pattern. As p_1 recursively divides the chain with segments of 2, the average cost of computing the next output is that of half a hash function evaluation. This makes the cost for relocation of pebbles the dominating portion of the total budget. It can be shown that the relocating pebbles will always arrive at their destinations "on time," if each relocating pebble moves two steps per round towards its destination. Since two consecutive pebbles need not be relocated simultaneously, we have at most $\frac{\log n}{2}$ "active" pebbles — still moving (or relocating) towards its destination — in each round and need no more than $\log n$ ($= \frac{\log n}{2}$ pebble $\times$ 2 step/pebble) hash evaluations per round. Based on these observations, Jakobsson presented a single-layer fractal algorithm that can traverse a hash chain of length n with a budget $\lceil \log n \rceil$ and total $\lceil \log n \rceil$ pebbles [9].

Coppersmith-Jakobsson algorithm. While the Jakobsson algorithm stresses simplicity over efficiency, the Coppersmith-Jakobsson algorithm takes the opposite approach; efficiency is improved at the expense of simplicity [3]. To consume leftover budgets completely, pebbles are partitioned into high-priority pebbles and low-priority pebbles. High-priority pebbles are relocated into already rather small segments, located close to the current output position. Low-priority pebbles, in turn, traverse larger distances, and further from the current output position. Therefore, the Coppersmith-Jakobsson algorithm has a double-layer fractal structure where the outer layer is created by low-priority pebbles and the inner layer by high-priority pebbles. Low-priority pebbles are only assigned those portions of the budget that remain after the current output has been computed and high-priority pebbles have exhausted their needs (i.e., arrived at their respective destinations). The role of low-priority pebbles is to soak up any computational leftovers and make sure high-priority pebbles stay inside small segments.

Assume that there are $\log n + \log(\log n + 1)$ pebbles. Just like the Jakobsson algorithm, each pebble p_i is located at position 2^i for $1 \le i \le \log n$. The remaining $\log(\log n + 1)$ extra pebbles are free at the outset and inserted when needed. In each round, "first things are done first." The current output link is computed and then any remaining budget is assigned to active high-priority pebbles, starting with the pebble with the lowest position (i.e., closest to the current output position). First then, any still remaining budget is assigned to active low-priority pebbles. While each pebble of the Jakobsson algorithm divides the chain with segments of fixed length, pebbles of the Coppersmith-Jakobsson algorithm are not assigned any predetermined task. Irrespective of the initial position, a high-priority pebble is relocated in the first segment (of length greater than two) that does not already contain an active pebble. Moreover, a high-priority pebble can become a low-priority pebble once it is reached by the current pointer (and vice versa). In other words, a pebble can belong to the inner layer of the fractal structure one time and to the outer layer another time. Based on the double-layer fractal structure with extra pebbles, the Coppersmith-Jakobsson algorithm can traverse a hash chain of length n with a budget $\lfloor \frac{1}{2} \log n \rfloor$ and total $\lceil \log n \rceil + \lceil \log(\log n + 1) \rceil$ pebbles, which can be shown almost optimal [3].

Our contribution. To accomplish better efficiency, the Coppersmith-Jakobsson algorithm combined two algorithms: a "greedy" algorithm that creates the inner layer of the fractal structure and a "sweeper" algorithm that sucks up all computational leftovers to build the outer layer. The greedy algorithm does not care about the future pebble movements and just fills up the first segment of length greater than two. Thanks to the outer layer of the fractal structure that is created by the sweep algorithm, the length of the hash chain looks relatively small in the greedy algorithm's view. The periods of the two fractal structures, however, do not match perfectly by the initial $\lceil \log n \rceil$ pebbles. Hence, $\lceil \log(\log n + 1) \rceil$ extra pebbles are required to harmonize them.

Our first observation on the Coppersmith-Jakobsson algorithm is that the greedy algorithm may not be the best choice for the entire traversal algorithm. If the greedy algorithm cares more about the whole picture (or the future pebble movements), we can reduce the burden of the sweeper algorithm, which may result in an improvement of the traversal algorithm. Our second observation is that we are able to use the extra pebbles more cleverly. The extra pebbles of the Coppersmith-Jakobsson algorithm are free in the beginning and inserted when needed. However, we can improve the traversal algorithm by storing whatever links in these extra pebbles. For example, if we store the first $\lceil \log(\log n + 1) \rceil$ links at the outset, we virtually reduce the total length of the hash chain by $\lceil \log(\log n + 1) \rceil$. Based on these observations, one could devise a hash chain traversal algorithm more efficient but more complex than the Coppersmith-Jakobsson algorithm. On the contrary, we aim for an algorithm more efficient and more straightforward than the Coppersmith-Jakobsson algorithm.

We introduce a new hash chain traversal algorithm that achieves both simplicity and efficiency. Surprisingly, pebbles in our algorithm are setup and recursively relocated exactly the same as those in the Jakobsson algorithm. Pebbles are setup at positions $2, 2^2, 2^3, ..., \frac{n}{2^2}, \frac{n}{2}, n$ and each pebble p_i at the initial position 2^i recursively divides the chain with segments of length 2^i. Therefore, our algorithm has a simple single-layer fractal structure and eliminates all the extra pebbles. The crux of our algorithm is that we can do this only with the budget $\lceil \frac{1}{2} \log n \rceil$. This seems a contradiction because the Jakobsson algorithm with the same pebble deployment strategy requires $\lceil \log n \rceil$ hash evaluations per chain link and the complex double-layer fractal structure and extra pebbles in the Coppersmith-Jakobsson algorithm are all to reduce the budget by half. Our algorithm uses the single-layer fractal structure but an improved stepping procedure. After a series of modifications and analyses, we found that the pebble relocation pattern of the Jakobsson algorithm is better than that of the Coppersmith-Jakobsson algorithm. However, instead of the stepping procedure of the Jakobsson algorithm (i.e., moving each active pebble two steps towards its destination in each round), we employ a different amortization principle; moving the leftmost active pebble (and repeatedly the next leftmost active pebble) towards its destination just as many steps as possible until the budget in the round is used up. Our analysis shows that active pebbles always arrive at their respective destinations on time with a budget $\lceil \frac{1}{2} \log n \rceil$. Therefore, our single-layer fractal algorithm can traverse a hash chain of length n with a budget $\lceil \frac{1}{2} \log n \rceil$ and total $\lceil \log n \rceil$ pebbles (without extra pebbles).

Related works. Influenced by amortization techniques proposed by Itkis and Reyzin [8], Jakobsson [9] introduced a hash chain traversal algorithm with a budget $\lceil \log n \rceil$ and $\lceil \log n \rceil$ pebbles. Coppersmith and Jakobsson [3] reduced the computational cost by half at the price of a more complex algorithm structure and extra pebbles: $\lfloor \frac{1}{2} \log n \rfloor$ hash evaluations per chain link and $\lceil \log n \rceil + \lceil \log(\log n + 1) \rceil$ pebbles. They also proved that their algorithm is near optimal by providing a theoretical lower bound for the most efficient algorithm possible (but not known to exist yet). Specifically, any traversal algorithm of hash chain of length n with α pebbles should have a budget b that is at least $\frac{1}{4\alpha} \log^2 n$. The optimal case is $\alpha = \frac{1}{2} \log n$, where the budget b is also $\frac{1}{2} \log n$. Thus, the Coppersmith-Jakobsson algorithm is (practically speaking) no more than a factor of two away from the optimal solution in terms of computation-times-storage complexity.

Sella [15] studied algorithms that traverse a hash chain of length n with a budget b, where b is a constant unrelated to n. One general algorithm and one specific algorithm were presented. The general algorithm for the budget $1 \leq b \leq \log n - 1$ has storage requirement of $(b+1) \cdot n^{1/(b+1)}$ and the specific algorithm for the budget $b = 1$ needs $2\sqrt{n}$ pebbles. The latter was also shown to be length-optimal; it can traverse the longest hash chain under given budget and storage. Kim [11] reduced the storage requirement of Sella's algorithm by $\frac{n^{1/(b+1)}-1}{n^{1/(b+1)}}$; total $(b+1) \cdot (n^{1/(b+1)}-1)$ pebbles are required. If the budget b comes near $\log n$, Kim's algorithm matches the memory and computation requirements of the Jakobsson algorithm but is more complex.

Jakobsson et al. [10] introduced a technique for traversal of hash trees (or Merkle trees) and proposed an efficient algorithm that generates a sequence of leaves along with their associated authentication paths. Szydlo [17] proposed an asymptotically optimal Merkle tree traversal algorithm that computes sequential tree leaves and authentication path data in time $2 \log N$ and space less than $3 \log N$ where N is the number of leaves. Berman et al. [2] investigated further tradeoffs between time and space requirements.

Efficient traversal algorithms for hierarchical (or multi-dimensional) chains were studied by Hu et al. [6]. They introduced the Sandwich-chain and Comb Skipchain in order to allow for both rapid generation and verification of intermediary links. Another interesting technique was taken by Fischlin [4], in which he shows how to augment the output from the hash chain traversal with a checksum in order to allow for faster verification of standard hash chain links.

On types of pebbles. In the Jakobsson algorithm [9], a pebble has a predetermined type τ and recursively divides the chain with segments of length τ. Meanwhile, the index i of a pebble p_i changes repeatedly by sorting. Consequently, we should say that p_i recursively divides the chain with segments of its type (rather than 2^i). However, for brevity's sake, we fix the index of each pebble and do not employ types. Note also that pebbles in the Coppersmith-Jakobsson algorithm [3] have neither types nor predetermined tasks.

On asymptotic complexity. The primary beneficiaries of hash chain traversal algorithms are applications with harsh constraints (e.g., small mobile devices, real-time systems, and heavily loaded servers). Hence, we avoid using asymptotic complexity notations such as $O(\cdot)$ and $\Theta(\cdot)$ because hidden constants are critical for the applications we consider. Actually, the Jakobsson algorithm [9], the Coppersmith-Jakobsson algorithm [3], and the theoretic optimal algorithm (not known to us yet) [3] have all the same complexity in the asymptotic sense.

2   Preliminaries

We mostly follow the terminology and basic definitions of previous works [3,9,15]. A *hash chain* H for a function $h : \{0,1\}^* \rightarrow \{0,1\}^l$ is a sequence of values $\langle v_0, v_1, \ldots, v_i, \ldots, v_n \rangle$, where v_n is a value chosen uniformly at random from $\{0,1\}^l$ and $v_i = h(v_{i+1})$. The function h is usually a cryptographic hash function or another publicly computable one-way function. We use the notation $h^0(x) = x$ and $h^i(x) = h(h^{i-1}(x))$ for $i \geq 1$. A single value v_i is referred to as a *link*. The *beginning link* v_0 and the *end link* v_n are the public key and the secret key, respectively. As v_0 is known publicly, the *length* of the hash chain $\langle v_0, v_1, \ldots, v_n \rangle$ is defined as n (not including v_0) or the number of links that are to be output. We assume that $n = 2^\kappa$ for some positive integer κ. A *hash chain traversal algorithm* for the hash chain $H = \langle v_0, v_1, \ldots, v_n \rangle$ computes and outputs links starting from v_1 and ending with v_n. The traversal algorithm is executed in n *rounds* and should output the i-th link v_i in round i.

Pictorially, we represent a hash chain as $(n+1)$ *points* that are equally spaced in a horizontal line, where points are placed from left to right, i.e., the i-th point is placed at position i and the $(i+1)$-th point on the right of the i-th point. Each point is associated with the link of its position; the i-th point is associated with the link v_i. The beginning link v_0 is also called the *leftmost* link of the chain and the end link v_n the *rightmost* link. A *segment* is a set of consecutive links (or sometimes the points associated with the links). Each segment has also its own leftmost link and rightmost link. Let $[i,j]$ denote a segment including points at $i, i+1, \ldots, j$ and $|[i,j]| = j - i + 1$ be the length of the segment. We introduce a *current pointer*, a kind of counter, that points the position i in round i. The link associated with the current pointer is called the *current (output) link*.

A *pebble* is a dynamic helper point that stores a specific link. If a pebble has position i, then it stores the associated link v_i whenever the value is available. We assume that the secret key v_n is stored in a pebble. Each pebble p_i also has its *starting position* (the position from which it starts), *current position* (the position at which it stays now), *destination* (the position to which it is going) and *status* (*ready, active,* or *arrived*). Here, a pebble that is *ready* has been assigned the starting position and destination but does not store its associated link (i.e., the link of the starting position) because it is not available at the moment. A pebble that is *active* has the starting position, destination, and its associated link, but has not yet arrived at the destination (i.e., still in motion towards the destination). A pebble is assigned status *arrived* if it is located at its destination and is still needed there (i.e., has not yet been reached by the current pointer).

We define the *budget b* as the number of computational units allowed per link of the sequence that is output. Here we only count hash function evaluations and not other computational operations associated with the algorithm execution. This is reasonable given the fact that the computational effort of performing one hash function evaluation far exceeds the remaining work per round. A traversal algorithm with a budget b and k pebbles *succeeds* in round i if it outputs the link v_i in round i with at most b hash evaluations and k stored links.

3   Algorithm

Setup. When a chain length n, a hash output length l, a description of hash function $h(\cdot)$, and κ ($= \log n$) pebbles are input, our setup algorithm chooses the end link v_n uniformly at random from $\{0, 1\}^l$ and locates pebbles with exponentially increasing distances. A pebble is modeled as an object that has four attributes: position, destination, value, and status. When a pebble is initialized, position is the same as destination. When a pebble is reached by the current pointer, position and destination are assigned a starting position and a destination, respectively. As a pebble moves to its destination, position is updated with its current position. The link associated with its position is stored in value whenever available. If the link is not available, value stores a special null value $\perp$. The attribute status can have one of three status values *ready, active*, and *arrived*. The setup algorithm assigns the *arrived* status to all pebbles and thus position and destination of each pebble have the same value. The setup algorithm is given in Algorithm 1, where $\leftarrow$ denotes the assignment operation, $\xleftarrow{R}$ the uniform random selection, and // comments. We say that a segment of length 2^β is in a *canonical form* if β pebbles are located in the segment and they store the 2^γ-th links of the segment where $1 \leq \gamma \leq \beta$.

Algorithm 1. Setup

INPUT: n, l, $h(\cdot)$ and p_i, where $n = 2^\kappa$ and $1 \leq i \leq \kappa$.

1. $v_n \xleftarrow{R} \{0, 1\}^l$;   // v_n is chosen uniformly at random from $\{0, 1\}^l$
2. $v_0 \leftarrow h^n(v_n)$;   // setup of the public key
3. $i \leftarrow 1$;
4. WHILE $i \leq \kappa$ DO   // setup of pebbles $p_1, p_2, \ldots, p_\kappa$
    p_i.position $\leftarrow 2^i$;   // pebbles are located in a canonical form
    p_i.destination $\leftarrow p_i$.position;
    p_i.value $\leftarrow h^{n-2^i}(v_n)$;   // p_i.value $= v_{2^i} = h^{n-2^i}(v_n)$
    p_i.status $\leftarrow$ *arrived*;
    $i \leftarrow i + 1$;

Each pebble p_i is located in position 2^i and stores its associated link v_{2^i} for $1 \leq i \leq \kappa$. The secret key v_n is stored in the rightmost pebble p_κ. Actually, the setup algorithm can be implemented with total n hash function evaluations, since p_i.value can be recycled in the computation of p_{i-1}.value. This is clear if we rewrite the setup algorithm as Algorithm 2.

Algorithm 2. Alternative setup

INPUT: n, l, $h(\cdot)$ and p_i, where $n = 2^\kappa$ and $1 \leq i \leq \kappa$.

1. $v_n \xleftarrow{R} \{0,1\}^l$;
2. $i \leftarrow \kappa$;   // setup of p_κ.
   p_i.position $\leftarrow n$;
   p_i.destination $\leftarrow p_i$.position;
   p_i.value $\leftarrow v_n$;
   p_i.status $\leftarrow$ *arrived*;
   $i \leftarrow i - 1$;
3. WHILE $i > 0$ DO   // setup of $p_{\kappa-1}, p_{\kappa-2}, \ldots, p_1$.
   p_i.position $\leftarrow \frac{p_{(i+1)}\cdot\text{position}}{2}$;
   p_i.destination $\leftarrow p_i$.position;
   p_i.value $\leftarrow h^{p_i \cdot \text{position}}(p_{(i+1)}.\text{value})$;
   p_i.status $\leftarrow$ *arrived*;
   $i \leftarrow i - 1$;
4. $v_0 \leftarrow h^2(p_1.\text{value})$;   // setup of the public key

---

The rightmost pebble p_κ requires no hash function evaluation. Pebbles (in the WHILE loop) $p_{\kappa-1}, p_{\kappa-2}, \ldots, p_1$ require $\frac{n}{2}, \frac{n}{2^2}, \ldots, 2$ hash function evaluations, where $\frac{n}{2} + \frac{n}{2^2} + \cdots + 2 = \frac{\frac{n}{2}\left(1 - \frac{1}{2^{\kappa-1}}\right)}{1 - \frac{1}{2}} = n - 2$. The public key v_0 needs 2 hash function evaluations. Therefore, total n hash function evaluations are required.

Traversal. The proposed traversal algorithm is given in Algorithm 3 and examples for $n = 16$ and 64 (or $\kappa = 4$ and 6) are given in Appendix A. In each round, the traversal algorithm first outputs the current output link and then spends the remaining budget to relocate pebbles for future pebble movements. After p_i has been reached by the current pointer, it will be assigned the new starting position of p_i.position $+ 3 \cdot 2^i$ and destination of p_i.destination $+ 2 \cdot 2^i$ to divide a segment of length 2^{i+1} (from p_i.position $+ 2^i + 1$ to p_i.position $+ 3 \cdot 2^i$) with two sub-segments of length 2^i (from p_i.position $+ 2^i + 1$ to p_i.position $+ 2 \cdot 2^i$ and from p_i.position $+ 2 \cdot 2^i + 1$ to p_i.position $+ 3 \cdot 2^i$). By this way, p_i recursively divides the chain with segments of length 2^i. If the link associated with the new starting position is not available, a special null value $\perp$ is stored in p_i.value and the status becomes *ready*. Since the new starting position of p_i corresponds to the destination of another pebble with index greater than i, the link of the new starting position will be available eventually. Once the link becomes available, p_i stores it and changes the status to *active*. Therefore, a newly arrived pebble always checks whether or not there is a *ready* pebble in the destination. When the round 2^i is finished, pebbles of the segment $[2^i + 1, 2^{i+1}]$ are located in a canonical form. According to Algorithm 3, the leftmost arrived pebble will always be located in the first even position from the current pointer inclusive.

Therefore, the link (say v_i) associated with the leftmost arrived pebble is output in even rounds and $h(v_i)$ in odd rounds.

To facilitate description, we define a utility function $\mathcal{P} : \{0, 1\}^* \times \mathbb{N} \to \mathbb{N}$.

- $\mathcal{P}(\alpha, \beta)$ returns the index of a pebble satisfying status $= \alpha$ and position $= \beta$.
- $\mathcal{P}(\alpha, 0)$ returns the index of the leftmost pebble satisfying status $= \alpha$.
- If there is no pebble satisfying the condition, $\mathcal{P}$ returns $\bot$.

In Algorithm 3, the utility function $\mathcal{P}$ is used to access pebbles of two categories; the leftmost arrived/active pebble and the arrived/ready pebble in a specific position. This allows us to simplify the implementation of $\mathcal{P}$ in various ways. For example, the first category can be dealt with ordinary pointers (rather than a function). However, the implementation details of the utility function $\mathcal{P}$ are not important to us, because the hash function evaluation is much more expensive than the evaluation of $\mathcal{P}$ with $\lceil \log n \rceil$ pebbles for all practical purposes.

Algorithm 3. Traversal

INPUT: n, $h(\cdot)$ and p_i, where $n = 2^\kappa$ and $1 \le i \le \kappa$.
  0. $b = \lceil \frac{\kappa}{2} \rceil$, $current = 0$;   // initialization of the budget and the current pointer
  1. IF $current = n$ THEN
          HALT;
      ELSE
          $current \leftarrow current + 1$;   // a new round begins
          $available \leftarrow b$;   // the remaining budget is set
  2. $i \leftarrow \mathcal{P}(arrived, 0)$;   // the leftmost arrived pebble is selected
      IF $current \bmod 2 = 1$ THEN   // if $current$ is odd
          OUTPUT $h(p_i.\text{value})$;
          $available \leftarrow available - 1$;
      ELSE   // if $current$ is even
          OUTPUT $p_i.\text{value}$;
          $p_i.\text{position} \leftarrow p_i.\text{position} + 3 \cdot 2^i$;   // a new starting position is assigned
          $p_i.\text{destination} \leftarrow p_i.\text{destination} + 2 \cdot 2^i$;   // a new destination is assigned
          $p_i.\text{value} \leftarrow \bot$;
          $p_i.\text{status} \leftarrow ready$;
          $j \leftarrow \mathcal{P}(arrived, p_i.\text{position})$;
          IF $j \ne \bot$ THEN   // if the link of the starting position is available
              $p_i.\text{value} \leftarrow p_j.\text{value}$;
          $p_i.\text{status} \leftarrow active$;
  3. $i \leftarrow \mathcal{P}(active, 0)$;   // the leftmost active pebble is selected
      IF $i = \bot$ THEN   // if there is no active pebble
          GO TO 1;
      WHILE $available > 0$ DO
          $p_i.\text{position} \leftarrow p_i.\text{position} - 1$;   // p_i moves one step towards its destination
          $p_i.\text{value} \leftarrow h(p_i.\text{value})$;   // p_i upates its associated link
          $available \leftarrow available - 1$;
          IF $p_i.\text{position} = p_i.\text{destination}$ THEN   // if p_i arrives at its destination
              $p_i.\text{status} \leftarrow arrived$;
              $j \leftarrow \mathcal{P}(ready, p_i.\text{destination})$;
              IF $j \ne \bot$ THEN
                  $p_j.\text{value} \leftarrow p_i.\text{value}$;
                  $p_j.\text{status} \leftarrow active$;
              GO TO 3;
      GO TO 1;   // this line is reached only if $available = 0$

---

4   Claims

Theorem 1. *The proposed hash chain traversal algorithm succeeds in round i, $1 \leq i \leq n$, for a length $n = 2^\kappa$, budget $b = \lceil \frac{\kappa}{2} \rceil$, and κ pebbles.*

Proof. For $\kappa = 1, 2, 3$, one can easily check the validity of the claim by executing the algorithm (like examples in Appendix A). Therefore, we assume $\kappa \geq 4$ ($b = \lceil \frac{\kappa}{2} \rceil \geq 2$) and prove Theorem 1 with a series of lemmata. Let segment-2 = $[1, 2]$ and segment-2^i = $[2^{i-1} + 1, 2^i]$ for $2 \leq i \leq \kappa$. Then, |segment-2| = 2 and |segment-2^i| = 2^{i-1} for $2 \leq i \leq \kappa$. The entire chain is composed of segment-2, segment-$2^2, \ldots,$ segment-2^κ. We begin with two trivial bootstrapping lemmata. Lemma 1 follows easily from $b \geq 2$ and Lemma 2 from p_2.position = 2^2 by the setup algorithm.

Lemma 1. *The traversal algorithm succeeds in* segment-2. □

Lemma 2. segment-2^2 *is in a canonical form when round $3 (= 2^1 + 1)$ begins.*

□

We define the *cumulative budget* of a segment as the sum of the budgets in the segment and the *cumulative expenditure* of a segment as the hash function evaluations required by the algorithm during the execution of the segment (i.e., while points in the segment are current). Let budget-2^i and expenditure-2^i be the cumulative budget and the cumulative expenditure in segment-2^i for $2 \leq i \leq \kappa - 1$. Then, the balance of segment-2^i can be defined by balance-2^i = budget-2^i − expenditure-2^i.

Lemma 3. *Assume that* segment-2^i *for $2 \leq i \leq \kappa - 1$ is in a canonical form when round $2^{i-1} + 1$ begins. Then, we have* balance-$2^i \geq 0$.

Proof. The cumulative budget budget-2^i = $2^{i-1}b$ is obtained simply by length 2^{i-1} times budget b. The cumulative expenditure expenditure-2^i consists of three parts. First, we need $2^{i-1} \cdot \frac{1}{2}$ hash evaluations to compute the output links; 1 hash evaluation in odd round and 0 hash evaluation in even round. Second, we need to relocate pebbles recursively in segment-2^i = $[2^{i-1} + 1, 2^i]$. When round $2^{i-1} + 1$ begins, pebbles in segment-2^i are in a canonical form. This means that $p_1, p_2, \ldots, p_{i-2}$ are located at positions $2^{i-1} + 2, 2^{i-1} + 2^2, \ldots, 2^{i-1} + 2^{i-2}$ and p_i at 2^i. Note that p_1 is the leftmost pebble, p_{i-2} is the midpoint pebble, p_i is the rightmost pebble, and p_{i-1} is not located in segment-2^i. Since p_j divides segment-2^i of length 2^{i-1} with sub-segments of length 2^j, p_i and p_{i-2} are not relocated in segment-2^i. p_{i-3} is relocated once ($= 2 - 1$) in segment-2^i, p_{i-4} three times ($= 2^2 - 1$), p_{i-5} seven times ($= 2^3 - 1$), and so on. Therefore, $2^{i-3}(2 - 1) + 2^{i-4}(2^2 - 1) + 2^{i-5}(2^3 - 1) + \cdots + 2(2^{i-3} - 1) = (i - 4)2^{i-2} + 2$ is the required number of hash evaluations.[1] Third, pebbles should be located in a canonical form in segment-2^{i+1} until round 2^i is finished. The worst case is

---

[1] This equation was derived for $i \geq 4$. However, it is also valid for $i = 2, 3$, because $(i - 4)2^{i-2} + 2 = 0$ for $i = 2, 3$.

that when round $2^{i-1} + 1$ begins, segment-2^{i+1} of length 2^i has only one pebble p_{i+1} at position 2^{i+1}. In that case, we need $2^i - 2$ hash evaluations. By summing up the three parts, we have the cumulative expenditure expenditure-$2^i = (2^{i-1} \cdot \frac{1}{2}) + ((i-4)2^{i-2} + 2) + (2^i - 2) = 2^{i-2}(i+1)$. From budget-$2^i$ and expenditure-2^i, we have balance-$2^i = 2^{i-1}b - 2^{i-2}(i+1) = 2^{i-2}(2b - i - 1) \geq 0$, where $2^{i-2} \geq 1$ and $2b - i - 1 = 2\lceil \frac{\kappa}{2} \rceil - i - 1 \geq 2 \cdot \frac{\kappa}{2} - i - 1 = \kappa - i - 1 \geq 0$ for $2 \leq i \leq \kappa - 1$. $\square$

From Lemma 3, we know that balance-$2^i \geq 0$. However, this does not guarantee the success of the algorithm because there might not be enough active pebbles to soak up computational leftovers in some rounds, which results in waste of budgets. The following lemma shows that pebbles are recursively relocated on time.

Lemma 4. *Assume that* segment-2^i *for* $2 \leq i \leq \kappa - 1$ *is in a canonical form when round* $2^{i-1} + 1$ *begins. Then, the traversal algorithm succeeds in* segment-2^i *and, before round* $2^i + 1$ *begins,* segment-2^{i+1} *gets to be in a canonical form.*

Proof. (Sketch)[2] When round $2^{i-1}+1$ begins, $p_1, p_2, \ldots, p_{i-2}$ are located at positions $2^{i-1}+2, 2^{i-1}+2^2, \ldots, 2^{i-1}+2^{i-2}$ and p_i at 2^i in segment-$2^i = [2^{i-1}+1, 2^i]$. As p_j for $1 \leq j \leq i - 2$ divides segment-2^i with sub-segments of length 2^j, p_j in round $2^i - 2^j$ outputs its stored link, leaves segment-2^i, and obtains a new starting position and destination in the next segment segment-2^{i+1}. Hence, pebbles leave segment-2^i in the order of p_{i-2}, p_{i-3}, $\ldots$, p_2, and p_1. Note that p_i leaves segment-2^i in round 2^i and then is relocated in segment-2^{i+2} (not segment-2^{i+1}). We assume the worst case that during round 2^i, p_i stays at its new starting position (in segment-2^{i+2}) and does not move towards its new destination, i.e., p_i does not consume any leftover budget during the execution of segment-2^i.

For $2 \leq j \leq i - 1$, let $S(p_{i-j})$ be the sub-segment of segment-2^i where p_{i-j} has the largest index during execution of segment-2^i. For example, $S(p_{i-2}) = [2^i - 2^{i-1} + 1, 2^i - 2^{i-2}]$, $S(p_{i-3}) = [2^i - 2^{i-2} + 1, 2^i - 2^{i-3}]$ and $S(p_{i-j}) = [2^i - 2^{i-j+1} + 1, 2^i - 2^{i-j}]$. Let $S'(p_{i-j}) = [2^i - 2^{i-j} + 1, 2^i]$ be the remaining sub-segment of segment-2^i after $S(p_{i-j})$ has been executed. Then, $|S(p_{i-j})| = |S'(p_{i-j})| = 2^{i-j}$. We also define $S(p_0) = [2^i - 1, 2^i]$ and $S'(p_0) = \phi$. During the execution of $S(p_{i-j})$, the traversal algorithm should (1) output 2^{i-j} links of $S(p_{i-j})$, which requires 2^{i-j-1} hash evaluations, (2) relocate pebbles recursively in $S(p_{i-j})$, which requires $(i - j - 3)2^{i-j-1} + 2$ hash evaluations, (3) relocate pebbles in $S'(p_{i-j})$, which requires $2^{i-j} - 2$ hash evaluations, and (4) move p_{i-j+1} to its destination, which requires 2^{i-j+1} hash evaluations. For $1 \leq i \leq \kappa - 1$ and $2 \leq j \leq i - 1$, the cumulative budget of $S(p_{i-j})$ is $2^{i-j}b \geq 2^{i-j-1}\kappa$, which is greater than $(1) + (2) + (3) = 2^{i-j-1}(i - j)$. Similarly, the cumulative budget of $S(p_0)$ is $2b \geq 4$, which is greater than $(1) + (2) + (3) = 1 + 0 + 0$. Therefore, the traversal algorithm succeeds in segment-2^i. However, it is not trivial that segment-2^{i+1} gets to be in a canonical form before the round 2^i+1 begins, because

---

[2] To prove Lemma 4, we should consider all fractal repetitions of segment-2^i. Instead of providing complex mathematical induction, we focus on the first level repetition of segment-2^i to give intuition. We leave the complete proof of Lemma 4 to readers.

the cumulative budget $2^{i-j}b$ is not always greater than $(1) + (2) + (3) + (4) = 2^{i-j-1}(i-j) + 2^{i-j+1}$.

Let balance-$S(p_{i-j}) = 2^{i-j}b - 2^{i-j-1}(i-j) - 2^{i-j+1}$ be the balance of the sub-segment $S(p_{i-j})$. When round $2^i - 2^{i-j+1} + 1$ begins, $S(p_{i-j})$ is in a canonical form and the pebble p_{i-j+1} of either *active* or *ready* status stays at the starting position in segment-2^{i+1}. If p_{i-j+1} does not arrive at its destination before round $2^i - 2^{i-j} + 1$, p_{i-j+1} has consumed computational leftovers of $S(p_{i-j})$ completely. Otherwise, there might be waste of budgets after the arrival of p_{i-j+1} at the destination. Suppose that for every j such that $2 \leq j \leq i - 1$, p_{i-j+1} has not arrived at the destination in segment-2^{i+1} during the execution of $S(p_{i-j})$. Then, there is no waste of budgets before the execution of $S(p_0)$ and by Lemma 3, p_1 can arrive at the destination on time, meaning that segment-2^{i+1} gets to be in a canonical form. If p_{i-j+1} for some $2 \leq j \leq i - 1$ has arrived at the destination in segment-2^{i+1} during the execution of $S(p_{i-j})$ (which implies balance-$S(p_{i-j}) \geq 0$), all the following pebbles, i.e., $p_{i-j}, p_{i-j-1}, \ldots, p_1$ also arrive at their respective destinations on time because balance-$S(p_{i-j-v}) = 2^{i-j-v}b - 2^{i-j-v-1}(i-j-v) - 2^{i-j-v+1} = 2^{-v}\text{balance-}S(p_{i-j}) + 2^{i-j-v-1}v \geq 0$ for $0 \leq v \leq i-j-1$ and balance-$S(p_0) = 2b - 3 \geq 0$. Consequently, segment-2^{i+1} gets to be in a canonical form.   □

By Lemma 1, the traversal algorithm succeeds in round 1 and round 2. By Lemma 2, segment-2^2 is in a canonical form when round $3(= 2^1 + 1)$ begins. By Lemma 4 (with Lemma 2), the traversal algorithm succeeds in segment-2^i for $2 \leq i \leq \kappa - 1$. In other words, the traversal algorithm succeeds in the first half of the chain, i.e., rounds $1, 2, \ldots, 2^{\kappa-1}$. Now we only have to show that the traversal algorithm succeeds in the second half of the chain, i.e., rounds $2^{\kappa-1} + 1, \ldots, 2^\kappa$.

Lemma 5. *If the traversal algorithm succeeds in the first half of the chain, i.e.,* segment-2, segment-$2^2, \ldots,$ segment-$2^{\kappa-1}$, *then it also succeeds in the second half of the chain, i.e.,* segment-2^κ.

Proof. Lemma 5 follows from the symmetry of the first half and the second half. When round 1 begins, the first half of the chain is in a canonical form by the setup algorithm. From Lemma 1, 2, and 4, the traversal algorithm succeeds in the first half and when round $2^{\kappa-1} + 1$ begins, the second half (i.e., segment-2^κ) is also in a canonical form. Therefore, the traversal algorithm succeeds in the second half, just like in the first half. Actually, the traversal in the second half is much easier because the second half can use all budgets for itself, i.e., there is no future pebble movements after segment-2^κ.   □

Theorem 1 follows from Lemma 1, 2, 4, and 5.   Q.E.D.

Acknowledgements. This research was supported by the KT and the BK21. We would like to thank the anonymous reviewers of CT-RSA 2009 for their comments and suggestions.

References

1. Anderson, R.J., Manifavas, C., Sutherland, C.: Netcard - a practical electronic-cash system. In: Lomas, M. (ed.) Security Protocols 1996. LNCS, vol. 1189, pp. 49–57. Springer, Heidelberg (1997)
2. Berman, P., Karpinski, M., Nekrich, Y.: Optimal trade-off for merkle tree traversal. Theor. Comput. Sci. 372(1), 26–36 (2007)
3. Coppersmith, D., Jakobsson, M.: Almost optimal hash sequence traversal. In: Blaze, M. (ed.) FC 2002. LNCS, vol. 2357, pp. 102–119. Springer, Heidelberg (2003)
4. Fischlin, M.: Fast verification of hash chains. In: Okamoto, T. (ed.) CT-RSA 2004. LNCS, vol. 2964, pp. 339–352. Springer, Heidelberg (2004)
5. Haller, N.: The s/key one-time password system. RFC 1760. Internet Engineering Task Force (1995)
6. Hu, Y.-C., Jakobsson, M., Perrig, A.: Efficient constructions for one-way hash chains. In: Ioannidis, J., Keromytis, A.D., Yung, M. (eds.) ACNS 2005. LNCS, vol. 3531, pp. 423–441. Springer, Heidelberg (2005)
7. Hu, Y.-C., Perrig, A., Johnson, D.B.: Ariadne: A secure on-demand routing protocol for ad hoc networks. Wireless Networks 11(1-2), 21–38 (2005)
8. Itkis, G., Reyzin, L.: Forward-secure signatures with optimal signing and verifying. In: Kilian, J. (ed.) CRYPTO 2001. LNCS, vol. 2139, pp. 332–354. Springer, Heidelberg (2001)
9. Jakobsson, M.: Fractal hash sequence representation and traversal. In: IEEE International Symposium on Information Theory, pp. 437–444. IEEE, Los Alamitos (2002); also available at Cryptology ePrint Archive, Report 2002/001, http://eprint.iacr.org/
10. Jakobsson, M., Leighton, T., Micali, S., Szydlo, M.: Fractal merkle tree representation and traversal. In: Joye, M. (ed.) CT-RSA 2003. LNCS, vol. 2612, pp. 314–326. Springer, Heidelberg (2003)
11. Kim, S.-R.: Improved scalable hash chain traversal. In: Zhou, J., Yung, M., Han, Y. (eds.) ACNS 2003. LNCS, vol. 2846, pp. 86–95. Springer, Heidelberg (2003)
12. Perrig, A., Canetti, R., Song, D.X., Tygar, J.D.: Efficient and secure source authentication for multicast. In: NDSS 2001. The Internet Society (2001)
13. Perrig, A., Canetti, R., Tygar, J.D., Song, D.X.: Efficient authentication and signing of multicast streams over lossy channels. In: IEEE Symposium on Security and Privacy, pp. 56–73. IEEE Computer Society, Los Alamitos (2000)
14. Rivest, R.L., Shamir, A.: Payword and micromint: Two simple micropayment schemes. In: Lomas, M. (ed.) Security Protocols 1996. LNCS, vol. 1189, pp. 69–87. Springer, Heidelberg (1997)
15. Sella, Y.: On the computation-storage trade-offs of hash chain traversal. In: Wright, R.N. (ed.) FC 2003. LNCS, vol. 2742, pp. 270–285. Springer, Heidelberg (2003)
16. Stubblebine, S.G., Syverson, P.F.: Fair on-line auctions without special trusted parties. In: Franklin, M.K. (ed.) FC 1999. LNCS, vol. 1648, pp. 230–240. Springer, Heidelberg (1999)
17. Szydlo, M.: Merkle tree traversal in log space and time. In: Cachin, C., Camenisch, J.L. (eds.) EUROCRYPT 2004. LNCS, vol. 3027, pp. 541–554. Springer, Heidelberg (2004)

Appendix A

We illustrate the proposed hash chain traversal algorithm with two examples. Fig. 1 shows execution of the algorithm for a hash chain of length $n = 16$, in which budget $b = 2$ and total 4 pebbles are required. Fig. 2 is for a hash chain of length $n = 64$, in which budget $b = 3$ and total 6 pebbles are required. Note that Fig. 2 contains rounds 1–32 and omits rounds 33–64. The latter is executed just as the former because of the symmetry of the two.

Legends in Fig. 1 and Fig. 2 are defined as follows.

C	current pointer
$\overline{P_i}$	pebble p_i with *active* status
P_i	pebble p_i with *arrived* status
(P_i)	pebble p_i with *ready* status
$P_{i,j}$	*active* pebble p_i and *arrived* pebble p_j at the same position
←--	trace of a pebble

For example, P_1 ←-- P_3 of round 2 in Fig. 1 is the result of a series of actions; (1) P_1 is assigned a new starting position 8 and destination 6, (2) P_1 is assigned its link and status *active* because P_3 is already at position 8, (3) P_1 moves from position 8 to position 7, and (4) P_1 moves from position 7 to position 6 (destination) and is assigned status *arrived*.

Round \ Position	1	2	3	4	5	6	7	8	9	10	11	12	13	14	15	16
Setup		P_1		P_2				P_3								P_4
1	C	P_1		P_2				P_3								P_4
2		C		P_2		P_1	←--	P_3								P_4
3			C	P_2		P_1		P_3								P_4
4				C		P_1		P_3						$\overline{P_2}$	←--	P_4
5					C	P_1		P_3					$\overline{P_2}$			P_4
6						C		P_3		$\overline{P_1}$	P_2					P_4
7							C	P_3		P_1	P_2					P_4
8								C		P_1	P_2					P_4
9									C	P_1	P_2					P_4
10										C	P_2		P_1	←--		P_4
11											C	P_2	P_1			P_4
12												C	P_1			P_4
13													C	P_1		P_4
14														C		P_4
15															C	P_4
16																C

Fig. 1. Execution of the proposed hash chain traversal for $n = 16$

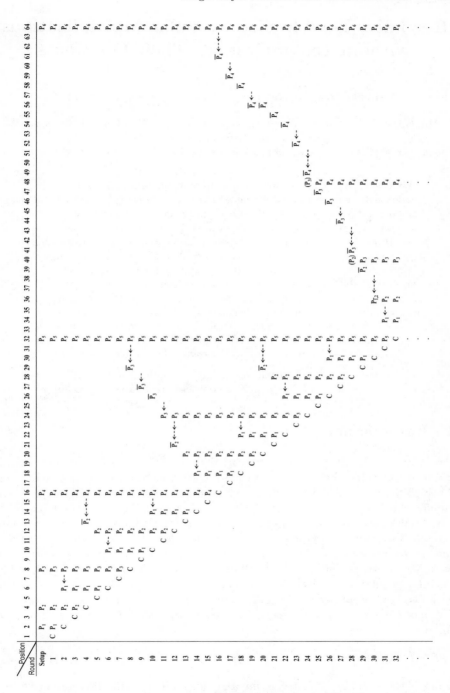

Fig. 2. Execution of the proposed hash chain traversal for $n = 64$

Recursive Double-Size Modular Multiplications without Extra Cost for Their Quotients

Masayuki Yoshino, Katsuyuki Okeya, and Camille Vuillaume

Hitachi, Ltd., Systems Development Laboratory, 292, Yoshida-cho, Totsuka-ku,
Yokohama, Japan
{masayuki.yoshino.aa,katsuyuki.okeya.ue,camille.vuillaume.ch}@hitachi.com

Abstract. A technique for computing the quotient ($\lfloor ab/n \rfloor$) of Euclidean divisions from the difference of two remainders ($ab \pmod{n} - ab \pmod{n+1}$) was proposed by Fischer and Seifert. The technique allows a 2ℓ-bit modular multiplication to work on most ℓ-bit modular multipliers. However, the cost of the quotient computation rises sharply when computing modular multiplications larger than 2ℓ bits with a recursive approach. This paper addresses the computation cost and improves on previous 2ℓ-bit modular multiplication algorithms to return not only the remainder but also the quotient, resulting in an higher performance in the recursive approach, which becomes twice faster in the quadrupling case and four times faster in the octupling case. In addition to Euclidean multiplication, this paper proposes a new 2ℓ-bit Montgomery multiplication algorithm to return both of the remainder and the quotient.

Keywords: modular multiplication, efficient implementation, RSA, arithmetic unit, low-end device, crypto-coprocessors, double-size technique.

1 Introduction

Cryptographic hardware accelerators are dedicated units for fast modular multiplication and typically process heavy modular exponentiation in low-end devices. Since they are penalized by a fixed operand size, Fischer and Seifert proposed an efficient 2ℓ-bit modular multiplication technique calling only the ℓ-bit arithmetic units based on Euclidean division which returns not only an ℓ-bit remainder but also an ℓ-bit quotient, i.e., on input a, b and n, both $ab \pmod{n}$ and $\lfloor ab/n \rfloor$ are returned. Such modular operation can be supported by most hardware architectures; indeed, the arithmetic units involved in the execution of a modular reduction can be easily enriched to simultaneously output a quotient with little extra hardware cost. Furthermore, software techniques for computing the *quotient* ($\lfloor ab/n \rfloor$) from the difference of two *remainders* ($ab \pmod{n} - ab \pmod{n+1}$) were also proposed [FS03]:

$$ab = qn + r = q(n+1) + (r-q) = (q-1)(n+1) + (r-q+n+1),$$

which is the keypoint of constructing a 2ℓ-bit modular multiplication algorithm.

The previous doubling techniques consist of two steps. The first step is to compute a 2ℓ-bit (not modular) multiplication step by step using the above

M. Fischlin (Ed.): CT-RSA 2009, LNCS 5473, pp. 340–356, 2009.
© Springer-Verlag Berlin Heidelberg 2009

relation $ab = qn + r$. Then, the second step reduces the 4ℓ-bit product with modulus $N = n_1 2^\ell + n_0$ using the equation $n_1 2^\ell = -n_0 \pmod{N}$: the 2ℓ-bit product $n_1 2^\ell$ is reduced to one negative ℓ-bit integer $(-n_0)$. Using successive reductions, the 4ℓ- bit product can be reduced to a 2ℓ-bit integer, that is, the *remainder* of double-size modular multiplication, which is the output of this algorithm.

Taking into account the above tricks, one may consider the natural question:

> *Question:* Is it possible to avoid the 2ℓ-bit quotient computation when the double-size technique is applied recursively ?

The motivation comes from the following fact: although double-size techniques output *remainders* only, the requirement for the underlying ℓ-bit multiplier is to return both of *quotients* and *remainders*, or simulate such a multiplier in software using two *remainders*.

This paper improves the double-size modular multiplication algorithms so as to return not only the remainder but also the quotient. Unlike the previous techniques which are penalized by the heavy cost of generating double-size quotients, our approaches are liberated from such additional computation: the new techniques preserve the quotient while processing modular multiplication, and return both a remainder and a quotient without any additional cost compared with previous techniques returning only a remainder. In a similar scenario with Chevallier-Mames et al., one of our schemes requires 6 calls for double-size computations, but only 36 calls in the $2^2\ell$-bit case, and only 216 calls in the $2^3\ell$-bit case. Furthermore, in order to speed up performances, this paper proposes several algorithms corresponding to different environments: with different modular multiplication algorithms, depending on whether or not precomputations are feasible, or depending on the modulus value. As a result, our proposals result in a performance improvement at least twice as fast as previous techniques in any environment.

The rest of this paper is organized as follows. Section 2 introduces previous techniques for both of *Euclidean* and *Montgomery* multiplications, and shows their issues for the recursive approach. Section 3 proposes a series of new algorithms to compute both a quotient and a remainder using *Euclidean* or *Montgomery* multiplication units. Section 4 shows another approach to compute the quadrupling case with precomputations, without calling the double-size techniques recursively. Section 5 compares the computational cost of our proposals to the previous, and finally we conclude the paper in Section 6.

2 Related Work for Quadrupling Modular Multiplication

2.1 Previous Double-Size Techniques

Double-size techniques strongly depend on the design of crypto coprocessors and especially the modular multiplication algorithm [FS03, CJP03, YOV07a, YOV07b].

Fischer and Seifert introduced the following two instructions based on the *Euclidean* division returning not only a remainder but also a quotient.

Definition 1. *For numbers,* $0 \le a, b, n, t$, EucMul^1 *instruction and* $\mathsf{EucMulInit}^1$ *instruction are defined as*

$$(\lfloor ab/n \rfloor, ab \pmod{n}) \leftarrow \mathsf{EucMul}(a, b, n),$$
$$(\lfloor (ab + t2^\ell)/n \rfloor, ab + t2^\ell \pmod{n}) \leftarrow \mathsf{EucMulInit}(a, b, t, n)$$

where a, b, n and t are ℓ-bit integers.

It is implicitly required in [FS03, CJP03] that both instructions can work with negative operands a, b and t: the processor is able to handle such operands with a signed representation and outputs correct results. Furthermore, non-reduced inputs such that $|a| > n$ are also acceptable.

Alg.1. Double-size modular multiplication 1

Input: $A = a_1 2^\ell + a_0$, $B = b_1 2^\ell + b_0$ and $N = n_1 2^\ell + n_0$;
Output: $AB \pmod{N}$;

$(q_1, r_1) \leftarrow \mathsf{EucMul}(a_1, b_1, n_1)$
$(q_2, r_2) \leftarrow \mathsf{EucMul}(q_1, n_0, 2^\ell)$
$(q_3, r_3) \leftarrow \mathsf{EucMul}(a_0 + a_1, b_0 + b_1, 2^\ell - 1)$
$(q_4, r_4) \leftarrow \mathsf{EucMul}(a_0, b_0, 2^\ell)$
$(q_5, r_5) \leftarrow \mathsf{EucMul}(2^\ell - 1, -q_2 + q_3 - q_4 + r_1, n_1)$
$(q_6, r_6) \leftarrow \mathsf{EucMul}(q_5, n_0, 2^\ell)$
Return $(-q_6 - r_2 + r_3 - r_4 + r_5)2^\ell + (r_2 + r_4 - r_6)$

One of the fastest double-size algorithms is introduced in Algorithm 1 (with six EucMul instruction calls) [CJP03]. Algorithm 1 applies a simple radix representation with base 2^ℓ such that $A = a_1 2^\ell + a_0$ and outputs a remainder only: $AB \pmod{N}$. Other algorithms proposed by Chevallier-Mames et al. use both EucMul and $\mathsf{EucMulInit}$ instructions or apply more sophisticated representations with fewer instruction calls. The number of calls are shown in Table 1 and the algorithms are introduced in Appendix A.

The previous double-size techniques consist of two steps. The first step is to compute a 2ℓ-bit (not modular) multiplication step by step using an ℓ-bit multiplier. Then, the second step reduces the 4ℓ-bit product with modulus N using the equation $n_1 2^\ell = -n_0 \pmod{N}$: the 2ℓ-bit product $n_1 2^\ell$ is reduced to one negative ℓ-bit integer $(-n_0)$. Using successive reductions, the 4ℓ-bit product can be reduced to a 2ℓ-bit integer, that is, the result of the double-size modular multiplication.

---

[1] In order to avoid misreading, this paper uses different names from the original Mult-ModDiv and MultModDivInit instructions [FS03]. Later, MonMul instruction computing Montgomery multiplications will be introduced in this paper.

Table 1. Call numbers of double-size techniques

Algorithm	Alg.1	Alg.10	Alg.11	Alg.12	Alg.13
EucMul	6	4	5	3	0
EucMulInit	0	1	0	0	0
MonMul	0	0	0	0	6
Calls	6	5	5	3	6

Unlike naive implementations of Euclidean multiplications, *Montgomery* multiplications are not affected by carry propagation delays and as a result, Montgomery multiplications have been widely implemented in cryptographic coprocessors [Mon85]. Yoshino et al. extended the double-size techniques using an ℓ-bit Euclidean instruction to an ℓ-bit Montgomery multiplication unit which is the following instruction [YOV07a, YOV08].

Definition 2. *For numbers,* $0 \le a, b, n$ *and positive constant* m, MonMul *instruction is defined as*

$$(ab - rm/n, r = abm^{-1} \pmod{n}) \leftarrow \mathsf{MonMul}(a, b, n)$$

where a, b *and* n *are* ℓ*-bit integers and* $\gcd(m, n) = 1$

The MonMul instruction has assumptions that are similar to those of EucMul and EucMulInit instructions: negative and non-reduced operands a, b such that $|a| > n$ are acceptable.

The double-size Montgomery multiplication consists of two steps, similarly to Algorithm 1. The first step is to compute a 2ℓ-bit multiplication step by step using the MonMul instruction. Then, the second step reduces the 4ℓ-bit product to a 2ℓ-bit integer based on the modulus equation. Only the 2ℓ-bit integer is returned by 2ℓ-bit Montgomery multiplications such as Algorithm 13.

Table 1 displays the cost of each double-size technique, including Algorithm 1 and other algorithms described in Appendix A.

2.2   Costly Quotient Computation for Size Quadrupling Approach

At first sight, it can be expected that from a recursive use of the best current doubling algorithm, one could straight-forwardly derive the best possible quadrupling technique. Unfortunately, this is not true. Originally, double-size techniques based on a single-size instruction outputting a quotient and a remainder returns only a double-size remainder. Therefore, a recursive approach requires additional computations for the double-size quotient. See for example Algorithm 2 deriving an quotient from the difference of two remainders (Step 3) [FS03]: in addition to one computation of a double-size modular multiplication for the remainder, extra computations are required for the quotient.

Currently, Algorithm 2 computing a quotient from two different remainders is the preferred method for the double-size approach; however, this fact means that the performance of the quadrupling approach is not as optimized as that of double-size techniques. The double-size technique using Algorithm 1 requires only six

Alg.2. Simulation of the EucMul instruction

Input: a, b, n with $0 \leq a, b, n$;
Output: $\lfloor ab/n \rfloor$, $ab \pmod{n}$;

$r \leftarrow ab \pmod{n}$
$r' \leftarrow ab \pmod{(n+1)}$
$q \leftarrow r - r'$
If $q < 0$, then $q \leftarrow q + n + 1$.
Return (q, r)

---

single-size instruction calls, however, the quadrupling approach using Algorithm 1 requires twelve double-size instruction calls, which is twice the number of calls compared with the doubling case. Unfortunately, a similar situation can be observed for all previous double-size techniques implemented in a recursive manner.

3   New Double-Size Techniques

This section proposes new solutions to compute a double-size quotient and remainder simultaneously without additional computational cost compared with previous double-size techniques. Unlike the previous techniques based on the equation of the 2ℓ-bit modular multiplication $R = AB \pmod{N}$ or the 2ℓ-bit Montgomery multiplication $R = ABM^{-1} \pmod{N}$, our approach calculates a quotient Q and a remainder R satisfying the equation $AB = QN + RM$ where $M = 1$ for Euclidean multipliers and $M = 2^{2\ell}$ for Montgomery multipliers.

Figure 1 illustrates the fact that positions of the remainders are different for each modular multiplication algorithm. Our technique computes the 4ℓ-bit product AB and divides it into the 2ℓ-bit quotient Q and remainder R step by step using ℓ-bit instructions.

The available approaches for our new double-size techniques depend on several conditions such as the type of coprocessors, the possibility of performing precomputations or the value of the modulus. This section introduces several algorithms to cover most of the above settings: Section 3.1–3.3 targets an *Euclidean* multiplication unit, and Section 3.4 shows an extension to a *Montgomery* multiplication unit.

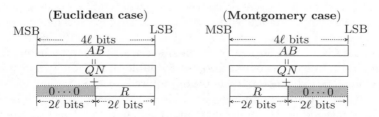

Fig. 1. Different Remainders Position in Equation $AB = QN + RM$

3.1 Simple ℓ-bit Width Representation Approach

Since the operand size of the EucMul instruction is limited to ℓ bits, double-size techniques need to split 2ℓ-bit integers into at most ℓ bits. One of the simplest splitting method is to apply the binary representation with ℓ-bit width, which does not require any precomputation.

Algorithm 3 introduced hereafter requires the same number of calls to the EucMul instruction as Algorithm 1.

Proposition 1. *Algorithm 3 requires* **_six_** *ℓ-bit EucMul instructions to compute 2ℓ-bit integers Q and R satisfying $AB = QN + R$ provided with 2ℓ-bit arbitrary integers A, B, N where $0 \le A$, $B < N$.*

Alg.3. Double-size EucMul instruction 1

Input: $A = a_1 2^\ell + a_0$, $B = b_1 2^\ell + b_0$ and $N = n_1 2^\ell + n_0$;
Output: $\lfloor AB/N \rfloor$ and $AB \pmod{N}$;

$(q_1, r_1) \leftarrow \mathsf{EucMul}(a_0, b_0, 2^\ell)$
$(q_2, r_2) \leftarrow \mathsf{EucMul}(a_0 + a_1, b_0 + b_1, n_1)$
$(q_3, r_3) \leftarrow \mathsf{EucMul}(a_1, b_1, n_1)$
$(q_4, r_4) \leftarrow \mathsf{EucMul}(r_3 - q_1, 2^\ell, n_1)$
$(q_5, r_5) \leftarrow \mathsf{EucMul}(q_3, n_0, n_1)$
$(q_6, r_6) \leftarrow \mathsf{EucMul}(q_2 - q_3 + q_4 - q_5, n_0, 2^\ell)$
Return $q_3 2^\ell + (q_2 - q_3 + q_4 - q_5)$ and $(q_1 - q_6 - r_1 + r_2 - r_3 + r_4 - r_5)2^\ell + (r_1 - r_6)$

Proof. Hereafter, the correctness of Algorithm 3 is proven.

First, 2ℓ-bit integers A, B, N expressed in binary are processed assuming the following ℓ-bit representation;

$$A = a_1 2^\ell + a_0,\ B = b_1 2^\ell + b_0,\ N = n_1 2^\ell + n_0.$$

Karatsuba can be applied to compute the multiplication AB in order to reduce computational cost.

$$AB = a_1 b_1 2^\ell (2^\ell - 1) + (a_1 + a_0)(b_1 + b_0)2^\ell - a_0 b_0 (2^\ell - 1) \tag{1}$$

Successively, each term of Equation (1) can be calculated with ℓ-bit EucMul instructions, outputting ℓ-bit quotients and remainders as the following equations.

$$a_0 b_0 = q_1 2^\ell + r_1 \tag{2a}$$
$$(a_1 + a_0)(b_1 + b_0) = q_2 n_1 + r_2 \tag{2b}$$
$$a_1 b_1 = q_3 n_1 + r_3 \tag{2c}$$

Then, Equations (2a), (2b) and (2c) can be reinjected in Equation (1).

$$AB = \{q_3 2^\ell + (q_2 - q_3)\}n_1 2^\ell + (r_3 - q_1)2^{2\ell} + (q_1 - r_1 + r_2 - r_3)2^\ell + r_1.$$

Similarly, one can change some of the terms in the above equation using the EucMul instruction in order to transform a 4ℓ-bit integer into multiples of the modulus $(n_1 2^\ell + n_0)$.

$$(r_3 - q_1)2^\ell = q_4 n_1 + r_4 \tag{3a}$$

$$q_3 n_0 = q_5 n_1 + r_5 \tag{3b}$$

$$(q_2 - q_3 + q_4 - q_5)n_0 = q_6 2^\ell + r_6 \tag{3c}$$

Finally, the product AB can be rewritten into the following equations, where underline related to EucMul instruction calls are depicted for easier comprehension.

$$
\begin{aligned}
AB &= \{q_3 2^\ell + (q_2 - q_3)\}n_1 2^\ell + (\underline{q_4 n_1 + r_4})2^\ell + (q_1 - r_1 + r_2 - r_3)2^\ell + r_1 (\because (3a)) \\
&= \{q_3 2^\ell + (q_2 - q_3 + q_4)\}(n_1 2^\ell + n_0) - \{q_3 2^\ell + (q_2 - q_3 + q_4)\}n_0 \\
&\quad + (q_1 - r_1 + r_2 - r_3 + r_4)2^\ell + r_1 \hspace{4cm} (\because (3b)) \\
&= \{q_3 2^\ell + (q_2 - q_3 + q_4 - \underline{q_5})\}(\underline{n_1}2^\ell + n_0) - (q_2 - q_3 + q_4 - q_5)n_0 \\
&\quad + (q_1 - r_1 + r_2 - r_3 + r_4 - \underline{r_5})2^\ell + r_1 \hspace{3cm} (\because (3c)) \\
&= \{q_3 2^\ell + (q_2 - q_3 + q_4 - q_5)\}(n_1 2^\ell + n_0) \\
&\quad + \{(q_1 - \underline{q_6} - r_1 + r_2 - r_3 + r_4 - r_5)\underline{2^\ell} + r_1 - \underline{r_6}\}. \hspace{2cm} (4)
\end{aligned}
$$

As a result, Equation (4) can be seen as $AB = QN + R$ with $Q = q_3 2\ell + (q_2 - q_3 + q_4 - q_5)$ and $R = (q_1 - q_6 - r_1 + r_2 - r_3 + r_4 - r_5)2^\ell + (r_1 - r_6)$.

□

Algorithm 3 was derived in the case that the instruction EucMul only is available; however, modern coprocessors may be equipped with not only the EucMul instruction but also a more advanced function, namely the EucMulInit instruction. Using both instructions, Algorithm 4 decreases the number of calls to the available instructions by one compared to Algorithm 3.

Proposition 2. *Algorithm 4 requires **four** ℓ-bit EucMul instructions and **one** ℓ-bit EucMulInit instruction to compute 2ℓ-bit integers Q and R satisfying $AB = QN + R$ provided with 2ℓ-bit arbitrary integers A, B, N where $0 \le A$, $B < N$.*

The correctness of Algorithm 4 is proven in Appendix B.

3.2   Sophisticated Representation Approach

Algorithms 3 and 4 manipulate integers in a simple radix representation with base 2^ℓ. Now this subsection introduces a representation change that may lead to further cost savings. The change, which is not as simple as the ℓ-bit width representation, needs to be performed prior to the execution of the corresponding double-size multiplication algorithm, and can be executed once and for all, especially when not only a single multiplication but a modular exponentiation is performed, as in RSA, DSA or DH [MOV96].

Alg.4. Double-size EucMul instruction 2

Input: $A = a_1 2^\ell + a_0$, $B = b_1 2^\ell + b_0$ and $N = n_1 2^\ell + n_0$;
Output: $\lfloor AB/N \rfloor$ and $AB \pmod N$;

$(q_1, r_1) \leftarrow \mathsf{EucMul}(a_0 + a_1, b_0 + b_1, n_1)$
$(q_2, r_2) \leftarrow \mathsf{EucMul}(a_1, b_1, n_1)$
$(q_3, r_3) \leftarrow \mathsf{EucMul}(a_0, b_0, 2^\ell)$
$(q_4, r_4) \leftarrow \mathsf{EucMulInit}(q_2, n_0, q_3 - r_2, n_1)$
$(q_5, r_5) \leftarrow \mathsf{EucMul}(q_1 - q_2 - q_4, n_0, 2^\ell)$
Return $q_2 2^\ell + (q_1 - q_2 - q_4)$ and $(q_3 - q_5 + r_1 - r_2 - r_3 - r_4)2^\ell + (r_3 - r_5)$

Proposition 3. *Algorithm 5 requires **five** ℓ-bit EucMul instructions to compute 2ℓ-bit integers Q and R satisfying $AB = QN + R$ provided with 2ℓ-bit arbitrary integers A, B, N where $0 \le A$, $B < N$.*

Let s be an integer satisfying $s^2 = N + \alpha$ and $s \le \lceil \sqrt{N} \rceil$. Then, the above binary representation can be changed using the ℓ-bit integer s; for instance $A = a_1 s + a_0$. Since $\alpha < 2s$ holds, α can be processed by the EucMul instruction.

Alg.5. Double-size EucMul instruction 3

Precomp: radix base s satisfying that $s^2 = N + \alpha$ and $s \le \lceil \sqrt{N} \rceil$;
Input: $A = a_1 s + a_0$, $B = b_1 s + b_0$, s, α;
Output: $\lfloor AB/N \rfloor$ and $AB \pmod N$;

$(q_1, r_1) \leftarrow \mathsf{EucMul}(a_0 + a_1, b_0 + b_1, s)$
$(q_2, r_2) \leftarrow \mathsf{EucMul}(a_1, b_1, s)$
$(q_3, r_3) \leftarrow \mathsf{EucMul}(a_0, b_0, s)$
$(q_4, r_4) \leftarrow \mathsf{EucMul}(\alpha, q_2, s)$
$(q_5, r_5) \leftarrow \mathsf{EucMul}(\alpha, q_1 - q_2 - q_3 + q_4 + r_2, s)$
Return $q_2 s + (q_1 - q_2 - q_3 + q_4 + r_2)$ and $(q_3 + q_5 + r_1 - r_2 - r_3 + r_4)s + (r_3 + r_5)$

The correctness of Algorithm 5 is proven in Appendix B.

3.3   Approach Using Specific Modulus

Optimal performances are reached when $\alpha = -1, 0, 1$ in Algorithm 5: the computational cost reduces to only 3 EucMul instructions. The existence of s can be guaranteed by modifying key generation, and [CJP03] argues that this can be achieved with simple algebraic techniques while preserving security of the modulus.

Other choices for the representation base may be an ℓ-bit integer t such that $t^2 = N + \alpha + \delta t$. If the key generation is modified, it is possible to select advantageous values for α and δ such as $\alpha = \delta = 1$.

Proposition 4. *Algorithm 6 requires **three** ℓ-bit EucMul instructions to compute 2ℓ-bit integers Q and R satisfying $AB = QN + R$ provided with 2ℓ-bit arbitrary integers A, B, N where $0 \le A$, $B < N$.*

Alg.6. Double-size EucMul instruction 4

Precomp: radix base t satisfying that $t^2 = N + \alpha + \delta t$;
Input: $A = a_1 t + a_0$, $B = b_1 t + b_0$, α, δ;
Output: $\lfloor AB/N \rfloor$ and $AB \pmod{N}$;

$(q_1, r_1) \leftarrow \mathsf{EucMul}(a_0 + a_1, b_0 + b_1, t)$
$(q_2, r_2) \leftarrow \mathsf{EucMul}(a_1, b_1, t)$
$(q_3, r_3) \leftarrow \mathsf{EucMul}(a_0, b_0, t)$
Return $q_2 t + (q_2 \delta + q_1 - q_2 - q_3 + r_2)$ and $\{q_1 \delta + q_2(\alpha + \delta^2 - \delta) - q_3(\delta - 1) + r_1 + r_2(\delta - 1) - r_3\}t + (q_1 + q_2(\delta - 1) - q_3 + r_2)\alpha + r_3$

The correctness of Algorithm 6 is proven in Appendix B.

3.4   Extension to Montgomery-Type Multiplication

Unlike naive implementations of *Euclidean* multiplications, Montgomery multiplications are not affected by carry propagation delays and as a result, Montgomery multiplications have been widely implemented for costly modular multiplications. On the other hand, the Montgomery quotient and remainder satisfy equations that are different from the case of Euclidean multiplication; more precisely, the relationship between product, quotient and remainder involves the so-called Montgomery constant M: $AB = QN + RM$. In practice, for efficient computations, M is set as $2^{2\ell}$ in the case of a 2ℓ-bit instruction [MOV96, YOV07a]. This paper follows the same pattern and assumes that $M = 2^{2\ell}$.

First, we propose Algorithm 7 based on an ℓ-bit width representation using the ℓ-bit Montgomery constant $m = 2^\ell$.

Proposition 5. *Algorithm 7 requires **six** ℓ-bit* MonMul *instructions to compute 2ℓ-bit integers Q and R satisfying $AB = QN + RM$ provided with 2ℓ-bit arbitrary integers A, B, N where $0 \le A$, B,N.*

Alg.7. Double-size MonMul instruction 1

Input: $A = a_1 + a_0 m$, $B = b_1 + b_0 m$ and $N = n_1 + n_0 m$ with $m = 2^\ell$;
Output: $(AB - RM)/N$ and $ABM^{-1} \pmod{N}$ with $M = 2^{2\ell}$;

$(q_1, r_1) \leftarrow \mathsf{MonMul}(a_0 + a_1, b_0 + b_1, n_1)$
$(q_2, r_2) \leftarrow \mathsf{MonMul}(a_1, b_1, n_1)$
$(q_3, r_3) \leftarrow \mathsf{MonMul}(a_0, b_0, m - 1)$
$(q_4, r_4) \leftarrow \mathsf{MonMul}(q_3 + r_2, 1, n_1)$
$(q_5, r_5) \leftarrow \mathsf{MonMul}(q_2, n_0, n_1)$
$(q_6, r_6) \leftarrow \mathsf{MonMul}(q_5, n_0, m - 1)$
Return $q_2 + (q_1 - q_2 + q_4 - q_5)m$ and $(-q_1 + q_2 - 2q_3 - q_4 - q_6 + r_1 - r_2 - r_3 + r_4 - r_5) + (q_3 + q_6 + r_3 + r_6)m$

The correctness of Algorithm 7 is in a similar fashion to Algorithm 3.

Not only the above simple ℓ-bit width binary representation, but also more sophisticated representations are applicable even for Montgomery multiplications; note that this direction has never been investigated by previous techniques for computing 2ℓ-bit Montgomery remainders [YOV07a, YOV07b, YOV08]. In fact, the following algorithm based on a more sophisticated representation enjoys higher performance, even for the doubling case.

Proposition 6. *Algorithm 8 requires **five** ℓ-bit* MonMul *instructions to compute 2ℓ-bit integers Q and R satisfying $AB = \overline{Q}N + RM$ provided with 2ℓ-bit arbitrary integers A, B, N where $0 \le A,\ B,\ N$.*

Let u be an integer satisfying $u^2 = N + \alpha m$ and $u \le \lceil \sqrt{n_1} + 1 \rceil m^{1/2}$ where $n_1 = \lfloor N/m \rfloor$. Since $u < m$ holds, a new representation with u such as $A = a_1 u + a_0 m$ is available for the MonMul instruction.

Alg.8. Double-size MonMul instruction 2

Precomp: radix base u satisfying that $u^2 = N + \alpha m$ and $u \le \lceil \sqrt{n_1} + 1 \rceil m^{1/2}$;
Input: $A = a_1 u + a_0 m$, $B = b_1 u + b_0 m$, u, α;
Output: $(AB - RM)/N$ and $ABM^{-1} \pmod N$;

$(q_1, r_1) \leftarrow$ MonMul$(a_0 + a_1, b_0 + b_1, u)$
$(q_2, r_2) \leftarrow$ MonMul(a_1, b_1, u)
$(q_3, r_3) \leftarrow$ MonMul(a_0, b_0, u)
$(q_4, r_4) \leftarrow$ MonMul(α, q_2, u)
$(q_5, r_5) \leftarrow$ MonMul$(\alpha, q_1 - q_2 - q_3 + q_4 + r_2, u)$
Return $q_2 + (q_1 - q_2 - q_3 + q_4 + r_2)u$ and $(q_3 + q_5 + r_1 - r_2 - r_3 + r_4) + (r_3 + r_5)u$

The correctness of Algorithm 8 is proven in a similar fashion to Algorithm 5.

4   Multiplication and Reduction Approach with Precomputation

Section 3 introduced a new double-size approach for outputting both a quotient and a remainder in order to accelerate long modular multiplication such as quadrupling. This section describes a different $k\ell$-bit modular multiplication approach which does not call double-size techniques recursively, and can be applied to arbitrary k such as $k = 3, 4, 5, \ldots$ It is easier to adjust the value k to a bit-length required by cryptographic standards than when using recursive approach, which only allows powers of two. For example, with a cryptographic processor which is able to process 512-bit modular multiplications one can select $k = 3$ for 1408-bit modular multiplications, which is a standard RSA key-length for credit card protocols [EMV]. However, in order to process this flexible approach efficiently, devices should be equipped with storage for precomputed values.

Proposition 7. *Algorithm 9 requires* ***twenty eight*** *ℓ-bit* EucMul *instructions to compute $\underline{4\ell\text{-bit}}$ integers R satisfying $R = AB$ (mod N) provided with $\underline{4\ell\text{-bit}}$ arbitrary integers A, B, N where $0 \le A, B, N$.*

Alg.9. Quadrupling-size classical modular multiplication

Precomp: elements $c_{i,j}$ satisfying that $\sum_{j=0}^{3} c_{i,j} 2^{j\ell} = 2^{i\ell}$ (mod N)($i = 4, 5, 6, 7$);
Input: $A = \sum_{i=0}^{3} a_i 2^{i\ell}$, $B = \sum_{i=0}^{3} b_i 2^{i\ell}$, $N = \sum_{i=0}^{3} n_i 2^{i\ell}$;
Output: AB (mod N);

Compute a multiplication according to Karatsuba.
  1. $(q_1, r_1) \leftarrow \mathsf{EucMul}(a_0, b_0, 2^\ell)$
  2. $(q_2, r_2) \leftarrow \mathsf{EucMul}(a_1, b_1, 2^\ell)$
  3. $(q_3, r_3) \leftarrow \mathsf{EucMul}(a_1 + a_0, b_1 + b_0, 2^\ell)$
  4. $(q_4, r_4) \leftarrow \mathsf{EucMul}(a_2 + a_0, b_2 + b_0, 2^\ell)$
  5. $(q_5, r_5) \leftarrow \mathsf{EucMul}(a_3 + a_1, b_3 + b_1, 2^\ell)$
  6. $(q_6, r_6) \leftarrow \mathsf{EucMul}(a_3 + a_2 + a_1 + a_0, b_3 + b_2 + b_1 + b_0, 2^\ell)$
  7. $(q_7, r_7) \leftarrow \mathsf{EucMul}(a_2, b_2, 2^\ell)$
  8. $(q_8, r_8) \leftarrow \mathsf{EucMul}(a_3, b_3, 2^\ell)$
  9. $(q_9, r_9) \leftarrow \mathsf{EucMul}(a_3 + a_2, b_3 + b_2, 2^\ell)$
For i from 0 to 7, compute s_i satisfying that $AB = \sum_{i=0}^{7} s_i 2^i$ (mod N)
  1. $s_0 \leftarrow r_1$
  2. $s_1 \leftarrow q_1 + r_3 - r_2 - r_1$
  3. $s_2 \leftarrow q_3 - q_2 - q_1 + r_4 + r_2$
  4. $s_3 \leftarrow q_4 + q_2 + r_6 - r_5 - r_4$
  5. $s_4 \leftarrow q_6 - q_5 - q_4 + r_7 + r_5$
  6. $s_5 \leftarrow q_7 + q_5 + r_9 - r_8 - r_7$
  7. $s_6 \leftarrow q_9 - q_8 - q_7 + r_8$
  8. $s_7 \leftarrow q_8$
For i from 0 to 3, compute $(q_{10+i}, r_{10+i}) \leftarrow \mathsf{EucMul}(s_{4+i}, c_{4+i,3}, n_3)$
For j from 0 to 2, do
  1. For i from 0 to 3, compute $(q_{14+i+4j}, r_{14+i+4j}) \leftarrow \mathsf{EucMul}(s_{4+i}, c_{4+i,3-j}, 2^\ell)$
For i from 0 to 2, compute $(q_{26+i}, r_{26+i}) \leftarrow \mathsf{EucMul}(q_{10} + q_{11} + q_{12} + q_{13}, n_{2-i}, 2^\ell)$
Return $\{s_3 + \sum_{i=0}^{3}(q_{14+i} + r_{10+i}) - q_{26}\}2^{3\ell} + \{s_2 + \sum_{i=0}^{3}(q_{18+i} + r_{14+i}) - q_{27} - r_{26}\}2^{2\ell} + \{s_1 + \sum_{i=0}^{3}(q_{22+i} + r_{18+i}) - q_{28} - r_{27}\}2^\ell + (s_0 + \sum_{i=0}^{3} r_{22+i} - r_{28})$

Algorithm 9 is designed for *classical* modular multiplications, however, it can be easily changed for *Montgomery* multiplications with ℓ-bit MonMul instructions. Furthermore, Algorithm 9 only treats the case of $k = 4$, but can be easily extended to general k.

Proposition 8. *There are algorithms to compute $k\ell$-bit modular multiplication runs in at most $(2k^2 - k)^2 \ell$-bit modular multiplications.*

---

[2] Roughly speaking, the approach consists of $k\ell$-bit (not modular) multiplication steps per ℓ-bits, and $k\ell$-bit modular reduction with precomputed value. Performance can be improved if a fast multiplication algorithm is applied on the first $k\ell$-bit multiplication step.

5 Comparison

This section evaluates the number of calls to the ℓ-bit instructions in 4ℓ-bit modular multiplication, which is the smallest operand size for recursive approach (with $2^2 = 4$).

Table 2 shows that the proposed double-size Euclidean multiplication is twice as fast as the previous double-size techniques from a viewpoint of the number of calls to ℓ-bit instructions. Our ℓ-bit width representation technique improves 72 calls to 36 calls if the EucMul instruction only is available (that is condition (a) in Table 2), and 50 calls to 25 calls if both EucMul and EucMulInit instructions are available (condition (b)). Furthermore, if precomputations are available (condition(c)), 50 calls are improved to 25 calls, and 18 calls are improved to 9 calls with specific moduli (condition (d)).

Table 2. Cost of quadrupling modular multiplication

	Previous techniques				Proposal			
Condition	(a)	(b)	(c)	(d)	(a)	(b)	(c)	(d)
Algorithm	Alg.1	Alg.10	Alg.11	Alg.12	Alg.3	Alg.4	Alg.5	Alg.6
EucMul	72	40	50	18	36	20	25	9
EucMulInit	0	10	0	0	0	5	0	0

Table 3 gives performance results of both the previous techniques and our proposals based on the ℓ-bit Montgomery multiplication unit. 72 calls are improved to 36 calls with the ℓ-bit width binary representation (condition (A)), and with the approach of using the more sophisticated representation, which requires a small precomputation, the number of calls is decreased from 72 to 25 (condition(B)).

Table 3. Cost of quadrupling *Montgomery* multiplication

	Previous	Proposal	
Condition	(A)	(A)	(B)
Algorithm	Alg.13	Alg.7	Alg.8
MonMul	72	36	25

In addition to the recursive approach, Section 5 proposed a multiplication-and-reduction technique which can directly compute $k\ell$-bit modular or Montgomery multiplications if storage for precomputations is available. In the 4ℓ-bit case, it has 28 calls to ℓ-bit instructions based on Euclid multiplication or Montgomery multiplication.

6 Conclusion

This paper addresses the issue of the high computation cost when getting a quotient from the difference of two remainders and improves 2ℓ-bit modular multiplication algorithms to return both a remainder and a quotient. Our technique results in a performance improvement for the recursive approach, and is at least twice as fast as previous schemes in the 4ℓ-bit case and four times faster in the 8ℓ-bit case.

Modular multiplications are one of the most time-critical parts in cryptosystems, however, their performance might be affected by other glue instructions which cannot take advantage of the coprocessor. Future work will consider this problem and deal with practical implementations.

References

[CJP03]   Chevallier-Mames, B., Joye, M., Paillierinst, P.: Faster double-size modular multiplication from euclidean multipliers. In: Walter, C.D., Koç, Ç.K., Paar, C. (eds.) CHES 2003. LNCS, vol. 2779, pp. 214–227. Springer, Heidelberg (2003)

[EMV]   EMV. EMV Issuer and Application Security Guidelines, Version 2.1 (2007), http://www.emvco.com/specifications.asp?show=4

[FS03]   Fischer, W., Seifert, J.-P.: Increasing the bitlength of a crypto-coprocessor. In: Kaliski Jr., B.S., Koç, Ç.K., Paar, C. (eds.) CHES 2002. LNCS, vol. 2523, pp. 71–81. Springer, Heidelberg (2003)

[Mon85]   Montgomery, P.L.: Modular Multiplication without Trial Division. Mathematics of Computation 44(170), 519–521 (1985)

[MOV96]   Menezes, A.J., van Oorschot, P.C., Vanstone, S.A.: Handbook of Applied Cryptography. CRC Press, Boca Raton (1996)

[Nis07a]   National Institute of Standards and Technology. NIST Special Publication 800-57 Recommendation for Key Management Part 1: General (Revised) (2007), http://csrc.nist.gov/CryptoToolkit/tkkeymgmt.html

[Nis07b]   National Institute of Standards and Technology. NIST Special Publication 800-78-1, Cryptographic Algorithms and Key Sizes for Personal Identity Verification (2007)
           http://csrc.nist.gov/CryptoToolkit/tkkeymgmt.html

[NM96]   Naccache, D., M'Raïhi, D.: Arithmetic Co-processors for Public-key Cryptography: The State of the Art. In: CARDIS, pp. 18–20 (1996)

[Pai99]   Paillier, P.: Low-cost double-size modular exponentiation or how to stretch your cryptoprocessor. In: Imai, H., Zheng, Y. (eds.) PKC 1999. LNCS, vol. 1560, p. 223. Springer, Heidelberg (1999)

[RSA78]   Rivest, R.L., Shamir, A., Adelman, L.M.: A Method for Obtaining Digital Signatures and Public-key Cryptosystems. Communications of the ACM 21(2), 120–126 (1978)

[YOV07a]   Yoshino, M., Okeya, K., Vuillaume, C.: Unbridle the bit-length of a crypto-coprocessor with montgomery multiplication. In: Biham, E., Youssef, A.M. (eds.) SAC 2006. LNCS, vol. 4356, pp. 188–202. Springer, Heidelberg (2007)

[YOV07b]   Yoshino, M., Okeya, K., Vuillaume, C.: Double-size bipartite modular multiplication. In: Pieprzyk, J., Ghodosi, H., Dawson, E. (eds.) ACISP 2007. LNCS, vol. 4586, pp. 230–244. Springer, Heidelberg (2007)

[YOV08]   Yoshino, M., Okeya, K., Vuillaume, C.: A Black Hen Lays White Eggs: Bipartite Multiplier Out of Montgomery One for On-Line RSA Verification. In: Grimaud, G., Standaert, F.-X. (eds.) CARDIS 2008. LNCS, vol. 5189, pp. 74–88. Springer, Heidelberg (2008)

A   Previous Double-Size Techniques

Algorithm using both the EucMul and the EucMulInit instructions requires only five calls to the instructions.

Alg.10. Double-size modular multiplication 2

Input: $A = a_1 2^\ell + a_0$, $B = b_1 2^\ell + b_0$ and $N = n_1 2^\ell + n_0$;
Output: AB (mod N);

$(q_1, r_1) \leftarrow \mathsf{EucMul}(a_1, b_1, n_1)$
$(q_2, r_2) \leftarrow \mathsf{EucMul}(a_0 + a_1, b_0 + b_1, 2^\ell - 1)$
$(q_3, r_3) \leftarrow \mathsf{EucMul}(a_0, b_0, 2^\ell)$
$(q_4, r_4) \leftarrow \mathsf{EucMulInit}(q_1, n_0, -q_2 + q_3 - r_1, n_1)$
$(q_5, r_5) \leftarrow \mathsf{EucMul}(q_4, n_0 + n_1, 2^\ell)$
Return $(r_2 - r_3 - r_4 + q_5)2^\ell + (r_3 + r_4 + r_5)$

In addition to Algorithm 1 and 10 with a simple base 2ℓ, Chevallier-Mames et al. presented another algorithm with a special radix s, which needs only five calls to the EucMul instruction.

Alg.11. Double-size modular multiplication 3

Precomp: $\alpha = s^2$ (mod N);
Input: $A = a_1 s + a_0$, $B = b_1 s + b_0$, $N = n_1 s + n_0$;
Output: AB (mod N);

$(q_1, r_1) \leftarrow \mathsf{EucMul}(a_0, b_0, s)$
$(q_2, r_2) \leftarrow \mathsf{EucMul}(a_0 + a_1, b_0 + b_1, s)$
$(q_3, r_3) \leftarrow \mathsf{EucMul}(a_1, b_1, s)$
$(q_4, r_4) \leftarrow \mathsf{EucMul}(\alpha, q_3, s)$
$(q_5, r_5) \leftarrow \mathsf{EucMul}(\alpha, -q_1 + q_2 - q_3 + q_4 + r_3, s)$
Return $(-r_1 + r_2 - r_3 + r_4 + q_1 + q_5)s + (r_1 + r_5)$

Furthermore, Chevallier-Mames et al. showed optimized algorithms with a specific modulus satisfying $t^2 = \alpha + \delta t \pmod{N}$ where α and δ are simple numbers such as $\alpha = \delta = 1$.

Alg.12. Double-size modular multiplication 4

Precomp: $t^2 = \alpha + \delta t \pmod{N}$;
Input: $A = a_1 t + a_0$, $B = b_1 t + b_0$, $N = n_1 t + n_0$;
Output: $AB \pmod{N}$;

$(q_1, r_1) \leftarrow \mathsf{EucMul}(a_0, b_0, t)$
$(q_2, r_2) \leftarrow \mathsf{EucMul}(a_0 + a_1, b_0 + b_1, t)$
$(q_3, r_3) \leftarrow \mathsf{EucMul}(a_1, b_1, t)$
Return $\alpha(-q_1 + q_2 - q_3 + r_3 + q_3\delta) + r_1 + t(-r_1 + r_2 - r_3 + q_1 + q_3(\alpha + \delta^2) + (-q_1 + q_2 - q_3 + r_3)\delta)$

Yoshino et al. proposed double-size Montgomery multiplication algorithms which call only six times the MonMul instruction.

Alg.13. Double-size Montgomery modular multiplication 1

Input: $A = a_1(m-1) + a_0 m$, $B = b_1(m-1) + b_0 m$, $N = n_1(m-1) + n_0 m$;
Output: $ABM^{-1} \pmod{N}$ where $M = m^2$;

$(q_1, r_1) \leftarrow \mathsf{MonMul}(a_1, b_1, n_1)$
$(q_2, r_2) \leftarrow \mathsf{MonMul}(q_1, n_0, (m-1))$
$(q_3, r_3) \leftarrow \mathsf{MonMul}(a_0 + a_1, b_0 + b_1, (m-1))$
$(q_4, r_4) \leftarrow \mathsf{MonMul}(a_0, b_0, (m-1))$
$(q_5, r_5) \leftarrow \mathsf{MonMul}((m-1), -q_2 + q_3 - q_4 + r_1, n_1)$
$(q_6, r_6) \leftarrow \mathsf{MonMul}(q_5, n_0, (m-1))$
Return $(q_2 + q_4 - q_6 - r_1 - r_2 + r_3 - r_4 + r_5)(m-1) + (r_2 + r_4 - r_6)m$

B   Proof of Correctness

B.1   Algorithm 4

Proof. Hereafter, the correctness of Algorithm 4 is proven.

First, 2ℓ-bit integers A, B, N presented with natural binary representation are converted to the following ℓ-bit representations; $A = a_1 2^\ell + a_0$, $B = b_1 2^\ell + b_0$, $N = n_1 2^\ell + n_0$. Karatsuba can be applied to compute the multiplication AB for less cost. $AB = a_1 b_1 2^\ell (2^\ell - 1) + (a_1 + a_0)(b_1 + b_0)2^\ell - a_0 b_0 (2^\ell - 1)$. Successively, each term of the above equation is represented in the following equations: $(a_1 + a_0)(b_1 + b_0) = q_1 n_1 + r_1$, $a_1 b_1 = q_2 n_1 + r_2$ and $a_0 b_0 = q_3 2^\ell + r_3$.

Then, the above equation is affected from changes of the above terms. $AB = \{q_2 2^\ell + (q_1 - q_2)\}n_1 2^\ell + (r_2 - q_3)2^{2\ell} + (q_3 + r_1 - r_2 - r_3)2^\ell + r_3$.

Similarly, one can change each representation with the following EucMulInit and EucMul instruction calls: $q_2 n_0 + (q_3 - r_2)2^\ell = q_4 n_1 + r_4$ and $(q_1 - q_2 - q_4)n_0 = q_5 2^\ell + r_5$. Finally, the product AB will be changed into the following equations, where underline related to EucMul instruction calls are depicted for easier comprehension.

$$AB = \{q_2 2^\ell + (q_1 - q_2)\}(n_1 2^\ell + n_0) - \{q_2 n_0 + (q_3 - r_2)2^\ell\}2^\ell$$
$$-(q_1 - q_2)n_0 + (q_3 + r_1 - r_2 - r_3)2^\ell + r_3$$
$$= \{q_2 2^\ell + (q_1 - q_2 - q_4)\}(n_1 2^\ell + n_0) - (q_1 - q_2 - q_4)n_0$$
$$+(q_3 + r_1 - r_2 - r_3 - r_4)2^\ell + r_3$$
$$= \{q_2 2^\ell + (q_1 - q_2 - q_4)\}(n_1 2^\ell + n_0)$$
$$+(q_3 - q_5 + r_1 - r_2 - r_3 - r_4)2^\ell + (r_3 - r_5)$$

As a result, $AB = QN + R$ where $Q = q_2 2\ell + (q_1 - q_2 - q_4)$ and $R = (q_3 - q_5 + r_1 - r_2 - r_3 - r_4)2^\ell + (r_3 - r_5)$ holds.

$\square$

B.2   Algorithm 5

Proof. $AB = (a_1 s + a_0)(b_1 s + b_0) = a_1 b_1 s(s-1) + (a_1 + a_0)(b_1 + b_0)s - a_0 b_0(s-1)$ Each term is converted based on the following equations. $\underline{(a_1 + a_0)(b_1 + b_0)s} = \underline{(q_1 s + r_1)s}$, $\underline{a_1 b_1 s(s-1)} = \underline{(q_2 s + r_2)s(s-1)}$ and $\underline{a_0 b_0(s-1)} = \underline{(q_3 s + r_3)(s-1)}$. Then, the equation is rewritten as follows.

$$AB = \{q_2 s + (q_1 - q_2 - q_3 + r_2)\}N + (q_3 + r_1 - r_2 - r_3)s + r_3$$
$$+\underline{q_2 \alpha s} + (q_1 - q_2 - q_3 + r_2)\alpha$$
$$= \{q_2 s + (q_1 - q_2 - q_3 + r_2)\}N + (q_3 + r_1 - r_2 - r_3)s + r_3$$
$$+\underline{(q_4 s + r_4)}s + (q_1 - q_2 - q_3 + r_2)\alpha$$
$$= \{q_2 s + (q_1 - q_2 - q_3 + q_4 + r_2)\}N + (q_3 + r_1 - r_2 - r_3 + r_4)s + r_3$$
$$+\underline{(q_1 - q_2 - q_3 + q_4 + r_2)\alpha}$$
$$= \{q_2 s + (q_1 - q_2 - q_3 + q_4 + r_2)\}N + (q_3 + r_1 - r_2 - r_3 + r_4)s + r_3 + \underline{q_5 s + r_5}$$
$$= \{q_2 s + (q_1 - q_2 - q_3 + q_4 + r_2)\}N + (q_3 + q_5 + r_1 - r_2 - r_3 + r_4)s + (r_3 + r_5).$$

As a result, $AB = QN + R$ where $Q = q_2 s + (q_1 - q_2 - q_3 + q_4 + r_2)$ and $R = (q_3 + q_5 + r_1 - r_2 - r_3 + r_4)s + (r_3 + r_5)$ proves that Algorithm 5 computes the quotient and remainder of 2ℓ-bit multiplications.

$\square$

B.3   Algorithm 6

Proof. $AB = (a_1 t + a_0)(b_1 t + b_0) = a_1 b_1 t(t-1) + (a_1 + a_0)(b_1 + b_0)t - a_0 b_0(t-1)$. All terms of the above equations are modified according to the following equations.

$(a_1 + a_0)(b_1 + b_0)t = (q_1 t + r_1)t = q_1 N + (q_1\delta + r_1)t + q_1\alpha,\ \underline{a_1 b_1}t(t-1) = \underline{(q_2 t + r_2)}t(t-1) = \{q_2 t + q_2(\delta-1) + r_2\}N + \{q_2(\alpha+\delta^2-\delta) + r_2(\delta-1)\}t + \{q_2(\delta-1)+r_2\}\alpha$ and $\underline{a_0 b_0}(t-1) = \underline{(q_3 t + r_3)}(t-1) = q_3 N + \{q_3(\delta-1)+r_3\}t + q_3\alpha - r_3$. Then, the equation is rewritten in the following equations.

$$AB = \{q_2 t + (q_2\delta + q_1 - q_2 - q_3 + r_2)\}N + (q_1 + q_2(\delta-1) - q_3 + r_2)\alpha + r_3$$
$$+\{q_1\delta + q_2(\alpha + \delta^2 - \delta) - q_3(\delta-1) + r_1 + r_2(\delta-1) - r_3\}t.$$

$\square$

B.4   Algorithm 9

Proof. The proof is conducted with general k, which includes $k = 4$.

First, A and B are represented with a_h and b_i $(h, i = 0, \cdots k)$ which are ℓ-bit pieces of A and B. $A = \sum_{h=0}^{k} a_h 2^{h\ell}$, $B = \sum_{i=0}^{k} b_i 2^{i\ell}$, $N = \sum_{j=0}^{k} n_j 2^{j\ell}$. Then, the $2k\ell$-bit product AB is rewritten as $AB = \sum_{h=0}^{k}\sum_{i=0}^{k} a_h b_i 2^{(h+i)\ell} = \sum_{h=0}^{k-1} s_i 2^{i\ell} + \sum_{h=k}^{2k-1} s_i 2^{i\ell}$. Now, precomputations are injected in the second term.

$$\sum_{h=k}^{2k-1} s_i 2^{i\ell} = \sum_{h=k}^{2k-1} s_i \sum_{j=0}^{k-1} c_{i,j} 2^{j\ell} = \sum_{h=k}^{2k-1} s_i (c_{i,k-1} 2^{(k-1)\ell} + \sum_{j=0}^{k-2} c_{i,j} 2^{j\ell})$$

$$= -q\sum_{j=0}^{k-2} n_j 2^{j\ell} + r2^{(k-1)\ell} + \sum_{h=k}^{2k-1} s_i \sum_{j=0}^{k-2} c_{i,j} 2^{j\ell}$$

where $s_i c_{i,k-1} = qn_{k-1} + r$ and $N = \sum_{j=0}^{k} n_j 2^{j\ell}$.

Then, the following equation for modular reduction holds.

$$AB \quad (\mathrm{mod}\ N) = \sum_{h=0}^{k-1} s_i 2^{i\ell} - q\sum_{j=0}^{k-2} n_j 2^{j\ell} + r2^{(k-1)\ell} + S\sum_{j=0}^{k-2} c_{i,j} 2^{j\ell}.$$

$\square$

Constant-Rounds, Almost-Linear Bit-Decomposition of Secret Shared Values

Tomas Toft[1,2,*]

[1] CWI Amsterdam, The Netherlands
[2] TU Eindhoven, The Netherlands

Abstract. Bit-decomposition of secret shared values – securely computing sharings of the binary representation – is an important primitive in multi-party computation. The problem of performing this task in a constant number of rounds has only recently been solved.

This work presents a novel approach at constant-rounds bit-decomposition. The basic idea provides a solution matching the big-$\mathcal{O}$-bound of the original while decreasing the hidden constants. More importantly, further solutions *improve* asymptotic complexity with only a small increase in constants, reducing it from $\mathcal{O}(\ell \log(\ell))$ to $\mathcal{O}(\ell \log^*(\ell))$ and even lower. Like previous solutions, the present one is unconditionally secure against both active and adaptive adversaries.

Keywords: Secret Sharing, Constant-rounds Multi-party Computation, Bit-decomposition.

1 Introduction

Secure multi-party computation (MPC) allows a number of parties to perform a computation based on private inputs, learning the result but revealing no other information than what this implies. The general solutions providing unconditional security phrase the computation as a circuit over secret shared inputs, however, the gates must be evaluated in an iterative fashion, implying that round complexity is equivalent to circuit depth.[1]

A clever choice of representation of the inputs can have great influence on possible constant-rounds solutions. Consider determining the sum of a number of private integer values. Their binary representation could be taken as input, but constant-rounds bitwise addition is expensive. A much simpler solution would be to choose a prime, p, greater than the maximal sum, and view the inputs as elements of $\mathbb{Z}_p$ – the $\mathbb{Z}_p$ computation simulates the task to be performed. However, if we wish to determine some property of the sum, say a particular bit, then we are stuck. This would be trivial given the binary representation.

---

[*] Supported by the research program Sentinels (`http://www.sentinels.nl`). Sentinels is being financed by Technology Foundation STW, the Netherlands Organization for Scientific Research (NWO), and the Dutch Ministry of Economic Affairs.
[1] With a linear secret sharing scheme only multiplication gates are counted regarding depth.

M. Fischlin (Ed.): CT-RSA 2009, LNCS 5473, pp. 357–371, 2009.

Being able to switch representation gives us the best of both worlds. An efficient means of extracting all bits in a constant number of rounds is therefore an important primitive (the other direction is trivial). Though this problem has already been solved, it is worth improving the solutions, as more efficient primitives implies better complexity for any application.

1.1 Setting and Goal

Formalizing the above, our setting is the following: A value x is secret shared among n parties $P_1, \ldots, P_n$ using an unconditionally secure, linear secret sharing scheme over the prime-field $\mathbb{Z}_p$. In addition to this, the parties have access to a secure, constant-rounds multiplication protocol for shared values, i.e. given sharings of a and b, the parties can obtain a sharing of the product. All primitives are assumed to allow concurrent execution. The above can be instantiated with Shamir's secret sharing scheme [Sha79] and the protocols of Ben-Or et al. [BGW88].

The parties wish to securely compute sharings of the binary representation of x in a constant number of rounds of interaction. I.e. they should obtain sharings of $x_{\ell-1}, \ldots, x_0 \in \{0,1\} \subseteq \mathbb{Z}_p$ – where ℓ is the bit-length of p – such that

$$x = \sum_{i=0}^{\ell-1} 2^i x_i$$

when viewed as occuring over the integers.

1.2 Related Work

Bit-decomposition has been considered in various settings. Algesheimer et al. presented a solution starting with an additive secret sharing, [ACS02]. This was not constant-rounds though, and only secure against semi-honest adversaries.

The first constant-rounds solution to the bit-decomposition problem is due to Damgård et al., [DFK+06]. Similar to here, unconditional security against active and adaptive adversaries was considered. The computation requires $\mathcal{O}(\ell \log(\ell))$ invocations of the multiplication protocol.

The [DFK+06] computation was later improved by a constant factor by Nishide and Ohta, [NO07] – the basic idea was the same, but it was observed that one of the two invocations of the most expensive primitive, bitwise addition, was unnecessary.

Toft later sketched a theoretically improved solution, [Tof07] also building on [DFK+06]. Complexity was reduced to $\mathcal{O}(\ell \log^{(c)}(\ell))$ for any constant integer, c, where $\log^{(c)}$ is the logarithm applied c times. The round complexity of that solution was $\mathcal{O}(c)$ and required multiple sequential applications of the bitwise addition protocol.

Concurrently and independently of Damgård et al., Schoenmakers and Tuyls considered bit-decomposition of Paillier encrypted values, i.e. the cryptographic setting, [ST06]. Though their main focus was not on constant round complexity, they noted that the techniques of [DFK+06] could be applied in that setting too.

1.3   Contribution

This work presents a practical, unconditionally secure,[2] constant-rounds bit-decomposition protocol with improved complexity compared to previous solutions. Security guarantees are equivalent to that of the primitives, i.e. when these are secure against active and adaptive adversaries, then so are the proposed protocols.

Complexity is $\mathcal{O}(\ell \log^{(*^{(c)})}(\ell))$ invocations of the multiplication protocol for any constant integer c, where $\log^{(*^{(1)})} = \log^*$, the iterated logarithm, while for $c > 1$, $\log^{(*^{(c)})}$ is the iteration of $\log^{(*^{(c-1)})}$.

$$\log^{(*^{(c)})}(\ell) = \begin{cases} 0 & \ell \leq 1 \\ 1 + \log^{(*^{(c)})}(\log^{(*^{(c-1)})}(\ell)) & \ell > 1 \end{cases}$$

Round complexity is $\mathcal{O}(c)$ implying a tradeoff between communication complexity and the number of rounds: decreasing big-$\mathcal{O}$-complexity comes at the cost of increased constants. However, even small c imply an essentially linear solution – $\log^*$ is in practice at most 5. Thus, the constants remain competitive for a solution which is "linear in practice." A comparison of this and previous work on constant-rounds bit-decomposition is seen in Table 1.

Table 1. Complexities of constant-rounds bit-decomposition protocols

	Rounds	Multiplications
[DFK+06]	$\mathcal{O}(1)$	$\mathcal{O}(\ell \log(\ell))$
[NO07]	$\mathcal{O}(1)$	$\mathcal{O}(\ell \log(\ell))$
[Tof07]	$\mathcal{O}(c)$	$\mathcal{O}(\ell \log^{(c)}(\ell) + c\ell)$
This paper	$\mathcal{O}(c)$	$\mathcal{O}(c\ell \log^{(*^{(c)})}(\ell))$

It is noted that the techniques presented here are also applicable in other settings, e.g. to MPC protocols based on Paillier encryption. The ideas may also be used to construct novel, non-constant-rounds variations, which may be preferable in practice.

1.4   An Overview of This Paper

Section 2 introduces preliminaries, including known sub-protocols. Section 3 presents the overall intuition and translation of the problem to post-fix comparison, which is introduced fully in Sect. 4. The following two sections both provide solutions to post-fix comparison, with Sect. 6 building on Sect. 5. Finally, Sect. 7 provides analysis and concluding remarks.

---

[2] Technically, security is perfect; there is, however, a negligible probability of "faults." The effect can be chosen, e.g. only expected constant-rounds, only unconditional security, imperfect correctness.

2 Preliminaries

This section elaborates on the setting introduced in Sect. 1.1. Moreover, additional primitives needed below are also listed.

2.1 Setting and Complexity

The basic setting of a linear secret sharing scheme with a multiplication protocol can be modelled as an incorruptible third party. This party allows the parties of the protocol to input values (share), perform arithmetic (when sufficiently many agree), and to output values (reconstruct). More formally, this trusted party is instantiated as an ideal functionality, e.g. similar to the arithmetic black-box of Damgård and Nielsen [DN03]. The present protocols then apply to any schemes shown equivalent to this ideal functionality. This also implies that in order to show security, it must simply be argued that any values revealed, do not reveal any information; sub-protocols are secure by assumption.

Similar to other work, interaction is considered the most important resource. As the primitives considered are abstract, the measures are phrased as invocations of the sub-protocols. For simplicity, as the multiplication protocol is both the most expensive and the most used primitive, only its invocations are counted. Complexity is therefore measured in *secure multiplications*. Note that secure computation of linear combinations follows directly from the linearity of the underlying scheme and is therefore considered costless.

Complexity is divided into two measures: Rounds (the overall number of times that messages are exchanged between all parties) and the communication size (the overall size of all messages). The former is represented by the number of sequential secure multiplications, while the latter consists of the overall number. Note that round complexity counts sequential multiplications, and *not* the actual number of messages. However, with abstract primitives it is not possible to be more precise. Moreover, with a constant-rounds multiplication protocol, the rounds-measures are equivalent under big-$\mathcal{O}$.

2.2 Basic Notation

Though protocols are considered, the notation used will be algorithmic in nature. $[a]$ will denote a sharing of a, with $a \leftarrow \mathsf{output}([a])$ indicating its reconstruction. Computation will be written in an infix notation, which eases readability. For example,

$$[y] \leftarrow c_1 \cdot [x_1] \cdot [x_2] + c_2 \cdot [x_3] \cdot [x_4]$$

denotes two invocations of the multiplication protocol followed by a linear computation. As the protocol invocations may be performed in parallel, this requires only a single round.

Certain intermediate values of the protocol will be bit-decomposed. To distinguish these, they will be written $[x]_B$. The notation is simply shorthand for a list bits, $[x_i]$. Any arithmetic operation taking such an input implicitly converts it to a field element. This is a linear computation and therefore costless.

Abusing notation, we sometimes write multiple shared values – such as a list of shared values – with the same notation as a single sharing, $[x]$. It will be clear from context when this occurs.

2.3 Known Primitives

Regarding the primitives needed, all are described in detail in both [DFK$^+$06] and [NO07]. The most basic is that of random bit generation – generating a random sharing of a uniformly random bit unknown to all parties. This is considered equivalent to two sequential multiplications.

The second primitive consists of comparison of secret shared values. An execution results in a sharing of $[b] \in \{0, 1\}$ which is 1 exactly when the first input is greater than the second. This work only needs a protocol for the case when the inputs are already bit-decomposed. Similar to arithmetic, this is also written with an infix notation, $\overset{?}{\geq}$. For efficiency, the present work considers a variation of the one used in [DFK$^+$06] and [NO07]; the basic idea is the same, hence the difference is only sketched.

Given two ℓ-bit inputs, $[a]_B$ and $[b]_B$, their bitwise XOR is computed in the standard way, $[a_i \oplus b_i] = [a_i] + [b_i] - 2[a_i b_i]$. Then, rather than performing a full prefix-OR on this, we stop after having determined and run brute-force prefix-OR on the first interesting block of size $\sqrt{\ell}$ (i.e. the first block containing a one). The associated block of $[a]_B$ is determined (in parallel with that of the XOR-list) and the answer to $[a]_B > [b]_B$ may be extracted as the dot product between these two list of length $\sqrt{\ell}$ – this is the bit of $[a]_B$ at the most significant differing bit-position, which is also the result.

Overall this requires six rounds (plus two rounds of preprocessing) in which the multiplication protocol is invoked $\ell + 5\ell + (5/2)(\ell + \sqrt{\ell}) + 2\ell + (5/2)(\ell + \sqrt{\ell}) + \sqrt{\ell} = 13\ell + 6\sqrt{\ell}$ times.[3]

In this work, we also need a comparison which does not only provide single bit output, but considers three cases – smaller, equal, or larger – encoded in two bits:

$$\text{comp}\left([a]_B, [b]_B\right) = \begin{cases} \left(\begin{bmatrix}1\\0\end{bmatrix}\right) & \text{if } [a]_B > [b]_B \\ \left(\begin{bmatrix}0\\0\end{bmatrix}\right) & \text{if } [a]_B = [b]_B \\ \left(\begin{bmatrix}0\\1\end{bmatrix}\right) & \text{if } [a]_B < [b]_B \end{cases}$$

This is essentially just a comparison that goes both ways – comparing both $[a]_B$ to $[b]_B$ and $[b]_B$ to $[a]_B$. This adds only $\ell + \sqrt{\ell}$ extra multiplications and require no additional rounds. The bulk of the work consists of determining the most significant bit-position where the inputs differ.

The final primitive is the generation of uniformly random, bitwise shared values of $\mathbb{Z}_p$, written $[r]_B \in_R \mathbb{Z}_p$. For ℓ-bit p, this is done by securely generating ℓ random bits, $[x]_B = ([x_{\ell-1}], \dots, [x_0])$, and verifying that $[x]_B < p$. This may of

---

[3] In difference to [DFK$^+$06] and [NO07], it is noted that the quadratic, brute force prefix-OR requires $5/2(k + \sqrt{k})$ multiplications on lists of size $\sqrt{k}$ rather than $5k$.

course fail; as in previous work, it is assumed that four attempts (run in parallel) are needed. This implies a complexity of seven rounds and $4(2\ell + (11\ell + 6\sqrt{\ell})) = 52\ell + 24\sqrt{\ell}$ multiplications; note the slightly reduced complexity as p is public.

3   Overall Intuition – Simplifying the Problem

The overall solution transforms the problem of bit-decomposition to one of postfix comparison described fully in Sect. 4 below. For clarity, we first introduce the overall idea through a naïve solution. Complexity is worse than previous ones, but the solution – seen as Protocol 1 – provides intuition needed for the following sections.

Protocol 1. The overall bit-decomposition

Input: $[x]$.
Output: The bit-decomposition, $[x]_B = ([x_{\ell-1}], [x_{\ell-2}] \dots, [x_0])$

    $[r]_B \in_R \mathbb{Z}_p$
    $c \leftarrow \mathsf{output}([x] + [r]_B)$
    $c' \leftarrow c + p$ ▷ Addition over the integers
    $[f] \leftarrow [r]_B \overset{?}{>} c$
5: **for** $i = 0, \dots, \ell$ **do**
      $[\tilde{c}_i] \leftarrow [f] \cdot (c'_i - c_i) + c_i$
    end for
    $[\tilde{c}]_B \leftarrow ([\tilde{c}_\ell], \dots, [\tilde{c}_0])$
    $[x \bmod 2^\ell] \leftarrow [x]$
10: **for** $i = \ell - 1, \dots, 1$ **do**
      $[u_i] \leftarrow [r \bmod 2^i]_B \overset{?}{>} ([\tilde{c} \bmod 2^i]_B)$
      $[x \bmod 2^i] \leftarrow ([\tilde{c} \bmod 2^i]_B) - [r \bmod 2^i]_B + 2^i[u_i]$
      $[x_i] \leftarrow 2^{-i} ([x \bmod 2^{i+1}] - [x \bmod 2^i])$
    end for
15: $[x_0] \leftarrow [x \bmod 2]$

Correctness: The protocol starts similarly to the one from [NO07], lines 1 through 8. $[x]$ is additively masked by a random $[r]_B$, and the binary representation of $\tilde{c} = x + r$ (over the integers) is computed securely. It is clear that $\tilde{c} \in \{c, c + p\}$, thus, if it can be determined which is the case, then for every bit-position, the relevant bit can be selected. This selection occurs in line 6, where $[f]$ is used to obliviously choose between the two bits. $[f]$ is 1 when $[r]_B > c$, i.e. exactly when an overflow has occurred in the computation of c. This is the case where $c + p$ should be chosen.

The remaining computation – lines 9 to 15 – extracts the actual bits. The main idea is to reduce $[x]$ securely modulo all powers of 2 and compute differences between neighbors. Clearly

$$(x \bmod 2^{i+1}) - (x \bmod 2^i) = 2^i x_i$$

which is easily mapped to a binary value.

p is ℓ-bit, so $\left[x \bmod 2^\ell\right]$ is computed correctly. Moreover, as we have $\tilde{c} - r = x$ over the integers, reducing both $[\tilde{c}]_B$ and $[r]_B$ modulo 2^i and subtracting provides the right result, at least if the computation had been modulo 2^i. Since it occurs modulo p, an underflow occurs when $\left[r \bmod 2^i\right]_B > \left[\tilde{c} \bmod 2^i\right]_B$. Line 12 compensates by adding 2^i in this case. Thus, the $[x_i]$ are correct. The needed reductions of $[\tilde{c}]_B$ and $[r]_B$ modulo powers of two are easily done. Both are bit-decomposed, and it is therefore merely a question of ignoring the top bits.

Complexity: Until the start of the first loop, a random element is generated, added to $[x]$, and then compared to the (public) outcome. Following this are the costless linear combinations of the loop as well as a bit of renaming, lines 8 and 9. All of this is linear and requires only a constant number of rounds.

In the actual bit extraction, $\ell - 1$ comparisons of length $1, \ldots, \ell - 1$ must be performed. This implies a quadratic number of multiplications, however, note that though the bits are extracted in a linear manner, this does not imply a linear number of rounds. Line 11 does not depend on previous iterations of the loop, and all other computation is costless.

Thus, overall complexity is $\mathcal{O}(\ell^2)$ multiplications and $\mathcal{O}(1)$ rounds. However, we have an immediate improvement, if the comparisons in the final loop can be performed more efficiently.

Security: Security follows directly from the security of the sub-protocols and the masking of $[x]$. $[r]$ is chosen uniformly at random from $\mathbb{Z}_p$, thus c is also uniformly random over $\mathbb{Z}_p$ and therefore reveals nothing.

4   The Post-fix Comparison Problem

Section 3 transformed the problem of bit-decomposition to that of performing $\ell - 1$ comparisons (line 11). In general this is quadratic, however, here the numbers are highly interdependent: they are reductions modulo all powers of two of the same initial values, $[\tilde{c}]_B$ and $[r]_B$, which leads to the formulation of the post-fix comparison problem.

Problem 1 (Post-fix Comparison). *Given two bitwise shared $\hat{\ell}$-bit values* $[a]_B = \sum_{i=0}^{\ell-1} 2^i [a_i]$ *and* $[b]_B = \sum_{i=0}^{\ell-1} 2^i [b_i]$, *compute*

$$[c_i] = \left[a \bmod 2^i\right]_B \overset{?}{>} \left[b \bmod 2^i\right]_B$$

for all $i \in \{1, 2, \ldots, \hat{\ell}\}$.

Recall from Sect. 3 that in addition to the post-fix comparison, Protocol 1 required only $\mathcal{O}(\ell)$ secure multiplications. Presenting a protocol which solves post-fix comparison more efficiently than $\mathcal{O}(\ell \log(\ell))$ ($\mathcal{O}(\ell \log^{(c)}(\ell))$) therefore implies an improvement over [DFK+06, NO07] ([Tof07]). The naïve solution above is of course much worse.

5   Solving Post-fix Comparison, $\mathcal{O}(\ell \log(\ell))$ Work

The $\mathcal{O}(\ell \log(\ell))$ bound is not beaten in one go. Rather, we start with a solution equalling it, which is used as a sub-protocol below. This not only introduces needed primitives, but also provides a gradual presentation of the ideas.

For this section, assume that we are given a post-fix comparison problem of size ℓ.[4] (Assume for simplicity that ℓ is a power of 2, this can always be achieved by padding with zeros.) Rather than considering the input as two separate numbers to be compared, we may write it as a string of bit-pairs,

$$\left(\begin{pmatrix} [a_{\ell-1}] \\ [b_{\ell-1}] \end{pmatrix}, \ldots, \begin{pmatrix} [a_0] \\ [b_0] \end{pmatrix} \right) \ .$$

Abusing notation, we write

$$\text{comp} \left(\begin{pmatrix} [a_{i+j-1}] \\ [b_{i+j-1}] \end{pmatrix}, \ldots, \begin{pmatrix} [a_j] \\ [b_j] \end{pmatrix} \right)$$

meaning the comparison of the numbers represented by the (sub-string) of bit-pairs.

The naïve solution to the post-fix comparison problem divided the inputs into all post-fixes, and processed these individually. Here, we split into blocks of length powers of two. The final result can then be constructed from these. Rather than simply solving Problem 1, we consider a slightly more difficult variation and require the output from the extended comparison, $\text{comp}(\cdot)$. This results in slightly more (shared) information which will be needed below.

The division into blocks is best viewed as a balanced, binary tree. Every node represents a comparison of some sub-block of bit-positions. The root node represents the full comparison, while the leaves represent (the comparison of) single bit positions. Internal nodes represent the comparison of either the first half (left child) or second half (right child) of its parent. I.e. if a node represents the comparison of $[a_{i_{2^k-1}}], \ldots, [a_{i_0}]$ and $[b_{i_{2^k-1}}], \ldots, [b_{i_0}]$, then its left (right) child represents $[a_{i_{2^k-1}}], \ldots, [a_{i_{2^{k-1}}}]$ and $[b_{i_{2^k-1}}], \ldots, [b_{i_{2^{k-1}}}]$ ($[a_{i_{2^{k-1}-1}}], \ldots, [a_{i_0}]$ and $[b_{i_{2^{k-1}-1}}], \ldots, [b_{i_0}]$). Such a tree is constructed by running Protocol 2 on the full input.

From this tree,

$$\begin{pmatrix} [c_i^\top] \\ [c_i^\perp] \end{pmatrix} = \text{comp} \left([a \bmod 2^i], [b \bmod 2^i] \right)$$

can be computed using Protocol 3. Note that $\|$ denotes the concatenation of lists; line 5 sets $[s]$ to an empty list, while line 8 appends the outcomes of the $\text{comp}(\cdot)$'s, i.e. pairs of bits. Executing the protocol on all i, $1 \leq i < \ell$, results in the desired output.

---

[4] Though $[\tilde{c}]_B$ is $(\ell + 1)$-bit, the extra bit added by the conversion to the post-fix comparison problem can be ignored.

Protocol 2. The tree construction protocol, treecomp($\cdot$), for solving post-fix comparison using $\mathcal{O}(\ell \log(\ell))$ multiplications.

Input: Two $\hat{\ell}$-bit, bitwise secret shared numbers, $[a_{\hat{\ell}-1}], \ldots, [a_0]$ and $[b_{\hat{\ell}-1}], \ldots, [b_0]$.
Output: A binary tree of results of sub-comparisons; each node represents a block of
   bit-positions.
   if $\hat{\ell} = 1$ **then**
      return $(\text{comp}([a_0], [b_0]), \bot, \bot)$
   end if
   $[d] \leftarrow \text{comp}(([a_{\hat{\ell}-1}], \ldots, [a_0]), ([b_{\hat{\ell}-1}], \ldots, [b_0]))$
5: $[c_{\text{left}}] \leftarrow \text{treecomp}(([a_{\hat{\ell}-1}], \ldots, [a_{\hat{\ell}/2}]), ([b_{\hat{\ell}-1}], \ldots, [b_{\hat{\ell}/2}]))$
   $[c_{\text{right}}] \leftarrow \text{treecomp}(([a_{\hat{\ell}/2-1}], \ldots, [a_0]), ([b_{\hat{\ell}/2-1}], \ldots, [b_0]))$
   return $([d], [c_{\text{left}}], [c_{\text{right}}])$

Protocol 3. Extracting one result for post-fix comparison from a treecomp($\cdot$)-tree.

Input: A bit-position, $1 \leq i < \ell$ and the output of treecomp($\cdot$) on two ℓ-bit, bitwise
   secret shared numbers, $[a_{\ell-1}], \ldots, [a_0]$ and $[b_{\ell-1}], \ldots, [b_0]$.
Output: Output equivalent to that of $\text{comp}([a \bmod 2^i], [b \bmod 2^i])$.
   if $i = \ell$ **then**
      return root.$[d]$
   end if
   node $\leftarrow$ leaf(i)
5: $[s] \leftarrow$ EmptyList
   while node is not root **do**
      if node is left child **then**
         $[s] \leftarrow [s] || (\text{sibling(node)}.[d])$
      end if
10:   node $\leftarrow$ parent(node)
   end while
   return comp $([s])$

Correctness: The intuition behind this section is that the division into blocks creates halves, quarters, eighths, etc. Any postfix of bit-positions, $i - 1, \ldots, 0$, can be "covered" by only a logarithmic number of disjoint blocks – at most one from each level of the tree is needed. Replacing the bits of a block with the output of comp($\cdot$) changes the numbers involved, but comparing these new numbers provides the same result. For example, for blocks $[a_\top]$ and $[a_\bot]$ ($[b_\top]$ and $[b_\bot]$) covering $[a]$ ($[b]$), we have

$$\text{comp}\left(\binom{[a_\top]}{[b_\top]} \middle\| \binom{[a_\bot]}{[b_\bot]} \right) = \text{comp}\left(\text{comp}\binom{[a_\top]}{[b_\top]} \middle\| \text{comp}\binom{[a_\bot]}{[b_\bot]} \right)$$

To see this, note that comparison simply extracts the most significant, differing bit-position. Replacing blocks with their output of comp essentially compresses strings to bit-pairs containing the relevant information, i.e. which was bigger (or whether they were equal). Thus, replacing the bits of the blocks of the cover with the outcome of their comparisons, results in a much smaller comparison problem with the same result.

Protocol 2 constructs all the covers needed for all positions, while Protocol 3 selects the covers needed for the i'th position. These are exactly the siblings of the left children on the path to the root. Finally, the log-length comparisons provide the right result.

Complexity: It is easily verified that Protocol 2 requires $\mathcal{O}(\ell \log(\ell))$ multiplications. Each of the $\log(\ell)$ levels of the tree is a subdivision of the original problem into disjoint blocks, which are then compared, i.e. each level requires a linear number of secure multiplications, which matches the stated bound. Round complexity is $\mathcal{O}(1)$; the only secure computation is the invocations of comp$(\cdot)$. As these are all independent of each other, they may be performed in parallel.

Protocol 3 traverses a path from a leaf to the root and gathers data. This is purely bookkeeping, the only multi-party computation is the comparison at the very end. The path from leaf to root is of length $\log(\ell)$, thus the bit-length of the values of $[s]$ is at most $\log(\ell)$, implying $\mathcal{O}(\log(\ell))$ multiplications.

Applying Protocol 2 on the input and Protocol 3 on the tree ℓ times – once for every bit-position i – is overall $\mathcal{O}(\ell \log(\ell))$. Only $\mathcal{O}(1)$ rounds are needed as the bit-positions may be processed concurrently.

Security: Security is easily argued. Neither of Protocols 2 and 3 output any information. They are simply deterministic applications of the primitives, which are secure and composable by assumption.

6   Solving Post-fix Comparison, $\mathcal{O}(\ell \log^*(\ell))$ and Better

In order to use the present ideas, while also decreasing complexity, the height of the tree of sub-comparisons of Sect. 5 must be reduced. However, if fan-out is increased (the block of each node split into more than two sub-blocks), each of the results of the post-fix comparison, $[c_i]$, may depend on more than one block from each level. There is no longer a single node per level which handles all less significant bits of ancestors on the path to the root. The solution is to increase fan-out and then "combine" the blocks represented by siblings, such that each output again depends on only one node per level.

Analogously to Sect. 5, initially each node will simply represent a distinct sub-block of that of its parent. For each node, the parts of the inputs it represents are compared. As each level simply divides $[a]$ and $[b]$ into sub-blocks, overall complexity is $\mathcal{O}(h\ell)$, where h is the height of the tree. The construction is analogous to Protocol 2 and is seen as Protocol 4; the main difference is the addition of the function, F, specifying the fan-out which depends on the size of the parent node.

Protocol 4. Extending the tree construction protocol, treecomp($\cdot$) to non-binary fan-out.

Input: Two $\hat{\ell}$-bit, bitwise secret shared numbers, $[a_{\hat{\ell}-1}], \ldots, [a_0]$ and $[b_{\hat{\ell}-1}], \ldots, [b_0]$, as well as a $F : \mathbb{N} \mapsto \mathbb{N}$ specifying the fan-out.

Output: A tree of results of sub-comparisons with fan-out specified by F; each node represents a block of bit-positions.

    if $\hat{\ell} = 1$ **then**

        return $(\text{comp}([a_0], [b_0]), \perp)$

    end if

    $[d] \leftarrow \text{comp}(([a_{\hat{\ell}-1}], \ldots, [a_0]), ([b_{\hat{\ell}-1}], \ldots, [b_0]))$

5: $f \leftarrow F(\hat{\ell})$

    $[c] \leftarrow \text{EmptyList}$

    for $i = 0, \ldots, f - 2$ **do**

$$[c] \leftarrow [c] \| \text{treecomp}\left(\left(\begin{bmatrix} a_{\hat{\ell}-i(\lceil \hat{\ell}/f \rceil)-1} \\ b_{\hat{\ell}-i(\lceil \hat{\ell}/f \rceil)-1} \end{bmatrix} \right), \ldots, \left(\begin{bmatrix} a_{\hat{\ell}-(i+1)(\lceil \hat{\ell}/f \rceil)} \\ b_{\hat{\ell}-(i+1)(\lceil \hat{\ell}/f \rceil)} \end{bmatrix} \right) \right)$$

    end for

10: $[c] \leftarrow [c] \| \text{treecomp}\left(\left(\begin{bmatrix} a_{\hat{\ell}-(f-1)(\lceil \hat{\ell}/f \rceil)-1} \\ b_{\hat{\ell}-(f-1)(\lceil \hat{\ell}/f \rceil)-1} \end{bmatrix} \right), \ldots, \left(\begin{bmatrix} a_0 \\ b_0 \end{bmatrix} \right) \right)$

    return $([d], [c])$

In order to ensure that only one node from each level is needed, each node must be "extended" to represent not only the comparison of its own block, but also those of its less significant siblings. This is best explained through an example. Consider node n_{32} on Fig. 1, this is the third child of n_3 and should therefore represent the comparison of the initial blocks of nodes n_{32}, n_{31}, and n_{30}. Once this is the case for all nodes, each post-fix will again depend on at most one node from every level. Continuing the example of Fig. 1: n_2, n_1, and n_0 are combined and stored in n_2 at the first level, while n_{32}, n_{31}, and n_{30} are combined and stored in n_{32} at the second. Together they represent everything up to and including the block of n_{32}.

Thus, once siblings are combined – $[d]$ is set to represent the comparison of not only their own block but that of all their less significant siblings – each of the ℓ output values can be computed. The task, which is a straight forward generalization of Protocol 3 of Sect. 5, is seen as Protocol 5. For all non-rightmost nodes on the path to the root, its right-neighbor sibling – which provides a cover for all the less significant bits of the parent – is included. Together, these provide the full cover.

The arguments of correctness and security are the same as in the previous section. Sub-strings are compressed resulting in a smaller comparison, and all interaction still occurs using only secure primitives. Regarding complexity, apart from the combining of the comparison results of siblings, this is $\mathcal{O}(h\ell)$. First each level is split into disjoint sub-blocks which are compared. Then ℓ comparisons of h-bit numbers are performed. It remains to explain how to choose F and combine siblings.

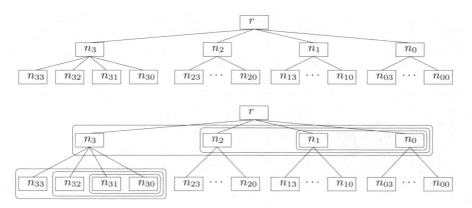

Fig. 1. Combining children with less significant siblings

Protocol 5. Extracting one result for post-fix comparison from a generalized treecomp($\cdot$)-tree.

Input: A bit-position, $1 \leq i < \ell$ and the output of the generalized treecomp($\cdot$) on two ℓ-bit, bitwise secret shared numbers, $[a_{\ell-1}], \ldots, [a_0]$ and $[b_{\ell-1}], \ldots, [b_0]$.
Output: Output equivalent to that of comp($[a \bmod 2^i], [b \bmod 2^i]$).

    if $i = \ell$ **then**
        return root.$[d]$
    end if
    node $\leftarrow$ leaf(i)
5: $[s] \leftarrow$ EmptyList
    while node is not root **do**
        if node is a non-rightmost child **then**
            $[s] \leftarrow [s]||(\text{right\_sibling(node)}.[d])$
        end if
10:    node $\leftarrow$ parent(node)
    end while
    return comp $([s])$

6.1 $\mathcal{O}(\ell \log^*(\ell))$

In order to see how the sub-results of the children can be combined efficiently, note that *combine* means *compare up to and including this block*. This is a sub-problem of the post-fix comparison problem, where only every $\hat{\ell}/F(\hat{\ell})$'th result is needed. Viewed differently, it is the full post-fix comparison problem on the numbers represented by the outcome of the comparisons of the children – as before, replacing a block with the outcome of its comparison does not change the result.

While it seems as if we are back to square one, this is not the case, as the size of the post-fix comparison problem is reduced. For a (parent) problem with $\hat{\ell}$-bit inputs, letting the children be of logarithmic-size – i.e. setting the fan-out

$F(x) = \lceil x/\log(x) \rceil$ – implies that the new problem is of size $\lceil \hat{\ell}/\log(\hat{\ell}) \rceil$. This can be solved using $\mathcal{O}(\hat{\ell})$ secure multiplications – linear in the parent – using the solution of Sect. 5. Doing this for *all* nodes is linear at each level of the tree, as each node still represents a distinct block of the inputs. Thus, overall this is $\mathcal{O}(h\ell)$.

The choice of F reduces the size of children in log-steps resulting in $h = \log^*(\ell)$, i.e. this solution requires only $\mathcal{O}(\ell \log^*(\ell))$ multiplications. Security again follows from the fact that the primitives are secure by definition and no information is output.

6.2 Decreasing Complexity Further

Further improvements are trivial. It is merely a question of choosing a greater fan-out and solving the post-fix comparison problems linearly in the parents. The choice of fan-out above ensured linearity with the solution of Sect. 5. Increasing fan-out to $F(x) = \lceil x/\log^*(x) \rceil$ provides post-fix comparison problems of size $\hat{\ell}/\log^*(\hat{\ell})$ for nodes of size $\hat{\ell}$. These can be solved using $\mathcal{O}(\hat{\ell})$ secure multiplications by applying the improved solution of Sect. 6.1. This implies a $\mathcal{O}(\ell \log^{(*^{(2)})}(\ell))$ solution.

The process can be repeated indefinitely, however, each iteration increases round complexity. Each new version executes its predecessor as a sub-computation in addition to the initial comparison for every block and the concluding comparison for every output. Thus, to ensure a constant number of rounds, only a constant number of improvements are possible.

It is simple to verify that the complexity of the this secure computation for post-fix comparison is as stated – $\mathcal{O}(c)$ rounds and $\mathcal{O}(\ell \log^{(*^{(c)})}(\ell))$ where c is constant. From Sect. 4 it now follows that constant-rounds bit-decomposition is possible in the same complexity.

Theorem 1. *For any integer c, there exists a $\mathcal{O}(c)$-rounds bit-decomposition protocol for arbitrary $[x] \in \mathbb{Z}_p$, which requires only $\mathcal{O}(\ell \log^{(*^{(c)})}(\ell))$ invocations of the multiplication protocol.*

7 Analysis and Conclusion

This paper has presented a novel solution to the problem of constant-rounds bit-decomposition. Rather than simulating bitwise addition, a random mask and the masked value (transformed to simulate that the addition occurred over the integers) are reduced modulo all powers of two. Subtraction in $\mathbb{Z}_{2^i}$ for all i is then simulated efficiently with $\mathbb{Z}_p$ arithmetic. The resulting complexity of $\mathcal{O}(\ell \log^{(*^{(c)})}(\ell))$ is essentially linear – $\log^*$ is "bounded" by 5 on any reasonable input. Only when the *bit-length*, ℓ, of the prime defining the field exceeds 2^{65536} is it larger. A detailed analysis allows the protocol to be compared to the previous solutions.

The initial transformation requires the generation of a random element and a comparison. As the latter has a public input, the initial XOR and the selection of the associated block are costless. This implies $(52\ell + 24\sqrt{\ell}) + (11\ell + 6\sqrt{\ell}) = 63\ell + 30\sqrt{\ell}$ multiplications in $7 + 5 = 12$ rounds.

The basic $\mathcal{O}(\ell\log(\ell))$ solution to the post-fix comparison problem first compares at each level (equivalent to one full ℓ-bit comparison per level); second, it performs a comparison for each of the ℓ paths. With a few observations, namely that

- a single bitwise XOR of the inputs suffices for *all* levels
- for the output of comp$(\cdot)$, the XOR is free as $\binom{1}{1}$ does not occur
- for bit-decomposition, the final comparison of the post-fix comparison problem does not have to be comp$(\cdot)$

it can be seen that this requires $1 + 5 + 5 = 11$ rounds and

$$\ell + \log(\ell) \cdot \left(13\ell + 7\sqrt{\ell}\right) + \ell \cdot \left(12\log(\ell) + 6\sqrt{\log(\ell)}\right) < 31\ell\log(\ell) + 8\ell$$

multiplications, where the inequality holds for $\ell \geq 4$.

The analysis of the improved solution is similar. The "inner" post-fix comparison protocol is bounded by two sequential invocations of comp$(\cdot)$ (for each parent). This adds 10 rounds as well as

$$2(13\ell + 7\sqrt{\ell}) = 26\ell + 14\sqrt{\ell}$$

multiplications per level. Thus, complexity for the general solution is $11 + 10c$ rounds and $(31 + 26c)\ell\log^{(*^{(c)})}(\ell) + 8\ell + 14c\sqrt{\ell}\log^{(*^{(c)})}(\ell)$ secure multiplications. For the full bit-decomposition protocol, this implies $23 + 10c$ rounds and

$$(31 + 26c)\ell\log^{(*^{(c)})}(\ell) + 71\ell + 14c\sqrt{\ell}\log^{(*^{(c)})}(\ell) + 30\sqrt{\ell}$$

multiplications overall.

Concluding, we focus on the $\mathcal{O}(\ell\log(\ell))$ and $\mathcal{O}(\ell\log^*(\ell))$ solutions, as these are closest to the competition and there is little practical gain from the theoretically better options. For fairness, the optimized comparison (and thus also random element generation) is used in all cases. The details are summarized in Table 2.

The proposed $\mathcal{O}(\ell\log(\ell))$ solution requires fewer rounds and multiplications than any of the other solutions of that complexity. In the long run, the better complexity of [Tof07] will provide fewer multiplications, however, this comes at

Table 2. Explicit complexities of constant-rounds bit-decomposition protocols

	Rounds	Multiplications
[DFK$^+$06]	38	$94\ell\log(\ell) + 63\ell + 30\sqrt{\ell}$
[NO07]	25	$47\ell\log(\ell) + 63\ell + 30\sqrt{\ell}$
[Tof07]	$38c$	$94\ell\log^{(c)}(\ell) + (130(c-1) + 63) \cdot \ell + 30\sqrt{\ell}$
This paper, $\mathcal{O}(\ell\log(\ell))$	23	$31\ell\log(\ell) + 71\ell + 30\sqrt{\ell}$
This paper, $\mathcal{O}(\ell\log^*(\ell))$	33	$57\ell\log^*(\ell) + 71\ell + 14\sqrt{\ell}\log^*(\ell) + 30\sqrt{\ell}$

a cost of increased round complexity. Moreover, for the $\mathcal{O}(\ell \log\log(\ell))$ solution, the turning point does not occur until $\ell \approx 2^{16}$, due to the large constants.

The proposed $\mathcal{O}(\ell \log^*(\ell))$ solution beats [Tof07] outright, however, both the solution of [NO07] and the present $\mathcal{O}(\ell \log(\ell))$ have better constants. When $\ell \approx 38$ this solution and the one of [NO07] use about the same number of secure multiplications, while the present $\mathcal{O}(\ell \log(\ell))$ solution is beaten around $\ell \approx 180$ bits. Thus, if additional rounds are acceptable, then the theoretically better solution may also win in practice.

Finally, it is noted that "hybrids" are also possible. The ideas of [Tof07] can be applied to [NO07], as well as be combined with the ideas of this paper. Though asymptotic complexity is not improved, such variations may be competitive in practice, as the c factor is on the linear rather than on the $\ell \log^{(c)}(\ell)$ term.

Acknowledgements

The author would like to thank Eike Kiltz and Berry Schoenmakers for discussions and suggestions. The anonymous referees are thanked for their input and comments.

References

[ACS02]   Algesheimer, J., Camenisch, J.L., Shoup, V.: Efficient computation modulo a shared secret with application to the generation of shared safe-prime products. In: Yung, M. (ed.) CRYPTO 2002. LNCS, vol. 2442, pp. 417–432. Springer, Heidelberg (2002)

[BGW88]   Ben-Or, M., Goldwasser, S., Wigderson, A.: Completeness theorems for noncryptographic fault-tolerant distributed computations. In: 20th Annual ACM Symposium on Theory of Computing, pp. 1–10. ACM Press, New York (1988)

[DFK⁺06]   Damgård, I.B., Fitzi, M., Kiltz, E., Nielsen, J.B., Toft, T.: Unconditionally secure constant-rounds multi-party computation for equality, comparison, bits and exponentiation. In: Halevi, S., Rabin, T. (eds.) TCC 2006. LNCS, vol. 3876, pp. 285–304. Springer, Heidelberg (2006)

[DN03]   Damgård, I.B., Nielsen, J.B.: Universally composable efficient multiparty computation from threshold homomorphic encryption. In: Boneh, D. (ed.) CRYPTO 2003. LNCS, vol. 2729, pp. 247–264. Springer, Heidelberg (2003)

[NO07]   Nishide, T., Ohta, K.: Multiparty computation for interval, equality, and comparison without bit-decomposition protocol. In: Okamoto, T., Wang, X. (eds.) PKC 2007. LNCS, vol. 4450, pp. 343–360. Springer, Heidelberg (2007)

[Sha79]   Shamir, A.: How to share a secret. Communications of the ACM 22(11), 612–613 (1979)

[ST06]   Schoenmakers, B., Tuyls, P.: Efficient binary conversion for paillier encrypted values. In: Vaudenay, S. (ed.) EUROCRYPT 2006. LNCS, vol. 4004, pp. 522–537. Springer, Heidelberg (2006)

[Tof07]   Toft, T.: Primitives and Applications for Multi-party Computation. PhD thesis, University of Aarhus (2007),
           http://www.daimi.au.dk/~ttoft/publications/dissertation.pdf

Local Sequentiality Does Not Help for Concurrent Composition

Andrew Y. Lindell

Aladdin Knowledge Systems and Bar-Ilan University, Israel
andrew.lindell@aladdin.com, lindell@cs.biu.ac.il

Abstract. Broad impossibility results have been proven regarding the
feasibility of obtaining protocols that remain secure under concurrent
composition when there is no honest majority. These results hold both
for the case of general composition (where a secure protocol is run many
times concurrently with arbitrary other protocols) and self composition
(where a single secure protocol is run many times concurrently). One
approach for bypassing these impossibility results is to consider more
limited settings of concurrency. In this paper, we investigate a restriction
that we call *local sequentiality*. In this setting, every honest party in
the multi-party network runs its protocol executions strictly sequentially
(thus, sequentiality is preserved locally, but not globally). Since security
is preserved under global sequential composition, one may conjecture
that it also preserved under local sequentiality. However, we show that
local sequentiality does *not* help. That is, any protocol that is secure
under local sequentiality is also secure under concurrent self composition
(when the scheduling is fixed). Thus, known impossibility results apply.

1 Introduction

In modern network settings, protocols must remain secure even when many protocol executions take place concurrently. Impossibility results have been proven,
showing that in the case of *no honest majority* and *no trusted setup*, large
classes of functions cannot be securely computed under concurrent composition
[8,7,9,19,20]. These results hold for both concurrent general composition (where a
secure protocol is run concurrently with arbitrary other protocols) and concurrent self composition (where a single secure protocol is run many times concurrently). In fact, these two types of composition have been shown to be (almost)
equivalent [20]. One suggestion for overcoming these impossibility results is to
introduce restrictions on the network or on the behavior of the honest parties.
Such restrictions come in two flavors:

1. *Assumptions on the network:* Here, an assumption is made regarding the network activity. For example, in the setting of m-bounded concurrent composition [1,18], it is assumed that at most m different concurrent executions overlap.
   Another assumption that has been used is that of timing, where it is assumed
   that parties have clocks that proceed at approximately the same rate [13,17].
   We note that restrictions of this type are true "assumptions", in that they cannot be enforced through local instructions provided to the honest parties.

M. Fischlin (Ed.): CT-RSA 2009, LNCS 5473, pp. 372–388, 2009.
© Springer-Verlag Berlin Heidelberg 2009

2. *Locally enforceable policies regarding the honest parties' behavior:* Here, the honest parties are provided with some policy that they can locally enforce (i.e., enforcing the policy requires no coordination with other honest parties or with a centralized entity).

Needless to say, the best case regarding the construction of secure protocols is to not assume any restriction whatsoever. However, as we have mentioned, this is not possible. We therefore aim to construct protocols that remain secure under *reasonable* network assumptions or realistic enforceable policy. (An important issue regarding network assumptions is what is the "damage" if an assumption does not hold due to some anomaly. Such cases should be carefully examined and dealt with.) In this paper, we consider a realistic restriction of the second type and ask if it suffices for bypassing the impossibility results.

Local sequentiality. We consider a restriction where honest parties run protocol executions in a strictly sequential manner (concluding one before beginning the next). Observe that since many different sets of parties participate in different executions, many protocol executions are run concurrently from the global perspective. Nevertheless, from a local perspective, sequentiality is enforced (of course, an adversary may act as it wishes; it is the honest parties who run protocols sequentially). This restriction seems promising since security in the stand-alone model implies security under *globally* sequential composition [6], where sequentiality is assumed with respect to all executions in the network (i.e., no two executions are ever run at the same time). Thus, one may hope that such a theorem also holds with respect to local sequentiality. Note, that although quite restrictive (and thus not very desirable), honest parties can easily enforce a policy of *local* sequentiality. This is in contrast to global sequentiality which is not at all realistic. Unfortunately, we show that local sequentiality does *not* guarantee security. In fact, we show that the known impossibility results for concurrent self composition all hold also for locally sequential self composition. This is proven in two steps. First, we show that any protocol that is secure under locally sequential self composition is also secure under concurrent self composition, as long as the adversary for concurrent self composition uses a fixed schedule. (Loosely speaking, an adversary uses a fixed schedule if the order in which the messages are sent in the different executions is fixed and does not depend on the adversary's input or view during the actual execution.) In order to prove this, we show that many concurrent executions by two parties can be simulated under local sequentiality by having many different parties run the executions (the concurrency is achieved on a global sense and yet suffices for the simulation). Second, we observe that the known impossibility results for concurrent self composition (as proven in [18,20] who build on [9,19]) all hold for adversaries that use fixed schedules. Broad impossibility results and lower bounds are therefore immediately obtained for the setting of local sequentiality as well. We stress that these impossibility results hold only when there is no honest majority or trusted setup phase (otherwise, security under concurrent composition can be achieved, see related work discussed below).

Related work. Secure computation was first studied in the stand-alone model, where it was shown that any multi-party functionality can be securely computed [28,15,4,11]. The study of concurrent composition of protocols was initiated by [14] in the context of witness indistinguishability, and was next considered by [12] in the context of non-malleability. Until recently, the majority of work on concurrent composition was in the context of concurrent zero-knowledge [13,27]. The concurrent composition of protocols for *general secure computation* was only considered much later. Specifically, the first definition and composition theorem for security under concurrent general composition was presented by [25] for the case that a secure protocol is executed once in an arbitrary network. The general case, where many secure protocol executions may take place (again, in an arbitrary network) was then considered in the definition (and composition theorem) of universal composability [7]. Here, it was also shown that any functionality can be securely realized assuming an honest majority [7], or assuming a trusted setup phase in the form of a common reference string [10]. However, in the case of no honest majority or trusted setup, broad impossibility results have been demonstrated for universal composability, concurrent general composition and concurrent self composition [8,7,9,19,20]. These impossibility results justify and provide motivation for considering restricted networks settings and weaker notions of security. Two examples of restrictions that have been considered are bounded concurrency [18,24,23] and timing [17], and an example of a weaker notion of security that has been studied can be found in [22,26,2].

2   Definitions

We denote the security parameter by n. A function $\mu(\cdot)$ is negligible in n (or just negligible) if for every polynomial $p(\cdot)$ there exists a value N such that for all $n > N$ it holds that $\mu(n) < 1/p(n)$. Let $X = \{X(n,a)\}_{n\in\mathbb{N},a\in\{0,1\}^*}$ and $Y = \{Y(n,a)\}_{n\in\mathbb{N},a\in\{0,1\}^*}$ be distribution ensembles. Then, we say that X and Y are computationally indistinguishable, denoted $X \overset{\text{c}}{\equiv} Y$, if for every probabilistic polynomial-time distinguisher D there exists a negligible function $\mu(\cdot)$ such that for every $a \in \{0,1\}^*$, $|\Pr[D(X(n,a)) = 1] - \Pr[D(Y(n,a)) = 1]| < \mu(n)$. When X and Y are equivalent distributions, we write $X \equiv Y$. A machine is said to run in polynomial-time if its number of steps is polynomial in the *security parameter*, irrespective of the length of its input.

2.1   Two-Party Concurrent Self Composition

We begin by defining concurrent self composition for two parties; the extension to local sequentiality is described below. The definition here is taken from [20] and follows the ideal/real paradigm of [16,3,21,6].

Two-party computation. A two-party protocol problem is cast by specifying a random process that maps pairs of inputs to pairs of outputs (one for each party). We refer to such a process as a functionality and denote it

$f : \{0,1\}^* \times \{0,1\}^* \rightarrow \{0,1\}^* \times \{0,1\}^*$, where $f = (f_1, f_2)$. That is, for every pair of inputs (x, y), the output-pair is a random variable $(f_1(x, y), f_2(x, y))$ ranging over pairs of strings. The first party (with input x) wishes to obtain $f_1(x, y)$ and the second party (with input y) wishes to obtain $f_2(x, y)$. We often denote such a functionality by $(x, y) \mapsto (f_1(x, y), f_2(x, y))$. In the context of concurrent composition, each party actually uses many inputs (one for each execution), and these may be chosen adaptively based on previous outputs.

Adversarial behavior. In this work we consider a malicious, static adversary that runs in polynomial time. Such an adversary controls one of the parties (who is called corrupted) and may then interact with the honest party while arbitrarily deviating from the specified protocol. Our definition does not guarantee any fairness. That is, the adversary always receives its own output and can then decide when (if at all) the honest party will receive its output. The scheduling of message delivery is decided by the adversary.

Security of protocols (informal). The security of a protocol is analyzed by comparing what an adversary can do in the protocol to what it can do in an ideal scenario that is trivially secure. This is formalized by considering an *ideal* computation involving an incorruptible *trusted third party* to whom the parties send their inputs. The trusted party computes the functionality on the inputs and returns to each party its respective output. Unlike in the case of stand-alone computation, here the trusted party computes the functionality many times, each time upon different inputs. Loosely speaking, a protocol is secure if any adversary interacting in the real protocol (where no trusted third party exists) can do no more harm than if it was involved in the above-described ideal computation. Since the trusted party computes the functionality and does nothing else, we will sometimes say that the parties interact with the functionality (rather than interacting with the trusted party computing the functionality).

Concurrent executions in the ideal model. An ideal execution with an adversary who controls P_2 proceeds as follows (when the adversary controls P_1 the roles are simply reversed):

Inputs: Party P_1 and P_2's inputs are respectively determined by probabilistic polynomial-time Turing machines M_1 and M_2, and initial inputs x and y to these machines. As we will see below, these Turing machines determine the values that the parties use as inputs in the protocol executions. These input values are computed from the initial input, the current session number and outputs that were obtained from executions that have already concluded. Note that the number of previous outputs ranges from zero (for the case that no previous outputs have yet been obtained) to some polynomial in n that depends on the number of sessions initiated by the adversary.

Session initiation: The adversary initiates a new session by sending a start-session message to the trusted party, who then sends (start-session, i) to the honest party, where i is the index of the session (i.e., this is the ith session to be started).

Honest party sends input to trusted party: Upon receiving (start-session, i) from the trusted party, the honest party P_1 applies its input-selecting machine M_1 to its initial input x, the session number i and its previous outputs, and obtains a new input x_i. That is, in the first session, $x_1 = M_1(x, 1)$. In later sessions, $x_i = M_1(x, i, \gamma_{i_1}, \ldots, \gamma_{i_j})$ where j sessions have already concluded and the outputs were $\gamma_{i_1}, \ldots, \gamma_{i_j}$. (We assume that the output γ_{i_k} explicitly contains the index of the session i_k from which it was output.) The honest party P_1 sends (i, x_i) to the trusted party.

Adversary sends input to the trusted party and receives output: Whenever the adversary wishes, it may send a message (i, y_i) to the trusted party, for any y_i of its choice. Upon sending this pair, it receives back $(i, f_2(x_i, y_i))$ where x_i is the value that P_1 previously sent the trusted party. (If i start-session messages have not yet been sent to the trusted party, then the (i, y_i) message from the adversary is ignored. In addition, once an input indexed by i has already been sent by the adversary, the trusted party ignores any subsequent such messages.)

Adversary instructs trusted party to answer honest party: When the adversary sends a message of the type (send-output, i) to the trusted party, the trusted party sends $(i, f_1(x_i, y_i))$ to the honest party P_1, where x_i and y_i are the respective inputs sent by P_1 and the adversary for this session. (If (i, x_i) and (i, y_i) have not yet been received by the trusted party, then this (send-output, i) message is ignored.)

Outputs: The honest party P_1 outputs the vector $(f_1(x_{i_1}, y_{i_1}), f_1(x_{i_2}, y_{i_2}), \ldots)$ of outputs that it received from the trusted party. Formally, whenever it receives an output, it writes it to its output-tape. Thus, the outputs do not appear in ascending order according to the session numbers, but rather in the order that they are received. The adversary may output an arbitrary (probabilistic polynomial-time computable) function of its auxiliary input z, the corrupted party P_2's input-selecting machine M_2, initial input y, and the outputs obtained from the trusted party.

Let $f : \{0,1\}^* \times \{0,1\}^* \rightarrow \{0,1\}^* \times \{0,1\}^*$ be a functionality, where $f = (f_1, f_2)$, and let $\mathcal{S}$ be a non-uniform probabilistic polynomial-time machine (representing the ideal-model adversary). Then, the ideal execution of f (with security parameter n, input-selecting machines $\overline{M} = (M_1, M_2)$, initial inputs (x, y), and auxiliary input z to $\mathcal{S}$), denoted CONC-IDEAL$_{f, \mathcal{S}, \overline{M}}(n, x, y, z)$, is defined as the output pair of the honest party and $\mathcal{S}$ from the above ideal execution.

We note that in the above-described interaction of the adversary with the trusted party, the adversary is allowed to obtain outputs in any order that it wishes, and can choose its inputs adaptively based on previous outputs. This is inevitable in a concurrent setting where the adversary can schedule the order in which all protocol executions take place.

Execution in the real model. We next consider the real model in which a real two-party protocol is executed (and there exists no trusted third party). Formally, a two-party protocol ρ is defined by two sets of instructions ρ_1 and ρ_2 for parties P_1 and P_2, respectively. A protocol is said to be polynomial-time if the running-time of each ρ_i in a *single execution* is bounded by a fixed polynomial in the security parameter n, irrespective of the length of the input.

Let f be as above and let ρ be a probabilistic polynomial-time two-party protocol for computing f. In addition, let $\mathcal{A}$ be a non-uniform probabilistic polynomial-time adversary that controls either P_1 or P_2. Then, the **real concurrent execution** of ρ (with security parameter n, input-selecting machines $\overline{M} = (M_1, M_2)$, initial inputs (x, y) to the parties, and auxiliary input z to $\mathcal{A}$), denoted CONC-REAL$_{\rho, \mathcal{A}, \overline{M}}(n, x, y, z)$, is defined as the output pair of the honest party and $\mathcal{A}$, resulting from the following process. The parties run concurrent executions of the protocol, where an honest party P_1 follows the instructions of ρ_1 in all of the executions; likewise, an honest P_2 always follows ρ_2. Thus, the parties play the same role in every execution. Now, the ith session is initiated by the adversary by sending a **start-session** message to the honest party. The honest party then applies its input-selecting machine on its initial input, the session number i and its previously received outputs, and obtains the input for this new session. Upon the conclusion of an execution of ρ, the honest party writes its output from that execution on its output-tape. The scheduling of all messages throughout the executions is controlled by the adversary. That is, the execution proceeds as follows. The adversary sends a message of the form (i, α) to the honest party. The honest party then adds the message α to the view of its ith execution of ρ and replies according to the instructions of ρ and this view. (Note that the honest party runs each execution of ρ obliviously to the other executions. Thus, this is stateless composition.) The adversary continues by sending another message (j, β), and so on. We note that there is no restriction on the scheduling allowed by the adversary.

Security as emulation of a real execution in the ideal model. Loosely speaking, a protocol is secure if for every real-model adversary $\mathcal{A}$ and pair of input-selecting machines (M_1, M_2), there exists an ideal model adversary $\mathcal{S}$ such that for all initial inputs x, y, the outcome of an ideal execution with $\mathcal{S}$ is computationally indistinguishable from the outcome of a real protocol execution with $\mathcal{A}$. Notice that the order of quantifiers is such that $\mathcal{S}$ comes after M_1 and M_2. Thus, $\mathcal{S}$ knows the *strategy* used by the honest parties to choose their inputs. However, $\mathcal{S}$ does not know the initial input of the honest party, nor the random tape used by its input-selecting machine (any "secrets" used by the honest parties are included in the initial input, not the input-selecting machine). Notice also that a special case of this definition is where the inputs are fixed ahead of time. In this case, the initial inputs are vectors where the ith value is the input for the ith session, and in the ith session the input-selecting machines M_1 and M_2 just output the ith value of the input vector. We now present the definition:

Definition 1 *A protocol ρ is said to* securely compute f under concurrent self composition *if for every real-model non-uniform probabilistic polynomial-time adversary $\mathcal{A}$ controlling party P_i ($i \in \{1, 2\}$) and every pair of probabilistic polynomial-time input-selecting machines $\overline{M} = (M_1, M_2)$, there exists an ideal-model non-uniform prob. polynomial-time adversary $\mathcal{S}$ controlling P_i, such that*

$$\left\{ \text{CONC-IDEAL}_{f, \mathcal{S}, \overline{M}}(n, x, y, z) \right\} \overset{\text{c}}{\equiv} \left\{ \text{CONC-REAL}_{\rho, \mathcal{A}, \overline{M}}(n, x, y, z) \right\}$$

where the ensembles are indexed by $n \in \mathsf{N}$ and $x, y, z \in \{0, 1\}^*$.

Definition 1 requires that the ideal-model simulator/adversary run in strict polynomial-time. However, a more liberal interpretation of "efficient simulation" is often allowed, in which case the simulator can run in *expected* polynomial-time.

Adaptively chosen inputs. Definition 1 allows the honest parties to choose their inputs adaptively, based on previously obtained outputs. Our impossibility result for locally sequential self composition relies inherently on this ability. Nevertheless, we claim that allowing inputs to be adaptively chosen is crucial for many (if not, most) applications.

Fixed roles. In our definition here, the parties are restricted to always using the same role (i.e., P_1 always runs ρ_1 and P_2 always runs ρ_2). A more general definition would allow P_1 and P_2 to run either ρ_1 or ρ_2 (the specific role taken by each can be negotiated upon initiating the execution). This property of the definition is inherited in the definition of locally sequential self composition (Section 2.2). Since our results for local sequentiality are negative, considering fixed roles only strengthens the results.

2.2 Locally Sequential Self Composition

In the setting of local sequentiality, we consider a multi-party network with parties $P_1, \ldots, P_m$ for some $m = \text{poly}(n)$. The parties run two-party protocols amongst themselves with the following limitation: if an honest party P_i is currently running a secure protocol ρ, then it ignores all incoming messages that do not belong to this execution of ρ. There are two possibilities regarding what P_i does with these incoming messages that are ignored: one possibility is to drop them and the other is to buffer them and deal with them one at a time after the current execution of ρ concludes. Our results hold for both of these possibilities, and for the sake of concreteness, we choose the latter option. We stress that although each individual party enforces a policy of sequentiality, protocol executions are run concurrently when considering the overall network. Specifically, P_1 and P_2 may run a protocol execution at the same time as P_3 and P_4. Furthermore, the specific interleaving between these executions is arbitrary and can be decided by the adversary. In addition, a corrupted P_5 can run two protocol executions at the same time with honest parties P_6 and P_7. (The adversary is not limited to sequentiality at all.)

The changes that are necessary to the definitions in Section 2.1 are as follows. First, the network now contains many parties. Therefore, in the definition of the ideal model, the start-session and send-output messages are modified so that they contain the identities of the parties. That is, in order to initiate a session between parties P_j and P_k, the adversary sends (start-session, j, k) to the trusted party, who then sends (start-session, i, j, k) to P_j and P_k, where this is the ith session that P_j and P_k run together. Likewise, all later messages (inputs, outputs and send-output) are all indexed by (i, j, k), and the input selecting machines also receive the identities of the participating parties, as well as the session index. Finally, the trusted party also checks that a (send-output, i, j, k) message has been received before it agrees to start the $i + 1^{\text{th}}$ session between P_j and P_k.

This completes the changes to the ideal model, which is denoted SEQ-IDEAL here. In the real model, denoted SEQ-REAL, we modify the honest parties' behavior so that they enforce local sequentiality (as described above). In addition, they index all messages with the identities involved and the session index. Everything else remains the same as in Section 2.1. We also impose fixed roles in this setting, and so every honest party is designated a role in that it runs either ρ_1 in every execution or ρ_2 in every execution (this just strengthens our results). We obtain the following definition:

Definition 2 (security under locally sequential self composition): *Let f be a two-party functionality, let $\rho = (\rho_1, \rho_2)$ be a polynomial-time protocol, and let $P_1, \ldots, P_m$ be parties in the network. Protocol ρ is said to* securely compute f under locally sequential self composition *if for every real-model non-uniform probabilistic polynomial-time adversary $\mathcal{A}$ controlling a subset $\mathcal{P} \subseteq \{P_1, \ldots, P_m\}$ of the parties and every series of probabilistic polynomial-time input-selecting machines $\overline{M} = (M_1, \ldots, M_m)$, there exists an ideal-model non-uniform probabilistic polynomial-time adversary $\mathcal{S}$ controlling the parties in $\mathcal{P}$, such that*

$$\left\{ \text{SEQ-IDEAL}_{f,\mathcal{S},\overline{M}}(n, \overline{x}, z) \right\} \stackrel{\text{c}}{\equiv} \left\{ \text{SEQ-REAL}_{\rho,\mathcal{A},\overline{M}}(n, \overline{x}, z) \right\}$$

where the ensembles are indexed by $n \in \mathsf{N}$ and $\overline{x}, z \in \{0,1\}^*$.

Locally sequential general composition. As a tool in order to prove our results, we refer to the notion of locally sequential general composition. In the setting of general composition, a secure protocol ρ for computing a functionality $\mathcal{F}$ is run alongside an arbitrary other protocol π. In sequential composition and in the real model, the arbitrary messages of π are sent in between executions of ρ (and thus, sequentiality is preserved). In order to define security, a hybrid model is considered where π is unchanged, but the executions of ρ are replaced by ideal calls to the functionality $\mathcal{F}$; this model is called the $\mathcal{F}$-hybrid model. Loosely speaking, a protocol ρ is said to be secure under locally sequential general composition if for every protocol ρ and every real-model adversary $\mathcal{A}$, there exists a *hybrid-model* adversary $\mathcal{S}$ such that the outputs of π^ρ in the real execution are indistinguishable from those generated by π in the $\mathcal{F}$-hybrid model. We denote a real execution of π with ρ in this model by SEQ-REAL$_{\pi^\rho,\mathcal{A}}(n, \overline{x}, z)$, and a hybrid execution of π with $\mathcal{F}$ by SEQ-HYBRID$_{\pi,\mathcal{A}}^{\mathcal{F}}(n, \overline{x}, z)$.

3   Locally Sequential Self Composition

In this section, we consider adversaries for the setting of concurrent self composition that have *fixed scheduling*. Intuitively, this means that the order in which the messages are sent in the different executions is fixed, and does not depend on the adversary's input or its view during the actual execution. Formally, a schedule for the setting of concurrent self composition is a series of indices $i_1, i_2, \ldots$ such that the jth message sent by the adversary to the honest party is (i_j, α) for some α. That is, the jth message belongs to session i_j, and this is the jth index in the schedule. We note that by convention, the first message sent by the adversary in any session is a start-session message. (This is without loss of generality because the honest party can be instructed to ignore any message (i, α) if at least i start-session messages were not received. Alternatively, the honest party can be instructed to send itself the appropriate number of start-session messages in the case that it receives a message (i, α) and less than i start-session messages were received.) Security for adversaries with fixed scheduling is formulated by requiring that for every real-model adversary $\mathcal{A}$ and every schedule I, there exists an ideal-model adversary/simulator $\mathcal{S}$, such that the usual requirement of indistinguishability holds.

We remark that the limitation of adversaries to fixed scheduling is severe and unrealistic. However, the known impossibility results for the setting of concurrent composition in [9,18,19,20] all use adversaries with fixed scheduling. Therefore, the relevant impossibility results can also be applied here. Furthermore, since our focus in this paper is impossibility, this is not a limitation.

We begin by showing an "equivalence" between locally sequential self composition and locally sequential general composition, and then use this to prove our main result.

3.1 Locally Sequential Self Versus General Composition

In this section, we show an equivalence between self and general composition in the setting of local sequentiality. An analogous equivalence for the concurrent setting was demonstrated in [20]; the proof here is almost the same.

Bit transmission. Informally speaking, a functionality enables bit transmission if it can be used by the parties to send bits to each other. For example, the "less than" functionality enables bit transmission, as follows. If P_1 wishes to send a bit to P_2, then P_2 fixes its input at a predetermined value (say 5). Now, if P_1 wishes to send P_2 the bit 0, then it uses input 4, and if it wishes to send P_2 the bit 1 then it uses input 6. In this way, P_2 will know which bit P_1 sends based on the output. More generally, P_1 can transmit a bit to P_2 if there exists an input y for P_2 and a pair of inputs x and x' for P_1 such that $f_2(x, y) \neq f_2(x', y)$. This suffices because P_1 and P_2 can decide that $f_2(x, y)$ should be interpreted as bit 0 and $f_2(x', y)$ as bit 1. Likewise, P_2 can transmit a bit to P_1 if the reverse holds. We say that a functionality enables bit transmission if it can be used by P_1 to transmit a bit to P_2 and vice versa. Thus, a functionality can only enable bit transmission if both parties receive output (it is impossible to transmit a bit to a party that receives no output). We now present the formal definition:

Definition 3 *A deterministic functionality* $f = (f_1, f_2)$ *enables bit transmission from* P_1 *to* P_2 *if there exists an input* y *for* P_2 *and a pair of inputs* x *and* x' *for* P_1 *such that* $f_2(x, y) \neq f_2(x', y)$. *Likewise,* $f = (f_1, f_2)$ *enables bit transmission from* P_2 *to* P_1 *if there exists an input* x *for* P_1 *and a pair of inputs* y *and* y' *for* P_2 *such that* $f_1(x, y) \neq f_1(x, y')$. *We say that a functionality enables bit transmission if it enables bit transmission from* P_1 *to* P_2 *and from* P_2 *to* P_1.

We are now ready to prove that in the setting of local sequentiality, security under self composition is equivalent to security under general composition (for functionalities that enable bit transmission).

Proposition 4 *Let $\mathcal{F}$ be a functionality that enables bit transmission. Then, a protocol ρ securely computes $\mathcal{F}$ under locally sequential self composition if and only if it securely computes $\mathcal{F}$ under locally sequential general composition.*

Proof Sketch: The proof of this proposition is almost identical to the proof of equivalence between security under concurrent self composition and concurrent general composition in [20]. We therefore only describe why the proposition holds here as well. The main observation is that when a functionality $\mathcal{F}$ enables bit transmission, it can be used by the parties to send any message to each other. Specifically, a message $\alpha = \alpha_1 \cdots \alpha_m$ of length m can be sent by party P to party P' by invoking m ideal calls to $\mathcal{F}$; in the ith call the bit α_i is transmitted. Furthermore, these m calls can be carried out *sequentially*, thereby preserving the requirement of local sequentiality. (Recall that in the setting of locally sequential general composition, arbitrary messages can only be sent in between executions of the secure protocol. Thus, the secure protocol can be invoked sequentially for the sake of bit transmission, and local sequentiality is still preserved.) We therefore have that security under self composition implies security under general composition. The other direction follows immediately from the fact that self composition is just a special case of general composition where the arbitrary protocol is "empty". ∎

Interchangeable roles. The model that we consider here is one where parties have fixed roles in the computation (e.g., a party who proves a statement in zero-knowledge must play the prover role in all executions). As shown in [20], in a setting where parties may adopt different roles, the equivalence stated in Proposition 4 holds for *all* functionalities (and not just for those enabling bit transmission). See [20] for more details.

3.2   Locally Sequential Composition and Concurrent Composition

We prove our impossibility results for locally sequential self composition by showing that security in this setting actually implies security in the setting of concurrent self composition (for adversaries with fixed scheduling). We then derive impossibility by applying the known results for concurrent self composition.

Theorem 5. *Let $\mathcal{F}$ be a functionality that enables bit transmission. If a protocol ρ securely computes $\mathcal{F}$ under locally sequential self composition (with many parties), then it securely computes $\mathcal{F}$ under concurrent self composition with fixed scheduling (and with two parties).*

Proof: We begin by providing the idea behind the proof of the theorem, and then proceed to the full proof. An execution of t concurrent sessions of a secure protocol with two parties P_1 and P_2 can be emulated by $2t$ parties $\widetilde{P}_1, \ldots, \widetilde{P}_{2t}$, each of which runs just a single execution. Specifically, the ith session between

P_1 and P_2 is actually run by $\widetilde{P}_{2i-1}$ and $\widetilde{P}_{2i}$ (where $\widetilde{P}_{2i-1}$ plays the role of P_1, and $\widetilde{P}_{2i}$ plays the role of P_2). Since the honest P_1 and P_2 run each execution obliviously of the others, there is no problem having different parties run each different execution. (Note that we rely heavily on the fact that stateless composition is used here.) Furthermore, since each pair runs only a single execution, local sequentiality is trivially preserved. One problem that arises according to this strategy, however, is that in the concurrent setting parties P_1 and P_2 use their previously received outputs in order to compute their new inputs. Therefore, in the setting of local sequentiality, outputs that are received by one honest party must be sent to all the other honest parties (in order to enable them to compute their inputs in the same way as P_1 and P_2 in the concurrent setting). In principle, this is achieved by using the *bit transmission* property of the functionality $\mathcal{F}$ (i.e., when an honest party receives its output from an execution of ρ, it sends this output to all other honest parties by running many executions of ρ and sending a single bit of the output in each such execution). However, this introduces a difficulty with respect to the scheduling of the delivery of these received outputs. In particular, the adversary may deliver some of these outputs and not others, and may also deliver them in different orders to different pairs. This concern can happen in the setting of local sequentiality, but not when just two parties run many concurrent executions. This problem is overcome by relying on the assumption that the adversary's schedule is fixed. In this way, it is possible to have the parties in the setting of local sequentiality only agree to run protocol executions in the same order as the concurrent adversary runs them. This completes the proof sketch; the full proof follows.

Let ρ be a protocol that is secure under locally sequential self composition with many parties. By Proposition 4 and the fact that $\mathcal{F}$ enables bit transmission, we have that ρ is also secure under locally sequential *general* composition. We now show that this implies that ρ is secure under concurrent self composition with *fixed scheduling*.

Let P_1 and P_2 be two parties with input-selecting machines M_1 and M_2, respectively, let $\mathcal{A}$ be an adversary who "attacks" P_1 and P_2 in the setting of concurrent self composition, and let I denote the schedule of messages for $\mathcal{A}$. Recall that I is a series of indices $i_1, i_2, \ldots$ such that the messages sent by $\mathcal{A}$ in a concurrent execution are $(i_1, \alpha_1), (i_2, \alpha_2), \ldots$ for α_j's of $\mathcal{A}$'s choice. Furthermore, in the first appearance of an index j in the series, it always holds that $\alpha_j =$ start-session. We observe that I fully defines the order in which sessions both begin and terminate. Therefore, the sessions from which outputs have already been received at the time that a new session begins are also fully determined. Recall that the input to every new session is based on already-received outputs.

In order to construct a simulator $\mathcal{S}$ for $\mathcal{A}$, we first define a setting of locally sequential general composition that emulates the concurrent executions of P_1 and P_2. Let $t = t(n)$ be an upper bound on the running time of $\mathcal{A}$, and thus on the length of the series I. Then, define parties $\widetilde{P}_1, \ldots, \widetilde{P}_{2t}$ who all run an arbitrary protocol π. Before describing π, we call the parties $(\widetilde{P}_1, \widetilde{P}_3, \ldots, \widetilde{P}_{2t-1}$ odd (i.e., they have odd indices), and we call the parties $(\widetilde{P}_2, \widetilde{P}_4, \ldots, \widetilde{P}_{2t})$ even.

We now present protocol π that works in the $\mathcal{F}$-hybrid model (the protocol specification includes the input-selecting machines M_1 and M_2 of P_1 and P_2):

Protocol π:

1. *Instructions for Party $\widetilde{P}_i$, for even i, $1 \leq i \leq 2t$:* Wait until a start-session message is received from party P_{i-1}. Until such a message is received, accept any message of the form (received-output, j, γ_j) from even parties P_j with $j < i$. When a start-session message is received, do the following. First, let $(j_1, \gamma_{j_1}), \ldots, (j_k, \gamma_{j_k})$ be the received-output values that were received (in the order that they were received). Then, check that according to the schedule I, the $\frac{i}{2}^{\text{th}}$ session between P_1 and P_2 (with adversary $\mathcal{A}$) begins after P_2 has already concluded and thus received output in sessions $j_1, j_2, \ldots, j_k$ (and only these sessions). Furthermore, check that this is the order that P_2 concludes these sessions in I. If one of these checks fails, halt and output $\bot$. Otherwise, compute $x \leftarrow M_2(x_i, \frac{i}{2}, \gamma_{j_1}, \ldots, \gamma_{j_k})$, where x_i is P_i's initial input. Then, send x to the ideal functionality $\mathcal{F}$ and receive back the output γ. Finally, send (received-output, i, γ) to all even parties P_j for $j > i$, write γ to the output tape and halt.

2. *Instructions for Party P_i, for odd i, $1 \leq i \leq 2t$:* The instructions here are the same as for an even i except that P_i receives outputs from odd parties, it checks that P_1 would have concluded the sessions and in this order, it applies M_1 (instead of M_2), and it sends its output to odd parties P_j, $j > i$.

This completes the description of π.

Notice that in the real model, when the ideal calls to $\mathcal{F}$ are replaced by executions of ρ, the parties do not respond to any non-ρ message that is received during the execution of ρ. (This is due to the local sequentiality requirement.)

Having described the protocol π, we now construct a real adversary $\widetilde{\mathcal{A}}$ such that the outcome of a locally sequential execution with $\widetilde{\mathcal{A}}$ is essentially *identical* to the outcome of an execution in the setting of concurrent self composition with $\mathcal{A}$ and schedule I. In other words, $\widetilde{\mathcal{A}}$ with $\widetilde{P}_1, \ldots, \widetilde{P}_{2t}$ essentially emulates the concurrent executions of $\mathcal{A}$ with P_1 and P_2 according to the schedule I.

We first define the corruption sets of $\mathcal{A}$ and $\widetilde{\mathcal{A}}$: if $\mathcal{A}$ controls P_1, then $\widetilde{\mathcal{A}}$ controls all the odd parties $\widetilde{P}_1, \widetilde{P}_3, \ldots, \widetilde{P}_{2t-1}$; if $\mathcal{A}$ controls P_2 then $\widetilde{\mathcal{A}}$ controls all the even parties $\widetilde{P}_2, \widetilde{P}_4, \ldots, \widetilde{P}_{2t}$. (This corruption pattern is fixed from here on.) Now, upon input z, the adversary $\widetilde{\mathcal{A}}$ internally invokes $\mathcal{A}$ with input z and perfectly emulates the setting of concurrent self composition with P_1 and P_2 (despite the fact that it runs in the setting of local sequentiality). This emulation is carried out as follows; for the sake of clarity our description and proof from here on is for the case that P_1 *is corrupted* (the other case is proven analogously):

1. *Session initiation:* When $\mathcal{A}$ sends the ith start-session message to P_2, then $\widetilde{\mathcal{A}}$ instructs $\widetilde{P}_{2i-1}$ to send a start-session message to $\widetilde{P}_{2i}$. (That is, for the 1st such message, $\widetilde{\mathcal{A}}$ instructs $\widetilde{P}_1$ to send a start-session message to $\widetilde{P}_2$; for the 2nd such message, $\widetilde{\mathcal{A}}$ instructs $\widetilde{P}_3$ to send a start message to $\widetilde{P}_4$, and so on.)

2. *Session execution:* Whenever $\mathcal{A}$ instructs P_1 to send P_2 a message (i, α) belonging to the ith session, $\widetilde{\mathcal{A}}$ instructs $\widetilde{P}_{2i-1}$ to send α to $\widetilde{P}_{2i}$. Likewise, when $\widetilde{P}_{2i-1}$ receives a message β back from $\widetilde{P}_{2i}$, adversary $\widetilde{\mathcal{A}}$ hands (i, β) to $\mathcal{A}$, as if it was received from P_2.

3. *Delivery of π-messages:* After $\widetilde{\mathcal{A}}$ sends $\widetilde{P}_{2i}$ the last message of session i, it immediately delivers all of the received-output messages that $\widetilde{P}_{2i}$ sends to all even parties.

4. *Conclusion:* When $\mathcal{A}$ outputs γ and halts, $\widetilde{\mathcal{A}}$ also outputs γ and halts.

Now, let x_1, x_2 be arbitrary values. Then, we denote by $\overline{x}_t(x_1, x_2)$ the length-$2t$ vector of the form $\overline{x}_t(x_1, x_2) = (x_1, x_2, x_1, x_2, \ldots, x_1, x_2)$; i.e., t repetitions of the pair (x_1, x_2). We claim that the execution with $\widetilde{\mathcal{A}}$ perfectly emulates the real execution of $\mathcal{A}$. That is, for every $x_1, x_2, z \in \{0, 1\}^*$ and every $n \in \mathsf{N}$,

$$\left\{ \text{SEQ-REAL}_{\pi^\rho, \widetilde{\mathcal{A}}}(n, \overline{x}_t(x_1, x_2), z) \right\} \equiv \left\{ \text{CONC-REAL}_{\rho, \mathcal{A}}(n, x_1, x_2, z) \right\} \qquad (1)$$

Actually, the format of the random variable SEQ-REAL differs from that of CONC-REAL; the first is a vector of $\widetilde{\mathcal{A}}$'s output together with the output of all the $2t$ participating parties $\widetilde{P}_1, \ldots, \widetilde{P}_{2t}$, whereas the second is the vector of $\mathcal{A}$'s output together with the output of P_1 and P_2. Rather than introducing special notation for this, our intention is that the output of $\widetilde{\mathcal{A}}$ together with the vector of outputs of the honest parties $\widetilde{P}_2, \widetilde{P}_4, \ldots, \widetilde{P}_{2t}$ is identically distributed to the output of $\mathcal{A}$ together with the output of the honest P_2. (Another way of stating this is that there exists an efficient function h that can be applied to SEQ-REAL so that $h(\text{SEQ-REAL})$ is identically distributed to CONC-REAL. This efficient function h merely involves "re-arranging" the outputs.)

Stated simply, Eq. (1) states that the output of $\widetilde{\mathcal{A}}$ and the honest parties in the setting of local sequentiality is *identically distributed* to the output of $\mathcal{A}$ and P_2 in the setting of concurrent self composition. In order to see this, notice that the distribution of messages sent by P_2 and $\widetilde{P}_{2i}$ in session i is identical, assuming that they begin with the same input. (This holds because the composition is stateless and so in each session, P_2 runs the protocol as if it was the only protocol being executed.) In addition, by the definition of π, we know that P_2 and $\widetilde{P}_{2i}$ also always use the same input in session i. This holds because **(a)** they both use the input-selecting machine M_2 to decide inputs, and **(b)** they both apply M_2 to the same series of messages $(x_2, \gamma_1, \ldots, \gamma_k)$. The latter holds due to the fact that the schedule of $\mathcal{A}$ is fixed and $\widetilde{P}_{2i}$ checks that it has all the outputs that P_2 would also have already received. (We remark that we don't actually need to use the fact that the schedule is fixed for this part of the proof; it can be proven in any case.) Now, since the messages sent by $\mathcal{A}$ to P_2 in the ith concurrent session are just forwarded by $\widetilde{\mathcal{A}}$ to $\widetilde{P}_{2i}$ in the setting of local sequentiality, the view of $\mathcal{A}$ and the resulting outputs are identical in both settings. Eq. (1) follows.

By the security of ρ under *locally sequential general composition*, we have that for every $\widetilde{\mathcal{A}}$ there exists an ideal-model adversary/simulator $\widetilde{\mathcal{S}}$ such that,

$$\left\{ \text{SEQ-HYBRID}_{\pi, \widetilde{\mathcal{S}}}^{\mathcal{F}}(n, \overline{x}, z) \right\} \stackrel{\mathrm{c}}{\equiv} \left\{ \text{SEQ-REAL}_{\pi^\rho, \widetilde{\mathcal{A}}}(n, \overline{x}, z) \right\} \qquad (2)$$

Eq. (2) holds for every $\overline{x}$, and in particular when $\overline{x}$ is of the form $\overline{x}_t(x_1, x_2)$, as in Eq. (1). It remains for us to show that for every $\widetilde{S}$ working in the $\mathcal{F}$-hybrid model for local sequentiality, there exists an adversary S working in the ideal model for $\mathcal{F}$ in the setting of concurrent self composition, such that for every $x_1, x_2, z \in \{0, 1\}^*$ and every $n \in \mathsf{N}$,

$$\{\text{CONC-IDEAL}_{\mathcal{F}, S}(n, x_1, x_2, z)\} \equiv \left\{\text{SEQ-HYBRID}_{\pi, \widetilde{S}}^{\mathcal{F}}(n, \overline{x}_t(x_1, x_2), z)\right\} \qquad (3)$$

Once again, the format of the random variable CONC-IDEAL differs to SEQ-HYBRID. As before, our intention is that Eq. (3) holds for the above-mentioned efficient function h applied to SEQ-HYBRID. (This suffices because we obtain the series of "equalities": CONC-IDEAL $\equiv h(\text{SEQ-HYBRID}) \stackrel{c}{\equiv} h(\text{SEQ-REAL}) \equiv \text{CONC-REAL}$, implying that CONC-IDEAL $\stackrel{c}{\equiv}$ CONC-REAL as required.)

We now describe the ideal-model adversary/simulator S for which Eq. (3) holds. For this part of the proof, it is crucial that $\mathcal{A}$ have a fixed schedule I (we do not know how to prove that the equation holds otherwise). Simulator S works as follows (as above, the corruption pattern is fixed so that S controls P_1 if $\widetilde{S}$ controls all the odd parties, and S controls P_2 if $\widetilde{S}$ controls all the even parties). Upon input z, the adversary S internally invokes $\widetilde{S}$ with input z and perfectly emulates the hybrid-model setting of local sequentiality for $\widetilde{S}$, while running in the ideal-model for concurrent self composition. This emulation is carried out as follows; as above, for the sake of clarity our description is for the case that P_1 is corrupted (the other case is analogous):

1. *Session initiation:* This step is carried out initially and after every time that P_2 receives output. Let $\overline{\gamma} = \gamma_1, \ldots, \gamma_k$ be the outputs that P_2 has already received in the concurrent execution with $\mathcal{A}$ (initially $\overline{\gamma}$ is empty). Then, S checks I to find the list of sessions which begin after P_2 has received the outputs in $\overline{\gamma}$, and before P_2 receives any more outputs. Let $i_1, \ldots, i_j$ be these sessions. Then, S sends j start-session messages to the trusted party (corresponding to sessions $i_1, \ldots, i_j$). Recall that upon receiving such a message, the trusted party sends (start-session, i) to P_2, where this is the ith start-session message to be received.

   Now, when $\widetilde{S}$ wishes to send a start-session$(i-1, i)$ message to the trusted party (where i is even), adversary S checks that P_i has already received the outputs that are defined by the schedule I and so by π. If not, S halts and outputs fail. Otherwise, S does nothing (start-session messages must have already been sent above).

   We note that by the construction of π, party P_i only accepts a start-session message from P_{i-1}. S can therefore just ignore any start message that is not of this form.

2. *Send input and receive output:* When $\widetilde{S}$ wishes to send an input message $(i-1, i, x)$ to the trusted party, S sends $(\frac{i}{2}, x)$ to the trusted party. When S receives the output $(\frac{i}{2}, \gamma)$ from the trusted party, it internally passes $\widetilde{S}$ the message $(i-i, i, \gamma)$, as if from $\widetilde{S}$'s trusted party.

As in the previous item, S can ignore all messages that are not of this form, because by the definition of π, honest parties will not participate in such executions.

3. *Instruction to send output to honest party:* When $\widetilde{S}$ wishes to send a message (send-output, $i-1, i$) to its trusted party, S does nothing. Rather, S sends send-output messages to the trusted party according to the order of outputs received in the schedule I. After sending such output, S proceeds to the "session-initiation" step.

4. *Conclusion:* When $\widetilde{S}$ outputs γ and halts, S also outputs γ and halts.

In order to show that Eq. (3) holds, we show that the inputs that the honest parties $\widetilde{P}_2, \widetilde{P}_4, \ldots$ send in SEQ-HYBRID are identically distributed to the inputs that P_2 sends in CONC-REAL. In particular, this means that the ith input sent by an honest party in SEQ-HYBRID is identically distributed to the ith input sent by P_2 in CONC-REAL. Observe that for every i, party $\widetilde{P}_{2i}$ obtains its input by computing $M_2(x_2, i, \overline{\gamma})$. Likewise, in the ith session, party P_2 obtains its input by computing $M_2(x_2, i, \overline{\gamma}')$. Thus, it suffices to show that $\overline{\gamma}' = \overline{\gamma}$; that is, the input computed by $\widetilde{P}_{2i}$ is based on the same set of received-outputs as the input computed by P_2 in the ith session. However, this follows immediately from the fact that both S and $\widetilde{P}_{2i}$ check that their list of received-outputs is as defined by the schedule I. That is, S sends a (start-session, i) messages as soon as the necessary outputs have already been received. Likewise, S ensures that the outputs are received by P_2 in the order defined by I. In the setting of SEQ-HYBRID, the same effect is achieved by the fact that each party $\widetilde{P}_{2i}$ checks its received outputs before computing its input (and if this check fails, it outputs $\perp$). Thus, as long as no party $\widetilde{P}_{2i}$ outputs $\perp$ in π, we have that the inputs sent are identically distributed. This suffices because in the real execution of SEQ-REAL with $\widetilde{\mathcal{A}}$, no honest party outputs $\perp$. Therefore, the probability that an honest party will output $\perp$ in the SEQ-HYBRID execution with $\widetilde{S}$ is at most negligible. (It is possible that $\widetilde{S}$ will instruct outputs to be sent in the "incorrect order". However, if sessions are then started, based on this order, we are guaranteed that an honest party will output $\perp$. Essentially, this means that $\widetilde{S}$ cannot behave in this way except with negligible probability.)

We have so far established that the inputs sent by the honest parties $\widetilde{P}_{2i}$ in SEQ-HYBRID are identically distributed to those sent by P_2 is CONC-IDEAL. Given this fact, the view of $\widetilde{S}$ in the emulation by S is identical to its view in a SEQ-HYBRID execution. This follows immediately from the fact that S forwards all messages to $\widetilde{S}$, as it expects to receive from the trusted party. Thus, Eq. (3) holds. Combining Equations (1), (2) and (3), the theorem follows. ∎

References

1. Barak, B.: How to Go Beyond the Black-Box Simulation Barrier. In: 42nd FOCS, pp. 106–115 (2001)
2. Barak, B., Sahai, A.: How To Play Almost Any Mental Game Over The Net. In: 46th FOCS, pp. 543–552 (2005)

3. Beaver, D.: Foundations of secure interactive computing. In: Feigenbaum, J. (ed.) CRYPTO 1991. LNCS, vol. 576, pp. 377–391. Springer, Heidelberg (1992)
4. Ben-Or, M., Goldwasser, S., Wigderson, A.: Completeness Theorems for Non-Cryptographic Fault-Tolerant Distributed Computation. In: 20th STOC, pp. 1–10 (1988)
5. Canetti, R.: Security and Composition of Multiparty Cryptographic Protocols. Theory of Cryptography Library, Record 98-18, version of June 4th (later versions do not contain the referenced material) (1998)
6. Canetti, R.: Security and Composition of Multiparty Cryptographic Protocols. Journal of Cryptology 13(1), 143–202 (2000)
7. Canetti, R.: Universally Composable Security: A New Paradigm for Cryptographic Protocols. In: 42nd FOCS, pp. 136–145 (2001)
8. Canetti, R., Fischlin, M.: Universally composable commitments. In: Kilian, J. (ed.) CRYPTO 2001. LNCS, vol. 2139, pp. 19–40. Springer, Heidelberg (2001)
9. Canetti, R., Kushilevitz, E., Lindell, Y.: On the Limitations of Universal Composable Two-Party Computation Without Set-Up Assumptions. In: Biham, E. (ed.) EUROCRYPT 2003. LNCS, vol. 2656, pp. 68–86. Springer, Heidelberg (2003)
10. Canetti, R., Lindell, Y., Ostrovsky, R., Sahai, A.: Universally Composable Two-Party and Multi-Party Computation. In: 34th STOC, pp. 494–503 (2002)
11. Chaum, D., Crepeau, C., Damgard, I.: Multi-party Unconditionally Secure Protocols. In: 20th STOC, pp. 11–19 (1988)
12. Dolev, D., Dwork, C., Naor, M.: Non-malleable cryptography. SIAM Journal on Computing 30(2), 391–437 (2000)
13. Dwork, C., Naor, M., Sahai, A.: Concurrent Zero-Knowledge. In: 30th STOC, pp. 409–418 (1998)
14. Feige, U., Shamir, A.: Witness Indistinguishability and Witness Hiding Protocols. In: 22nd STOC, pp. 416–426 (1990)
15. Goldreich, O., Micali, S., Wigderson, A.: How to Play any Mental Game – A Completeness Theorem for Protocols with Honest Majority. In: 19th STOC, pp. 218–229 (1987)
16. Goldwasser, S., Levin, L.A.: Fair computation of general functions in presence of immoral majority. In: Menezes, A., Vanstone, S.A. (eds.) CRYPTO 1990. LNCS, vol. 537, pp. 77–93. Springer, Heidelberg (1991)
17. Kalai, Y., Lindell, Y., Prabhakaran, M.: Concurrent General Composition of Secure Protocols in the Timing Model. In: the 37th STOC, pp. 644–653 (2005)
18. Lindell, Y.: Bounded-Concurrent Secure Two-Party Computation Without Setup Assumptions. In: 35th STOC, pp. 683–692 (2003)
19. Lindell, Y.: General Composition and Universal Composability in Secure Multi-Party Computation. In: 44th FOCS, pp. 394–403 (2003)
20. Lindell, Y.: Lower Bounds and Impossibility Results for Concurrent Self Composition. Journal of Cryptology 21(2), 200–249 (2008)
21. Micali, S., Rogaway, P.: Secure Computation. In: Feigenbaum, J. (ed.) CRYPTO 1991. LNCS, vol. 576, pp. 392–404. Springer, Heidelberg (1992)
22. Pass, R.: Simulation in Quasi-Polynomial Time, and Its Application to Protocol Composition. In: Biham, E. (ed.) EUROCRYPT 2003. LNCS, vol. 2656, pp. 160–176. Springer, Heidelberg (2003)
23. Pass, R.: Bounded-Concurrent Secure Multi-Party Computation with a Dishonest Majority. In: The 36th STOC (to appear, 2004)
24. Pass, R., Rosen, A.: Bounded-Concurrent Secure Two-Party Computation in a Constant Number of Rounds. In: 44th FOCS, pp. 404–413 (2003)

25. Pfitzmann, B., Waidner, M.: Composition and Integrity Preservation of Secure Reactive Systems. In: 7th ACM Conference on Computer and Communication Security, pp. 245–254 (2000)
26. Prabhakaran, M., Sahai, A.: New Notions of Security: Universal Composability Without Trusted Setup. In: The 36th STOC (to appear, 2004)
27. Richardson, R., Kilian, J.: On the concurrent composition of zero-knowledge proofs. In: Stern, J. (ed.) EUROCRYPT 1999. LNCS, vol. 1592, p. 415. Springer, Heidelberg (1999)
28. Yao, A.: How to Generate and Exchange Secrets. In: 27th FOCS, pp. 162–167 (1986)

Breaking and Repairing Damgård *et al.* Public Key Encryption Scheme with Non-interactive Opening

David Galindo

University of Luxembourg
david.galindo@uni.lu

Abstract. We show a simple chosen-ciphertext attack against a public key encryption scheme with non-interactive opening (PKENO) presented by Damgård, Kiltz, Hofheinz and Thorbek in CT-RSA 2008. In a PKENO scheme a receiver can convincingly reveal to a verifier what the result of decrypting a ciphertext C is, without interaction and without compromising the confidentiality of non-opened ciphertexts. A special interesting feature of PKENO is that a verifier can even ask for opening proofs on invalid ciphertexts. Those opening proofs will convince the verifier that the ciphertext was indeed invalid. We show that one of the schemes by Damgård *et al.* does not achieve the claimed security goal. Next we provide a fix for it. The repaired scheme presents essentially no overhead and is proven secure under the Decisional Bilinear Diffie-Hellman assumption in the standard model.

Keywords: identity-based encryption, public key encryption, non-interactive proofs, standard model.

1 Introduction

The primitive *public key encryption with non-interactive opening* (PKENO) allows a receiver Bob to reveal the plaintext m obtained in decrypting any given ciphertext C under Bob's public key pk_B to a verifier Alice. By using PKENO Bob can do so convincingly and without interaction. More precisely, Bob runs a proving algorithm Prove on inputs its secret key sk_B and the intended ciphertext C, thereby generating a proof π. On the other hand, Alice runs a verification algorithm Ver on inputs Bob's public key pk_B, ciphertext C, the plaintext m that purportedly was output by the decryption algorithm, and an opening proof π. The verification algorithm outputs 1 if C was indeed an encryption of m, and 0 otherwise. A special interesting feature of PKENO is that Bob can also convince Alice of the fact that a given ciphertext C is *invalid*, i.e. it is rejected by the decryption algorithm.

PKENO was recently introduced in [DT07, DHKT08] as a means to enable publicly-verifiable decryption. For instance, Damgård and Thorbek [DT07] use PKENO in multiparty computation to prove that a given party did not follow the protocol, in the sense that it sent encrypted fake information. Damgård, Kiltz,

M. Fischlin (Ed.): CT-RSA 2009, LNCS 5473, pp. 389–398, 2009.
© Springer-Verlag Berlin Heidelberg 2009

Hofheinz and Thorbek [DHKT08] present two constructions of PKENO schemes. The first proposal is a generic construction that takes as atomic primitives an identity-based encryption scheme (IBE) and a one-time signature scheme. This construction resembles Canetti, Halevi and Katz IBE-to-PKE transformation [CHK04] (where PKE stands for public key encryption). The second proposal is a concrete scheme based on a chosen-ciphertext secure pairing-based key encapsulation mechanism by Boyen, Mei and Waters [BMW05].

The security of a PKENO scheme is roughly defined as follows. First, a PKENO must be *indistinguishable against chosen-ciphertext and prove attacks* (IND-CCPA security). This attack model is obtained by augmenting the standard IND-CCA game, namely giving the adversary additional access to a prove oracle, that on input a ciphertext returns an opening proof. Secondly it must be difficult for a malicious receiver to forge a proof on any ciphertext sent to him, i.e. it must be infeasible to produce a proof convincing a verifier that the ciphertext opens to a different result than what is obtained from the decryption algorithm.

In this paper we show that the second scheme in [DHKT08] is insecure. In particular, we exhibit an attack that breaks the confidentiality of any honestly-generated ciphertext $C^\star$ by having access to a proving oracle on a related but different ciphertext C. Next we propose a fix that precludes the above attacks. We show that the new scheme is nearly as efficient as the original one and that it is secure under under the Decisional Bilinear Diffie-Hellman assumption in the standard model.

2   Preliminaries

In this section we review the definitions and tools we need to present our results. We start by fixing some notation.

2.1   Notation

If x is a string, then $|x|$ denotes its length, while if S is a set then $|S|$ denotes its size. If k is a natural number, then 1^k denotes the string of k ones. If S is a set then $s_1, \ldots, s_n \xleftarrow{\$} S$ denotes the operation of picking n elements s_i of S independently and uniformly at random. We write $\mathcal{A}(x, y, \ldots)$ to indicate that $\mathcal{A}$ is an algorithm with inputs $x, y, \ldots$ and by $z \leftarrow \mathcal{A}(x, y, \ldots)$ we denote the operation of running $\mathcal{A}$ with inputs $(x, y, \ldots)$ and letting z be the output. We use the abbreviation PPT to refer to a probabilistic polynomial time algorithm [Gol01].

2.2   Public Key Encryption Scheme with Non-interactive Opening

A PKENO scheme PKENO = (Gen, Enc, Dec, Prove, Ver) is a tuple of five PPT algorithms:

- Gen is a probabilistic algorithm taking as input a security parameter 1^k. It returns a public key pk and a secret key sk. The public key includes the description of the set of plaintexts $\mathcal{M}_{pk}$.

- Enc is a probabilistic algorithm taking as inputs a public key pk and a message $m \in \mathcal{M}_{pk}$. It returns a ciphertext C.
- Dec is a deterministic algorithm that takes as inputs a ciphertext C and a secret key sk. It returns a message $m \in \mathcal{M}_{pk}$ or the special symbol $\perp$ meaning that the ciphertext is invalid.
- Prove is a probabilistic algorithm taking as inputs a ciphertext C and a secret key sk. It returns a proof π.
- Ver is a deterministic algorithm taking as inputs a public key pk, a ciphertext C, a plaintext m and a proof π. It returns a result $res \in \{0,1\}$ meaning accepted and rejected proof respectively. In particular $1 \leftarrow \mathsf{Ver}(pk, C, \perp, \pi)$ must be interpreted as the verifier being convinced that C is an invalid ciphertext.

For correctness and completeness it is required that for a honestly generated key pair $(pk, sk) \leftarrow \mathsf{Gen}(1^k)$, the following holds:

- **Correctness.** For all messages $m \in \mathcal{M}_{pk}$, we have $\Pr[\mathsf{Dec}(\mathsf{Enc}(pk, m)) = m] = 1$.
- **Completeness.** For all ciphertexts C we have that

$$\Pr\left[\, 1 \leftarrow \mathsf{Ver}\big(\, pk, C, \mathsf{Dec}(sk, C), \mathsf{Prove}(sk, C)\,\big)\,\right] = 1,$$

i.e. the verification algorithm accepts with overwhelming probability.

Definition 1 (IND-CCPA security). *Let us a consider the following game between a challenger and an adversary $\mathcal{A}$:*

Setup. *The challenger runs* $\mathsf{Gen}(1^k)$ *and gives pk to* $\mathcal{A}$.
Phase 1. *The adversary issues queries of the form:*
 a) decryption query to an oracle $\mathsf{Dec}(sk, \cdot)$
 b) proof query to an oracle $\mathsf{Prove}(sk, \cdot)$.
 These queries may be asked adaptively, that is, they may depend on the answers to previous queries.
Challenge. *At some point,* $\mathcal{A}$ *outputs two equal-length messages* $m_0, m_1 \in \mathcal{M}_{pk}$. *The challenger chooses a random bit β and returns* $C^\star \leftarrow \mathsf{Enc}(pk, m_\beta)$.
Phase 2. *As in Phase 1, with the restriction that decryption or proof queries on $C^\star$ are not allowed.*
Guess. *The adversary $\mathcal{A}$ outputs a guess* $\beta' \in \{0,1\}$. *The adversary wins the game if $\beta = \beta'$.*

Define $\mathcal{A}$'s advantage as $\mathsf{Adv}^{ind-ccpa}_{\mathsf{PKENO},\mathcal{A}}(1^k) = \big|\Pr[\beta' = \beta] - 1/2\big|$. *A scheme* PKENO *is called* indistinguishable against chosen-ciphertext and prove attacks (IND-CCPA secure) *if for every adversary $\mathcal{A}$,* $\mathsf{Adv}^{ind-ccpa}_{\mathsf{PKENO},\mathcal{A}}(\cdot)$ *is negligible.*

Definition 2 (Soundness). *Let us a consider the following game between a challenger and an adversary $\mathcal{A}$:*

Setup. *The challenger runs* $\mathsf{Gen}(1^k)$ *and gives* (pk, sk) *to* $\mathcal{A}$.

Challenge. *The adversary chooses a message* $m \in \mathcal{M}_{pk}$ *and gives it to an encryption oracle which returns* $C \leftarrow \mathsf{Enc}(pk, m)$.

Define $\mathcal{A}$*'s advantage in forging a proof by* $\mathsf{Adv}^{\mathsf{forge}}_{\mathsf{PKENO},\mathcal{A}}(1^k) = \Pr[1 \leftarrow \mathsf{Ver}(pk, C, m', \pi') \wedge m' \neq m]$. *A scheme* PKENO *is said to satifsy* computational proof soundness *if for every adversary* $\mathcal{A}$, $\mathsf{Adv}^{\mathsf{forge}}_{\mathsf{PKENO},\mathcal{A}}(\cdot)$ *is negligible.*

Definition 3 (PKENO security). *A PKENO scheme is said to be* secure *if it has* IND-CCPA *security* and *computational proof soundness.*

2.3   Pairing Assumptions

Parameter Generation Algorithms for Bilinear Groups. (Symmetric) Pairing-based schemes are parameterized by a *pairing parameter generator*. This is a PPT algorithm $\mathcal{G}$ that on input 1^k returns the description of an multiplicative cyclic group $\mathbb{G}_1$ of prime order p, where $2^k < p < 2^{k+1}$, the description of a multiplicative cyclic group $\mathbb{G}_T$ of the same order, and a non-degenerate bilinear pairing $e\colon \mathbb{G}_1 \times \mathbb{G}_1 \to \mathbb{G}_T$. See [BF03, BSS05] for a description of the properties of such pairings. We use $\mathbb{G}_1^*$ to denote $\mathbb{G}_1 \setminus \{0\}$, i.e. the set of all group elements except the neutral element. We shall use $\mathbb{P} = (\mathbb{G}_1, \mathbb{G}_T, p, e)$ as shorthand for the description of bilinear groups.

The BDDH Assumption. Let $\mathbb{P}$ be the description of pairing groups. Consider the following problem first considered by Joux [Jou00] and later formalized by Boneh and Franklin [BF03]: given $(g, g^a, g^b, g^c, W) \in \mathbb{G}_1^4 \times \mathbb{G}_T$ as input, output yes if $W = e(g, g)^{abc}$ and no otherwise. More formally, to a parameter generation algorithm for pairing-groups $\mathcal{G}$ and an adversary $\mathcal{B}$ we associate the following experiment.

> **Experiment** $\mathbf{Exp}^{\mathsf{bddh}}_{\mathcal{G},\mathcal{A}}(1^k)$
> $\mathbb{P} \xleftarrow{\$} \mathcal{G}(1^k)$
> $a, b, c, w \xleftarrow{\$} \mathbb{Z}_p^*$
> $\beta \xleftarrow{\$} \{0, 1\}$
> If $\beta = 1$ then $W \leftarrow e(g, g)^{abc}$ else $W \leftarrow e(g, g)^w$
> $\beta' \xleftarrow{\$} \mathcal{A}(1^k, \mathbb{P}, g, g^a, g^b, g^c, W)$
> If $\beta \neq \beta'$ then return 0 else return 1

We define the advantage of $\mathcal{B}$ in the above experiment as

$$\mathsf{Adv}^{\mathsf{bddh}}_{\mathcal{G},\mathcal{A}}(k) = \left| \Pr\left[\mathbf{Exp}^{\mathsf{bddh}}_{\mathcal{G},\mathcal{A}}(1^k) = 1 \right] - \frac{1}{2} \right|.$$

We say that the *Bilinear Decision Diffie-Hellman (BDDH) assumption relative to generator* $\mathcal{G}$ holds if $\mathsf{Adv}^{\mathsf{bddh}}_{\mathcal{G},\mathcal{A}}$ is a negligible function in k for all PPT algorithm $\mathcal{A}$.

2.4   Collision-Resistant Hashing

Let us consider a family of hash functions $\mathcal{CR} = \{\mathsf{CR}_s : \Sigma_s \to \mathbb{Z}_{p(s)}\}_{s \in S(k)}$, where k is a security parameter; Σ_s and $S(k)$ are finite sets such that $|\sigma|$ and $|s|$ are polynomially bounded as functions of k for any $\sigma \in \Sigma_s$ and $s \in S(k)$; $p(s)$ are prime numbers with $|p(s)|$ polynomially bounded in the security parameter. We say that $\mathcal{CR}$ is *collision resistant* if for any PPT algorithm $\mathcal{A}$ we have

$$\Pr[\ \mathsf{CR}_s(x) = \mathsf{CR}_s(y) \text{ and } x \neq y : s \xleftarrow{\$} S(k)\ ;\ (x, y) \longleftarrow \mathcal{A}(s)]$$

is negligible in k (cf. [Gol01] for a definition of the class of negligible functions).

3   Damgård *et al.* Scheme

Let $\mathbb{P} = (\mathbb{G}_1, \mathbb{G}_T, p, e)$ the description of a bilinear group. Let $\mathsf{TCR} : \mathbb{G}_1 \to \mathbb{Z}_p$ be a (target) collision-resistant hash function. Let (E, D) be a chosen-ciphertext secure symmetric encryption scheme (see [Gol01] for a definition) such that its keys' space is $\mathbb{G}_T$.

Damgård, Hofheinz, Kiltz and Thorbek's second scheme presented in [DHKT08] is depicted in Figure 1. $\mathcal{M}_{pk}$ is simply defined as the set of messages of the symmetric encryption scheme. Here we complete the description originally given by the authors in order to describe how the verification algorithm behaves when it takes inputs of the form $(sk, C, \perp, \emptyset)$.

3.1   Attack

Let us call a ciphertext $C = (c_1, c_2, c_3)$ *consistent* if $e(g, c_2) = e(c_1, X_1^t X_2)$, where $t = \mathsf{TCR}(c_1)$. Notice that the consistency of a ciphertext *does not depend* at any rate on the component c_3. Furthermore, the algorithm $\mathsf{Prove}(sk, \cdot)$ on input C always return a non-trivial proof $\pi \neq \emptyset$ as long as C is consistent.

The above suggests the following straightforward attack. Given the target ciphertext $C^\star = (c_1^\star, c_2^\star, c_3^\star)$, where $c_3^\star \leftarrow E(K^\star, m_\beta)$ for unknown β, the attacker submits $C = (c_1^\star, c_2^\star, c_3) \neq C^\star$ to the proving oracle for random $c_3 \neq c_3^\star$. C is consistent since it was obtained by changing the third component of $C^\star$, and the latter is consistent by definition. The proving oracle returns $(d_1^\star, d_2^\star) \leftarrow \mathsf{Prove}(sk, C)$, which is a valid opening proof for $C^\star$. The attacker thereby obtains the symmetric encryption key $K^\star$ by computing $e(c_1^\star, d_1^\star)/e(c_2^\star, d_2^\star)$. Finally, one recovers β from $m_\beta \leftarrow \mathsf{D}(K^\star, c_3^\star)$. This attack succeeds with probability 1.

It is symptomatic of the latter attack the fact that if one looks at scheme in Figure 1 as a standard public key encryption scheme (i.e. by skipping the **Prove** and **Ver** algorithms), then the security reduction given in [BMW05] for the resulting PKE scheme can not answer decryption queries related to a ciphertext of the form $(c_1^\star, c_2, c_3)$. Therefore the security proof in [BMW05] does not suffice to simulate the proving oracle in the scheme in [DHKT08], as the authors thereby claimed.

One way to try to overcome this attack is making the consistency of the whole ciphertext depend also on the third component. This is the simple idea behind

$\mathsf{Gen}(1^k)$
  $\mathbb{P} \xleftarrow{\$} \mathcal{G}(1^k)$
  $x_1, x_2, y \xleftarrow{\$} \mathbb{Z}_p$
  $X_1 \leftarrow g^{x_1}$; $X_2 \leftarrow g^{x_2}$; $Y \leftarrow e(g,g)^y$
  $pk \leftarrow (1^k, \mathbb{P}, \mathsf{E}, \mathsf{D}, \mathsf{TCR}, X_1, X_2, Y)$
  $sk \leftarrow (pk, x_1, x_2, y)$
  output (pk, sk)

$\mathsf{Dec}(sk, C)$
  parse C as (c_1, c_2, c_3)
  $t \leftarrow \mathsf{TCR}(c_1)$
  if $c_1^{x_1 t + x_2} \neq c_2$
    output $\perp$
  else $K \leftarrow e(c_1, g^y)$
    output $m \leftarrow \mathsf{D}(K, c_3)$

$\mathsf{Ver}(pk, C, m, \pi)$
  parse C as (c_1, c_2, c_3)
  $t \leftarrow \mathsf{TCR}(c_1)$
  if $m = \perp$ and $\pi = \emptyset$
    if $e(g, c_2) = e(X_1^t X_2, c_1)$
      output 0 and 1 otherwise
  if $\pi \neq \emptyset$
    if $e(g, c_2) = e(X_1^t X_2, c_1)$ and
      $e(g, d_2) = Y \cdot e(X_1^t X_2, d_1)$
      $K \leftarrow e(c_1, d_1)/e(c_2, d_2)$
      $m' \leftarrow \mathsf{D}(K, c_3)$
      if $m' = m$ output 1
      else output 0
    else output 0

$\mathsf{Enc}(pk, m)$
  $r \xleftarrow{\$} \mathbb{Z}_p$, ; $c_1 \leftarrow g^r$
  $t \leftarrow \mathsf{TCR}(c_1)$; $c_2 \leftarrow (X_1^t X_2)^r$
  $K \leftarrow Y^r \in \mathbb{G}_T$
  $c_3 \leftarrow \mathsf{E}(K, m)$
  output $C \leftarrow (c_1, c_2, c_3)$

$\mathsf{Prove}(sk, C)$
  parse C as (c_1, c_2, c_3)
  $t \leftarrow \mathsf{TCR}(c_1)$
  if $c_1^{x_1 t + x_2} \neq c_2$
    output $\emptyset$
  else $s \xleftarrow{\$} \mathbb{Z}_p$; $d_2 \leftarrow g^s$
    $d_1 \leftarrow g^y \cdot (X_1^t X_2)^s$
    output $\pi \leftarrow (d_1, d_2)$

Fig. 1. Damgård *et al.* [DHKT08] scheme

our proposal. However to be able to simulate decryption and proof queries one needs to slightly change the underlying encryption scheme. Fortunately a set of techniques in the same work [BMW05] resolves the question.

4   Repaired Scheme

Let $\mathbb{P} = (\mathbb{G}_1, \mathbb{G}_T, p, e)$ the description of a bilinear group. The new scheme uses a hash function $\mathsf{H} : \{0,1\}^n \rightarrow \mathbb{G}_1$ introduced by Chaum, Evertse and van de Graaf [CEvdG87] and rediscovered recently by Waters [Wat05]. On input of an integer n polynomially-bounded in k, the randomized hash key generator chooses $n+1$ random groups elements $h_0, \ldots, h_n \in \mathbb{G}_1$ and returns $h = (h_0, h_1, \ldots, h_n)$ as the hash function key. The hash function $\mathsf{H} : \{0,1\}^n \rightarrow \mathbb{G}_1^*$ is evaluated on a n-bit string $\boldsymbol{t} = (t_1, \ldots, t_n) \in \{0,1\}^n$ as the product $\mathsf{H}(\boldsymbol{t}) = h_0 \prod_{i=1}^n h_i^{t_i}$. In addition

$\mathsf{Gen}(1^k)$
  $\mathbb{P} \overset{\$}{\leftarrow} \mathcal{G}(1^k)$
  $\alpha, y_0, \ldots, y_n \overset{\$}{\leftarrow} \mathbb{Z}_p$
  $h_0 \leftarrow g^{y_0}, \ldots, h_n \leftarrow g^{y_n} \; ; \; Y \leftarrow e(g, g)^\alpha$
  $h \leftarrow (h_0, \ldots, h_n)$
  $pk \leftarrow (1^k, \mathbb{P}, \mathsf{E}, \mathsf{D}, \mathsf{CR}, h, Y)$
  $sk \leftarrow (pk, \alpha, y_0, \ldots, y_n)$
  output (pk, sk)

$\mathsf{Dec}(sk, C)$
  parse C as (c_0, c_1, c_2)
  $t \leftarrow \mathsf{CR}(c_0, c_1)$
  parse t as $(t_0, \ldots, t_n)$
  $t \leftarrow y_0 + \sum_{i=1}^n y_i t_i \mod p$
  if $c_1^t \neq c_2$
    output $\perp$
  else $K \leftarrow e(c_1, g^\alpha)$
  output $m \leftarrow \mathsf{D}(K, c_0)$

$\mathsf{Ver}(pk, C, m, \pi)$
  parse C as (c_0, c_1, c_2)
  $t \leftarrow \mathsf{CR}(c_0, c_1)$
  parse t as $(t_0, \ldots, t_n)$
  if $m = \perp$ and $\pi = \emptyset$
    if $e(g, c_2) = e(\mathsf{H}(t), c_1)$
      output 0
    else output 1
  if $\pi \neq \emptyset$
    if $e(g, c_2) = e(\mathsf{H}(t), c_1)$ and
      $e(g, d_1) = Y \cdot e(\mathsf{H}(t), d_1)$
      $K \leftarrow e(c_1, d_1)/e(c_2, d_2)$
      $m' \leftarrow \mathsf{D}(K, c_0)$
      if $m' = m$ output 1
      else output 0
    else output 0

$\mathsf{Enc}(pk, m)$
  $r \overset{\$}{\leftarrow} \mathbb{Z}_p, \; ; \; c_1 \leftarrow g^r$
  $K \leftarrow Y^r \in \mathbb{G}_T$
  $c_0 \leftarrow \mathsf{E}(K, m)$
  $t \leftarrow \mathsf{CR}(c_0, c_1) \; ; \; c_2 \leftarrow \mathsf{H}(t)^r$
  output $C \leftarrow (c_0, c_1, c_2)$

$\mathsf{Prove}(sk, C)$
  parse C as (c_0, c_1, c_2)
  $t \leftarrow \mathsf{CR}(c_0, c_1)$
  parse t as $(t_0, \ldots, t_n)$
  $t \leftarrow y_0 + \sum_{i=1}^n y_i t_i \mod p$
  if $c_1^t \neq c_2$
    output $\emptyset$
  else $s \overset{\$}{\leftarrow} \mathbb{Z}_p$
  $d_1 \leftarrow g^\alpha \cdot \mathsf{H}(t)^s \; ; \; d_2 \leftarrow g^s$
  output $\pi \leftarrow (d_1, d_2)$

Fig. 2. Repaired scheme

the scheme uses a collision-resistant hash function $\mathsf{CR} : \mathbb{G}_1 \times \{0,1\}^l \rightarrow \{0,1\}^n$. Let (E, D) be a one-time *chosen-plaintext* secure symmetric encryption scheme (see [Gol01] for a definition) such that its keys' space is $\mathbb{G}_T$.

The new scheme is displayed in Figure 2, being the plaintext space equal to that of the encryption scheme (E, D). Its key generation, encryption and decryption algorithms are identical to those of a PKE scheme by Boyen, Mei and Waters in [BMW05]. That scheme has in turn a strong resemblance to Water's identity-based encryption scheme [Wat05]. Indeed, the public and private keys are essentially identical to the master public and secret keys of Waters' IBE, and the n-bit string t plays the role of the recipient's identity. The intuition on why the scheme is secure is simple and is as follows. Given the target ciphertext $C^\star = (c_0^\star, c_1^\star, c_2^\star)$, where $c_0^\star \leftarrow E(K^\star, m_\beta)$ for unknown β, it is infeasible to compute a different ciphertext $C = (c_0, c_1, c_2)$ which is simultaneously consistent and such that $c_1 = c_1^\star$ (i.e. such that C gives raise when decrypted to $K = K^\star$). Consistency now means that $e(g, c_2) = e(H(t), c_1)$, with $t \leftarrow CR(c_0, c_1)$. Infeasibility holds under the assumptions that the hash function CR is collision-resistant and that the BBDH problem is hard.

Theorem 1. *The scheme from Figure 2 is* IND-PCCA *secure if the DBDH assumption holds and* CR *is target collision-resistant.*

Sketch of the proof. The proof uses exactly the same techniques as the security proof given for the encryption scheme in [BMW05] does. As noted before a ciphertext in the new scheme is essentially an IBE ciphertext where the identity t is determined from the first two elements. The crucial part in our simulation is to answer proof queries, since decryption queries are answered as in [BMW05]. We next explain how to answer those queries.

For all proof queries involving a ciphertext $C = (c_0, c_1, c_2)$ the simulator first checks that the ciphertext is consistent. This amounts to checking $e(g, c_2) = e(H(t), c_1)$, with $t \leftarrow CR(c_0, c_1)$. If the ciphertext is consistent, that is $c_0 = g^r$ and $c_2 = H(t)^r$ for some $r \in \mathbb{Z}_p$, the simulator creates a Waters IBE-like private key (d_1, d_2) for the identity t as the opening. We give the main details in the next paragraph on how (d_1, d_2) is built.

Let (g, g^a, g^b, g^c, W) be the input of the BDDH problem. The simulator defines $Y = e(g^a, g^b)$, which implies that α is implicitly defined as ab. The hash function H is indeed a *programmable hash function* [HK08]. Therefore the simulator can use an alternative generation algorithm that on inputs (g, g^b) produces a public hash key $\widetilde{h} = (\widetilde{h}_0, \cdots, \widetilde{h}_n) \in \mathbb{G}^{n+1}$ together with a secret trapdoor information st, such that h (as obtained by the original PKENO key generation algorithm Gen) and $\widetilde{h}$ are identically distributed. Then with non-negligible probability on k the simulator computes $F(t), J(t) \in \mathbb{Z}_p$ by using the trapdoor st such that $H(t) = (g^b)^{F(t)} g^{J(t)}$, where $t = CR(c_0, c_1)$ and (c_0, c_1, c_2) is any consistent proof query made by the attacker. At the same time, the simulator is able to compute $J(t^\star) \in \mathbb{Z}_p$ such that $H(t^\star) = g^{J(t^\star)}$, where $t^\star = CR(c_0^\star, c_1^\star)$ and $(c_0^\star, c_1^\star, c_2^\star)$ is the challenge ciphertext (see [HK08] for further details). Thanks to the properties of the programmable hash function, proof queries with respect to consistent ciphertexts $C = (c_0, c_1, c_2)$ are answered as

$$\pi = \left((g^a)^{\frac{-J(t)}{F(t)}} \cdot H(t)^r, (g^a)^{\frac{-1}{F(t)}} g^r \right)$$

for random r chosen by the simulator. It is a bit cumbersome but straightforward to check that π thereby obtained is a correct proof (see [BMW05] for hints).

When the attacker outputs the pair of plaintexts m_0, m_1 to be challenged on, the simulator creates the challenge ciphertext $C^\star = (c_0^\star, c_1^\star, c_2^\star)$ which can be viewed as an Waters identity-based encryption under the identity $t^\star = CR(c_0^\star, c_1^\star)$. This is done by setting $c_1^\star = g^c$, and $c_1^\star = E(K, m_\beta)$ for random $\beta \overset{\$}{\leftarrow} \{0, 1\}$ and $K \overset{\$}{\leftarrow}$

$\mathbb{G}_T$. Thanks again to the properties of the programmable hash function $c_2^\star$ is easily constructed as $(g^c)^{J(t^\star)}$. Since CR is collision resistant, the adversary will not be able to make any consistent ciphertext nor proof queries related to the *identity* $t^\star$. Thus, we can answer *all* proof queries on consistent ciphertexts $C \neq C^\star$. Finally, the security of our scheme boils down to the security of the underlying IBE security, i.e. Waters' scheme, and the collision-freeness of CR.   □

Theorem 2. *The scheme in Section 4 satisfies proof soundness unconditionally.*

Proof. Let $C = (\, c_0 = E(Y^r, m),\ c_1 = g^r,\ c_2 = \mathsf{H}(t)^r\,)$ with $t = \mathsf{CR}(c_0, c_1)$ be an honestly generated ciphertext and let $\pi = (d_1, d_2)$ be a proof for C. If the checking if $e(g, c_2) = e(\mathsf{H}(t), c_1)$ and $e(g, d_2) = Y \cdot e(\mathsf{H}(t), d_1)$ is passed, then the verifier concludes there exist $r', s' \in \mathbb{Z}_p$ such that $c_2 = \mathsf{H}(t)^{r'}$ and $g^r = c_1 = g^{r'}$; $d_1 = g^\alpha \cdot \mathsf{H}(t)^{s'}$ and $d_2 = g^{s'}$. Since $g^r = g^{r'}$ implies $r = r'$, the key K' recovered by the verification algorithm

$$K' \leftarrow e(c_1, d_1)/e(c_2, d_2) = e(g^r, g^\alpha \cdot \mathsf{H}(t)^{s'})/e(\mathsf{H}(t)^r, g^{s'}) = e(g^r, g^\alpha) = K.$$

Therefore in the verification algorithm the message m' obtained is always the encrypted message m thanks to the properties of the bilinear pairing, which hold unconditionally.

That the adversary can not either forge proofs on invalid ciphertexts can be seen by using similar arguments.   □

5   Conclusion

We have shown that a PKENO scheme presented in [DHKT08] is insecure. Next we have proposed a fix and proved the resulting scheme secure under the BDDH assumption in the standard model. The resulting scheme is essentially as efficient as the original one and it is based on previous plain PKE scheme by Boyen, Mei and Waters [BMW05]. In particular encryption requires one exponentiation in $\mathbb{G}_T$, two exponentiations in $\mathbb{G}$ and an average of $n/2$ group operations in $\mathbb{G}$, which amount to less than one exponentiation in $\mathbb{G}$. In contrast, Damgård *et al.* scheme requires one exponentiation in $\mathbb{G}_T$, one exponentiation in $\mathbb{G}$ and one double-exponentiation in $\mathbb{G}$. The decryption, prove and verification algorithms have the same cost in both our scheme and Damgård *et al.* We leave as an open problem the construction of an IND-CCPA secure PKENO scheme without pairings.

References

[BF03]   Boneh, D., Franklin, M.K.: Identity-Based encryption from the Weil pairing. SIAM Journal of Computing 32(3), 586–615 (2003); this is the full version of an extended abstract of the same title presented at in: Kilian, J. (ed.) CRYPTO 2001. LNCS, vol. 2139, p. 213. Springer, Heidelberg (2001)

[BMW05]   Boyen, X., Mei, Q., Waters, B.: Direct chosen ciphertext security from identity-based techniques. In: ACM Conference on Computer and Communications Security 2005, pp. 320–329 (2005)

[BSS05]   Blake, I.F., Seroussi, G., Smart, N.P.: Advances in Elliptic Curve Cryptography. London Mathematical Society Lecture Note Series, vol. 317. Cambridge University Press, Cambridge (2005)

[CEvdG87]   Chaum, D., Evertse, J.-H., van de Graaf, J.: An improved protocol for demonstrating possession of discrete logarithms and some generalizations. In: Price, W.L., Chaum, D. (eds.) EUROCRYPT 1987. LNCS, vol. 304, pp. 127–141. Springer, Heidelberg (1988)

[CHK04]   Canetti, R., Halevi, S., Katz, J.: Chosen-ciphertext security from identity-based encryption. In: Cachin, C., Camenisch, J.L. (eds.) EUROCRYPT 2004. LNCS, vol. 3027, pp. 207–222. Springer, Heidelberg (2004)

[DHKT08]   Damgård, I., Hofheinz, D., Kiltz, E., Thorbek, R.: Public-key encryption with non-interactive opening. In: Malkin, T.G. (ed.) CT-RSA 2008. LNCS, vol. 4964, pp. 239–255. Springer, Heidelberg (2008)

[DT07]   Damgård, I., Thorbek, R.: Non-interactive proofs for integer multiplication. In: Naor, M. (ed.) EUROCRYPT 2007. LNCS, vol. 4515, pp. 412–429. Springer, Heidelberg (2007)

[Gol01]   Goldreich, O.: Foundations of Cryptography - Basic Tools. Cambridge University Press, Cambridge (2001)

[HK08]   Hofheinz, D., Kiltz, E.: Programmable hash functions and their applications. In: Wagner, D. (ed.) CRYPTO 2008. LNCS, vol. 5157, pp. 21–38. Springer, Heidelberg (2008)

[Jou00]   Joux, A.: A one round protocol for tripartite Diffie-Hellman. In: Bosma, W. (ed.) ANTS 2000. LNCS, vol. 1838, pp. 385–394. Springer, Heidelberg (2000)

[Wat05]   Waters, B.: Efficient identity-based encryption without random oracles. In: Cramer, R. (ed.) EUROCRYPT 2005. LNCS, vol. 3494, pp. 114–127. Springer, Heidelberg (2005)

Strengthening Security of RSA-OAEP

Alexandra Boldyreva

Georgia Institute of Technology, Atlanta, GA, USA
sasha@gatech.edu

Abstract. OAEP is one of the few standardized and widely deployed public-key encryption schemes. It was designed by Bellare and Rogaway as a scheme based on a trapdoor permutation such as RSA. RSA-OAEP is standardized in RSA's PKCS #1 v2.1 and is part of several standards. RSA-OAEP was shown to be IND-CCA secure in the random oracle model under the standard RSA assumption. However, the reduction is not tight, meaning that the guaranteed level of security is not very high for a practical parameter choice. We first observe that the situation is even worse because the analysis was done in the single-query setting, i.e. where an adversary gets a single challenge ciphertext. This does not take into account the fact that in reality an adversary can observe multiple ciphertexts of related messages. The results about the multi-query setting imply that the guaranteed concrete security can degrade by a factor of q, which is the number of challenge ciphertexts an adversary can get. We re-visit a very simple but not well-known modification of the RSA-OAEP encryption which asks that the RSA function is only applied to a part of the OAEP transform. We show that in addition to the previously shown fact that security of this scheme is tightly related to the hardness of the RSA problem, security does not degrade as the number of ciphertexts an adversary can see increases. Moreover, this scheme can be used to encrypt long messages without using hybrid encryption. We believe that this modification to the RSA-OAEP is easy to implement, and the benefits it provides deserves the attention of standard bodies.

1 Introduction

BACKGROUND AND MOTIVATION. OAEP is one of the few standardized and widely deployed public-key encryption schemes. It was designed by Bellare and Rogaway [5] as a scheme based on a trapdoor permutation such as RSA. RSA-OAEP is standardized in RSA's PKCS #1 v2.1 and is part of the ANSI X9.44, IEEE P1363, ISO 18033-2 and SET standards. The scheme is parameterized by k_0, k_1. The encryption algorithm of OAEP[F] takes a public key f, which is an instance of a trapdoor permutation family F, and a message M, picks k_0-bit string r at random, pads M with k_1 zeros to get M' and computes the ciphertext $C = f(s \parallel t)$ for $s = G(r) \oplus M'$ and $t = H(s) \oplus r$, where G and H are hash functions. OAEP[F] was proven to be IND-CPA secure assuming F is a one-way trapdoor permutation family [5] and IND-CCA secure assuming F is partial one-way [12], both in the random oracle (RO) model, i.e., where G and

M. Fischlin (Ed.): CT-RSA 2009, LNCS 5473, pp. 399–413, 2009.

H are modeled as random oracles [4]. Partial one-wayness is a stronger property than one-wayness and it asks that given the result of applying a random instance of the function family to a random point x it be hard to compute the first part of x. RSA is believed to be one-way, so under this assumption the result of [5] implies that OAEP[RSA] (RSA-OAEP) is IND-CPA in the RO model. In [12] it was shown that one-wayness of RSA also implies partial one-wayness, therefore RSA-OAEP is IND-CCA under the standard RSA assumption (stating that RSA is one-way), in the RO model.

While the concrete security reduction showing OAEP is IND-CCA secure assuming partial one-wayness of the underlying permutation family is tight, the concrete bound showing RSA-OAEP is IND-CCA under the RSA assumption is quite loose, due to the "lossy" reduction from partial one-wayness to one-wayness of RSA. Such a loose concrete security bound implies that it may be easier to break the scheme than to invert RSA, and to maintain reasonable security guarantees one would need to use the scheme with a larger security parameter. It was shown in [16] that keys of length about 4-5 thousand bits are necessary, i.e. at least 4 times larger than the standard 1024-bit keys, and this means decryption will be about $64 = 4^3$ times slower than before (since decryption requires a modulo exponentiation whose complexity is cubic in the length of the security parameter). This is basically impractical.

Moreover, we note that the definitions of security of encryption in [5,12] only consider an adversary given a single challenge ciphertext. In reality, of course, an adversary can observe multiple ciphertexts of possibly related messages. Such mismatch was studied in [3,2], who defined security in the "multi-query" setting where the adversary can see multiple challenge ciphertexts on messages of its choice[1]. The result of [3] implies that security (IND-CPA or IND-CCA) in the single-query setting implies security in the multi-query setting, however, concrete security degrades as the number of queries increases, and this loss cannot be avoided in general. However it is possible for some specific constructions, e.g. [3] shows that IND-CPA security of the ElGamal encryption scheme [11] stays tightly related to security of the decisional Diffie-Hellman problem regardless of how many queries an adversary makes. Concrete security in the multi-query setting of RSA-OAEP has not been explicitly addressed before our work.

Interestingly, an extremely simple modification to the the RSA-OAEP scheme permits several concrete security improvements. Unlike most of alternative constructions that have been suggested [17,9,15], the modification we study does not change the transform construction. The modified scheme differs from OAEP in that it uses trapdoor permutations of particular structure. Informally, they just leave the last part of the input (t-part of the output of the OAEP transform) in the clear. The scheme can be immediately instantiated with the RSA family if we apply an RSA function only to the s-part of the OAEP transform output, or to a portion of the s-part. This modification has been suggested under the name OAEP++

---

[1] In fact, [3] considers what they call a "multi-user" setting which also allows the adversary to see multiple challenge ciphertexts under multiple public keys. We do not consider multiple public keys in this work.

by Kobara and Imai in [14] in order to improve concrete security of OAEP. They show that RSA-OAEP++ is IND-CCA secure in the RO model under the standard RSA assumption and the reduction is tight. However, they only consider the single-query setting. The result of [3] implies that in the practical multi-query setting the concrete security bound is worse by a factor of q, i.e. security may degrade as an adversary observes more ciphertexts of possibly related messages. We note that this modification has been also suggested in [8] for an orthogonal reason of showing some positive results about non-malleability of OAEP when one or both ROs are instantiated with existing functions. The paper [8] neither considers the multi-query setting nor provides concrete security bounds.

OUR CONTRIBUTIONS. We show that this simple modification has even more advantages. We prove that concrete IND-CCA security of the modified RSA-OAEP scheme stays tightly related to one-wayness of RSA regardless of how many challenge ciphertexts an adversary sees (is independent of parameter q). The proof is in Section 5 and it uses the self-reducibility property of RSA. There we explain why does not the same idea apply to the original RSA-OAEP scheme. Hence, the modified RSA-OAEP provides significantly better security guarantees than the original version, for very practical parameter sizes, which results in a very efficient scheme.

Additionally, the modified RSA-OAEP scheme can be used to encrypt long messages without using symmetric encryption in the hybrid encryption construct. For that the function G in the transform is made variable-output-length, i.e. its output size is of the length of the message plus the zero padding of length k_1. For a fixed-output-length hash $G'(\cdot)$ one can efficiently construct $G(\cdot)$ as $G'(\langle 0 \rangle \| \cdot) \| G'(\langle 1 \rangle \| \cdot) \ldots \| G'(\langle l \rangle \| \cdot)$, where $\langle i \rangle$ means the binary representation of the counter $i \in \mathbb{N}$. The function H in the transform needs to be variable-input-length, which is not a problem. The RSA function is applied to the first k (e.g. 1024 bits) of the s-part of the OAEP transform. The proof of security stays virtually the same. This scheme yields more compact ciphertexts for long messages than the one obtained through the use of hybrid encryption because there is no need to encrypt the symmetric key.

We hope the standard bodies will pay attention to the modified RSA-OAEP as the advantages it offers seem to be well worth a very simple modification to the standard scheme.

MORE RELATED WORK. After it was realized by [12] that IND-CCA security of RSA-OAEP is not tight there appeared several alternative encryption schemes using different transforms before applying the RSA function. These include OAEP+ [17], SAEP+ [9], REACT [15]. Another alternative, which was proposed in [18] is the simplest construction and is known as Simple RSA or RSA-KEM. IND-CCA security of all of these schemes are tightly related to the hardness of the RSA problem, in the RO model and in the single-query setting. The latter two schemes, unlike the former two, can also be shown to have an improved security reduction in the multi-query setting (though it was not formally proved). We think it is important to show that the standardized RSA-OAEP scheme has similar properties, with the help of a very simple modification that is easy to implement, because it appears very hard to replace the standardized schemes with completely different constructions.

Improving the concrete security bounds is very important. Many papers besides the aforementioned work of [3] focused on this issue. For example, Coron [10] showed a new proof with improved security reduction for the RSA-based Full-Domain Hash signature scheme and his technique has been widely used since then. Abe et al. [1] improved the time bound in the security proofs of some of RSA-based encryption schemes by considering 4-round Feistel network transformation.

2   Preliminaries

NOTATION AND CONVENTIONS. We denote by $\{0,1\}^*$ the set of all binary strings of finite length. We will refer to members of $\{0,1\}^*$ as strings. If X, Y are strings then $X \parallel Y$ denotes the concatenation of X and Y. If S is a set then $X \xleftarrow{\$} S$ denotes that X is selected uniformly at random from S. If $k \in \mathbb{N}$ then 1^k denotes the string consisting of k consecutive "1" bits. If A is a randomized algorithm and $n \in \mathbb{N}$, then the notation $X \xleftarrow{\$} A(X_1, X_2, \ldots, X_n)$ denotes that X is assigned the outcome of the experiment of running A on inputs $X_1, X_2, \ldots, X_n$. When describing algorithms, if X is a variable and Y is a string, then $X \leftarrow Y$ denotes that X is assigned the value of Y.

All algorithms we consider are possibly randomized unless indicated otherwise. By convention, the running-time of an algorithm is measured relative to the bit-length of the input and refers to both the actual running-time and program size, including that of any overlying experiment, according to some fixed RAM model of computation. k denotes the security parameter. All algorithms we consider run in time polynomial in k.

SYNTAX OF PUBLIC-KEY ENCRYPTION. A public-key encryption (PKE) scheme $\mathcal{PE} = (\mathcal{K}, \mathcal{E}, \mathcal{D})$ with associated message space MsgSp, which may depend on the security parameter k, consists of three algorithms. The key-generation algorithm $\mathcal{K}$ on input 1^k returns a public key pk and matching secret key sk. The encryption algorithm $\mathcal{E}$ takes pk and a plaintext M to return a ciphertext. The deterministic decryption algorithm $\mathcal{D}$ takes sk and a ciphertext C to return a plaintext. The consistency condition requires that for all $k \in \mathbb{N}$ and all $M \in \mathrm{MsgSp}(k)$ the probability of $\mathcal{D}_{sk}(C) = M$ is 1, where the probability is over the experiment

$$(pk, sk) \xleftarrow{\$} \mathcal{K}(1^k) \; ; \; C \xleftarrow{\$} \mathcal{E}_{pk}(M) \; .$$

SECURITY OF PKE. We recall the notions of security of public-key encryption (PKE). We only consider the definitions addressing chosen-ciphertext attack (as opposed to a weaker version for chosen-plaintext attack). We present two variants of the standard IND-CCA definition. In the first one the adversary is given a single challenge ciphertext, and in the second definition the adversary can see multiple challenge ciphertexts. We then show the relation between the definitions.

Definition 1. [Single- and Multi-query CCA Security of PKE] *Let* $\mathcal{PE} = (\mathcal{K}, \mathcal{E}, \mathcal{D})$ *be a PKE scheme. Let the left or right selector be the map* LR *defined by* $\mathrm{LR}(M_0, M_1, b) = M_b$ *for all equal-length strings* M_0, M_1, *and for any* $b \in \{0, 1\}$. *For an adversary* A *and* $b \in \{0, 1\}$ *define the experiment:*

$$\textbf{\textit{Experiment}} \ \mathbf{Exp}_{\mathcal{PE},A}^{\text{ind-cca}}(1^k)$$

$$b \xleftarrow{\$} \{0, 1\}$$

$$(pk, sk) \xleftarrow{\$} \mathcal{K}(1^k)$$

$$d \xleftarrow{\$} A^{\mathcal{E}_{pk}(\mathrm{LR}(\cdot, \cdot, b)), \mathcal{D}_{sk}(\cdot)}$$

$$\textit{If } b = d \textit{ then return 1 else return 0}$$

It is mandated the LR *encryption oracle (also known as the challenge oracle) is queried on pairs of messages in* $\mathrm{MsgSp}(k)$ *and of equal length and the decryption oracle is not queried on the outputs of the* LR *encryption oracle.*

For an adversary A *who is allowed to make a single query to its challenge oracle (we will refer to such an adversary a single-query adversary) define the single-query(sq)-cca-advantage,* $\mathbf{Adv}_{\mathcal{PE},A}^{\text{ind-cca-sq}}(k)$ *as*

$$2 \cdot \Pr\left[\mathbf{Exp}_{\mathcal{PE},A}^{\text{ind-cca}}(1^k) = 1 \right] - 1 \ .$$

We define the multi-query(mq)-cca-advantage, $\mathbf{Adv}_{\mathcal{PE},A}^{\text{ind-cca-mq}}(k)$ *the exact same way, but for the adversary* A *who can query its challenge oracle an arbitrary number of times. We will refer to such* A *a multi-query adversary.*

A scheme $\mathcal{PE}$ *is said to be IND-CCA secure in the single- (resp. multi-) query setting if the single-query (resp, multi-query) -cca-advantage of any polynomial-time adversary is negligible.* ∎

It is shown by using a hybrid argument in [3] that for any $k \in \mathbb{N}$, a scheme $\mathcal{PE}$ and any multi-query adversary A making q queries to its challenge oracle there exists a single-query adversary B so that

$$\mathbf{Adv}_{\mathcal{PE},A}^{\text{ind-cca-mq}}(k) \leq q \cdot \mathbf{Adv}_{\mathcal{PE},B}^{\text{ind-cca-sq}}(k) \ , \tag{1}$$

where the running time of B is that of A plus $O(\log q)$, and B does the same number of decryption oracle queries as A.

It was also shown in [3] that the above bound is tight and cannot be improved in general. But for specific schemes, e.g. ElGamal, the concrete security in the multi-query setting is basically the same as in the single-query setting.

In this paper we are interested in improving the bound in concrete security treatment of the popular RSA-OAEP scheme in the multi-query setting. Accordingly we recall the computational assumptions used in the analyses of the scheme.

COMPUTATIONAL ASSUMPTIONS. A *trapdoor-permutation generator* is an algorithm $\mathcal{F}$ that on input 1^k returns the description of a permutation and its inverse f, f^{-1}. The trapdoor property means that for every instance f there exist a function f^{-1} with the same domain and range so that $f(f^{-1}) \equiv f^{-1}(f) \equiv \mathrm{ID}$, the identity function.

Definition 2. [One-wayness] *A trapdoor permutation generator $\mathcal{F}$ is called one-way if for every $k \in \mathbb{N}$ and every adversary I its advantage $\mathbf{Adv}_{\mathcal{F},I}^{\mathrm{owf}}(k)$ defined as*

$$\Pr\left[\, x = x' \;:\; (f, f^{-1}) \overset{\$}{\leftarrow} \mathcal{F}(1^k)\,;\; x \overset{\$}{\leftarrow} \{0,1\}^k\,;\; x' \overset{\$}{\leftarrow} I(1^k, f, f(x)) \,\right]$$

is negligible. ∎

Definition 3. [Partial-Domain One-wayness] *A trapdoor permutation generator $\mathcal{F}$ is called partial-domain one-way for $k \in \mathbb{N}$ and some extra parameter $k' \le k$, whch can be a linear function of k, if for every $k \in \mathbb{N}$ and every adversary I its advantage $\mathbf{Adv}_{\mathcal{F},I}^{\mathrm{pd-owf}}(k,k')$ defined as*

$$\Pr\left[\, x[1\ldots k'] = x' \;:\; (f, f^{-1}) \overset{\$}{\leftarrow} \mathcal{F}(1^k)\,;\; x \overset{\$}{\leftarrow} \{0,1\}^k\,;\; x' \overset{\$}{\leftarrow} I(1^k, f, f(x)) \,\right]$$

is negligible, where $x[1\ldots k']$ denotes the first k' bits of x. ∎

An *RSA trapdoor permutation generator* is an algorithm $\mathcal{F}$ that on input 1^k returns $(N,e),(N,d)$ where N is the product of two random distinct $\lfloor k/2 \rfloor$-bit primes and $ed \equiv 1 \bmod \phi(N)$. (Here $\phi(\cdot)$ is Euler's phi function.)

The standard assumption is that the RSA trapdoor permutation generator is one-way, and the reasonable security level requires k to be at least 1024 bits. It was shown in [12] that under this assumption RSA is also partial one-way. But the concrete reduction in [12] is not tight showing that a much larger RSA modulus is required to guarantee reasonable level of the stronger notion of partial one-wayness.

3   RSA-OAEP and Its Security

OAEP ENCRYPTION. The OAEP encryption [5] is parameterized by k_0, k_1 and k_2 (that can be linear functions of k, but typically $k_0 = k_1 = 128$ and $k_2 = k$) and makes use of a trapdoor permutation generator $\mathcal{F}$ with domain and range $\{0,1\}^{k_2}$ and two random oracles

$$G\colon \{0,1\}^{k_0} \to \{0,1\}^{k_2-k_0} \quad \text{and} \quad H\colon \{0,1\}^{k_2-k_0} \to \{0,1\}^{k_0}\,.$$

The message space is $\{0,1\}^{k_2-k_0-k_1}$. The scheme $\mathrm{OAEP}[\mathcal{F}] = (\mathcal{K}, \mathcal{E}, \mathcal{D})$ is defined as follows:

- The key generation algorithm $\mathcal{K}(1^k)$ picks a pair $(f, f^{-1}) \overset{\$}{\leftarrow} \mathcal{F}(1^{k_2})$ and returns f as pk and f^{-1} as sk.
- The encryption algorithm $\mathcal{E}(pk, M)$ picks $r \overset{\$}{\leftarrow} \{0,1\}^{k_0}$, computes $s \leftarrow G(r) \oplus (M \parallel 0^{k_1})$, $t \leftarrow H(s) \oplus r$ and $C \leftarrow f(s\|t)$ and returns C.
- The decryption algorithm $\mathcal{D}(sk, C)$ computes $s \parallel t \leftarrow f^{-1}(C)$, $r \leftarrow t \oplus H(s)$ and $M \leftarrow s \oplus G(r)$. If the last k_1 bits of M are zeros, then it returns the first $k_2 - k_0 - k_1$ bits of M, otherwise it returns $\perp$.

SECURITY OF OAEP. The encryption scheme OAEP[$\mathcal{F}$] is IND-CCA secure in the RO model if the underlying trapdoor permutation generator $\mathcal{F}$ is partial-domain one-way [12]. The concrete security results in [12] are done for the single-query IND-CCA security. We "translate" them into the the multi-query IND-CCA security using the result from [3] recalled in Equation 1.

Theorem 1. [12,3] *Let $\mathcal{F}$ be a trapdoor permutation generator with domain and range $\{0,1\}^k$. Let OAEP[$\mathcal{F}$] be the encryption scheme defined above. Then for any adversary A making q_e challenge oracle and q_d decryption oracle queries, q_H, q_G queries to RO oracles H and G, there exist an adversary B s.t.*

$$\mathbf{Adv}_{\mathcal{F},B}^{\mathrm{pd-owf}}(k, k_2 - k_0) \geq \frac{\mathbf{Adv}_{OAEP[\mathcal{F}],A}^{\mathrm{ind\text{-}cca\text{-}mq}}(k)}{2 q_e q_H} - \frac{1}{q_e q_H}\left(\frac{q_d q_G + q_d + q_G}{2^{k_0}} + \frac{q_d}{2^{k_1}}\right),$$

and the running time of B is that of A plus $q_G \cdot q_H \cdot (T_F(k) + O(1)) + O(\log q_e)$, where $T_F(k)$ is the time needed for evaluating a random instance of $\mathcal{F}$. ∎

As we can see the reduction is not particularly tight, but the situation becomes even worse if we use RSA, pretty much the only practical trapdoor permutation. It is believed to be one-way, and it was shown in [12] that under this assumption it is partial one-way as well, but the reduction is not tight. The concrete result is as follows.

Theorem 2. [12,3] *Consider the RSA trapdoor permutation generator with domain and range $\{0,1\}^k$. Let OAEP[RSA] be the encryption scheme defined above. Then for any adversary A making q_e challenge oracle and q_d decryption oracle queries, q_H, q_G queries to RO oracles H and G there exist an adversary B s.t.*

$$\mathbf{Adv}_{RSA,B}^{\mathrm{owf}}(k) \geq \frac{(\mathbf{Adv}_{OAEP[RSA],A}^{\mathrm{ind\text{-}cca\text{-}mq}}(k))^2}{4 q_e}$$
$$- \frac{1}{q_e}\left(\frac{q_d q_G + q_d + q_G}{2^{k_0}} + \frac{q_d}{2^{k_1}} + \frac{32}{2^{k-2k_0}}\right),$$

and the running time of B is 2 times that of A plus $q_H \cdot (q_H + 2q_G) \cdot O(k^3) + O(\log q_e)$. ∎

Such a loose concrete security bound implies that to maintain reasonable security guarantees, i.e. so that it not much harder to break the scheme than to invert 1024-bit RSA, one would need to use the scheme with a larger security parameter. It is shown in [16] show that keys of length about 4-5 thousand bits are necessary, i.e. at least 4 times larger that the standard 1024-bit keys, and this means decryption will be about $64 = 4^3$ times slower than before (since decryption requires a modulo exponentiation whose complexity is cubic in the length of the parameters). This is basically impractical. Note that the this estimate is for $q_e = 1$, i.e. when a single challenge ciphertext is considered. If we take into account the maximum number of queries to the challenge oracle an adversary makes – q_e, then to have reasonable security guarantees in the practical multi-query settings the RSA parameters should be even larger, making the scheme's algorithms prohibitively slow.

4   Known Concrete Security Improvements

Interestingly an extremely simple modification to the scheme permits several concrete security improvements. The modified scheme differs from OAEP[F] in that it uses trapdoor permutations of particular structure, which leave the last part of the input in the clear. Let $\mathcal{F}$ be a generator producing trapdoor permutations with domain and range $\{0,1\}^k$. Define a new generator $\mathcal{F}_k$ first to run $\mathcal{F}$; let (f, f^{-1}) be its output of $\mathcal{F}$, and define the first output of $\mathcal{F}_k$ as $f_p(x) \equiv f(x[1,\ldots,k]) \parallel \mathrm{ID}(x[k+1,\ldots,p]) = f(x[1,\ldots,k]) \parallel x[k+1,\ldots,p]$ for any inputs x of length $p \geq k$, where $x[1,\ldots,k]$ denotes the first k bits of x. The second output, the inverse permutation, is defined straight-forwardly. With regard to the OAEP construction we will be interested in cases when $p = k_2$ and $k \leq k_2 - k_0$, so that applying $\mathcal{F}_k$ to the output of the OAEP transform leaves the t-part in the clear. This modification has been suggested under the name OAEP++ by Kobara and Imai in [14] in order to improve concrete security of OAEP. This modification has also been previously suggested in [8] for an orthogonal reason of showing some positive results about non-malleability of OAEP when one or both ROs are instantiated with existing functions. The paper [8] does not provide concrete security bounds.

It is basically straightforward to see that if $\mathcal{F}$ is one-way, then $\mathcal{F}_k$ is partial one-way, in that it is infeasible to recover first k bits of the preimage. With respect to RSA, we get that RSA_k, applying RSA to only the first k bits of the input, is partial-one-way under the standard RSA assumption. That immediately implies that OAEP[$\mathcal{F}_k$], when $k \leq k_2 - k_0$ is IND-CCA in the RO model, if $\mathcal{F}$ is one-way, and we get that OAEP[RSA_k] is IND-CCA in the RO model under the standard RSA assumption[2]. For the concrete security result we can use the bound of Theorem 1.

But as shown in [14] we can get rid of factor q_h. This is possible for the modified scheme for the following reason. The proof of the original scheme constructs an adversary B breaking partial one-wayess of $\mathcal{F}$ using the IND-CCA adversary A for OAEP[F]. B needs to partially invert its input $y = f(s \parallel t)$, i.e. find s. This input y is given to A as the challenge ciphertext. The proof argues that the only way A can win the IND-CCA game is by querying the random oracle H on s at some point. While B cannot check which of the RO queries A made is the correct value B is looking for (since B does not know the second part t to verify this), it can just pick one query at random. This is where the factor q_h, the number of RO queries, is coming from. For the modified scheme, the proof from [12] applies without a single change, except we can note that B will now be able to select the correct s out of A's RO queries because t is in the clear. B just checks if $f(s_i \parallel t) = y$ for all queries to the random oracle H that A makes. Here is the improved security result, which also takes into account the multi-query setting (not considered in [14]).

---

[2] This was previously observed in [8]. The reduction in [14] does not use this observation and the proof is done "from scratch".

Theorem 3. [14,3] *Let $\mathcal{F}$ be a trapdoor permutation family with domain and range $\{0,1\}^k$. Let $\mathcal{F}_k$ be a trapdoor permutation generator producing permutations with domain and range $\{0,1\}^p$ for $p \geq k$ as defined above. Let $OAEP[\mathcal{F}_k] = (\mathcal{K}, \mathcal{E}, \mathcal{D})$ be the encryption scheme defined in Section 3 so that $p = k_2$ and $k \leq k_2 - k_0$. Then for any adversary A making q_e challenge oracle queries, q_d decryption oracle queries, q_H, q_G queries to RO oracles H and G there exist an adversary B s.t.*

$$\mathbf{Adv}^{\mathrm{owf}}_{\mathcal{F},B}(k) \geq \frac{\mathbf{Adv}^{\mathrm{ind\text{-}cca\text{-}mq}}_{OAEP[\mathcal{F}_k],A}(k)}{2q_e} - \frac{1}{q_e}\left(\frac{q_d q_G + q_d + q_G}{2^{k_0}} + \frac{q_d}{2^{k_1}}\right),$$

and the running time of B is that of A plus $q_G \cdot q_H \cdot (T_{\mathcal{F}_k}(k) + O(1)) + O(\log q_e)$, where $T_{\mathcal{F}_k}(k)$ is the time needed for evaluating a random instance of $\mathcal{F}_k$. ∎

The RSA instantiation result is immediate if we use RSA in place of $\mathcal{F}$ and RSA_k in place of $\mathcal{F}_k$ above.

5   Improving the Security in the Multi-query Setting

We show that security in the multi-query setting does not have to degrade as more messages are encrypted by each user (when an adversary does multiple queries to the challenge encryption oracle), i.e. we can get rid of the factor q_e in the bound of Theorem 3 when $OAEP(RSA_k)$ is used. Hence, the modified scheme provides significantly better security guarantees than the original version, for very practical parameter sizes. The following theorem states the improvement result.

Theorem 4. *Let RSA be a trapdoor permutation generator with domain and range $\{0,1\}^k$. Let RSA_k be a trapdoor permutation generator with domain and range $\{0,1\}^p$ for $p \geq k$ as defined in Section 4. Let $OAEP[RSA_k]$ be the encryption scheme defined in Section 3 so that $k_2 = p$ and $k \leq k_2 - k_0$. Then for any adversary A attacking IND-CCA security of the scheme making at most q_e queries to its challenge oracle, q_d decryption oracle queries, q_H, q_G queries to RO oracles H and G, there exist an adversary B s.t.*

$$\mathbf{Adv}^{\mathrm{owf}}_{RSA,B}(k) \geq \frac{\mathbf{Adv}^{\mathrm{ind\text{-}cca\text{-}mq}}_{OAEP[RSA_k],A}(k)}{2} - \left(\frac{q_d q_G + q_e q_d + q_e q_G}{2^{k_0}} + \frac{q_d}{2^{k_1}}\right),$$

and the running time of B is that of A plus $(q_e + q_G \cdot q_H) \cdot O(k^3) + O(\log q_e)$. ∎

What does the improvement mean in practice? The current belief is that 1024-bit RSA provides 80 bits of security, so for any adversary B with reasonable resources $\mathbf{Adv}^{\mathrm{owf}}_{RSA,B}(k) \leq 2^{-80}$ (and there are indications that this estimate is outdated in that it does not take into account newer attacks and growing computing power, and the bound is likely to be lower [13]). Now assume an adversary manages to obtain 2^{20} ciphertexts of chosen messages. This is about the number of TLS connections that were required to mount the well-known attack on RSA-PKCS1 by Bleichenbacher [7] (though his attacks needed that many chosen ciphertexts). Then according to Theorem 3 the bound on $\mathbf{Adv}^{\text{ind-cca-mq}}_{\text{OAEP}[\text{RSA}_k],A}(k)$ is only about 2^{-59}, which not a strong security level. Theorem 4 implies that in fact security of the scheme does not degrade as an adversary mounts more chosen-plaintext attacks and stays tightly related to the assumed security level of the underlying RSA problem.

Proof. We show how to modify the proof of security of RSA-OAEP in the single-query setting from [12], which assumes an adversary A attacking IND-CCA security of OAEP[F] (in the RO model). In our case we consider a special case of the scheme, OAEP[RSA_k] (i.e. when RSA is applied to the first k bits of the OAEP output, leaving the t-part in the clear). This will allow us to use self-reducibility of RSA and to incorporate the RSA challenge into multiple challenge ciphertexts, which the adversary is allowed to see in the multi-query setting.

Following [12] we use the game-playing technique of [6,19] and consider a sequence of experiments or games, associated with the adversary A. For the most part the proof is a simple extension of the proof in [12]. For $i \in \mathbb{N}$ we let $\Pr[\mathsf{G}_0]$ denote the probability that Game i outputs 1.

Game 0 corresponds to $\mathbf{Exp}^{\text{ind-cca}}_{\text{OAEP}[\text{RSA}_k],A}(1^k)$, the standard multi-query IND-CCA experiment (c.f. Definition 1 for the multi-query case). Each of q_e challenge ciphertexts is generated according to the definition of encryption of OAEP[RSA_k] as follows. For $1 \le i \le q_e$, to encrypt $M_{i,b}$ first r_i^* is chosen at random from $\{0,1\}^{k_0}$. Then $C_i \leftarrow f_k(s_i\|t_i)$, where $s_i = G(r_i^*) \oplus M_{i,b}\|0^{k_1}$ and $t_i = r_i^* \oplus H(s_i)$. Decryption oracle queries are answered according to the decryption algorithm of OAEP[RSA_k]. By construction and Definition 1 we get

$$\frac{1}{2} + \frac{1}{2} \cdot \mathbf{Adv}^{\text{ind-cca-mq}}_{\text{OAEP}[\text{RSA}_k],A}(k) = \Pr[\mathsf{G}_0] .$$

Game 1 is different from Game 0 in that it moves the computation of the random coins, $r_1^+, \ldots, \ldots, r_{q_e}^+$, used in the challenge ciphertexts explicitly up front, together with the computations of $g_1^+, \ldots, g_{q_e}^+$. By computation we mean choosing the values at random from the corresponding domains ($\{0,1\}^{k_0}$ and $\{0,1\}^{k_2-k_0}$ resp.) and storing the results. Further in the game r_i^+ is used in place of r_i^* and g_i^+ is used in place of $G(r_i^+)$, for all $1 \le i \le q_e$. I.e. each challenge ciphertext has the form $f_k(s_i^*\|t_i^*)$, where $s_i^* = (M_{i,b}\|0^{k_1}) \oplus g_i^+, t_i^* = h_i^+ \oplus r_i^*$ for $r_i^* = r_i^+$ and $h_i^+ = H(s_i^*)$. And whenever A queries the random oracle G on r_i^+ for any $1 \le i \le q_e$, it is given back g_i^+. These changes do not affect the distribution of the view of A compared to that in Game 0, because $(r_1^*, G(r_1^*), \ldots, r_{q_e}^*, G(r_{q_e}^*))$ and $(r_1^+, g_1^+, \ldots, r_{q_e}^+, g_{q_e}^+)$ have the same distribution, since G is a random oracle:

$$\Pr[\mathsf{G}_1] = \Pr[\mathsf{G}_0] .$$

Game 2 differs from Game 1 only in that the queries to the random oracle G on points $r_1^*, \ldots, r_{q_e}^*$ made by the adversary or by the decryption oracle are answered at random independently from the values $g_1^+, \ldots, g_{q_e}^+$ used to compute the challenge ciphertexts. Hence the challenge ciphertexts are independent from the challenge bit b (since they are uniformly distributed, independent of the rest of A's view) and

$$\Pr[\mathsf{G}_2] = \frac{1}{2} \ .$$

Similarly to [12] we can argue that the view of A and thus its outputs have the same distribution in Games 1 and 2 unless A queries G oracle on either of the points $r_1^*, \ldots, r_{q_e}^*$ (directly or making the decryption oracle make this query). Let us denote the probability of such event in this game $\Pr[\mathsf{AskG}_2]$, and such an event is defined similarly in the following games.

$$\Pr[\mathsf{G}_2] - \Pr[\mathsf{G}_1] \leq \Pr[\mathsf{AskG}_2] \ .$$

Game 3 is different from Game 2 in that it moves the computation of $s_1^+, \ldots, s_{q_e}^+$ and $h_1^+, \ldots, h_{q_e}^+$) explicitly up front. By computation we mean choosing the values at random from the corresponding domains ($\{0,1\}^{k_2 - k_0}$ and $\{0,1\}^{k_0}$ resp.) and storing the results. Further s_i^+ is used in place of s_i^* and h_i^+ is used in place of $H(s_i^+)$, for all $1 \leq i \leq q_e$. I.e. each challenge ciphertext has the form $f_k(s_i^* \| t_i^*)$, where $s_i^* = s_i^+$, $t_i^* = h_i^+ \oplus r_i^*$ for $r_i^* = r_i^+$ and $h_i^+ = H(s_i^*)$. And whenever A queries the random oracle H on s_i^+ for any $1 \leq i \leq q_e$, it is given back h_i^+. These changes do not affect the distribution of the view of A compared to that in Game 2, because we replaced each quadruple $(s_i^*, H(s_i^*), g_i^+, b)$ with another having the same distribution, since H is a random oracle:

$$\Pr[\mathsf{AskG}_3] = \Pr[\mathsf{AskG}_2] \ .$$

In Game 4, the difference with Game 3 is only in that the queries to the random oracle H made by the adversary or by the decryption oracle on points $s_1^+, \ldots, s_{q_e}^+$ are answered at random independently from the values $h_1^+, \ldots, h_{q_e}^+$ used in the challenge ciphertexts.

Similarly to [12] we can argue that the view of A and thus its outputs have the same distribution in Games 3 and 4 unless A or the decryption oracle queries the H oracle on either of the points $s_1^+, \ldots, s_{q_e}^+$. Let's denote the probability of such event in this game $\Pr[\mathsf{AskH}_4]$, and such an event is defined similarly in the following games.

$$\Pr[\mathsf{AskG}_4] - \Pr[\mathsf{AskG}_3] \leq \Pr[\mathsf{AskH}_4]$$

and

$$\Pr[\mathsf{AskG}_4] \leq \frac{q_e(q_G + q_d)}{2^{k_0}} \ .$$

Game 5 is similar to Game 4 except the way the challenge ciphertexts are generated. In this game they are simply picked at random, independently from everything else. We can argue similarly to the proof in [12] that this does not change the view and the outputs of A. The reason is that f_k is a permutation and in Game 4 it was applied to uniformly distributed points $s_i^* \parallel t_i^*$, where $s_i^* = s_i^+$ and $t_i^* = h_i^+ \oplus r_i^+$.

$$\Pr[\mathsf{AskH}_5] = \Pr[\mathsf{AskH}_4] .$$

Games 6–8 deal with answering decryption oracle queries which were simulated perfectly before that. The definitions of the games and their analysis done in [12] hold for our modified scheme and are independent of the number of the challenge encryption oracle queries A does, but we describe them for completeness. For a k_2-bit ciphertext C we call its last k_0 bits t, and the fist $k_2 - k_0$ bits of $f_k^{-1}(C)$ we call s. We call r the result of xoring t with $H(s)$.

Game 6 is like Game 5 except the decryption oracle rejects all ciphertexts for which the underlying r-value has not been previously queried to the G oracle by the adversary. The views of A in Games 5 and 6 are different only if A queries a valid ciphertext without querying the underlying r-value to G oracle. A ciphertext is valid if the last k_1 bits of $s \oplus G(r)$ are zeros. But if r has not been queried, then $G(r)$ is an independent random string and validity will be satisfied with probability at most 2^{-k_1}. For q_d decryption queries we get

$$\Pr[\mathsf{AskH}_6] - \Pr[\mathsf{AskH}_5] \leq \frac{q_d}{2^{k_1}} .$$

Game 7 is like Game 6 except that the decryption oracle rejects all ciphertexts for which the underlying s-value has not been previously queried to the H oracle by the adversary. The views of A in Games 6 and 7 are different only if A queries a valid ciphertext without querying the underlying s-value to H oracle when the query r was made to the G oracle. Since $r = H(s) \oplus t$, $H(s)$ was not previously defined, it is random and independent. Hence the probability that r was queried is at most $q_G/2^{k_0}$. And for q_d decryption queries we get

$$\Pr[\mathsf{AskH}_7] - \Pr[\mathsf{AskH}_6] \leq \frac{q_d q_G}{2^{k_0}} .$$

In the last Game 8 the decryption oracle queries, for which either of the corresponding r and s values has not been queried, are rejected. The other ciphertexts are decrypted by using a simple plaintext extractor who expects all previously made G and H queries made by A and returns the matching plaintext. Namely, to decrypt C, if there exist stored r_i, s_j for $1 \leq i \leq q_G$ and $1 \leq j \leq q_H$ so that $H(s_j) \oplus t = r_i$, the last k_0 of $f_k(s_j \parallel t) = C$ and the last k_1 bits of $s_i \oplus G(r)$ are zeros, then return the rest of $s_i \oplus G(r)$. If no such r_i, s_j are found, then return $\perp$. The view of the adversary does not change and thus

$$\Pr[\mathsf{AskH}_8] = \Pr[\mathsf{AskH}_7] .$$

Putting it all together we get

$$\frac{1}{2} \cdot \mathbf{Adv}_{\mathrm{OAEP}[F_k],A}^{\mathrm{ind\text{-}cca\text{-}mq}}(k) = \Pr[\mathsf{G}_0] - \frac{1}{2}$$

$$= \Pr[\mathsf{G}_1] - \frac{1}{2}$$

$$\leq \Pr[\mathsf{G}_2] - \frac{1}{2} + \Pr[\mathsf{AskG}_2]$$

$$\leq \Pr[\mathsf{AskG}_2]$$

$$= \Pr[\mathsf{AskG}_3]$$

$$\leq \Pr[\mathsf{AskG}_4] + \Pr[\mathsf{AskH}_4]$$

$$\leq \frac{q_e(q_G + q_d)}{2^{k_0}} + \Pr[\mathsf{AskH}_5]$$

$$\leq \frac{q_e(q_G + q_d)}{2^{k_0}} + \frac{q_d}{2^{k_1}} + \Pr[\mathsf{AskH}_6]$$

$$\leq \frac{q_e(q_G + q_d) + q_d q_G}{2^{k_0}} + \frac{q_d}{2^{k_1}} + \Pr[\mathsf{AskH}_7]$$

$$= \frac{q_e(q_G + q_d) + q_d q_G}{2^{k_0}} + \frac{q_d}{2^{k_1}} + \Pr[\mathsf{AskH}_8] \ .$$

We now claim that there exists an adversary B such that

$$\Pr[\mathsf{AskH}_8] \leq \mathbf{Adv}_{\mathrm{RSA},B}^{\mathrm{owf}}(k) \ . \tag{2}$$

This is where we use self-reducibility of RSA to improve tightness of the reduction. To justify Equation (2) we construct B as follows. B is given an RSA public key (N, e) and a challenge $y = x^e \mod \mathbb{N}$ for a random $x \in \mathbb{Z}_N^*$. B picks q_e values at random from $\mathbb{Z}_N^*$, let us call them $v_1, \ldots, v_{q_e}$; and q_e values at random from $\{0, 1\}^{p-k}$, let us call them $w_1, \ldots, w_{q_e}$. B runs A on public key (N, e), answers its RO queries with random and independent values (and records all queries and answers). To answer the decryption oracle queries B checks if the corresponding G and H queries were made, and in this case a simple plaintext extractor we described above is used; otherwise, the ciphertexts are rejected (B returns $\perp$). For $1 \leq i \leq q_e$ for an i-th query to the challenge oracle made by A, B returns $(yv_i^e \mod N) \parallel w_i$.

We claim that B simulates the view of A in Game 8 perfectly. (Except for the mismatch between the sets $\mathbb{Z}_{\mathbb{N}}^*$ and $\{0, 1\}^k$, which is usually ignored. For the possible simple resolutions of this issue see [12].) The challenge ciphertexts are random and independent strings, and the decryption queries are answered according to the simple plaintext extractor algorithm that uses the recorded queries to the random oracles and the (random) answers. Event AskH_8 means that A made a query h to the random oracle G so that $h[1, \ldots, k]^e = yv_j^e ($ mod $N)$ for some $1 \leq j \leq q_e$. B searches for such query and outputs hv_j^{-1} mod N, which is y^d mod N, i.e. it breaks one-wayness of RSA.

The running time of the constructed adversary is greater than that of the adversary in Theorem 3 by the time to prepare q_e ciphertexts which needs q_e RSA applications, and the justification of the former can be found in [12].

REMARK. We comment on why does not the above proof showing the security improvement work for the unmodified OAEP[RSA] scheme. The reason is that in the original scheme the RSA permutation is applied to the whole output $s \parallel t$ of the OAEP transform. The tight security of OAEP[RSA] is only shown assuming partial-domain one-wayness of RSA. In the proof above the adversary B given a challenge y could still use self-reducibility of RSA and generate challenge ciphertexts for A as $yv_1^e, \ldots, yv_{q_e}^e \mod N$. In A's view, these ciphertexts have the right distribution (in Game 8) unless A queries the H oracle on any of the underlying s values (the first part of $y^d v_1, \ldots, y^d v_{q_e}$). But if this happens, B cannot compute y^d, as it does not know the remaining part of the transform.

6   Encrypting Long Messages with Modified RSA-OAEP

We observe that the modified RSA-OAEP scheme can be used to encrypt long messages without employing symmetric encryption in the hybrid encryption construct. For that the function G in the transform is made variable-output-length, i.e. it's output is the length of the message plus the zero padding of length k_1. For a fixed-output-length hash $G'(\cdot)$ one can efficiently construct $G(\cdot)$ as $G'(\langle 0 \rangle \parallel \cdot) \parallel G'(\langle 1 \rangle \parallel \cdot) \ldots \parallel G'(\langle l \rangle \parallel \cdot)$, where $\langle i \rangle$ means the binary representation of the counter $i \in \mathbb{N}$. In the RO model G is a random oracle if G' is. The function H in the transform needs to be variable-input-length, which is not a problem, since most of the hash functions are. The RSA function is applied to the first k (e.g. 1024 bits) of the s-part of the OAEP transform. The proof of security stays virtually the same. This scheme yields more compact ciphertexts for long messages than the one obtained through the use of hybrid encryption because there is no need to encrypt the symmetric key.

7   Conclusions

We re-visited a previously suggested slight modification of the well-known and practical RSA-OAEP encryption. We showed that this scheme has extra advantages, namely its IND-CCA security remains tightly related (in the RO model) to hardness of the RSA problem, even in the multi-query setting. Additionally, this scheme can be used for encryption of long messages without employing the hybrid encryption method and symmetric encryption. We believe the modification is very simple to implement and may be considered by the standard bodies.

Acknowledgments

We thank David Cash and Eike Kiltz for useful discussions, and Adam O'Neill and anonymous reviewers for their comments on the draft. Alexandra Boldyreva is supported in part by NSF CAREER award 0545659 and NSF Cyber Trust award 0831184.

References

1. Abe, M., Kiltz, E., Okamoto, T.: Chosen ciphertext security with optimal cipher-text overhead. In: Pieprzyk, J. (ed.) Asiacrypt 2008. LNCS, vol. 5350, pp. 355–371. Springer, Heidelberg (2008)
2. Baudron, O., Pointcheval, D., Stern, J.: Extended notions of security for multicast public key cryptosystems. In: Welzl, E., Montanari, U., Rolim, J.D.P. (eds.) ICALP 2000. LNCS, vol. 1853, pp. 499–511. Springer, Heidelberg (2000)
3. Bellare, M., Boldyreva, A., Micali, S.: Public-key encryption in a multi-user setting: Security proofs and improvements. In: Preneel, B. (ed.) EUROCRYPT 2000. LNCS, vol. 1807, pp. 259–274. Springer, Heidelberg (2000)
4. Bellare, M., Rogaway, P.: Random oracles are practical: A paradigm for designing efficient protocols. In: Ashby, V. (ed.) ACM CCS 1993, Fairfax, Virginia, USA, November 3–5, pp. 62–73. ACM Press, New York (1993)
5. Bellare, M., Rogaway, P.: Optimal asymmetric encryption. In: De Santis, A. (ed.) EUROCRYPT 1994. LNCS, vol. 950, pp. 92–111. Springer, Heidelberg (1995)
6. Bellare, M., Rogaway, P.: The security of triple encryption and a framework for code-based game-playing proofs. In: Vaudenay, S. (ed.) EUROCRYPT 2006. LNCS, vol. 4004, pp. 409–426. Springer, Heidelberg (2006)
7. Bleichenbacher, D.: Chosen ciphertext attacks against protocols based on the RSA encryption standard PKCS #1. In: Krawczyk, H. (ed.) CRYPTO 1998. LNCS, vol. 1462, pp. 1–12. Springer, Heidelberg (1998)
8. Boldyreva, A., Fischlin, M.: On the security of OAEP. In: Lai, X., Chen, K. (eds.) ASIACRYPT 2006. LNCS, vol. 4284, pp. 210–225. Springer, Heidelberg (2006)
9. Boneh, D.: Simplified OAEP for the RSA and rabin functions. In: Kilian, J. (ed.) CRYPTO 2001. LNCS, vol. 2139, pp. 275–291. Springer, Heidelberg (2001)
10. Coron, J.-S.: On the exact security of full domain hash. In: Bellare, M. (ed.) CRYPTO 2000. LNCS, vol. 1880, pp. 229–235. Springer, Heidelberg (2000)
11. ElGamal, T.: A public key cryptosystem and a signature scheme based on discrete logarithms. IEEE Transactions on Information Theory 31, 469–472 (1985)
12. Fujisaki, E., Okamoto, T., Pointcheval, D., Stern, J.: RSA-OAEP is secure under the RSA assumption. Journal of Cryptology 17(2), 81–104 (2004)
13. Kaliski, B.: TWIRL and RSA key size. RSA Laboratories (2003)
14. Kobara, K., Imai, H.: OAEP++: A very simple way to apply OAEP to deterministic OW-CPA primitives. Cryptology ePrint Archive, Report 2002/130 (2002), http://eprint.iacr.org/
15. Okamoto, T., Pointcheval, D.: REACT: Rapid enhanced-security asymmetric cryptosystem transform. In: Naccache, D. (ed.) CT-RSA 2001. LNCS, vol. 2020, pp. 159–175. Springer, Heidelberg (2001)
16. Pointcheval, D.: How to encrypt properly with RSA. RSA Laboratories' Crypto-Bytes 5(1), 9–19 (Winter/Spring 2002)
17. Shoup, V.: OAEP reconsidered. In: Kilian, J. (ed.) CRYPTO 2001. LNCS, vol. 2139, pp. 239–259. Springer, Heidelberg (2001)
18. Shoup, V.: A proposal for an ISO standard for public-key encryption. ISO/IEC JTC 1/SC27 (2001)
19. Shoup, V.: Sequences of games: a tool for taming complexity in security proofs. cryptology eprint archive, report 2004/332 (2004), http://eprint.iacr.org/

Fault Attacks on RSA Public Keys:
Left-To-Right Implementations Are Also Vulnerable

Alexandre Berzati[1,3], Cécile Canovas[1], Jean-Guillaume Dumas[2],
and Louis Goubin[3]

[1] CEA-LETI/MINATEC, 17 rue des Martyrs, 38054 Grenoble Cedex 9, France
{alexandre.berzati,cecile.canovas}@cea.fr
[2] Université de Grenoble, Laboratoire Jean Kuntzmann, umr CNRS 5224, BP 53X,
51 rue des mathématiques, 38041 Grenoble, France
Jean-Guillaume.Dumas@imag.fr
[3] Versailles Saint-Quentin University,
45 Avenue des Etats-Unis, 78035 Versailles Cedex, France
Louis.Goubin@prism.uvsq.fr

Abstract. After attacking the RSA by injecting fault and corresponding countermeasures, works appear now about the need for protecting
RSA public elements against fault attacks. We provide here an extension
of a recent attack [BCG08] based on the public modulus corruption. The
difficulty to decompose the *"Left-To-Right"* exponentiation into partial
multiplications is overcome by modifying the public modulus to a number
with known factorization. This fault model is justified here by a complete study of faulty prime numbers with a fixed size. The good success
rate of this attack combined with its practicability raises the question of
using faults for changing algebraic properties of finite field based cryptosystems.

Keywords: RSA, fault attacks, *"Left-To-Right"* exponentiation, number theory.

1 Introduction

Injecting faults during the execution of cryptographic algorithms is a powerful
way to recover secret information. Such a principle was first published by Bellcore
researchers [BDL97, BDL01] against multiple public key cryptosystems. Indeed,
these papers provide successful applications including RSA in both standard and
CRT modes. This work was completed, and named Differential Fault Analysis
(DFA), by E. Biham and A. Shamir with applications to secret key cryptosystems [BS97]. The growing popularity of this kind of attack, in the last decade,
was based on the ease for modifying the behavior of an execution [BECN+04] and
the difficulty for elaborating efficient countermeasures [BOS03, Wag04, Gir05b].

Many applications against the RSA cryptosystem, based on fault injection,
have been published. The first ones dealt with the perturbation of the private

M. Fischlin (Ed.): CT-RSA 2009, LNCS 5473, pp. 414–428, 2009.

key or temporary values during the computation [BDL97, BDJ$^+$98, BDL01]. The perturbation of public elements was considered as a real threat when J-P. Seifert published an attack on the RSA signature check mechanism [Sei05, Mui06]. This paper first mentions the possibility of modifying the public modulus N such that the faulty one is prime or easy to factor. Then, E. Brier *et al.* extended this work to the full recovery of the private exponent d for various RSA implementations [BCMCC06]. Both works are based on the assumption that the fault occurs before performing the RSA modular exponentiation. A. Berzati *et al.* first address the issue of modifying the modulus during the exponentiation [BCG08].Still this work was limited to an application against *"Right-To-Left"* type exponentiation algorithms.

In this paper we aim to generalize the previous attack to *"Left-To-Right"* type exponentiations. Under the fault assumption that the modulus can become a number with a known factorization, we prove that it is possible to recover the whole private exponent. We provide a detailed study of this fault model, based on number theory, to show its consistency and its practicability for various kinds of perturbation. Finally, we propose an algorithm to recover the whole private exponent that is efficient either in terms of fault number or in computational time.

2   Background

2.1   Notations

Let N, the public modulus, be the product of two large prime numbers p and q. The length of N is denoted by n. Let e be the public exponent, coprime to $\varphi(N) = (p-1) \cdot (q-1)$, where $\varphi(\cdot)$ denotes Euler's totient function. The public key exponent e is linked to the private exponent d by the equation $e \cdot d \equiv 1 \bmod \varphi(N)$. The private exponent d is used to perform the following operations.

RSA Decryption: Decrypting a ciphertext C boils down to compute $\tilde{m} \equiv C^d \bmod N \equiv C^{\sum_{i=0}^{n-1} 2^i \cdot d_i} \bmod N$ where d_i stands for the i-th bit of d. If no error occurs during computation, transmission or decryption of C, then $\tilde{m}$ equals m.

RSA Signature: The signature of a message m is given by $S \equiv \dot{m}^d \bmod N$ where $\dot{m} = \mu(m)$ for some hash and/or deterministic padding function μ. The signature S is validated by checking that $S^e \equiv \dot{m} \bmod N$.

2.2   Modular Exponentiation Algorithms

Binary exponentiation algorithms are often used for computing the RSA modular exponentiation $\dot{m}^d \bmod N$ where the exponent d is expressed in a binary form as $d = \sum_{i=0}^{n-1} 2^i \cdot d_i$. Their polynomial complexity with respect to the input length make them very interesting to perform modular exponentiation.

The Algorithm 1 describes a way to compute modular exponentiations by scanning bits of d from least significant bits (LSB) to most significant bits

Algorithm 1. *"Right-To-Left"* modular exponentiation	**Algorithm 2.** *"Left-To-Right"* modular exponentiation
INPUT: m, d, N OUTPUT: $A \equiv m^d$ mod N	INPUT: m, d, N OUTPUT: $A \equiv m^d$ mod N
1 : $A:=1$; 2 : $B:=m$; 3 : **for** i **from** 0 **upto** $(n-1)$ 4 : **if** $(d_i == 1)$ 5 : $A := (A \cdot B)$ mod N; 6 : **end if** 7 : $B := B^2$ mod N; 8 : **end for** 9 : **return** A;	1 : $A:=1$; 2 : **for** i **from** $(n-1)$ **downto** 0 3 : $A := A^2$ mod N; 4 : **if** $(d_i == 1)$ 5 : $A := (A \cdot m)$ mod N; 6 : **end if** 7 : **end for** 8 : **return** A;

(MSB). That is why it is usually referred to as the *"Right-To-Left"* modular exponentiation algorithm. This is that specific implementation that is attacked in [BCG08] by corrupting the public modulus of RSA.

The dual algorithm that implements the binary modular exponentiation is the *"Left-To-Right"* exponentiation described in Algorithm 2. This algorithm scans bits of the exponent from MSB to LSB and is lighter than *"Right-To-Left"* one in terms of memory consumption.

3   Modification of the Modulus and Extension Attempt

3.1   Previous Work

J-P. Seifert first addressed the issue of corrupting RSA public key elements [Sei05, Mui06]. This fault attack aims to make a signature verification mechanism accept false signatures by modifying the value of the public modulus N. No information about the private exponent d is revealed with this fault attack. Its efficiency is linked to the attacker's ability to reproduce the fault model chosen for the modification of the modulus.

Seifert's work inspired the authors of [BCMCC06] who first used the public modulus perturbation to recover the whole private key d. The attacker has to perform a perturbation campaign to gather a large enough number of (message, faulty signature) pairs. As in Seifert's attack, the fault on the modulus is induced before executing the exponentiation. Three methods based on the use of Chinese Remainder Theorem and the resolution of quite small discrete logarithms are proposed in [BCMCC06] and [Cla07] to recover the private exponent from the set of gathered pairs.

A new fault attack against *"Right-To-Left"* exponentiation has been presented lately [BCG08]. This work completes the state-of-the-art by allowing the attacker to use other fault models for recovering the private exponent. The details of this work are presented below.

3.2 Public Key Perturbation during RSA Execution: Case of the "*Right-To-Left*" Algorithm

Fault Model. In J.P Seifert and E. Brier *et al.*'s proposals [Sei05, BCMCC06] the fault is provoked before the exponentiation so that the whole execution is executed with the faulty modulus, $\hat{N}$.

The attack presented by A. Berzati *et al.* [BCG08] extends the fault model by allowing the attacker to inject the fault during the execution of the "*Right-To-Left*" exponentiation. The modification of N is supposed to be a transient random byte modification. It means that only one byte of N is set to a random value. The value of the faulty modulus $\hat{N}$ is not known by the attacker. However, the time location of the fault is a parameter known by the attacker and used to perform the cryptanalysis. This fault model has been chosen for its simplicity and practicability in smart card context [Gir05a, BO06]. Furthermore, it can be easily adapted to 16-bit or 32-bit architectures.

Faulty Computation. Let $d = \sum_{i=0}^{n-1} 2^i \cdot d_i$ be the binary representation of d. The output of a RSA signature can be written as:

$$S \equiv \dot{m}^{\sum_{i=0}^{n-1} 2^i \cdot d_i} \bmod N \tag{1}$$

We consider that a fault has occurred j steps before the end of the exponentiation, during the computation of a square. According to the fault model described, all subsequent operations are performed with a faulty modulus $\hat{N}$. We denote by $A \equiv \dot{m}^{\sum_{i=0}^{(n-j-1)} 2^i \cdot d_i} \bmod N$ the internal register value and by $\hat{B}$ the result of the faulty square:

$$\hat{B} \equiv \left(\dot{m}^{2^{(n-j-1)}} \bmod N \right)^2 \bmod \hat{N} \tag{2}$$

Hence, the faulty signature $\hat{S}$ can be written as:

$$\hat{S} \equiv A \cdot \hat{B}^{\sum_{i=(n-j)}^{n-1} 2^{[i-(n-j)]} \cdot d_i} \bmod \hat{N} \tag{3}$$

$$\equiv [(\dot{m}^{\sum_{i=0}^{(n-j-1)} 2^i \cdot d_i} \bmod N) \tag{4}$$
$$\cdot (\dot{m}^{2^{(n-j-1)}} \bmod N)^{\sum_{i=(n-j)}^{n-1} 2^{[i-(n-j)+1]} \cdot d_i}] \bmod \hat{N}$$

From the previous expression of $\hat{S}$, one can first notice that the fault injection splits the computation into a correct (computed with N) and a faulty part (computed with $\hat{N}$). A part of d is used during the faulty computation. This is exactly the secret exponent part that will be recovered in the following analysis.

Attack Principle. From both correct signature S and faulty one $\hat{S}$ (obtained from the same message m), the attacker can recover the isolated part of the private key $d_{(1)} = \sum_{i=n-j}^{n-1} 2^i \cdot d_i$. Indeed, he tries to find simultaneously candidate values for the faulty modulus $\hat{N}'$ (according to the random byte fault assumption) and for the part of the exponent $d'_{(1)}$ that satisfies:

$$\hat{S} \equiv \left(S \cdot \dot{m}^{-d'_{(1)}} \bmod N \right) \cdot \left(\dot{m}^{2^{(n-j-1)}} \bmod N \right)^{2^{[1-(n-j)]} \cdot d'_{(1)}} \bmod \hat{N}' \qquad (5)$$

According to [BCG08], the pair $(d'_{(1)}, \hat{N}')$ that satisfies (5) is the right one with a probability very close to 1. Then, the subsequent secret bits will be found by repeating this attack using the knowledge of the already found most significant bits of d and a signature faulted earlier in the process. In terms of fault number, the whole private key recovery requires an average of (n/l) faulty signatures, where l is the average number of bits recovered each time. As a consequence, this few number of required faults makes the attack both efficient and practicable.

3.3   Application to the *"Left-To-Right"* Modular Exponentiation

In this section, we try to apply the previously explained fault attack to the *"Left-To-Right"* implementation of RSA. Under the same fault model, we wanted to know what does prevent an attacker from reproducing the attack against the dual implementation.

We denote by A the internal register value just before the modification of the modulus N:

$$A \equiv \dot{m}^{\sum_{i=j}^{n-1} 2^{i-j} \cdot d_i} \bmod N \qquad (6)$$

Hence, knowing that the first perturbed operation is a square, the faulty signature $\hat{S}$ can be written as:

$$\hat{S} \equiv \left(\left(\left(A^2 \cdot \dot{m}^{d_{j-1}} \right)^2 \cdot \dot{m}^{d_{j-2}} \right)^2 \ldots \right)^2 \cdot \dot{m}^{d_0} \bmod \hat{N} \qquad (7)$$

$$\equiv A^{2^j} \cdot \dot{m}^{\sum_{i=0}^{j-1} 2^i \cdot d_i} \bmod \hat{N}$$

By observing (7), one can notice that the perturbation has two consequences on the faulty signature $\hat{S}$. First, it splits the computation into a correct part (*i.e:* the internal register value A) and a faulty one, like for the perturbation of the *"Right-To-Left"* exponentiation [BCG08]. The other one is the addition of j cascaded squares of the local variable A, computed modulo $\hat{N}$. This added operation defeats the previous attack on the *"Right-To-Left"* exponentiation [BCG08] because of the difficulty to compute square roots in RSA rings.

Our idea for generalizing the previous attack to *"Left-To-Right"* exponentiation is to take advantage of the modulus modification to change the algebraic properties of the RSA ring. In other words, if $\hat{N}$ is a prime number, then it is possible to compute square roots in polynomial time. Moreover, it is actually sufficient that $\hat{N}$ is B-smooth with B small enough to enable an easy factorization of $\hat{N}$, then the Chinese Remainder Theorem enables also to compute square roots in polynomial time. We show next anyway that the number of primes $\hat{N}$ is sufficient to provide a realistic fault model.

4   Fault Model

According to the previous section, the square root problem can be overcome by perturbing the modulus N such that $\hat{N}$ is prime. In this section we will study the consistency and the practicability of such a fault model. Even though this model has already been adopted in Seifert's attack [Mui06, Sei05], we propose next further experimental evidences of the practicability of this model.

4.1 Theoretical Estimations

Let us first estimate the number of primes with a fixed number of bits. From [Dus98, Theorem 1.10], we have the following bounds for the number of primes π below a certain integer x:

$$\pi(x) \geq \frac{x}{\ln(x)}\left(1 + \frac{1}{\ln(x)} + \frac{1.8}{\ln^2(x)}\right), \text{ for } x \geq 32299. \tag{8}$$

$$\pi(x) \leq \frac{x}{\ln(x)}\left(1 + \frac{1}{\ln(x)} + \frac{2.51}{\ln^2(x)}\right), \text{ for } x \geq 355991.$$

Then, for numbers of exactly t bits such that $t \geq 19$ bits, the number of primes is $\pi_t = \pi(2^t) - \pi(2^{t-1})$. By using the previous bounds (8), the probability that a t-bit number is prime, $pr_t = \frac{\pi_t}{2^{t-1}}$, satisfies:

$$pr_t > Inf(t) = \frac{0.480t^5 - 1.229t^4 + 0.0265t^3 - 7.602t^2 + 9.414t - 3.600}{t^3(t-1)^3\ln^3(2)} \tag{9}$$

$$pr_t < Sup(t) = \frac{0.480t^5 - 1.229t^4 + 2.157t^3 - 11.862t^2 + 13.674t - 5.02}{t^3(t-1)^3\ln^3(2)}$$

For instance, if $t = 1024$ bits:

$$Inf(1024) = \frac{1}{709.477} \text{ and } Sup(1024) = \frac{1}{709.474}$$

Therefore around one 1024-bit number out of 709 is prime; and among the 2048-bit numbers, more than one out of 1419 is prime.

Consider now a set of k randomly selected numbers of exactly t bits and let PN be the random variable expressing the expected number of primes in this set. This variable follows a binomial law $\mathcal{B}(k, pr_t)$. Then we can give the following confidence interval of primes (with a and b integer bounds):

$$\Pr[a \leq PN \leq b] = \sum_{i=a}^{b}\binom{k}{i}pr_t^i(1 - pr_t)^{k-i} \tag{10}$$

For example, we construct the following set $\mathcal{N}$ according to a random byte fault model. In other words, if $\oplus$ is the bit by bit exclusive OR, then[1]:

$$\mathcal{N} = \{N \oplus R_8 \cdot 2^{8i}, \ R_8 = 0 .. 255, \ i = 0 .. \left(\frac{n}{8} - 1\right)\}$$

---

[1] For the sake of clarity we assume that a byte fault can take 2^8 values. In fact, it can take only $2^8 - 1$. Indeed, the error can not be null otherwise the value of N is unchanged and the fault can not be exploited.

Then the cardinality of $\mathcal{N}$ is

$$|\mathcal{N}| = 256 \cdot \frac{n}{8} = 32 \cdot n$$

Would the set $\mathcal{N}$ be composed of randomly selected values, then the proportion of primes in $\mathcal{N}$ would follow (9). Hence, we can set $k = |\mathcal{N}|$ and compute the corresponding average and bounds with an approximation of pr_t. For $n = 1024$, according to (9), we can estimate pr_{1024} and thus the average number of faulty primes is $32 \cdot 1024/709.47 \approx 46.186$. Equation (10) combined with the estimation of pr_{1024} shows also that the number of primes in a set is comprised between $[18, 80]$ in 99.999% of the cases. For $n = 2048$, the average number of primes is 46.176 and comprised between 18 and 80 in 99.999% of the cases. Obviously $\mathcal{N}$ is not a set of randomly chosen elements; howbeit, empirical evidence shows that such sets behave quite like random sets of elements, as shown below.

4.2   Experimental Results

We have computed such sets for randomly selected RSA moduli and counted the number of primes in those sets. The repartition seems to follow a binomial rule (as expected) and we have the following experimental data to support our belief (see Figure 1).

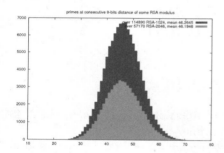
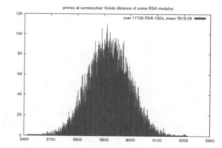

(a) Primes at consecutive 8-bit distance of some RSA modulus

(b) Primes at consecutive 16-bit distance of some RSA modulus

Fig. 1. Experimental distribution of primes among faulty RSA moduli

As shown in Table 1 it was anyway *never* the case that no prime was found in a set $\mathcal{N}$ (more than that we always found more than 18 primes in such a set). This experimental lower-bound equals to the one obtained by considering a random set. The same observation can be done for the upper-bound. Hence, our obtained results confirm our theoretical analysis.

The presented results can be extended to other fault models. The Table 1 presents also theoretical expected results when 16-bit or 32-bit architectures are targeted. For $t = 1024$ with 16-bit architecture the average number of primes is 5911.83 and is between $[5520, 6320]$ in 99.999% of the cases.

Table 1. Experimental counts of primes in $\mathcal{N}$

Architecture	n bits	$\lvert\mathcal{N}\rvert$	$\lvert\mathcal{N}\rvert \cdot pr_n$	# of exp.	# of primes		
					Min.	Avg.	Max.
8-bit	1024	2^{15}	46.186	114890	18	46.26	79
8-bit	2048	2^{16}	46.176	57170	22	46.19	80
16-bit	1024	2^{22}	5911.83	17725	5621	5919.08	6212
32-bit	1024	2^{37}	$\approx 1,94 \cdot 10^8$				

4.3 Consequences

This study strengthen J-P. Seifert's assumption [Sei05, Mui06] of considering only prime modification of the modulus. We have showed that our fault model can be considered as a random modification of the public modulus. Then, an average of 709 faults on N will be required to obtain a prime $\hat{N}$ in the case of a 1024-bit RSA.

Additional Remark. By carefully studying the experimental results, one can notice that, for a given modulus N, the byte location of the fault influences the number of prime found in the subset. Thus, if the attacker has the ability of setting the byte location of the fault, he can increase his chances to get a prime faulty modulus and therefore, dramatically reduce the number of faulty signatures required to perform the attack.

4.4 The Algorithm of Tonelli and Shanks

The algorithm of Tonelli and Shanks [Coh93] is a probabilistic and quite efficient algorithm used to compute square roots modulo P, where P is a prime number. The principle of the algorithm is based on the isomorphism between the multiplicative group $(\mathbb{Z}/P\mathbb{Z})^*$ and the additive group $\mathbb{Z}/(P-1)\mathbb{Z}$. Suppose $P-1$ is written as:

$$P - 1 = 2^e \cdot r, \quad \text{with } r \text{ odd.} \tag{11}$$

Then, the cyclic group G of order 2^e is a subgroup of $\mathbb{Z}/(P-1)\mathbb{Z}$. Let z be a generator of G, if a is a quadratic residue modulo N, then:

$$a^{(P-1)/2} \equiv (a^r)^{2^{e-1}} \equiv 1 \bmod P \tag{12}$$

Noticing that $a^r \bmod P$ is a square in G, then it exists an integer $k \in [\![0 : 2^e - 1]\!]$ such that

$$a^r \cdot z^k = 1 \text{ in } G \tag{13}$$

And so, $a^{r+1} \cdot z^k = a$ in G. Hence, the square root of a, is given by

$$a^{1/2} \equiv a^{(r+1)/2} \cdot z^{k/2} \bmod P \tag{14}$$

Both main operations of this algorithm are:

- Finding the generator z of the subgroup G,
- Computing the exponent k.

The whole complexity of this algorithm is that of finding k, $\mathcal{O}\left(\ln^4 P\right)$ binary operations or $\mathcal{O}\left(\ln P\right)$ exponentiations. The details of the above algorithm are described in [Coh93]. In practice, on a Pentium IV 3.2GHz, the GIVARO[2] implementation of this algorithm takes on average 7/1000 of a second to find a square root for a 1024-bit prime modulus.

4.5   Smooth Modulus

As in [Mui06], what we really need for the faulty modulus is only to be easily factorable. Indeed, one can compute square roots modulo non prime modulus as long as the factorization is known. The idea is first to find square roots modulo each prime factors of $\hat{N}$. Then to lift them independently to get square roots modulo each prime power. And finally to combine them using the Chinese Remainder Theorem (see e.g. [Sho05, §13.3.3] for more details). The number of square roots increases but since they are computed on comparatively smaller primes, the overall complexity thus remains $\mathcal{O}\left(\ln^4 \hat{N}\right)$ binary operations. In the following we thus consider only prime faulty moduli.

5   Cryptanalysis

The purpose of our fault attack against the *"Left-To-Right"* exponentiation is similar to the attack against the *"Right-To-Left"* one [BCG08]. The modulus N is transiently modified to a prime value during a squaring, j_k steps before the end of the exponentiation. Then, from a correct/faulty signature pair $(S, \hat{S}_k)$, the attack aims to recover the part of private exponent $d_{(k)} = \sum_{i=0}^{j_k-1} 2^i \cdot d_i$ isolated by the fault. By referring to [BCG08], the following analysis can be easily adapted for faults that first occurs during a multiplication.

Dictionary of Prime Modulus. The first step consists in computing a dictionary of prime faulty modulus candidates $(\hat{N}_i)$. The attacker tests all possible values obtained by modifying N according to a chosen fault model. Then, candidate values for $\hat{N}$ are tested using the probabilistic Miller-Rabin algorithm [Rab80]. According to our study (see Sect. 4.1), for a random byte fault assumption, the faulty modulus dictionary will contain 46 entries in average either for a 1024-bit or a 2048-bit RSA. The size of the dictionary depends on the fault model (see Table 1).

---

[2] GIVARO is an open source C++ library over the GNU Multi-Precision Library. It is available on http://packages.debian.org/fr/sid/libgivaro-dev

Computation of Square Roots. For each entry $\hat{N}_i$ of the modulus dictionary, the attacker chooses a candidate value for the searched part of the private exponent $d'_{(k)}$. Now he can compute[3]:

$$R_{(d'_{(k)}, \hat{N}_i)} \equiv \hat{S}_k \cdot \dot{m}^{-d'_{(k)}} \bmod \hat{N}_i \tag{15}$$

For the right pair $(d_{(k)}, \hat{N})$, $R_{(d_{(k)}, \hat{N})}$ is expected to be a multiple quadratic residue (*i.e:* a j_k-th quadratic residue, see Sect. 3.3). As a result, if $R_{(d'_{(k)}, \hat{N}_i)}$ is not a quadratic residue, the attacker can directly deduce that the candidate pair $(d'_{(k)}, \hat{N}_i)$ is a wrong one. The quadratic residuosity test can be done in our case because all precomputed candidate values for the faulty modulus are prime numbers. The test is based on Fermat's theorem:

$$\text{If } \left(R_{(d'_{(k)}, \hat{N}_i)}\right)^{(\hat{N}_i - 1)/2} \equiv 1 \bmod \hat{N}_i \tag{16}$$

$$\text{then } R_{(d'_{(k)}, \hat{N}_i)} \text{is a quadratic residue modulo } \hat{N}_i$$

If the test is satisfied then the attacker can use the Tonelli and Shanks algorithm (see Sect. 4.4) to compute the square roots of $R_{(d'_{(k)}, \hat{N}_i)}$. Therefore, to compute the j_k-th square root of $R_{(d'_{(k)}, \hat{N}_i)}$, this step is expected to be repeated j_k-times. But, when one of the j_k quadratic residuosity test fails, the current candidate pair is directly $(d'_{(k)}, \hat{N}_i)$ rejected and the square root computation is aborted. The attacker has to choose another candidate pair.

Final Modular Check. The purpose of the two first steps is to cancel the effects on the faulty signature due to the perturbation. Now, from the j_k-th square root of $R_{(d'_{(k)}, \hat{N}_i)}$ the attacker will simulate an error-free end of execution by computing:

$$S' \equiv \left(\left(R_{(d'_{(k)}, \hat{N}_i)}\right)^{1/2^{j_k}} \bmod \hat{N}_i\right)^{2^{j_k}} \cdot \dot{m}^{d'_{(k)}} \bmod N \tag{17}$$

Finally, he checks if the following equation is satisfied:

$$S' \equiv S \bmod N \tag{18}$$

As in the *"Right-To-Left"* attack [BCG08], when this latter condition is satisfied, it means that the candidate pair is very probably the searched one (see Sect. 6.3). Moreover, the knowledge of the already found least significant bits of d is used to reproduce the attack on the subsequent secret bits. As a consequence, the attacker has to collect a set of faulty signatures $\hat{S}_k$ by injecting the fault

---

[3] This computation is possible only when d'_k is invertible in $\mathbb{Z}/\mathbb{Z}\hat{N}_i$; in our case all the considered N_i are primes and Euclid's algorithm always computes the inverse.

at different steps j_k before the end of the exponentiation. Moreover, multiple faulty signature $\hat{S}_{k,f}$ have to be gathered for a given step j_k to take into account the probability for having a faulty signature $\hat{S}_k$ computed under a prime $\hat{N}$, that is to say exploitable by the cryptanalysis. This set $(\hat{S}_{k,f}, j_k)_{k,f}$ is sorted in descending fault location. If faults are injected regularly, each sorted pair is used to recover a l-bit part of the exponent such that for the k-th pair $(\hat{S}_{k,f}, j_k)$, the recovered part of d is $d_{(k)} = \sum_{i=0}^{j_k-1} 2^i \cdot d_i = \sum_{i=0}^{k \cdot l - 1} 2^i \cdot d_i$. These results can be applied for faults that are not injected regularly (*i.e:* $j_k - j_{k-1} = l_k < l_{max}$). The attack algorithm is given in more details next.

Algorithm 3. DFA against *"Left-To-Right"* modular exponentiation

---

INPUT: N, $\dot{m}$, the correct signature S, the size of the dictionary D_{length},

        the set of pairs $(\hat{S}_{k,f}, j_k)_{0 \leq k < n/l,\ 1 \leq f \leq \mu(F_n)}$

OUTPUT: the private exponent d

1: *//Computation of the dictionary of prime faulty modulus candidates*
2: Dict $= Build\_Prime\_Dict(N, D_{length})$;
3: *//Initialization*
4: $d := 0$;
5: *//All the faulty signatures are tested*
6: **for** k **from** 0 **upto** $\lfloor n/l \rfloor$
7:   **for** f **from** 1 **upto** $\mu(F_n)$
8:     **for** $d_{(k)}$ **from** 0 **upto** $2^l - 1$
9:       $d' := d_{(k)} \cdot 2^{j_k} + d$;
10:       **for** i **from** 1 **upto** D_{length}
11:         $R := \hat{S}_{k,f} \cdot \dot{m}^{-d'} \bmod Dict[i]$;
12:         *//The function computes j_k square roots and returns 0 when a test fails*
13:         $R := Test\_And\_Tonelli(R, j_k, Dict[i])$;
14:         *//If a test fails, then we have to test another candidate pair*
15:         **if** $(R == 0)$
16:           **break**;
17:         **else**
18:           $S' := R^{2^{j_k}} \cdot \dot{m}^{d'} \bmod N$
19:           *//Final check*
20:           **if** $(S' == S \bmod N)$
21:             *//The attack continues for the subsequent l-bit part of d*
22:             $d := d'$;
23:             **goto** *line_6*;
24:           **end if**;
25:         **end if**;
26:       **end for**;
27:     **end for**;
28:   **end for**;
29: **end for**;
30: **return** d;

---

6   Performance

6.1   Fault Number

Our fault model is based on the modification of the modulus N such that its corresponding faulty value is prime. In Section 4.1, we have shown that the probability for a t-bit number to be prime, pr_t, can be bounded. Now, let the number of fault to make $\hat{N}$ prime be the random variable F_t. This random variable follows a geometric probability law. Hence the average number of faults to make $\hat{N}$ prime is:

$$\frac{1}{Sup(t)} < \mu\left(F_t\right) = \frac{1}{pr_t} < \frac{1}{Inf(t)} \tag{19}$$

For large values of t (*i.e*: at least 1024 or 2048-bit RSA), we can use the pinching (or sandwich) theorem to approximate this value asymptotically :

$$\mu\left(F_t\right) \sim \frac{t \cdot \ln^3(2)}{0.480} \sim \frac{t}{1.441} \tag{20}$$

From a given faulty signature, the attacker can recover a l-bit part of d. There are at most n/l such parts for an RSA of size n. This shows that the average number of faults required for a whole private key satisfies:

$$\text{Number of faults} = \mathcal{O}\left(\frac{n^2}{1.441 \cdot l}\right) \text{ tries} \tag{21}$$

This number can be dramatically reduced if the attacker has the ability to chose the byte location of the fault (see Sect. 4.1) or if the fault model is larger (*i.e*: smooth modulus, different architectures targeted ...).

6.2   Computational Complexity

We now give the overall complexity of the attack. The size of the dictionary, D_{length}, is let as an attack parameter since the attacker can fix a limit if the chosen fault model requires more resources than he can get. According to our previous analysis (see Sect. 4.1), $D_{length} = 46$ for a random byte fault assumption.

Theorem 1. *Algorithm 3 is correct and its average complexity for a random byte fault perturbation of the modulus satisfies:*

$$C_{attack} = \mathcal{O}\left(\frac{2^{8+l} \cdot n^3 \cdot (n+l)}{16 \cdot l}\right) \text{ exponentiations}$$

Proof. Correctness as been shown in section 5. Now for the complexity, the attacker has to test all possible candidate pairs $(d'_{(k)}, \hat{N}_i)$. The number of pairs depends on the size of the dictionary of prime modulus denoted by D_{length} and the window recovery length l:

$$|(d'_{(k)}, \hat{N}_i)| = 2^l \cdot D_{length} \tag{22}$$

For each pair the attacker first computes $R_{(d'_{(k)},\hat{N}_i)}$ (see (15)) by executing a modular exponentiation of the message and a multiplication.

Then, he performs a series of at most j_k quadratic residuosity tests and, for each success, a square root is computed. By noticing that the probability to fail in the test follows a geometric probability law, the average number of performed tests[4] is $\frac{1}{\Pr[\text{Test fails}]} = 2$. As a consequence, the average complexity of this step is:

$$C_{Square\ roots}(k) = \mathcal{O}\left(2 \cdot C_{Test} + C_{Tonelli\ \&\ Shanks}\right) \qquad (23)$$
$$= \mathcal{O}\left(j_k \cdot n\right) \text{ exponentiations}$$

The last step of the attack is the final check (see (17)). It requires to compute j_k modular squares and a modular exponentiation of the message followed by a multiplication. The latter computation is also bounded by $\mathcal{O}(j_k \cdot n)$ exponentiations.

Now in the case of a fixed size dictionary the average number of primes of this dictionary for a byte modification of the modulus is $N_{faults\ per\ blocs} = \frac{2^8 n/8}{D_{length}}$.

Then, the attack has to test all of the gathered faulty signatures in order to recover the whole exponent. Hence, as j_k is bounded by $k \cdot l$, the overall computational complexity is bounded by:

$$C_{attack} = \sum_{k=0}^{n/l} N_{faults\ per\ blocs} \cdot C_{Square\ roots}(k) \cdot 2^l \cdot D_{length} \qquad (24)$$
$$= \mathcal{O}\left(\frac{2^{8+l} \cdot n^3 \cdot (n+l)}{16 \cdot l}\right)$$

The presented attack is thus longer than the *"Right-To-Left"* one [BCG08], the principal reason being the extra number of faulty pairs to analyze in order to get a prime modulus.

6.3   False-Acceptance Probability

As defined in [BCG08], the false-acceptance probability is the probability for a wrong pair $(d'_{(k)}, \hat{N}_i)$ to satisfy (18). In our case, the computation of the final check is done in $\mathbb{Z}/N\mathbb{Z}$ and requires extra squares. As a consequence the false-acceptance probability given in [BCG08] has to be adapted by replacing the search space for $\hat{N}$ by the dictionary length D_{length}:

$$0 < \Pr[F.A] < min\left(\frac{(N-1)\cdot 2^l \cdot D_{length}}{N \cdot (2^l \cdot D_{length} - 1)}, \frac{2^l \cdot D_{length}}{N}\right) \qquad (25)$$

Moreover, because of the quadratic residuosity tests (see Sect. 5), false candidates can be rejected before computing the final check. Hence, the final check will not

---

[4] The test fails when tested value is not a quadratic residue. But all the $\hat{N}_i$ are prime. Let be z_i a generator in $\mathbb{Z}/\hat{N}_i\mathbb{Z}$, all the elements of the group can be expressed as a power of z_i. Hence one element out of 2 is a power of z_i^2 and a quadratic residue.

always be done. The probability that a wrong pair pass all the j_k tests is given by:

$$\Pr \left[R_{(d'_{(k)}, \hat{N}_i)} \text{is a } j_k\text{-times quadratic residue} \right] \qquad (26)$$

$$= \prod_{i=0}^{j_k-1} \Pr \left[\left(R_{(d'_{(k)}, \hat{N}_i)} \right)^{1/2^i} \text{ is a quadratic residue} \right]$$

$$= \frac{1}{2^{j_k}}$$

This probability indicates that, for recovering the k-th part of d, only one out of 2^{j_k} wrong pairs will pass all the quadratic residuosity tests. Eventually, the false-acceptance probability can be upper-bounded:

$$\Pr[F.A] < min \left(\frac{1}{2^{j_k}}, \frac{(N-1) \cdot 2^l \cdot D_{length}}{N \cdot (2^l \cdot D_{length} - 1)}, \frac{2^l \cdot D_{length}}{N} \right) \qquad (27)$$

This expression first shows that because of the last term $\frac{2^l \cdot D_{length}}{N}$, the false-acceptance probability is highly negligible for commonly used RSA length. Furthermore, one can advantageously notice that the final check can be avoided when the number of consecutive quadratic residuosity tests to pass is large enough (*i.e*: $2^{j_k} > D_{length} \cdot 2^l$).

7   Conclusion

In this paper, we generalize the fault attack presented in [BCG08] to *"Left-To-Right"* implementation of RSA by assuming that the faulty modulus can be prime. Although this model has been already used [Sei05], this paper provides a detailed theoretical analysis in fault attack context. Furthermore this analysis proves that such a fault model is not only practicable but extendable to different architectures. This emphases the need for protecting RSA public elements during the execution.

More generally the use of a faulty prime modulus to compute square roots in polynomial time raises the question of using faults for changing algebraic properties of the underlying finite domain. This paper provides an element of answer that may be completed by future fault exploitations.

References

[BCG08]   Berzati, A., Canovas, C., Goubin, L.: Perturbating RSA Public Keys: An Improved Attack. In: Oswald, E., Rohatgi, P. (eds.) CHES 2008. LNCS, vol. 5154, pp. 380–395. Springer, Heidelberg (2008)

[BCMCC06]   Brier, É., Chevallier-Mames, B., Ciet, M., Clavier, C.: Why One Should Also Secure RSA Public Key Elements. In: Goubin, L., Matsui, M. (eds.) CHES 2006. LNCS, vol. 4249, pp. 324–338. Springer, Heidelberg (2006)

[BDJ⁺98]   Bao, F., Deng, R.H., Jeng, A., Narasimhalu, A.D., Ngair, T.: Breaking
Public Key Cryptosystems on Tamper Resistant Devices in the Presence
of Transient Faults. In: Christianson, B., Lomas, M. (eds.) Security Protocols 1997. LNCS, vol. 1361, pp. 115–124. Springer, Heidelberg (1998)

[BDL97]   Boneh, D., DeMillo, R.A., Lipton, R.J.: On the Importance of Checking
Cryptographic Protocols for Faults. In: Fumy, W. (ed.) EUROCRYPT
1997. LNCS, vol. 1233, pp. 37–51. Springer, Heidelberg (1997)

[BDL01]   Boneh, D., DeMillo, R.A., Lipton, R.J.: On the Importance of Eliminating Errors in Cryptographic Computations. Journal of Cryptology 14(2),
101–119 (2001)

[BECN⁺04]   Bar-El, H., Choukri, H., Naccache, D., Tunstall, M., Whelan, C.:
The Sorcerer's Apprentice Guide to Fault Attacks. Cryptology ePrint
Archive, Report 2004/100 (2004)

[BO06]   Blömer, J., Otto, M.: Wagner's Attack on a secure CRT-RSA Algorithm
Reconsidered. In: Breveglieri, L., Koren, I., Naccache, D., Seifert, J.-P.
(eds.) FDTC 2006. LNCS, vol. 4236, pp. 13–23. Springer, Heidelberg
(2006)

[BOS03]   Blömer, J., Otto, M., Seifert, J.-P.: A New CRT-RSA Algorithm Secure
Against Bellcore Attack. In: ACM Conference on Computer and Communication Security (CCS 2003), pp. 311–320. ACM Press, New York
(2003)

[BS97]   Biham, E., Shamir, A.: Differential Fault Analysis of Secret Key Cryptosystems. In: Kaliski Jr., B.S. (ed.) CRYPTO 1997. LNCS, vol. 1294,
pp. 513–525. Springer, Heidelberg (1997)

[Cla07]   Clavier, C.: De la sécurité physique des crypto-systèmes embarqués. PhD
thesis, Université de Versailles Saint-Quentin (2007)

[Coh93]   Cohen, H.: A Course in Computational Algebraic Number Theory.
Springer, New York (1993)

[Dus98]   Dusart, P.: Autour de la fonction qui compte le nombre de nombres
premiers. PhD thesis, Université de Limoges (1998)

[Gir05a]   Giraud, C.: DFA on AES. In: Dobbertin, H., Rijmen, V., Sowa, A. (eds.)
AES 2005. LNCS, vol. 3373, pp. 27–41. Springer, Heidelberg (2005)

[Gir05b]   Giraud, C.: Fault-Resistant RSA Implementation. In: Breveglieri, L.,
Koren, I. (eds.) Fault Diagnosis and Tolerance in Cryptography, pp.
142–151 (2005)

[Mui06]   Muir, J.A.: Seifert's RSA Fault Attack: Simplified Analysis and Generalizations. Cryptology ePrint Archive, Report 2005/458 (2006)

[Rab80]   Rabin, M.O.: Probabilistic algorithm for testing primality. Journal of
Number Thoery 12(1), 128–138 (1980)

[Sei05]   Seifert, J.-P.: On Authenticated Computing and RSA-Based Authentication. In: ACM Conference on Computer and Communications Security
(CCS 2005), pp. 122–127. ACM Press, New York (2005)

[Sho05]   Shoup, V.: A Computational Introduction to Number Theory and Algebra. Cambridge University Press, Cambridge (2005)

[Wag04]   Wagner, D.: Cryptanalysis of a provably secure CRT-RSA algorithm. In:
Proceedings of the 11th ACM Conference on Computer Security (CCS
2004), pp. 92–97. ACM Press, New York (2004)

Fault Analysis Attack against
an AES Prototype Chip Using RSL

Kazuo Sakiyama, Tatsuya Yagi, and Kazuo Ohta

The University of Electro-Communications
Department of Information and Communication Engineering
1-5-1 Chofugaoka, Chofu, Tokyo 182-8585, Japan
{saki,tatsuya_popo,ota}@ice.uec.ac.jp

Abstract. This paper reports a successful Fault Analysis (FA) attack against a prototype AES (Advanced Encryption Standard) hardware implementation using a logic-level countermeasure called Random Switching Logic (RSL). The idea of RSL was proposed as one of the most effective countermeasures for preventing Differential Power Analysis (DPA) attacks. The RSL technique was applied to AES and a prototype ASIC was implement with a 0.13-μm standard CMOS library. Although the main purpose of using RSL is to enhance the DPA resistance, our evaluation results for the ASIC reveal that the DPA countermeasure of RSL can negatively affect the resistance against FA attacks. We show that the circuits using RSL has a potential vulnerability against FA attacks by increasing the clock frequency.

Keywords: Fault Analysis, Random Switching Logic, AES, Clock-based Attack.

1 Introduction

Secure systems have to be resistant against various malicious attacks in order to prevent leakage of information or an unexpected use of the system. Recent threats to cryptographic devices are considered Side-Channel Attacks (SCAs) which are attacks that observe in a non-intrusive way computational timing, power variants or electromagnetic radiation of the device. By simple observation or mathematical processing of the observed physical phenomena, one can retrieve secret data out of the device.

Among SCAs, passive attacks are based on measuring physical characteristics leaking from side-channels of the embedded system. Timing Analysis (TA) checks the computation time. If the execution time varies with the data or the key used in the computations, this can be detected by the attacker [1]. Simple Power Analysis (SPA) measures the power fluctuations during cryptographic operations and guesses the actual types of computations. Furthermore, in [2], Kocher, Jaffe and Jun introduced Differential Power Analysis (DPA), a type of differential SCA that also considers effects correlated to data values. Brier, Clavier and Oliver introduced Correlation Power Analysis (CPA) that aims at enhancing DPA by

M. Fischlin (Ed.): CT-RSA 2009, LNCS 5473, pp. 429–443, 2009.
© Springer-Verlag Berlin Heidelberg 2009

improving the Hamming weight model [3]. Electromagnetic Analysis (EMA) and Acoustic Analysis (AA) were also introduced as effective SCA examples [4].

For silicon devices, one of the most straightforward attacks is a physical attack directly to the silicon. This is a powerful method if such a probing point is available. For instance, it is easy to retrieve secret information by probing the data on a bus line on the silicon. The fault induction attack or Fault Analysis (FA) attack reported in [5] is a technique that works by disturbing the device by inducing errors during a computation. These attacks are named active attacks after the technique.

Our research in the paper uses the FA attack based on the clock pin of a cryptographic device to induce a fault. By applying the clock-based FA attack against a prototype AES (Advanced Encryption Standard [6]) hardware, we will show that it is possible to retrieve the secret key from a device protected by a DPA countermeasure called RSL.

The remainder of this paper is as follows. Section 2 describes previous work for DPA countermeasures and reviews the RSL technique and its application to AES. In Sect. 3, we show our evaluation method and explain the details about the clock-based FA attacks against the prototype AES hardware using RSL. The evaluation results are discussed in Sect. 4. Section 5 concludes the paper and describes future work.

2   Countermeasures against DPA

In order to resist SCAs, a lot of different countermeasures have been proposed in recent years. The countermeasures are classified mainly into two categories depending on the type of the countermeasure; algorithm level and logic (or gate) level. In [7], Coron proposed countermeasures against DPA for elliptic curve cryptography [8,9]. Those countermeasures utilize random bits when performing a computational sequence in scalar multiplication, and hence they are considered as algorithm-level countermeasures. On the other hand, logic-level countermeasures try to prevent an information leakage from a device independent of cryptographic algorithms. Hereafter we will focus on the logic-level countermeasures and discuss the previous work.

2.1   Logic-Level Countermeasures

Tiri and Verbauwhede proposed a design methodology using a complementary logic called WDDL (Wave Dynamic Differential Logic) [10]. WDDL employs a dual-rail circuit style so that the same power consumption can be consumed regardless the input values, e.g. an OR operation $a \vee b$ is operated with an AND operation $(\neg a) \wedge (\neg b)$. The difficulty in implementing WDDL circuits is that a pair of the complementary logics has to be balanced physically (e.g. the wire load capacitance should be balanced) so that we can observe the same power consumption for any input signals. Masked-AND proposed by Trichina [11] can mask input and output data of AND gates with random bits to resist against power analysis attacks. However, the vulnerability of the proposed Masked-AND gate was reported in [12].

MDPL (Masked Dual-rail Precharge Logic) proposed by Popp and Mangard is designed for solving the physical constraints in WDDL although it is used in a dual-rail circuit [13]. However, security problems are already pointed out in terms of DPA attacks even on their implemented prototype ASIC [14,15,16,17]. To date, the RSL technique in [18,19] seems to surmount the security problems by improving the specification, $e.g.$ applying re-masking operation for each RSL gate with different random bits. The details are described in the next subsection.

2.2 Random Switching Logic (RSL)

As a logic-level countermeasure against DPA attacks, RSL was proposed by Suzuki, Saeki and Ichikawa in 2004 [20]. RSL is a technique for masking intermediate values by using random data, and can be used in a single-rail circuit style. The original RSL was based on the use of a full-custom cell library to realize NAND, NOR and XOR operations. In 2007, the authors mapped the RSL operations on FPGA slices and implemented AES using the RSL technique [18]. As a result, they showed that the FPGA implementation of AES with RSL is secure against DPA attacks. Furthermore, in order to develop an ASIC prototype of AES with RSL in a general design environment, RSL was modified so that it can be realized with a standard cell library [19]. Consequently, they implemented a prototype ASIC of AES with RSL based on a 0.13-μm CMOS library. The prototype ASIC is mounted on an evaluation platform called SASEBO [21] and available publicly for the purpose of evaluating the side-channel resistance. In this paper, we use the prototype ASIC for evaluating security of the RSL technique. Next, we review RSL in detail.

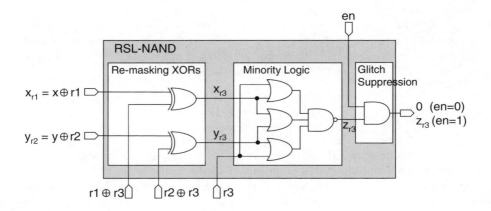

Fig. 1. Block diagram for RSL-NAND with re-masking and glitch suppression

Figure 1 shows a NAND gate based on RSL (RSL-NAND) that is composed of re-masking XORs, an inverted majority (minority) logic gate and a glitch suppression gate. The RSL-NAND has six 1-bit inputs and one 1-bit output. RSL-NAND has two major features. One is the re-masking mechanism to enhance the resistance against SCAs. The other is the glitch suppression gate controlled by an enable signal. The details of the functionality are described as follows.

Random Re-Masking. Two inputs, x_{r1} and y_{r2} are masked with random bits, $r1$ and $r2$, respectively, *i.e.*

$$x_{r1} = x \oplus r1, \quad y_{r2} = y \oplus r2. \tag{1}$$

In order to re-mask these input data with a new random bit $r3$, we input two bits, $r1 \oplus r3$ and $r2 \oplus r3$ to the re-masking XORs. The operation of re-masking is described as

$$x_{r1} \oplus (r1 \oplus r3) = x \oplus r3 = x_{r3}, \quad y_{r2} \oplus (r2 \oplus r3) = y \oplus r3 = y_{r3}. \tag{2}$$

Minority Logic. The minority logic evaluates three inputs, x_{r3}, y_{r3} and $r3$ and outputs 1-bit value. If two or more inputs are ones, it outputs zero and if not, it outputs one. By utilizing the minority gate, we can calculate a NAND operation of x and y re-masked with $r3$ as

$$z_{r3} = \neg\big((x_{r3} \vee r3) \wedge (y_{r3} \vee r3) \wedge (x_{r3} \vee y_{r3})\big) = \neg(x \wedge y) \oplus r3. \tag{3}$$

Glitch Suppression. The last remaining input is the enable signal en that is used for suppressing glitch propagations. The glitch is a short-period electrical pulse caused by different arrival time of input signals of a (composite) gate. It is known that the glitch propagation in a hardware chip leaks information via side channel as reported in [22]. To avoid the glitch propagation, the output signal of RSL-NAND is controlled by a 1-bit enable signal as

$$\begin{cases} z_{r3} & (en = 1) \\ 0 & (en = 0). \end{cases} \tag{4}$$

Namely, the output signal is forced to zero (*i.e.* precharge to zero) until all input signals are fixed and the glitch disappears. This is because we need to consider the arrival time of input signals and the delay of the minority gate when determining the timing delay of value switching of the enable signals. Thus, we can lower the risk of glitch-based attacks. The details of enable signals will be explained in the next subsection.

2.3   Combinational Logic Using the RSL Gates

In order to explain how the RSL gates work in an actual hardware design, we use a general architecture illustrated in Fig. 2. The architecture consists of three functional blocks, the RSL gates, a Random Number Generator (RNG) and a controller for enable signals. In this example case, we focus on 1-bit datapath starting from a flip-flop $regD1$ and ending in a flip-flop $regD2$. Here, we assume that the input data, x_{r1} is already masked with $r1$ before arriving at $regD1$.

The RSL gate labeled as $RSL1$ receives two masked inputs, x_{r1} and y_{r2} and outputs z_{r3} when the enable signal $en1 = 1$ by following the Eqs. (1)(2)(3)(4). Likewise, $RSL2$ and $RSL3$ perform RSL-NAND operations and when $en2 = 1$ and $en3 = 1$, the output of $RSL3$ will be as follows:

$$w_{r7} = \neg\Big(\neg\big(\neg(x \wedge y) \wedge s\big) \wedge u\Big) \oplus r7 \quad (en1 = en2 = en3 = 1). \tag{5}$$

The delay control for the enable signals has an important role to suppress the glitch propagations. We discuss the mechanism of the controller hereafter.

The order of asserting enable signals must be from $en1$ to $en3$. This can be explained by considering the direction of the signal propagation. For instance, if $en2$ is high before $en1$ is asserted and then $en1$ becomes high (e.g. $(en1, en2) = (0, 1) \rightarrow (1, 1)$), a glitch may occur in $RSL2$ depending on the arrival time of the input signals for $RSL2$. This glitch cannot be suppressed in $RSL2$ and propagates to $RSL3$ because $en2$ is already high. Therefore, in order to avoid such glitch propagation, $en1$ has to be high at least before $en2$ becomes high.

On the other hand, the order of negating those enable signals have to be opposite, i.e. in the order from $en3$ to $en1$. If negating $en1$ earlier than $en2$ (e.g. and $en1 = 0$ and $en2 = 1$), there may happen a glitch in $RSL2$ and propagate to $RSL3$ as well. The asserting or negating order is summarized as

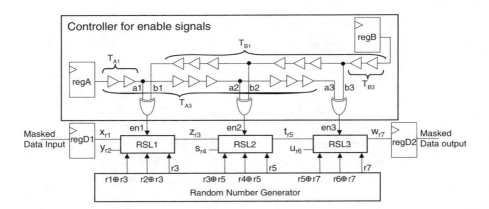

Fig. 2. General hardware architecture using several RSL gates; Random Number Generator (RNG) and a controller for generating enable signals are necessary

follows: $(en1, en2, en3) = (0,0,0) \rightarrow (1,0,0) \rightarrow (1,1,0) \rightarrow (1,1,1) \rightarrow (1,1,0) \rightarrow$ $(1,0,0) \rightarrow (0,0,0) \rightarrow \cdots$.

In order to facilitate the control of the enable signals, two flip-flops of $regA$ and $regB$ are used in the controller block. At each positive edge of the clock, they change values as $(regA, regB) = (0,0) \rightarrow (1,0) \rightarrow (1,1) \rightarrow (0,1) \rightarrow (0,0) \rightarrow \cdots$. To explain the reason, we use a simple example as follows.

Suppose $(regA, regB) = (0,0)$ at the first cycle of the clock. If we change the values of those flip-flops as $(regA, regB) = (1,0)$ in the next cycle, $en1$ starts being high after the time of T_{A1}, where T_{A1} is the delay time from the output of $regA$ to the input of the XOR gate generating $en1$. Then, the signals $en2$ and $en3$ are asserted in this order with the delay times of T_{A2} and T_{A3}, respectively as shown in Fig. 3. Moreover, if $(regA, regB) = (1,1)$ are set at the next cycle, $en3$ starts being low in the time of T_{B3}. Likewise, $en2$ and $en1$ are negated in this order. Eventually, we can obtain the enable signals as illustrated in Fig. 3.

Figure 3 also shows a data flow from $regD1$ to $regD2$. The data x_{r1} is determined by $regD1$ at a positive edge of the clock and an RSL-NAND operation is performed in $RSL1$ with y_{r2}. After all signals are fixed in $RSL1$, $e1$ is asserted

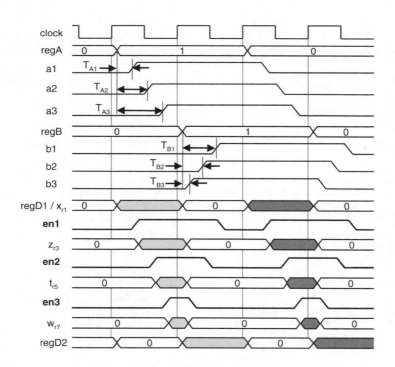

Fig. 3. Timing waveform of the enable signals generated by the controller and data flow corresponding to the architecture shown in Fig. 2

and the next NAND operation is performed in $RSL2$. In the same way, the signal w_{r7} is determined and stored in $regD2$ at the next positive edge of the clock. Note that we need two cycles to send a data from $regD1$ to $regD2$.

3   Evaluation Method for AES Hardware Using RSL

Based on the specification of RSL reviewed in the previous section, we discuss the security of a circuit using the RSL gates in terms of FA attacks. We introduce a clock-based FA attack that can induce a fatal error in a circuit implemented with the RSL gates by increasing the clock frequency.

3.1   Attack Model

Through the observation in Sect. 2, we notice that the RSL technique has a potential risk for FA attacks when increasing the clock frequency. In order to explain this, we consider a case of providing a high clock frequency to the circuit using the RSL gates in Fig. 2. Figure 4 shows a waveform when providing a clock at a high clock frequency such as $T_{clock} \leq T_{A3}$. This type of the clock is denoted as "fast clock" in this paper. Here T_{clock} is the clock period. For a normal operation case (*i.e.* $T_{clock} > T_{A3}$), the enable signal $en3$ should be high at the positive edge of the clock every two cycles (see Fig. 3). In contrast, this is not true for the case of providing a fast clock. As can be seen from the Fig. 4, the enable signal $en3$ is low at every positive edge of the clock and the register $regD2$ cannot store the correct data from the signal w_{r7}. As a result, the register $regD2$ stays zero as long as the fast clock is provided.

This attack model is based on the clock-based FA attacks. In this model, we put the following assumption for attackers' ability:

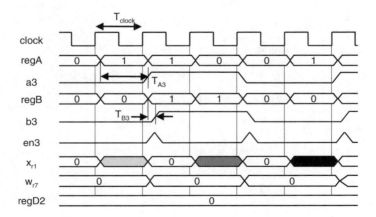

Fig. 4. Waveform corresponding to the architecture illustrated in Fig. 2 when providing a fast clock such that $T_{clock} \leq T_{A3}$

– to operate target devices several times.
– to increase the clock frequency of a target device.

Under the assumption, it is possible to force register values in a circuit using RSL to be zeroes.

As mentioned in [19], the RSL technique was applied for non-linear functions in AES except the SUBBYTES functions in the key scheduling process. Accordingly, we can illustrate a block diagram that focuses on the final round of the encryption of an AES hardware with RSL (denoted as RSL-AES) as shown in Fig. 5. When providing a fast clock to RSL-AES, we assume that the output of the SUBBYTES function becomes zeroes and one of the 128-bit inputs for the ADDROUNDKEY function becomes zeroes after the SHIFTROWS transformation. Here, we assume that the key scheduling block is correctly working and the correct 10th round key, K10 is input as another 128-bit input of the ADDROUNDKEY block.

Thus, the ADDROUNDKEY function is expected to output 10th round key. The 10th round key is XORed with random bits `r_unmask` since the output data of the RSL gates is masked with random bits by the re-masking mechanism, and therefore data from the final RSL gates needs to be unmasked. The unmasking is operated at the final step in Fig. 5. Note that we assume that the final unmasking block is allocated after the ADDROUNDKEY function since a XOR operation in the ADDROUNDKEY function could be the DPA selection function [2]. In this model, we can obtain 128-bit output of `dout` as

$$\text{dout} = \text{K10} \oplus \text{r\_unmask}, \tag{6}$$

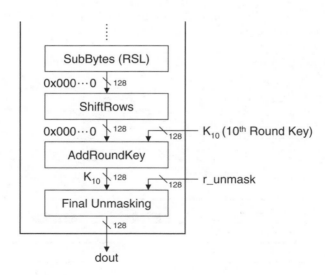

Fig. 5. Final round of AES encryption with a fault induction to the SUBBYTES block that is implemented with RSL.

where r_unmask is 128-bit data used for unmasking the final result. If r_unmask is 128-bit random data, we have no clue for guessing K10. Therefore, this attack model seems useless. However, we will show that the attack model is applicable in the case of the prototype ASIC in the following subsections.

3.2 SASEBO-R

We use the SASEBO-R (Side-channel Attack Standard Evaluation Board, type-R) [21] in order to verify our attack model. The SASEBO-R is designed to develop evaluation schemes against physical attacks. The board has two hardware chips, a prototype ASIC that has several cryptographic functions and a Xilinx FPGA that controls the ASIC chip. The experiment setting with SASEBO-R is shown in Fig. 6. Note that we did not use an oscilloscope that is necessary for SPA or DPA attacks in general.

There are several cryptographic hardware in the ASIC chip and an RSL-AES is also implemented [23]. To perform one of the cryptographic operations, the control FPGA sends all necessary signals to the ASIC chip such as a clock signal, control signals, plaintext, *etc.* After performing a cryptographic operation,

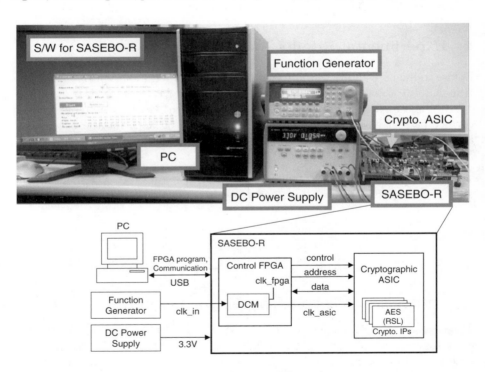

Fig. 6. Evaluation environment with the SASEBO-R

FPGA receives the result (*e.g.* ciphertext in the case of encryption) from the ASIC chip. The FPGA can be programmed and controlled by software in a PC via USB interface.

3.3 Clock-Based Fault Induction

The clock signal, the most important signal in this paper, is provided from a function generator to the FPGA on the SASEBO-R although we normally use a 24-MHz on-board oscillator. The function generator sent a clock signal `clk_in` to the DCM (Digital Clock Manager) module in the control FPGA. In the default setting of the DCM, it generates two clock signals, `clk_fpga` and `clk_asic`, which are synchronized. Note that we set the DCM so that the clock frequency of `clk_in` can be the same as that of `clk_fpga` through the paper.

By changing the setting of the DCM module, it can operate as a frequency multiplier for `clk_asic`, *i.e.* the DCM module in this setting can provide a clock `clk_asic` whose frequency is multiple of `clk_fpga`.

For FA attacks in this paper, we set the DCM such as $f_{asic} = 6 \times f_{fpga}$ since our concern is to provide a fast clock only to the ASIC chip. In other words, it is expected that a fault is induced only in the ASIC chip in this setting.

4 Experimental Results

Before inducing a fault, we perform several 128-bit AES encryptions using the RSL-AES at a normal clock frequency, *i.e.* $f_{fpga} = f_{asic} = 24$ MHz.

Although attackers have no idea about the secret key at this moment, one correct pair of plaintext and ciphertext needs to be kept for the subsequent analysis. One of the obtained results is summarized in Table 1.

Table 1. Experimental results of 128-bit AES encryption using RSL-AES in the default setting; the clock frequency of f_{fpga} and f_{asic} are both 24 MHz

No.	Plaintext and Ciphertext values (in hex.)
plaintext	9d e6 2f 88 12 fa e8 11 61 1a f3 80 f7 77 fd ef
ciphertext	c9 d7 ed 00 e5 07 c9 37 21 bf b0 9b eb 6b 9d 21

4.1 Determination of Clock Frequency to Induce Fault

Then, we set the DCM such that $f_{asic} = 6 \times f_{fpga} = 144$ MHz and increased the clock frequency of $f_{clk\_in}(= f_{fpga})$ gradually from 10 MHz to 27 MHz. Here, we used a fixed plaintext and operated AES encryptions with the same secret key.

As shown in the Table 2, we observed a fixed value when providing f_{clk_in} from 10 MHz to 15 MHz. The observed fixed value is exactly the same as the

Table 2. Experimental results of 128-bit AES encryption using RSL-AES when increasing the input clock frequency from 10 MHz to 24 MHz under the condition of $f_{asic} = 6 \times f_{fpga}$. The same secret key and plaintext are used as the case of Table 1.

$f_{clk\_in} = f_{fpga}$	f_{asic}	Output results
10 MHz $\sim$ 15 MHz	60 MHz $\sim$ 90 MHz	a fixed value
15 MHz $\sim$ 27 MHz	90 MHz $\sim$ 162 MHz	different values for each encryption

ciphertext observed at the normal clock setting, which means that RSL-AES operated correctly. On the contrary, in the case for setting f_{clk_in} from 15 MHz to 27 MHz, the output data was different for each encryption. Here, we guessed that this case was in our expected faulty situation because the random behavior of the output data could be explained by our attack model in Eq. (6). Accordingly, we decided to use $f_{clk\_in} = f_{fpga} = 24$ MHz and $f_{asic} = 144$ MHz.

4.2 Encryption Results of RSL-AES in the Faulty Situation

For further analysis, we collected several encryption results with $f_{fpga} = 24$ MHz and $f_{asic} = 144$ MHz). As a result, we could obtain the faulty ciphertexts as shown in Table 3.

From Table 3, we noticed that 1-byte data in the first column (denoted as dout15) is the same as the data in the fifth column (dout11) for each encryption result. More precisely, we observed

$$\texttt{dout15} \oplus \texttt{dout11} = 0 \times 00$$

for all faulty encryption results. Motivated by this finding, we continued the same operation for dout15 and other douts. As a result, we could see

Table 3. Experimental results of 128-bit AES encryption in the faulty situation ($f_{fpga} = 24$ MHz, f_{asic} 144 MHz). The same secret key is used as the case of Table 1.

No.	Observed Output Value (in hex.)															
1	c5	5a	67	1d	c5	0b	9e	e0	f0	51	71	74	fe	34	01	d2
2	65	fa	c7	bd	65	ab	3e	40	50	f1	d1	d4	5e	94	a1	72
3	46	d9	e4	9e	46	88	1d	63	73	d2	f2	f7	7d	b7	82	51
4	24	bb	86	fc	24	ea	7f	01	11	b0	90	95	1f	d5	e0	33
5	8c	13	2e	54	8c	42	d7	a9	b9	18	38	3d	b7	7d	48	9b
6	fc	63	5e	24	fc	32	a7	d9	c9	68	48	4d	c7	0d	38	eb
7	2c	b3	8e	f4	2c	e2	77	09	19	b8	98	9d	17	dd	e8	3b
8	df	40	7d	07	df	11	84	fa	ea	4b	6b	6e	e4	2e	1b	c8
$\vdots$					$\vdots$											
dout	15	14	13	12	11	10	09	08	07	06	05	04	03	02	01	00

$$\mathtt{dout15} \oplus \mathtt{dout14} = 0 \times 9f,$$
$$\mathtt{dout15} \oplus \mathtt{dout13} = 0 \times a2,$$
$$\mathtt{dout15} \oplus \mathtt{dout12} = 0 \times d8,$$
$$\mathtt{dout15} \oplus \mathtt{dout11} = 0 \times 00, \tag{7}$$

$$\vdots$$

$$\mathtt{dout15} \oplus \mathtt{dout00} = 0 \times 17.$$

for all faulty encryption results. By considering our attack model expressed in Eq. (6), the results indicate that r_unmask has some relation between bytes, because we know that K10 is fixed for all encryptions.

In order to clarify the relation of r_unmask in bytes, we XORed two 128-bit output values listed in Table 3. The results are listed in Table 4. We found a very interesting property; each result has the same value in bytes.

4.3   Cryptanalysis for the Experimental Results

By applying our attack model, the results of XORing two faulty output values can be expressed using Eq. (6) as

$$\mathtt{dout\_no1} \oplus \mathtt{dout\_no2} = \mathtt{r\_unmask\_no1} \oplus \mathtt{r\_unmask\_no2}, \tag{8}$$

where dout_no1 and dout_no2 are 128-bit output values observed with two different encryptions in the faulty situation, and r_unmask_no1 and r_unmask_no2 are different 128-bit data used for unmasking in the encryptions. Although several possibilities can be considered for the concealed specification of r_unmask, we guess that the same 8-bit random data is used in r_unmask in bytes, *i.e.*

$$\mathtt{r\_unmask} = \mathtt{rnd[7:0]} \parallel \mathtt{rnd[7:0]} \parallel \cdots \mathtt{rnd[7:0]},$$

from the results summarized in Table 4 and the derived equation expressed in Eq. (8).

Table 4. XOR results of two output data listed in Table 3

XOR(No., No.)	XORed Results (in hex.)															
(1, 2)	a0	a0	a0	a0	a0	a0	a0	a0	a0	a0	a0	a0	a0	a0	a0	a0
(1, 3)	83	83	83	83	83	83	83	83	83	83	83	83	83	83	83	83
(1, 4)	e1	e1	e1	e1	e1	e1	e1	e1	e1	e1	e1	e1	e1	e1	e1	e1
(1, 5)	49	49	49	49	49	49	49	49	49	49	49	49	49	49	49	49
(1, 6)	39	39	39	39	39	39	39	39	39	39	39	39	39	39	39	39
(1, 7)	e9	e9	e9	e9	e9	e9	e9	e9	e9	e9	e9	e9	e9	e9	e9	e9
(1, 8)	1a	1a	1a	1a	1a	1a	1a	1a	1a	1a	1a	1a	1a	1a	1a	1a
$\vdots$							$\vdots$									
dout	15	14	13	12	11	10	09	08	07	06	05	04	03	02	01	00

In this way, we can guess that 10th round key K10 is masked with 8-bit (same) random bits. In our guess, we only need 2^8 different candidates of K10 for all possible values of rnd[7:0] because

$$K10 = \text{dout} \oplus (\text{rnd}[7:0] \parallel \text{rnd}[7:0] \parallel \cdots \parallel \text{rnd}[7:0]).$$

To confirm our guess, the corresponding 2^8 secret-key candidates are prepared and verified with the correct pair of the plaintext and the ciphertext as shown in Table 1. As a result, we could find one correct secret-key from the 2^8 candidates. The retrieved round key and secret key are listed in Table 5.

Table 5. Retrieved round key and secret key from the ASIC chip

No.	Output value (in hex.)
Round key (final round)	3d a2 9f e5 3d f3 66 18 08 a9 89 8c 06 cc f9 2a
Retrieved secret key	00 01 02 03 04 05 06 07 08 4e bb fa 4d 0f ef 53

4.4 Efficiency of our FA Attack

We needed hundreds of the AES encryptions to understand the fault effect and to retrieve the secret key. However, once we know the architecture of the RSL-AES, only two encryptions are needed; one is for collecting one correct pair of plaintext and ciphertext at the normal clock frequency and the other is for obtaining one output generated in the faulty situation using a fast clock. This fact implies that:

- we can obtain information related to the hardware architecture as well as the secret key by this FA attack, and
- the knowledge about the hardware architecture makes it easy to retrieve secret information from the device.

5   Conclusions and Future Work

We presented a successful fault analysis attack against an AES hardware implementation with a DPA countermeasure by increasing the clock frequency. In this paper, we evaluated an AES with the RSL technique embedded in a prototype ASIC on the SASEBO-R. For the purpose of suppressing the propagation of the glitches, RSL uses enable signals that have different timing delays. This fundamental mechanism of RSL was intended to prevent DPA attacks. However, we showed that the glitch suppression mechanism worked negatively for our clock-based FA attack that uses a faster clock. In other words, the DPA countermeasure resulted in lowering the ASIC's tolerance to FA attacks.

As a countermeasure for our introduced clock-based FA attacks, a detection circuit for illegal clock inputs will be effective to avoid a high-frequency clock input. In this case, a special ability, *e.g.* invasive attacks may be required for attackers in order to bypass the detector. However, once a fault is induced in some way and a part of hardware specification is known, attackers can take advantage of the knowledge for attacking the same type of hardware chips. That is, a major lesson learned from this research is that we need to check side effects carefully when implementing a new countermeasure. Our future work includes a further security evaluation for other cryptographic implementations using DPA countermeasures.

Acknowledgements

The authors would like to thank Akashi Satoh and Research Center for Information Security, National Institute of Advanced Industrial Science and Technology (AIST) for providing the SASEBO boards. We deeply appreciate the CT-RSA2009 reviewers for their helpful comments.

References

1. Kocher, P.: Timing attacks on implementations of Diffie-Hellman, RSA, DSS and other systems. In: Koblitz, N. (ed.) CRYPTO 1996. LNCS, vol. 1109, pp. 104–113. Springer, Heidelberg (1996)
2. Kocher, P., Jaffe, J., Jun, B.: Differential power analysis. In: Wiener, M. (ed.) CRYPTO 1999. LNCS, vol. 1666, pp. 388–397. Springer, Heidelberg (1999)
3. Brier, E., Clavier, C., Oliver, F.: Correlation power analysis with a leakage model. In: Joye, M., Quisquater, J.-J. (eds.) CHES 2004. LNCS, vol. 3156, pp. 16–29. Springer, Heidelberg (2004)
4. Shamir, A., Tromer, E.: Acoustic cryptanalysis on noisy people and noisy machines. Preliminary proof-of-concept presentation,
   http://www.wisdom.weizmann.ac.il/~tromer/acoustic/
5. Boneh, D., DeMillo, R.A., Lipton, R.J.: On the importance of checking cryptographic protocols for faults (extended abstract). In: Fumy, W. (ed.) EUROCRYPT 1997. LNCS, vol. 1233, pp. 37–51. Springer, Heidelberg (1997)
6. FIPS Pub. 197: Specification for the AES (November 2001),
   http://csrc.nist.gov/pub-lications/fips/fips197/fips-197.pdf
7. Coron, J.-S.: Resistance against differential power analysis for elliptic curve cryptosystems. In: Koç, Ç.K., Paar, C. (eds.) CHES 1999. LNCS, vol. 1717, pp. 292–302. Springer, Heidelberg (1999)
8. Miller, V.: Use of elliptic curves in cryptography. In: Williams, H.C. (ed.) CRYPTO 1985. LNCS, vol. 218, pp. 417–426. Springer, Heidelberg (1986)
9. Koblitz, N.: Elliptic curve cryptosystem. Math. Comp. 48, 203–209 (1987)
10. Tiri, K., Verbauwhede, I.: A logic level design methodology for a secure DPA resistant ASIC or FPGA implementation. In: Proceedings of Design, Automation and Test in Europe Conference (DATE 2004), pp. 246–251 (2004)
11. Trichina, E.: Combinational logic design for AES subbyte transformation on masked data. Technical report, Cryptology ePrint Archive: Report 2003/236 (2003)
12. Mangard, S., Popp, T., Gammel, B.M.: Side-channle leakage of masked cmos gates. In: Menezes, A. (ed.) CT-RSA 2005. LNCS, vol. 3376, pp. 351–365. Springer, Heidelberg (2005)

13. Popp, T., Mangard, S.: Masked dual-rail pre-charge logic: DPA-resistance without routing constraints. In: Rao, J.R., Sunar, B. (eds.) CHES 2005. LNCS, vol. 3659, pp. 172–186. Springer, Heidelberg (2005)
14. Popp, T., Kirschbaum, M., Zefferer, T., Mangard, S.: Evaluation of the masked logic style MDPL on a prototype chip. In: Paillier, P., Verbauwhede, I. (eds.) CHES 2007. LNCS, vol. 4727, pp. 81–94. Springer, Heidelberg (2007)
15. Tiri, K., Schaumont, P.: Changing the odds against masked logic. In: Biham, E., Youssef, A.M. (eds.) SAC 2006. LNCS, vol. 4356, pp. 134–146. Springer, Heidelberg (2007)
16. Schaumont, P., Tiri, K.: Masking and dual-rail logic don't add up. In: Paillier, P., Verbauwhede, I. (eds.) CHES 2007. LNCS, vol. 4727, pp. 95–106. Springer, Heidelberg (2007)
17. Gierlichs, B.: DPA-resistance without routing constraints? In: Paillier, P., Verbauwhede, I. (eds.) CHES 2007. LNCS, vol. 4727, pp. 107–120. Springer, Heidelberg (2007)
18. Suzuki, D., Saeki, M., Ichikawa, T.: Random switching logic: A new countermeasure against DPA and second-order DPA at the logic level. IEICE Transaction on Fundamentals E90-A(1), 160–169 (2007)
19. Suzuki, D., Saeki, M.: Satoh A. A design methodology for a DPA-resistant cryptographic LSI with RSL techniques (I). In: Symposium Record of Symposium on Cryptography and Information Security (SCIS 2008), 6 pages (2008)
20. Suzuki, D., Saeki, M., Ichikawa, T.: Random switching logic: A countermeasure against DPA based on transition probability. Technical report, Cryptology ePrint Archive: Report 2004/346 (2004)
21. Research Center for Information Security (RCIS). Side-channel attack standard evaluation board (SASEBO),
    http://www.rcis.aist.go.jp/special/SASEBO/index-en.html
22. Mangard, S., Pramstaller, N., Oswald, E.: Successfully attacking masked AES hardware implementations. In: Rao, J.R., Sunar, B. (eds.) CHES 2005. LNCS, vol. 3659, pp. 157–171. Springer, Heidelberg (2005)
23. Research Center for Information Security (RCIS). Side-channel attack standard evaluation board (SASEBO),
    http://www.rcis.aist.go.jp/special/SASEBO/CryptoLSI-en.html

Evaluation of the Detached Power Supply as Side-Channel Analysis Countermeasure for Passive UHF RFID Tags

Thomas Plos

Institute for Applied Information Processing and Communications (IAIK)
Graz University of Technology, Inffeldgasse 16a, 8010 Graz, Austria
Thomas.Plos@iaik.tugraz.at

Abstract. Radio-frequency identification (RFID) is an emerging technology that has found its way into many applications, even in security related areas. Integration of cryptographic algorithms into RFID tags is necessary and the implementation of them needs to be secure against side-channel analysis (SCA) attacks. RFID tags operating in the ultra-high frequency (UHF) range are susceptible to so-called parasitic-backscatter attacks, which can be applied from a distance. In this article, we evaluate the efficiency of the detached power-supply countermeasure by applying it to a smart card and performing differential power analysis (DPA) attacks. Consecutively, we discuss the suitability of this countermeasure for protecting passive UHF tags from parasitic-backscatter attacks. The results show that the non-ideal properties of the analog switches used by the detached power supply decrease the effectiveness of this countermeasure. Moreover, we have identified side-channel leakage at the I/O pin of the smart card as a considerable problem for the detached power-supply approach. We conclude that utilizing the detached power supply to protect passive UHF tags from parasitic-backscatter attacks is feasible, if the integration interval is sufficiently long and the analog switches have adequate properties. However, longer integration intervals also increase the power loss of the tag, resulting in reduced read ranges.

Keywords: Differential Power Analysis, Side-Channel Analysis, Detached Power Supply, Parasitic Backscatter, RFID, UHF.

1 Introduction

Over the last years, radio-frequency identification (RFID) technology has become increasingly important and is used in many applications like ticketing, car immobilizers, electronic passports, toll collection, and supply-chain management. The move of RFID technology from simple identification towards more sophisticated applications has increased its need for security. Integrating cryptographic security into RFID systems in an efficient and reliable manner is a challenging task and still an active field of research.

In an RFID system, a tag and a reader communicate contactlessly by means of a radio-frequency field. The tag is a small circuit that consists of a microchip

M. Fischlin (Ed.): CT-RSA 2009, LNCS 5473, pp. 444–458, 2009.

attached to an antenna and receives its data and possibly the clock signal from the field. Passive tags, which are much more prevalent, also obtain their power from the field. Active tags are supplied by an extra battery [6]. The microchip of passive tags contains an analog front-end and a digital circuit. Modulating and demodulating the data, generating the clock signal and extracting the power from the field is done by the analog front-end. Processing the data and computing the appropriate responses is handled by the digital circuit. The complexity of the digital circuit highly depends on the application and ranges from a simple state machine to a microcontroller with sensors and non-volatile memory.

The power consumption of passive tags must not exceed a specific limit, since they are supplied by the field and have to achieve certain read ranges. More-over, passive tags are cost sensitive and produced in high volume, limiting also the available chip area. These constraints make it fairly difficult to integrate strong security into passive tags. As an attacker will always try to break the weakest link, an overall-secure RFID system requires the tags to be secure as well. Although the research community has initially believed that integrating standardized cryptographic algorithms into passive tags is infeasible, numerous publications in the recent years have shown the contrary. The most prominent attempts to bring standardized cryptographic algorithms to passive tags are, for example, the integration of symmetric schemes like the Advanced Encryption Standard (AES) [5, 15], asymmetric schemes like Elliptic Curve Cryptography (ECC) [1, 24, 9], and coupon-based schemes like GPS [17, 10].

Standardized cryptographic algorithms are secure in a mathematical and cryptanalytical sense. However, when integrating the algorithms into real de-vices, they can be vulnerable to implementation attacks. An important imple-mentation attack is side-channel analysis. There, an attacker measures physi-cal properties of a device during the execution of the cryptographic algorithm and tries to reveal secrets stored on the device. A powerful side-channel anal-ysis technique is differential power analysis (DPA) introduced by Kocher *et al.* in 1998 [14]. This technique exploits the fact that the power consumption of CMOS devices is dependent on the processed data and operations. DPA attacks use a simplified model for the power consumption, predict intermediate values computed by the cryptographic algorithm, and combine the predicted results with the measured power consumption by means of statistical methods. Since this technique can be applied to a large number of measurements, even a small data-dependent leakage in the power consumption is sufficient for a successful DPA attack. When using the electromagnetic (EM) emissions of a cryptographic device instead of its power consumption, as shown by Gandolfi *et al.* [7], DPA attacks are named differential electromagnetic analysis (DEMA) attacks.

Passive tags are also susceptible to side-channel analysis. Hutter *et al.* [11] have performed DPA and DEMA attacks on passive tags in the high frequency (HF) range. Oren and Shamir [20] and Plos [21] have conducted attacks on the EM emissions of passive tags in the ultra-high frequency (UHF) range. For UHF tags, it has turned out that the power reflected by the tag's antenna contains data-dependent information. Oren and Shamir have named this effect *parasitic backscatter*, which allows to perform successful DEMA attacks from a distance of one meter and more. Hence, such parasitic-backscatter attacks pose a serious threat for UHF tags.

Integrating countermeasures into cryptographic devices makes side-channel analysis techniques like DPA and DEMA attacks less effective. Principally, countermeasures can be divided into hiding and masking [16]. The goal of hiding is to decouple the power consumption of a cryptographic device from its internally processed data values. Masking breaks the link between the intermediate values and the values actually computed by the device. Unfortunately, integrating countermeasures usually also increases the power consumption and the design complexity of a device.

A countermeasure recently proposed by Shamir [22, 23] for protecting UHF tags from parasitic-backscatter attacks is the detached power supply. It is a hardware countermeasure based on hiding at the architectural level. The decoupling of the power consumption is accomplished by using two capacitors. At any time, one capacitor powers the digital circuit of the tag and the other capacitor is charged by the tag's analog front-end. By periodically switching the capacitors, energy is transferred from the analog front-end to the digital circuit without a direct physical connection. The simplicity and the fact that its application requires no extensive modification of existing chip designs makes it interesting for RFID tags. So far, no information is available about the efficiency of this countermeasure. Although there exists a practical implementation of a similar approach by Corsonello *et al.* [2] using a three-phase charge pump, results illustrating how well this countermeasure prevents side-channel analysis are missing.

In this work, we evaluate the efficiency of the detached power supply and discuss its suitability for protecting passive UHF tags from parasitic-backscatter attacks. It is the first article that presents practical results of the detached power-supply countermeasure with respect to side-channel analysis. We have implemented the detached power supply with discrete components and applied it to an unprotected smart card with an integrated AES. Using a smart card has been necessary since commercially-available passive UHF tags do not yet contain standardized cryptographic algorithms. By performing DPA attacks, we have analyzed a basic version of the detached power supply and an enhanced version which uses an additional discharge phase. The results have shown that even the enhanced version of the detached power supply is still vulnerable to power analysis because of the non-ideal properties of the analog switches. Moreover, it has turned out that there is a strong side-channel leakage at the I/O pin of the smart card, regardless of whether the detached power has been used or not. In order to address this problem, we suggest a simple countermeasure that prevents the side-channel leakage at output pins. Finally, an estimation is provided about the required capacitor size and the power-consumption overhead caused by integrating the detached power supply into a passive UHF tag.

Our results show that utilizing the detached power supply to protect passive UHF tags from parasitic-backscatter attacks is feasible, if the integration interval is sufficiently long and the analog switches have adequate properties. However, longer integration intervals also increase the additional power loss of the tags, resulting in reduced read ranges.

This article is organized as follows. Section 2 describes the principle of the detached power supply. Section 3 illustrates the practical implementation of the countermeasure and the measurement setup. The results of the side-channel analysis are presented in Section 4. A suggestion for a simple countermeasure to prevent side-channel leakage at output pins is shown in Section 5. In Section 6, the costs of integrating the detached power supply into passive UHF tags is discussed. Conclusions are drawn in Section 7.

2   Description of the Detached Power Supply

Initially, the detached power supply has been intended to protect smart cards from power analysis [22]. However, the countermeasure is much more suitable for preventing parasitic-backscatter attacks on UHF tags [23]. First, the detached power supply does not protect from side-channel analysis that uses the direct emissions of the device. Second, manipulating the capacitors (*e.g.* interconnecting the pins of the capacitors) makes the countermeasure completely useless. In case of the parasitic-backscatter attack, where a passive attacker remotely measures the power reflected from a tag, both arguments are of no importance. Moreover, the power consumption of passive tags is much lower compared to smart cards. The lower power consumption allows capacitors to be smaller and they can directly be integrated into the chip of the tag.

In the following, the principle of the detached power supply is explained in more detail. After the description of the basic version, an enhanced version with an additional discharge phase is presented.

2.1   Basic Version of the Detached Power Supply

The basic version of the detached power-supply [23] comprises: two capacitors, four switches, and two diodes. Figure 1(a) presents a schematic diagram and a time lapse of the switches' states. Shaded areas in the sequence diagram indicate the intervals in which a certain switch is closed. A complete cycle consists of four phases. In the first phase the switches S_1 and S_4 are closed and S_2 and S_3 are opened. This causes C_1 to be charged by the analog front-end, while C_2, which has been charged in a previous phase, powers the tag's digital circuit. During the second phase, which is rather short in time, S_1 is opened and S_2 is closed. Now, the digital circuit is powered by both capacitors. The diodes D_1 and D_2 prevent charge equalization between the two capacitors. In the next phase, S_4 is opened and S_3 is closed. The digital circuit is powered by C_1 and C_2 is charged by the analog front-end, the opposite way around than in the first phase. In the fourth and last phase, which is again a short phase where both capacitors power the digital circuit, S_3 is opened and S_4 is closed. After the fourth phase, the cycle is completed by opening S_2 and closing S_1 and starts from the beginning. The toggling between C_1 and C_2 is done at the switching frequency f_S.

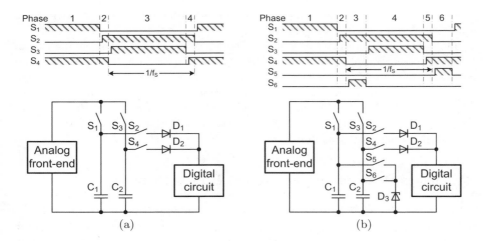

Fig. 1. Sequence diagrams comprising the states of the switches during the particular phases and a schematic overview of the circuits used for the basic version (a) and the enhanced version (b) of the detached power supply

Switching between particular phases can be assigned to a fixed number of clock cycles or triggered by the voltage of the discharging capacitor when reaching a certain threshold. Regardless of the switching strategy, an attacker can still get information about the total amount of charge consumed by the digital circuit during a discharge phase. This information can be minimized by selecting larger capacitors allowing to make the discharge phase longer.

2.2   Enhanced Version of the Detached Power Supply

In order to remove the remaining information leakage present in the basic version of the detached power supply, it is suggested in [22] to discharge each capacitor to a fixed voltage level before reconnecting it to the analog front-end. This enhancement requires two additional switches and a voltage-limiting element for properly discharging the capacitors. An example for a simple voltage-limiting element is a Zener diode. Figure 1(b) shows the schematic overview of the circuit and a sequence diagram comprising the states of the switches during the particular phases. In contrast to the basic version, the enhanced version needs six instead of four phases for a complete cycle. During each of the two additional phases, one capacitor is switched in parallel to the Zener diode D_3 to get further discharged to a predefined voltage level. All other phases are the same as in the basic version. In the third phase C_2 is connected to the Zener diode via S_6, in the sixth phase C_1 via S_5. As a result, the charge stored in the concerning capacitor is no longer related to the charge consumed by the digital circuit while it has been supplied by the capacitor. Theoretically, this makes power-analysis attacks completely useless.

3   Implementation of the Detached Power Supply and the Measurement Setup

Evaluating the effectiveness of the detached power supply requires its implementation and furthermore application to a physical device. Since passive UHF tags that are commercially available do not yet have integrated standardized cryptographic algorithms, we have decided to use an unprotected smart card instead. This allows to draw conclusions for passive UHF tags, because the countermeasure operates independently of the circuit it protects. Both the simple and the enhanced version of the detached power supply have been implemented as an analog circuit using discrete components. The switches of each circuit are handled by a microcontroller. For better flexibility, the microcontroller is connected to a PC via a serial interface to adjust certain parameters like the switching frequency or activation/deactivation of the detached power supply. Figure 2 illustrates how the detached power supply and the smart card are combined to protect against power analysis. The detached power supply is integrated into the supply line of the smart card. All other pins like clock, reset, and I/O are left untouched.

The utilized smart card runs at a clock frequency of 3.57 MHz and consumes about 5 mA at 5 V. A software version of the AES using a key length of 128 bits is implemented on the smart card. The AES is a block cipher operating on 128-bit blocks of input data. Encrypting a single block of data takes the smart card less than 3 800 clock cycles. A detailed description of the AES can be found in [19].

Crucial components of the detached power supply are the analog switches which should ideally have: high switching speed, low cross talk, high off isolation, and low on resistance. After testing several switches, it has pointed out that USB high-speed multiplexers are a good solution since they have excellent electrical properties. The multiplexers can operate up to some hundreds of MHz, have an off isolation of about 100 dB at 100 kHz, and have a maximum on resistance of 10 Ω.

Standard ceramic types with a value of 0.1 µF have been selected for the capacitors. Taking into account the smart card's average power consumption of about 5 mA and allowing the capacitors to be discharged from 5 to 3.5 V, results in a minimum switching frequency of about 36 kHz. This means that a single 0.1 µF capacitor can power the smart card over a time of approximately 100 clock cycles.

Further components of the circuit are low voltage-drop Shottky diodes that prevent charge equalization between the two capacitors while they are temporarily switched in parallel to the smart card. A Zener diode has been chosen as voltage-limiting element for the enhanced version of the detached power supply. The breakdown voltage of the Zener diode depends on the deployed switching frequency of the detached power supply and needs to be below the minimum voltage reached by the capacitor after being discharged by powering the smart card.

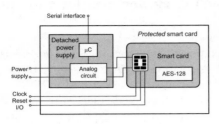

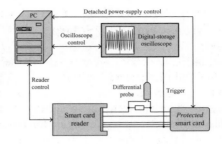

Fig. 2. Overview of the smart card protected with the detached power-supply countermeasure

Fig. 3. Measurement setup used to determine the power consumption of the protected smart card

Besides implementing the detached power supply and applying it to a suitable cryptographic device, building a measurement setup is necessary for the evaluation process. Figure 3 shows a measurement setup that allows automated acquisition of the power traces. Main components of the measurement setup are: the protected smart card, a smart-card reader, a PC, a differential probe, and a digital-storage oscilloscope. The smart-card reader, the digital-storage oscilloscope, and the microcontroller that is responsible for managing the switches of the detached power supply are controlled by a MATLAB script running on the PC. A photo of the actual measurement setup containing the protected smart card, the smart-card reader, and the differential probe is presented in Figure 4.

The measurement cycle for retrieving a single power trace always follows the same scheme. After receiving an appropriate command from the smart-card reader, the protected smart card starts encrypting the incoming data block. When the encryption begins, the protected smart card releases a trigger event that causes the digital-storage oscilloscope to record a power trace. Determining the power consumption is achieved by measuring the voltage drop across a $1\,\Omega$ resistor in the supply line of the protected smart card with a differential probe. The measurement cycle is finished by transferring the power trace to the PC, where further analysis work is conducted.

4   Results of the Side-Channel Analysis

This section presents the results obtained by analyzing the power consumption of the smart card equipped with the detached power supply described in Section 3. DPA attacks have been carried out using the correlation coefficient ρ to detect linear dependencies between the measured power consumption and the formed hypotheses. As power model, the Hamming-weight model has been selected. In that way, attacking the first key byte of the AES implemented on the smart card without using any countermeasures, has led to a correlation coefficient of 0.55 for the correct hypothesis. Based on a given ρ, the rule of thumb stated in [16] allows to determine the number of power traces n that is needed for a successful attack with high probability ($> 99.99\%$):

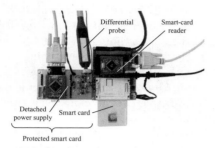

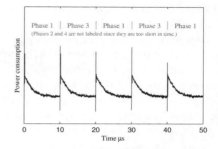

Fig. 4. Photo of the actual measurement setup containing the protected smart card, a smart-card reader, and a differential probe

Fig. 5. Power trace of the protected smart card with the basic version of the detached power supply using a switching frequency of 100 kHz

$$n = 3 + 8 \frac{3.719^2}{ln^2(\frac{1+\rho}{1-\rho})} \tag{1}$$

Substituting 0.55 for ρ in (1) computes to about 75. Hence, 75 power traces are required on average for a successful DPA attack on the smart card. This value is later used as a reference to better quantify the impact on the number of needed power traces when the countermeasure is activated. Consecutively, the results of the DPA attacks applied to the smart card with activated detached power supply are presented. Both basic and enhanced version of the countermeasure are examined and compared with each other.

4.1   Results of the Basic Version of the Detached Power Supply

After analyzing the side-channel leakage of the smart card itself, measurements with the basic version of the detached power supply as countermeasure have been conducted. Various switching frequencies starting from 640 kHz to 36 kHz have been examined. It has shown that analyzing power traces resulting from measurements with activated detached power supply is costlier since they require additional pre-processing steps. An example of such a power trace is given in Figure 5. The traces are afflicted with a variable offset, making it necessary to align them vertically. Moreover, in our implementation the switching of the capacitors is controlled by an extra microcontroller and occurs independently from the operation of the smart card. This makes standard DPA attacks based on power traces highly inefficient. However, transforming the power traces from the time domain into the frequency domain as suggested by Gebotys [8], has solved this problem as well.

When applying the pre-processing techniques described above, power-analysis attacks have been successful if the basic version of the detached power-supply countermeasure has been activated. As expected in theory, decreasing the switching frequency has lowered the side-channel leakage. Using a switching frequency of 100 kHz, which equals to an integration over about 36 clock cycles of the smart card's power consumption, has led to a maximum ρ of 0.1 for the correct hypothesis. Selecting 36 kHz, which is the lowest possible switching frequency for our setup equaling to an integration over 100 clock cycles, has resulted in a maximum ρ of 0.04. According to (1), approximately 2 800 measurements are necessary on average for an attack when integrating over 36 clock cycles and more than 17 200 measurements if integrating over 100 clock cycles.

A side-channel leakage that is not decreased by lowering the switching frequency could be the indicator for inadequate analog switches. We have observed for example that analog switches with an insufficiently large isolation resistance at higher frequencies make the detached power-supply countermeasure useless. Data dependencies modulated on harmonics of the smart card's clock frequency pass the analog switches more or less unhampered. Hence, the choice of suitable switches is essential.

4.2   Results of the Enhanced Version of the Detached Power Supply

Evaluating the efficiency of the enhanced version of the detached power supply is of much more interest. As a vulnerability to power analysis of the basic version is already stated in theory, it is not so for the enhanced version. The microcontroller that is responsible for activating the analog switches of the detached power supply has been configured such that an additional discharge phase of 10 clock cycles is introduced. During this phase, one of the two capacitors is further discharged to a fixed voltage level before reconnecting to the power supply to get rid of the remaining side-channel leakage.

Two switching frequencies have been examined, 100 kHz and 36 kHz. By applying the same pre-processing and analysis techniques as described in Section 4.1, a maximum correlation coefficient ρ of 0.022 has been obtained for a switching frequency of 100 kHz. According to (1), this corresponds to 57 100 measurements needed for a successful attack with high probability. Figure 6 shows the correlation coefficient as function of the number of measurements. The correct hypothesis is printed in black, incorrect hypotheses are printed in gray. Compared to the basic version of the detached power supply using the same switching frequency, the maximum correlation coefficient is reduced from 0.1 to 0.022 and the number of needed power traces is increased from 2 800 to 57 100. A certain vulnerability to side-channel analysis is still present. However, when decreasing the switching frequency from 100 kHz to 36 kHz, the maximum-achievable correlation coefficient is lowered as well. We have limited the number of measurements per experiment to 100 000 and therefore have not been able to perform a successful attack on the smart card when using a switching frequency of 36 kHz.

The remaining side-channel leakage is explained with the non-ideal properties of the analog switches. In contrast to ideal switches whose resistance is assumed to be zero when activated, our analog switches have a maximum on resistance R_{ON} of about 10 Ω. This resistance acts in series to the capacitor during the

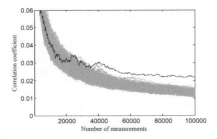

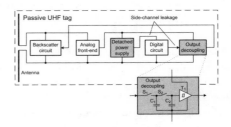

Fig. 6. Correlation coefficient as a function of the number of measurements for the enhanced version of the detached power supply using a switching frequency of 100 kHz

Fig. 7. Schematic of the decoupling principle for the output pin and its combination with the detached power supply to protect passive UHF tags

discharge phase. Consequently, the discharge speed of the capacitor is limited by the time constant τ which is the product of R_{ON} and the value of the capacitor. The time constant indicates the time required to discharge a capacitor to about 37 % of its initial charge. A shorter time constant is achieved by selecting more appropriate switches with lower R_{ON} or by using smaller capacitors.

We have verified the influence of R_{ON} by performing simulations with a computer program that is intended for analyzing the behavior of electronic circuits. In the simulations, two identical capacitors, one charged to 4.0 V and the other to 3.9 V, are discharged over a 10 Ω resistor to a constant voltage of 3.3 V. The discharge time is selected to be equal to an interval of 10 clock cycles of our smart card. After discharging the capacitors, one capacitor has a remaining voltage of 3.352 V and the other of 3.344 V. Hence, the initial voltage difference of 100 mV has not completely disappeared, but has been reduced to 8 mV. This clarifies that even under ideal conditions as they can be found in the simulation, a non-zero R_{ON} has a noticeable effect.

Besides the influence of the R_{ON} of real switches, there is another aspect that needs to be considered when using the detached power supply to protect a cryptographic device against power analysis. As already mentioned in [22], data-dependent information can not only leak through the power line of a device but also through its I/O pins. We have measured the voltage variations at the I/O pin of our smart card while encrypting data to evaluate whether such attacks pose a threat or not. The deployed smart card operates conform to the ISO 7816 standard [12] and uses one pin for serially receiving and transmitting data in a half-duplex mode. During the processing of data, the smart card keeps its I/O pin in high-impedance mode (tri-state). We have detected rather strong data-dependent leakage at the smart card's I/O pin, regardless whether the detached power supply has been activated or not. A maximum correlation coefficient ρ of about 0.15 has been obtained by only measuring the voltage variations at the I/O pin and by consecutively applying a DPA attack.

5   A Suggestion for Preventing Side-Channel Leakage at Output Pins

As the results of the previous section illustrate, decorrelating the power consumption of a device alone is not enough. Due to coupling effects on the chip, data-dependent information can also leak by the device's I/O pins. An example for such coupling effects is crosstalk, where the switching activity of one wire influences the signal of neighboring wires [13]. There exist various techniques for minimizing crosstalk like reducing the wire lengths, and increasing the distance between critical signal wires or introducing additional ground planes between them. This measures need to be applied during the chip-design phase and probably increase the design complexity and the resulting chip size. In addition, with shrinking CMOS technology the impact of the crosstalk effect is increased [25].

The digital circuit of UHF tags requires at least one output pin which controls a so-called backscatter circuit that is used to transmit data to the reader. This backscatter circuit is connected in parallel to the antenna and varies the power consumption of the tag which in turn influences the power reflected by the antenna. In contrast to the parasitic backscatter, this is an intended effect. Consequently, side-channel leakage coupled into the output pin can propagate through the backscatter circuit to the tag's antenna. As proposed in [22], the most obvious solution to avoid the side-channel leakage is to temporarily disconnect or to ground the output pin during the execution of cryptographic operations. However, this is not possible if the tag needs to transmit data and to execute cryptographic operations in parallel.

A decoupling principle similar to the detached power-supply approach can be applied to address the side-channel leakage at the digital circuit's output pin. By using the output-decoupling principle shown in Figure 7, data present at the output pin of the digital circuit is transmitted to the backscatter circuit without direct physical connection at any time. Figure 7 also illustrates how to combine the detached power supply and the decoupling principle to protect passive UHF tags. The components required for the decoupling of the output are the switches S_1 and S_2, the capacitors C_1 and C_2, and a Schmitt trigger T_1. In a first phase, S_2 is opened and S_1 is closed. This causes C_1 to be charged to the voltage at the output pin of the digital circuit. The voltage across C_2 is buffered by the Schmitt trigger and converted to a distinct digital voltage level at its output. In a second phase, which is rather short in time, S_1 is opened and S_2 is closed. Since the value of C_2 is much smaller than that of C_1, the resulting voltage across the parallel connection of the two capacitors is approximately equal to the initial voltage across C_1. After the second phase, the cycle is completed and continues from the beginning. The duration of a cycle defines the maximum-achievable data rate and needs to be selected properly to meet the requirements of the application.

In terms of additional hardware costs, the decoupling of the output pin is rather cheap. The size of the capacitors and the switches can be kept small because of the low currents, and the Schmitt trigger only requires a handful of additional transistors. Another advantage of this approach is its simplicity. No extensive redesign of the tag chip is required to address potential coupling effects at the output pin of the digital circuit.

6   Discussing the Costs of Integrating the Detached Power Supply into Passive UHF Tags

Integrating the detached power supply into passive UHF tags introduces additional costs in terms of chip size and power consumption that need to be considered. The low power consumption of passive UHF tags, which is typically in the range of some microamperes, allows to deploy small capacitors for the detached power supply. For a rough estimation of the required capacitor size, we have taken the experimental results from [15], describing the digital circuit of a passive UHF tag with an integrated AES module. As stated by the authors, their implementation on a 0.18 μm CMOS process has a power consumption of about 2.6 μA at a supply voltage of 1.8 V. The clock frequency of the AES module is around 420 kHz. When integrating over 100 clock cycles and allowing the capacitors to be discharged to a voltage of 1 V, each of the two capacitors for the detached power supply needs to have a value of at least 0.78 nF. Although capacitors in the size of 1 nF are not unusual for current UHF tags [4], integrating another two capacitors of that size directly into the chip of a tag could be too costly. On-chip capacitors have the advantage that it is more difficult for an attacker to manipulate them. However, in order to protect the tags from parasitic backscatter attacks, which are conducted remotely and without modification of the tag, cheaper external capacitors can be deployed as well.

Another important aspect for passive UHF tags is the increased power consumption caused by the detached power supply. Real switches have an on resistance that is different from zero and charging a capacitor always results in extra thermal power loss. When using the enhanced version of the detached power supply, the power loss is further increased due to the additional discharge phase of the capacitors. Generally, the power loss of a switched-capacitor circuit is proportional to its output current and inversely proportional to the switching frequency and the size of the capacitors [18]. Consequently, the power loss introduced by the detached power supply is reduced when selecting larger capacitors, increasing the switching frequency, and using a device with low power consumption. The requirement to utilize high switching frequencies conflicts with the results in Section 4, which illustrate that the switching frequency needs to be lowered to obtain a better resistance against power analysis. For the example of the passive UHF tag with the integrated AES module mentioned before, we have calculated a power loss of approximately 22 % when integrating over 100 clock cycles. There is a square-root relation between the power consumption of the tag and the maximum read range [3]. Doubling the power consumption of the tag results in a read range that is decreased by a factor of $\sqrt{2}$. Hence, a power loss of 22 % reduces the read range by a factor of 1.13.

7 Conclusion

In this work, we have evaluated the effectiveness of the detached power-supply countermeasure to prevent power analysis and discussed its suitability for protecting passive UHF tags from parasitic-backscatter attacks. It is the first article that presents concrete results about the efficiency of this countermeasure with respect to side-channel analysis. DPA attacks have been applied to a smart card protected with a basic version of the detached power supply and an enhanced version that uses an additional discharge phase. Due to the non-ideal properties of the deployed analog switches, even the enhanced version shows a susceptibility to power analysis. Moreover, we have depicted that the side-channel leakage of I/O pins poses a serious problem when utilizing the detached power supply. In order to address this issue, a simple decoupling principle for output pins has been presented. Additionally, we provided an estimation concerning the required capacitor size and the power-consumption overhead introduced by integrating the detached power supply into a passive UHF tag.

We conclude that the detached power supply significantly reduces the side-channel leakage in the power consumption of a cryptographic device, if the integration interval is sufficiently long and the utilized analog switches have adequate properties. Using this countermeasure to protect passive UHF tags from parasitic backscatter attacks is feasible. However, longer integration intervals also increase the power loss caused by the detached power supply. A higher power loss results in reduced read ranges of the tags. Combining the detached power supply with other countermeasures, for example on algorithmic level, is indispensable if more sophisticated attacks need to be prevented as well. Such attacks involve manipulating the capacitors and measuring the direct emissions close to the tag chip.

Acknowledgements

This work has been funded by the European Commission through Contract Number IST-FP6-033546 (Project BRIDGE) under the Sixth Framework Programme and by the Austrian Science Fund (FWF) Project Number P18321.

References

[1] Batina, L., Guajardo, J., Kerins, T., Mentens, N., Tuyls, P., Verbauwhede, I.: Public-Key Cryptography for RFID-Tags. In: Workshop on RFID Security 2006 (RFIDSec 2006), Graz, Austria, July 12-14 (2006)

[2] Corsonello, P., Perri, S., Margala, M.: A New Charge-Pump Based Countermeasure Against Differential Power Analysis. In: Proceedings of the 6th International Conference on ASIC (ASICON 2005), vol. 1, pp. 66–69. IEEE, Los Alamitos (2005)

[3] Curty, J.-P., Declercq, M., Dehollain, C., Joehl, N.: Design and Optimization of Passive UHF RFID Systems. Springer, Heidelberg (2007)

[4] Facen, A., Boni, A.: Power supply generation in cmos passive uhf rfid tags. Research in Microelectronics and Electronics 2006, Ph. D., pp. 33–36 (June 2006)

[5] Feldhofer, M., Dominikus, S., Wolkerstorfer, J.: Strong Authentication for RFID Systems using the AES Algorithm. In: Joye, M., Quisquater, J.-J. (eds.) CHES 2004. LNCS, vol. 3156, pp. 357–370. Springer, Heidelberg (2004)

[6] Finkenzeller, K.: RFID-Handbook, 2nd edn. Carl Hanser Verlag (2003) ISBN 0-470-84402-7

[7] Gandolfi, K., Mourtel, C., Olivier, F.: Electromagnetic Analysis: Concrete Results. In: Koç, Ç.K., Naccache, D., Paar, C. (eds.) CHES 2001. LNCS, vol. 2162, pp. 251–261. Springer, Heidelberg (2001)

[8] Gebotys, C.H., Ho, S., Tiu, C.C.: EM Analysis of Rijndael and ECC on a Wireless Java-Based PDA. In: Rao, J.R., Sunar, B. (eds.) CHES 2005. LNCS, vol. 3659, pp. 250–264. Springer, Heidelberg (2005)

[9] Hein, D., Wolkerstorfer, J., Felber, N.: ECC is Ready for RFID - A Proof in Silicon. In: Selected Areas in Cryptography, 15th International Workshop, SAC 2008, Sackville, Canada, August 14-15 (2008); revised Selected Papers. LNCS (September 2008)

[10] Hofferek, G., Wolkerstorfer, J.: Coupon Recalculation for the GPS Authentication Scheme. In: Grimaud, G., Standaert, F.-X. (eds.) CARDIS 2008. LNCS, vol. 5189, pp. 162–175. Springer, Heidelberg (2008)

[11] Hutter, M., Mangard, S., Feldhofer, M.: Power and EM Attacks on Passive 13.56 MHz RFID Devices. In: Paillier, P., Verbauwhede, I. (eds.) CHES 2007. LNCS, vol. 4727, pp. 320–333. Springer, Heidelberg (2007)

[12] International Organisation for Standardization (ISO). ISO/IEC 7816: Identification cards - Integrated circuit(s) cards with contacts (1989)

[13] Kirkpatrick, D.A., Sangiovanni-Vincentelli, A.L.: Techniques For Crosstalk Avoidance In The Physical Design Of High-performance Digital Systems. In: IEEE/ACM International Conference on Computer-Aided Design, 1994E, pp. 616–619 (November 1994)

[14] Kocher, P.C., Jaffe, J., Jun, B.: Differential Power Analysis. In: Wiener, M. (ed.) CRYPTO 1999. LNCS, vol. 1666, pp. 388–397. Springer, Heidelberg (1999)

[15] Man, A.S., Zhang, E.S., Lau, V.K., Tsui, C., Luong, H.C.: Low Power VLSI Design for a RFID Passive Tag baseband System Enhanced with an AES Cryptography Engine. In: Proceedings of 1st Annual RFID Eurasia, Istanbul, Turkey, September 5-6, 2007, pp. 1–6. IEEE, Los Alamitos (2007)

[16] Mangard, S., Oswald, E., Popp, T.: Power Analysis Attacks – Revealing the Secrets of Smart Cards. Springer, Heidelberg (2007)

[17] McLoone, M., Robshaw, M.J.B.: Public Key Cryptography and RFID Tags. In: Abe, M. (ed.) CT-RSA 2007. LNCS, vol. 4377, pp. 372–384. Springer, Heidelberg (2006)

[18] Midya, P.: Efficiency analysis of switched capacitor doubler. In: IEEE 39th Midwest symposium on Circuits and Systems, 1996, vol. 3, pp. 1019–1022 (August 1996)

[19] National Institute of Standards and Technology (NIST). FIPS-197: Advanced Encryption Standard (November 2001), http://www.itl.nist.gov/fipspubs/

[20] Oren, Y., Shamir, A.: Remote Password Extraction from RFID Tags. IEEE Transactions on Computers 56(9), 1292–1296 (2007)

[21] Plos, T.: Susceptibility of UHF RFID Tags to Electromagnetic Analysis. In: Malkin, T.G. (ed.) CT-RSA 2008. LNCS, vol. 4964, pp. 288–300. Springer, Heidelberg (2008)

[22] Shamir, A.: Protecting Smart Cards from Passive Power Analysis with Detached Power Supplies. In: Paar, C., Koç, Ç.K. (eds.) CHES 2000. LNCS, vol. 1965, pp. 71–77. Springer, Heidelberg (2000)
[23] Shamir, A.: Method and Apparatus for Protecting RFID Tags from Power Analysis. Patent Number WO 2008/019246 A2 (February 2008), http://www.freepatentsonline.com/
[24] Tuyls, P., Batina, L.: RFID-Tags for Anti-counterfeiting. In: Pointcheval, D. (ed.) CT-RSA 2006. LNCS, vol. 3860, pp. 115–131. Springer, Heidelberg (2006)
[25] Yang, Z., Mourad, S.: Deep Submicron On-chip Crosstalk. In: Proceedings of the 16th IEEE Instrumentation and Measurement Technology Conference, 1999. IMTC 1999, vol. 3, pp. 1788–1793 (1999)

Securing RSA against Fault Analysis by Double Addition Chain Exponentiation

Matthieu Rivain

Oberthur Technologies & University of Luxembourg
m.rivain@oberthur.com

Abstract. Fault Analysis is a powerful cryptanalytic technique that enables to break cryptographic implementations embedded in portable devices more efficiently than any other technique. For an RSA implemented with the Chinese Remainder Theorem method, one faulty execution suffices to factorize the public modulus and fully recover the private key. It is therefore mandatory to protect embedded implementations of RSA against fault analysis.

This paper provides a new countermeasure against fault analysis for exponentiation and RSA. It consists in a *self-secure* exponentiation algorithm, namely an exponentiation algorithm that provides a direct way to check the result coherence. An RSA implemented with our solution hence avoids the use of an extended modulus (which slows down the computation) as in several other countermeasures. Moreover, our exponentiation algorithm involves 1.65 multiplications per bit of the exponent which is significantly less than the 2 required by other self-secure exponentiations.

1 Introduction

The *physical cryptanalysis* gathers different cryptanalytic techniques taking advantage of the physical properties of cryptographic implementations. Among these, one mainly identifies *side channel analysis* [27,26] that physically observes cryptographic computations and *fault analysis* [8,6] that physically disturbs them. The latter consists in exploiting the faulty outputs resulting from erroneous computations in order to retrieve information on the secret key. Fault analysis has been introduced first against RSA and other public key schemes [8] and then against DES [6]. Several works followed that improved fault analysis and generalized it to other algorithms.

A straightforward way to protect any algorithm against fault analysis is by performing twice the computation and by checking that the same result is obtained. In case of inconsistency, an error message is returned thus preventing the exposure of the faulty result. A variant consists in verifying an encryption by a decryption (or *vice versa*). These countermeasures are suitable for fast algorithms such as block ciphers, but when a public key cryptosystem such as RSA must be implemented, a doubling of the execution time becomes prohibitive. That is why, securing RSA against fault analysis constitutes a challenging issue

M. Fischlin (Ed.): CT-RSA 2009, LNCS 5473, pp. 459–480, 2009.

of embedded cryptography. Several methods have been proposed so far but the number of secure and practical solutions is still quite restricted.

In this paper, we provide a new countermeasure against fault analysis for exponentiation and RSA that constitutes an efficient alternative to the existing solutions. First we introduce preliminaries about RSA, fault analysis and the existing countermeasures (Sect. 2). Then we describe our self-secure exponentiation algorithm (Sect. 3) and the resulting secure RSA-CRT algorithm (Sect. 4). Afterward we analyze the security of our solution (Sect. 5) and we address its resistance *vs* side channel analysis (Sect. 6). Finally, we give an analysis of the time and memory complexities of our solution and we compare them to previous solutions in the literature (Sect. 7).

2   RSA and Fault Analysis

2.1   The RSA Cryptosystem

RSA is nowadays the most widely used public key cryptosystem [33]. An RSA public key is composed of a public modulus N which is the product of two large secret primes p and q and of a public exponent e which is co-prime with the Euler's totient of N namely $\varphi(N) = (p-1) \cdot (q-1)$. The corresponding RSA private key is composed of the public modulus N and the secret exponent d that is defined as the inverse of e modulo $\varphi(N)$.

An RSA signature (or deciphering) s of a message $m < N$ is obtained by computing: $s = m^d \bmod N$. The signature verification (or message ciphering) is the inverse operation that can be performed publicly since, according to Euler's Theorem, we have: $m = s^e \bmod N$.

For efficient implementation of RSA, one makes often use of the Chinese Remainder Theorem (CRT). This theorem implies that $m^d \bmod N$ can be computed from $m^d \bmod p$ and $m^d \bmod q$. The RSA-CRT hence consists in performing the two following exponentiations: $s_p = m^{d_p} \bmod p$ and $s_q = m^{d_q} \bmod q$, where $d_p = d \bmod (p-1)$ and $d_q = d \bmod (q-1)$. By Fermat's little Theorem, we have $s_p = m^d \bmod p$ and $s_q = m^d \bmod q$. Therefore, once s_p and s_q have been computed, s can be recovered from s_p and s_q by applying a so-called *recombination step*: $s = \mathsf{CRT}(s_p, s_q)$. Two methods exist for CRT recombination: the one from Gauss and the one from Garner. The less memory consuming is the Garner's recombination that is defined as $\mathsf{CRT}(s_p, s_q) = s_q + q \cdot \big(i_q \cdot (s_p - s_q) \bmod p \big)$, where $i_q = q^{-1} \bmod p$. The whole RSA-CRT is around 4 times faster than the straightforward RSA which makes its use very common, especially in the context of low resource devices were computation time is often critical.

2.2   Fault Analysis against RSA

The most powerful fault attack against RSA is known as the *Bellcore attack* [8] that targets a CRT implementation. It consists in corrupting one of the two CRT exponentiations, say the one modulo p. The RSA computation thus results in a faulty signature $\tilde{s}$ that is correct modulo q (*i.e.* $\tilde{s} \equiv s \bmod q$) and corrupted modulo p (*i.e.* $\tilde{s} \not\equiv s \bmod p$). This implies that the difference $\tilde{s} - s$ is a multiple of q but is not a multiple of p, and hence we have $\gcd(\tilde{s} - s, N) = q$. Therefore, a pair signature/faulty signature provides a way to factorize N and consequently to fully break RSA. Actually, a pair message/faulty signature suffices to mount the attack since we have $\gcd(\tilde{s}^e - m, N) = q$ [22]. This way, RSA is broken with a single faulty computation.

RSA implemented in straightforward mode (*i.e.* without CRT) is also vulnerable to fault analysis. Several attacks have been published that assume either a faulty exponent [3], a faulty modulus [5,12,35] or a faulty intermediate power [8,9,34]. These attacks require several faulty signatures to fully recover the key but still constitute practical threats.

Another kind of fault attacks known as *safe-error attacks* can be distinguished from the ones addressed above. Depending on the algorithm, a fault injection may have no effect for some secret key values and may cause a corruption for others. In that case, simply observing wether the computation was corrupted or not reveals information on the secret key. Such attacks are especially threatening since they bypass classical fault analysis countermeasures that return an error in case of fault detection. Among these attacks, two categories can be distinguished: the *C-safe-error attacks* [41] that target dummy operations and the *M-safe-error attacks* that target registers allocations [40,24]. Our countermeasure provides an error detection mechanism and does not aim to thwart safe-error attacks. However, as discussed in Sect. 5.2, these last can be simply prevented.

Securing RSA Against Fault Analysis. A simple way to protect RSA against fault analysis is by verifying the signature s before returning it, namely by performing the following check: $m \overset{?}{=} s^e \bmod N$. This method offers a perfect security against differential fault analysis since a faulty signature is systematically detected. This countermeasure is efficient as long as e is small, but in the opposite case, it implies to perform two exponentiations which doubles the time complexity of RSA. This overhead is clearly prohibitive in the context of low resource devices. Moreover, depending on the context, the public exponent e may not be available (*e.g.* the Javacard API for RSA signature [37]). That is why, many works in the last decade have been dedicated to the search of alternative solutions. We review hereafter the main proposals that can be divided into two families: the *extended modulus based countermeasures* and the *self-secure exponentiations*.

2.3 Extended Modulus Based Countermeasures

We present hereafter different countermeasures that all rely on the use of an extended modulus in order to add redundancy in the computation.

Shamir's Trick and Variants. A first solution to protect RSA with CRT has been proposed by Shamir [36]. It consists in performing the two CRT exponentiations with extended moduli $p \cdot t$ and $q \cdot t$ where t is a small integer. Namely, one computes $s_p^* = m^{d \bmod \varphi(p \cdot t)} \bmod p \cdot t$ and $s_q^* = m^{d \bmod \varphi(q \cdot t)} \bmod q \cdot t$. The consistency of the computation is then checked by verifying that s_p^* mod t equals s_q^* mod t. If no error is detected, the algorithm returns $\mathsf{CRT}(s_p^*$ mod p, s_q^* mod $q)$. Under its simplest form, this countermeasure does not protect the CRT recombination which enables a successful fault attack [2]. Several works have proposed variants of Shamir's countermeasure in order to deal with this issue [2,7,14].

Vigilant Scheme. In [38], Vigilant proposed another countermeasure based on a modulus extension. The modulus is multiplied by $t = r^2$ for a small random number r. The message is then formatted as follows: $\hat{m} = \alpha m + \beta \cdot (1+r)$ mod Nt where (α, β) is the unique solution in $\{1, \cdots, Nt\}^2$ of the system $\alpha \equiv 1$ mod N, $\alpha \equiv 0$ mod t, $\beta \equiv 0$ mod N and $\beta \equiv 1$ mod t. Then, the exponentiation $s_r = \hat{m}^d$ mod Nt is performed. As shown in [38], s_r equals $\alpha m^d + \beta \cdot (1 + dr)$ mod Nt. Therefore, the signature can be recovered from s_r since it satisfies $s = s_r$ mod N and the consistency of the computation can be verified by checking $s_r \equiv 1 + dr$ mod t. This method can be extended to protect RSA-CRT (see [38] for details).

Security Considerations. The security of an extended modulus based countermeasure is not perfect. For instance, if a faulty message $\widetilde{m}$ satisfies $\widetilde{m} \equiv m$ mod t and $\widetilde{m} \not\equiv m$ mod N, then the exponentiation of this message results in a faulty signature that is not detected. The non-detection probability of an extended modulus based countermeasure is roughly about 2^{-k} where k denotes the bit-length of the modulus extension t. Therefore, the greater k, the more secure the countermeasure. However, the greater k, the slower the exponentiation (see Sect. 7.3). This kind of countermeasure hence offers a time/security tradeoff. A usual choice for k is 64 bits which provides a fair security. However, depending on the application, one may choose $k = 32$ (low security, more efficient exponentiation) or $k = 80$ (strong security, less efficient exponentiation).

2.4   Self-secure Exponentiations

For the countermeasures presented hereafter, the redundancy is not included in the modular operations anymore but at the exponentiation level. Namely, the exponentiation algorithm provides a direct way to check the consistency of the computation.

Giraud Scheme. The Giraud Scheme [18] relies on the use of the Montgomery powering ladder. It uses the fact that this exponentiation algorithm works with a pair of intermediate variables (a_0, a_1) storing values of the form $(m^\alpha, m^{\alpha+1})$. At the end of the exponentiation the pair (a_0, a_1) equals (m^{d-1}, m^d) and the consistency of the computation can be verified by checking wether $a_0 \cdot m$ equals a_1. If a fault is injected during the computation, the coherence between a_0 and a_1 is lost and the fault is detected by the final check.

Boscher et al. Scheme. The scheme by Boscher et al. [10] is based on the *right-to-left square-and-multiply-always* algorithm [15] which was originally devoted to thwart *simple side channel analysis* (see Sect. 6.1). In [10], the authors observe that this algorithm computes a triplet (a_0, a_1, a_2) that equals $(m^d, m^{2^l-d-1}, m^{2^l})$ at the end of the algorithm, where l denotes the bit-length of d. The principle of their countermeasure is hence to check that $a_0 \cdot a_1 \cdot m$ equals a_2 at the end of the exponentiation. Once again, in case of fault injection, the relation between the a_i's is broken and the fault is detected by the final check.

The main drawback of these two countermeasures is that they both impose the use of an exponentiation algorithm that performs 2 modular multiplications per bit of the exponent while other exponentiation algorithms require an average of 1.5 multiplications per bit of the exponent (and sometimes less).

In the next section, we propose a new self-secure exponentiation. Our method requires around 1.65 multiplications per bit of the exponent in average and hence constitutes an efficient alternative to the existing countermeasures.

3   A New Self-secure Exponentiation Based on Double Addition Chains

3.1   Basic Principle

In the following, we shall call *double exponentiation* an algorithm taking as inputs an element m and a pair of exponents (a, b), and computing the pair of powers (m^a, m^b).

The core idea of our method is to process a double exponentiation to compute the pair $(m^d, m^{\varphi(N)-d})$ modulo N. Then, the consistency of the computation is verified by performing the following check:

$$m^d \cdot m^{\varphi(N)-d} \stackrel{?}{\equiv} 1 \bmod N \ . \tag{1}$$

If no error occurs during the computation then, due to Euler's Theorem, this check is positive. In that case, the algorithm returns $m^d \bmod N$. On the other hand, if the computation is corrupted, then the result of this check is negative with high probability. In that case, the algorithm returns an error message.

In order to construct a self-secure exponentiation based on aforementioned principle, we need a double exponentiation algorithm. We propose hereafter such an algorithm that is well suited for implementation constrained in memory. Our solution is based on the building of an *addition chain*. This notion, as well as the ensued notion of *addition chain exponentiation* are briefly introduced in the next section (see [25] for more details).

3.2   Addition Chain Exponentiations

At first, we give the definition of an addition chain.

Definition 1. *An* addition chain *for an integer a is a sequence $x_0, x_1, \cdots, x_n$ with $x_0 = 1$ and $x_n = a$ that satisfies the following property: for every k there exist indices $i, j < k$ such that $x_k = x_i + x_j$.*

An addition chain $(x_i)_i$ for an integer a provides a way to evaluate any element m to the power a. Let $m_0 = m$. For k from 1 to n, one computes $m_k = m_i \cdot m_j$

where $i, j < k$ are such that $x_k = x_i + x_j$. By induction, the sequence $(m_k)_k$ satisfies: $m_k = m^{x_k}$ for every $k \leq n$ which leads to $m_n = m^{x_n} = m^a$. Such an addition chain exponentiation may require an important amount of memory to store the intermediate powers required for the computation of subsequent powers. This can make the exponentiation unpractical, especially in the context of low resource devices. Therefore, the minimum number of variables required to store the intermediate powers is an important parameter of the addition chain exponentiation. This parameter that directly results from the addition chain will be called the *memory depth* of the chain in the following.

In this paper, an addition chain $x_0, x_1, \cdots, x_n$ with $(x_{n-1}, x_n) = (a, b)$ is called a *double addition chain* for the pair (a, b). A double addition chain for a pair (a, b) provides a way to perform the double exponentiation $m \mapsto (m^a, m^b)$ for any element m.

Remark 1. What we call here double exponentiation shall not be confused with multi-exponentiations (also known as simultaneous exponentiations) that compute a product of powers $\prod_i m_i^{a_i}$ (see for instance [32]). What we call double addition chain is also called *addition sequence* in the general case where possibly more than two powers must be computed [11,19]. Addition sequences have not been so much investigated. In [11], the authors propose some heuristics but these are not suitable for implementations constrained in memory.

3.3   A Heuristic for Double Addition Chains

In this section, we propose a heuristic to compute a double addition chain with a memory depth of 3 for any pair of natural integers (a, b). This provides us with a double exponentiation algorithm that is well suited for implementations constrained in memory.

Without loss of generality, we assume $a \leq b$. The chain involves a pair of intermediate results (a_i, b_i) that are initialized to $(0, 1)$ and that equal (a, b) once all the additions have been performed. In order to have a memory depth of 3, one single additional variable is used that keeps the value 1 (this amounts to keep the element m in a register for the resulting exponentiation). Therefore, at the ith step of the chain, one can either increment a_i or b_i by 1, double a_i or b_i, or add a_i and b_i together.

To construct such a chain, we start from the pair (a, b) and go down to the pair $(0, 1)$ by applying the inverse operations. Namely, we define a sequence $(\alpha_i, \beta_i)_i$ such that $(\alpha_0, \beta_0) = (a, b)$ and $(\alpha_n, \beta_n) = (0, 1)$ for some $n \in \mathbb{N}$, and where, for every i, the pair $(\alpha_{i+1}, \beta_{i+1})$ is obtained from (α_i, β_i) by decrementing, by dividing by two and/or by subtracting an element to the other one. In order to limit the memory required to the storage of the chain, we have to restrict the set of possible operations. Our heuristic is the following one:

$$(\alpha_{i+1}, \beta_{i+1}) = \begin{cases} (\alpha_i, \beta_i/2) & \text{if } \alpha_i \leq \beta_i/2 \text{ and } \beta_i \bmod 2 = 0 \\ (\alpha_i, (\beta_i - 1)/2) & \text{if } \alpha_i \leq \beta_i/2 \text{ and } \beta_i \bmod 2 = 1 \\ (\beta_i - \alpha_i, \alpha_i) & \text{if } \alpha_i > \beta_i/2 \end{cases} \quad (2)$$

Proposition 1. *If $\alpha_0, \beta_0 \in \mathbb{N}^*$ are such that $\alpha_0 \leq \beta_0$ then the sequence $(\alpha_i, \beta_i)_i$ satisfies the following properties:*

1. *For every i, we have $\alpha_i \leq \beta_i$.*
2. *There exists $n \in \mathbb{N}$ such that $(\alpha_n, \beta_n) = (0, 1)$.*

Proof. The first property is straightforward: it is true for $i = 0$ and it is preserved by every step. The second one is demonstrated as follows. For every i such that $\alpha_i > 0$, we have $\alpha_{i+1} \leq \beta_{i+1} \leq \beta_i$ and $\alpha_{i+1} + \beta_{i+1} < \alpha_i + \beta_i$. This implies that there exists $n' \in \mathbb{N}$ such that $\alpha_{n'} > 0$ and $\alpha_{n'+1} \leq 0$. From (2), one deduces $\alpha_{n'} = \beta_{n'} > 0$ and $\alpha_{n'+1} = 0$. Denoting x the natural integer such that $(\alpha_{n'+1}, \beta_{n'+1}) = (0, x)$, we finally get $(\alpha_{n'+\lceil \log x \rceil}, \beta_{n'+\lceil \log x \rceil}) = (0, 1)$. $\diamond$

At this point, we need a binary representation for the sequence of additions to perform for the processing of the sequence $(a_i, b_i)_i$. Let us denote by n the natural integer satisfying $(\alpha_n, \beta_n) = (0, 1)$. We define τ and ν as the n-bit vectors whose coordinates satisfy:

$$\tau_i = \begin{cases} 0 & \text{if } \alpha_{n-i} \leq \beta_{n-i}/2 \\ 1 & \text{if } \alpha_{n-i} > \beta_{n-i}/2 \end{cases} \tag{3}$$

and

$$\nu_i = \beta_{n-i} \bmod 2 . \tag{4}$$

The sequence $(a_i, b_i)_i$ can be computed from τ and ν by initializing (a_0, b_0) to $(0, 1)$ and by iterating:

$$(a_{i+1}, b_{i+1}) = \begin{cases} (a_i, 2b_i) & \text{if } \tau_{i+1} = 0 \text{ and } \nu_{i+1} = 0 \\ (a_i, 2b_i + 1) & \text{if } \tau_{i+1} = 0 \text{ and } \nu_{i+1} = 1 \\ (b_i, a_i + b_i) & \text{if } \tau_{i+1} = 1 \end{cases}$$

One can verify that $(a_i, b_i) = (\alpha_{n-i}, \beta_{n-i})$ holds for every i which yields $(a_n, b_n) = (a, b)$.

Let us remark that the whole sequence ν is not necessary for processing this addition chain (and the resulting exponentiation). Indeed, only the bits ν_i for which τ_i equals 0 are required. Therefore, the exponentiation algorithm shall make use of a single compressed sequence ω in order to avoid memory loss. We simply define ω as the sequence obtained from τ by inserting every bit ν_i for which $\tau_i = 0$ between τ_i and τ_{i+1}. In the sequel, we shall denote by n^* the bit-length of ω. Moreover, when we will need to make appear the relationship between the pair (a, b) and ω, we will use the notation $\omega(a, b)$.

The sequence $\omega(a, b)$ thus constitutes the binary representation of the double addition chain for the pair of exponents (a, b). To process the relying double exponentiation one must pre-compute ω. This is done by computing the pair (α_i, β_i) for every $i \in \{1, \cdots, n\}$. The following algorithm details such a computation. It makes use of two registers R_0 and R_1 that store the intermediate results α_i and β_i. It makes also use of a Boolean variable γ such that α_i is stored in $R_{\gamma \oplus 1}$ and β_i is stored in R_γ.

Algorithm 1. Double addition chain computation – ChainCompute

INPUT: A pair of natural integers (a, b) s.t. $a \leq b$
OUTPUT: The chain $\omega(a, b)$

---

1. $R_0 \leftarrow a$; $R_1 \leftarrow b$; $\gamma \leftarrow 1$; $j \leftarrow n^*$
2. **while** $(R_{\gamma \oplus 1}, R_\gamma) \neq (0, 1)$ **do**
3.   **if** $(R_\gamma / 2 > R_{\gamma \oplus 1})$
4.   **then** $\omega_{j-1} \leftarrow 0$; $\omega_j \leftarrow R_\gamma \bmod 2$; $R_\gamma \leftarrow R_\gamma / 2$; $j \leftarrow j - 2$
5.   **else** $\omega_j \leftarrow 1$; $R_\gamma \leftarrow R_\gamma - R_{\gamma \oplus 1}$; $\gamma \leftarrow \gamma \oplus 1$; $j \leftarrow j - 1$
6. **end while**
7. **return** ω

Remark 2. The length n^* is *a priori* unknown before the computation of the chain. However, as shown in Sect. 7.2, it is upper bounded by $2.2\lceil \log b \rceil$ (with high probability). For a practical implementation of Algorithm 1, one may use a buffer of $2.2\lceil \log b \rceil$ bits to store ω, initializing j by the final bit index of this buffer.

The following algorithm describes the resulting double modular exponentiation algorithm. It makes use of two registers R_0 and R_1 that store the intermediate results m^{a_i} and m^{b_i} and one more register to hold m. It makes also use of a Boolean variable γ such that m^{a_i} is stored in $R_{\gamma \oplus 1}$ and m^{b_i} is stored in R_γ.

Algorithm 2. Double modular exponentiation – DoubleExp

INPUT: An element $m \in \mathbb{Z}_N$, a chain $\omega(a, b)$ s.t. $a \leq b$, a modulus N
OUTPUT: The pair of modular powers $(m^a \bmod N, m^b \bmod N)$

---

1. $R_0 \leftarrow 1$; $R_1 \leftarrow m$; $\gamma \leftarrow 1$
2. **for** $i = 1$ to n^* **do**
3.   **if** $(\omega_i = 0)$ **then**
4. $R_\gamma \leftarrow R_\gamma^2 \bmod N$; $i \leftarrow i + 1$
5.   **if** $(\omega_i = 1)$ **then** $R_\gamma \leftarrow R_\gamma \cdot m \bmod N$
6.   **if** $(\omega_i = 1)$ **then**
7. $R_{\gamma \oplus 1} \leftarrow R_{\gamma \oplus 1} \cdot R_\gamma \bmod N$; $\gamma \leftarrow \gamma \oplus 1$
8. **end for**
9. **return** $(R_{\gamma \oplus 1}, R_\gamma)$

3.4   The Secure Exponentiation Algorithm

Following the principle described in Sect. 3.1, Algorithm 2 provides a way to perform a modular exponentiation secure against fault analysis. The resulting secure modular exponentiation is depicted in the following algorithm.

Algorithm 3. Secure modular exponentiation

INPUT: A message m, a secret exponent d, a modulus N and its Euler's totient $\varphi(N)$
OUTPUT: The modular power $m^d \bmod N$

---

1. $\omega \leftarrow \mathsf{ChainCompute}(d,\ 2\varphi(N) - d)$
2. $(s, c) \leftarrow \mathsf{DoubleExp}(m,\ \omega,\ N)$
3. **if** $s \cdot c \bmod N \neq 1$ **then return** "error";   **else return** s

---

Remark 3. For the chain computation (Step 1), $\varphi(N) - d$ is replaced by $2\varphi(N) - d$ in order to fit the constraint $a \leq b$ imposed by the chain computation algorithm. This does not affect the result of the double exponentiation in Step 2 since we have $m^{\varphi(N)-d} \equiv m^{2\varphi(N)-d} \bmod N$.

4   A New Secure RSA-CRT Algorithm

For an RSA computation, the secure modular exponentiation proposed above can be extended to be performed in CRT mode. Two double exponentiations are performed separately in order to compute the pairs (s_p, c_p) and (s_q, c_q) where $c_p = m^{p-1-d_p} \bmod p$ and $c_q = m^{q-1-d_q} \bmod q$. Then the signature s is recovered from s_p and s_q by CRT recombination and its value is checked modulo p (resp. q) using c_p (resp. c_q) according to (1).

Algorithm 4. Secure RSA-CRT

INPUT: A message m, the secret exponents d_p and d_q, the secret primes p and q
OUTPUT: The modular power $m^d \bmod p \cdot q$

---

1. $\omega_p \leftarrow \mathsf{ChainCompute}(d_p,\ 2(p-1) - d_p)$
2. $(s_p, c_p) \leftarrow \mathsf{DoubleExp}(m \bmod p,\ \omega_p,\ p)$
3. $\omega_q \leftarrow \mathsf{ChainCompute}(d_q,\ 2(q-1) - d_q)$
4. $(s_q, c_q) \leftarrow \mathsf{DoubleExp}(m \bmod q,\ \omega_q,\ q)$
5. $s \leftarrow \mathsf{CRT}(s_p, s_q)$
6. **if** $(s \cdot c_p \bmod p \neq 1$ **or** $s \cdot c_q \bmod q \neq 1)$ **then return** "error" **else return** s

---

Remark 4. We assume that $m \bmod p$ (resp. $m \bmod q$) cannot be corrupted before the beginning of the double exponentiation. This is mandatory for the security of Algorithm 4, since such a corruption would not be detected and would enable the Bellcore attack. In practice, this can be ensure by computing a cyclic redundancy code for $m \bmod p$ (resp. $m \bmod q$) at the beginning of the RSA-CRT algorithm. Then, at the beginning of the double exponentiation algorithm, $m \bmod p$ (resp. $m \bmod q$) is recomputed from m and its integrity is checked once it has been loaded in two different registers (m and R_1 in Algorithm 2). Any corruption occurring after this check shall be detected by the final check.

Remark 5. The chains ω_p and ω_q can be either computed on-the-fly as depicted in Algorithm 4 (Steps 1 and 3) or pre-computed and stored in non-volatile memory. The first solution has the advantages of preserving the classical RSA-CRT parameters and of enabling the exponent blinding countermeasure (see Sect. 6.2). The second solution has the advantage of avoiding the timing and memory overhead induced by the chain computations.

5   Security against Fault Analysis

In this section, we analyze the security of our method against fault analysis. We start with two remarks of practical purpose, then we investigate the detection probability of a fault injection and finally we address safe-error attacks.

Remark 6. In Algorithms 3 and 4, we assume that the integrity of the chain computation parameters is checked before executing the chain computation algorithm. This avoids any attack that would corrupt d (resp. d_p, d_q) before the computation of $2\varphi(N) - d$ (resp. $2(p-1) - d_p$, $2(q-1) - d_q$).

Remark 7. Some papers claim that coherence checks using conditional branches should be avoided to strengthen fault analysis security [42,14]. The argument behind this assertion is that the coherence check could be easily skipped by corrupting the status register. An alternative solution to direct coherence checking is to use an *infection procedure* that renders the erroneous signature harmless in case of fault detection [42]. However, most of the proposed countermeasures have security flaws due to ineffective infection methods (for instance [7,14] have been broken in [39,4]). Moreover, the infection procedure can also be skipped as it has been practically demonstrated in [23]. In [16], a simple solution is proposed that performs a coherence check without conditional branches in a way that is secure against operations skipping. We suggest to use this solution for the coherence checks performed in Algorithm 3 (Step 3) and Algorithm 4 (Step 6).

5.1   Fault Detection

We analyze hereafter the different fault attacks that can be attempted on our secure exponentiation algorithm and we investigate the corresponding detection probability. We only focus on transient faults, namely faults whose effect lasts for one computation. Permanent fault attacks are easily thwarted by the addition of some cyclic redundancy codes to check the parameters integrity.

We use the generic notation M to denote the involved modulus that may equal N (for a straightforward RSA), p or q (for a RSA-CRT) and we denote by $\mathrm{ord}_M(m)$ the order of an element m in $\mathbb{Z}_M^*$. When the fault causes the corruption of an intermediate variable v, we denote the corrupted variable by $\widetilde{v}$ and the error by ε such that $\widetilde{v} = v + \varepsilon$. We analyze here the condition about ε for a non-detection and we bound the probability $\mathcal{P}$ of non-detection in the *uniform fault model i.e.* assuming that ε is uniformly distributed.

For our analysis, the following lemma shall be useful (see the proof in Appendix A).

Lemma 1. *Let M be an integer greater than* 30. *Let m be a random variable uniformly distributed over $\mathbb{Z}_M^*$ and let u be a random variable uniformly distributed over $\{1, \cdots, \varphi(M)\}$ and independent of m. We have:*

$$\mathrm{P}\left(\mathrm{ord}_M(m)|u\right) < \frac{2}{M^{1/3}} \ . \tag{5}$$

For the sake of simplicity, we approximate hereafter a uniform distribution over $\mathbb{Z}_M$ by a uniform distribution over $\mathbb{Z}_M^*$. This approximation is sound in our context since M is a large prime or an RSA modulus.

Corruption of one of the two exponents. Among the exponents a and b, one equals d and the other one equals $\varphi(M) - d$. On the one hand, if $\varphi(M) - d$ is corrupted, then the result of the exponentiation remains correct (*i.e.* it equals $m^d \bmod M$) and the attack failed whatever the result of the final check (which is however very likely to detect the fault). On the other hand, if d is corrupted, we show hereafter that the final check will detect the error with high probability.

In fact, the error is not detected if and only if we have $m^{\tilde{d}} \cdot m^{\varphi(M)-d} \equiv 1 \bmod M$ that is $m^\varepsilon \equiv 1 \bmod M$. This occurs if and only if ε is a multiple of the order of m. Therefore, the probability of non-detection can be expressed as $\mathcal{P} = \mathrm{P}\left(\mathrm{ord}_M(m)|\varepsilon\right)$. Hence, the lower the order of m, the higher the probability of non-detection. Since a potential attacker does not know $\varphi(M)$, he cannot chose m in a way that affect its order. For this reason, m can be considered uniformly distributed over $\mathbb{Z}_M$. Therefore, in the uniform fault model, Lemma 1 implies $\mathcal{P} < 2/M^{1/3}$.

Remark 8. The bound provided by Lemma 1 is not tight at all but it is sufficient to show that $\mathcal{P}$ is negligible. For instance, if M satisfies $\log M \geq 244$, which is necessary (but not sufficient) for the security of RSA (even for RSA-CRT where $\log N = 2 \log M$), $\mathcal{P}$ is strictly lower than 2^{-80} which is negligible.

Corruption of the message or an intermediate power. From the definition of the double addition chain given in Sect. 3.3, one can see that for every $i \in \{1, \cdots, n\}$, the pair (a_n, b_n) can be expressed as a linear transformation of the triplet $(a_i, b_i, 1)$. Let us denote by α_i^a, β_i^a, δ_i^a the three coefficients of the expression of a_n, namely $a_n = \alpha_i^a a_i + \beta_i^a b_i + \delta_i^a$. By analogy, we denote by α_i^b, β_i^b, δ_i^b the coefficients in the expression of b_n.

If the message m is corrupted at the ith step of the exponentiation, this last returns the following pair of powers: $\left(m^a(m^{-1} \cdot \tilde{m})^{\delta_i^a}, m^b(m^{-1} \cdot \tilde{m})^{\delta_i^b}\right)$ modulo M. The error is not detected if and only if we have $(m^{-1} \cdot \tilde{m})^{\delta_i^a + \delta_i^b} \equiv 1 \bmod M$, that is $(1 + \varepsilon \cdot m^{-1})^{\delta_i^a + \delta_i^b} \equiv 1 \bmod M$. This occurs if and only if the order of $m' = 1 + \varepsilon \cdot m^{-1}$ divides $\delta_i^a + \delta_i^b$. Therefore, the probability of non-detection can be expressed as $\mathcal{P} = \mathrm{P}\left(\mathrm{ord}_M(m')|\delta_i^a + \delta_i^b\right)$. Following the same reasoning, a corruption of the intermediate power m^{a_i} (resp. m^{b_i}) is not detected with a probability $\mathcal{P} = \mathrm{P}\left(\mathrm{ord}_M(m')|\alpha_i^a + \alpha_i^b\right)$ where $m' = 1 + e \cdot m^{-a_i}$ (resp. $\mathcal{P} = \mathrm{P}\left(\mathrm{ord}_M(m')|\beta_i^a + \beta_i^b\right)$ where $m' = 1 + \varepsilon \cdot m^{-b_i}$).

Since a and b are unknown to the attacker, this one cannot chose the value of $\delta_i^a + \delta_i^b$, $\alpha_i^a + \alpha_i^b$ or $\beta_i^a + \beta_i^b$ since these directly ensue from a and b. That is why, we make the heuristic assumption that $\mathcal{P}$ equals $\mathrm{P}\left(\mathrm{ord}_M(m')|u\right)$ where u is uniformly distributed over $\{1, \cdots, \varphi(M)\}$. In the uniform fault model, we have the uniformity of m' that holds from the one-to-one relationship between ε and m' for every $m \neq 0$. Consequently, Lemma 1 implies $\mathcal{P} < 2/M^{1/3}$ and p is negligible.

Corruption of the chain. A faulty chain $\widetilde{w}$ results in a faulty pair of powers $(m^{\widetilde{a}}, m^{\widetilde{b}})$. The error is not detected if and only if the order of m divides $\widetilde{a} + \widetilde{b}$, hence the non-detection probability can be expressed as $\mathcal{P} = \mathrm{P}\left(\mathrm{ord}_M(m)|\widetilde{a} + \widetilde{b}\right)$.

As shown in Sect. 7.2, the expected bit-length of the chain ω yielding a pair of l-bit exponents (a, b) is of $2l$. This suggests an almost bijective relationship between the chains space and the exponents pairs space. In the uniform fault model, we can therefore consider that $\widetilde{a}$ and $\widetilde{b}$ are uniformly distributed which, by Lemma 1, implies $\mathcal{P} < 2/M^{1/3}$.

Corruption of the modulus. If the modulus M is corrupted at the ith step of the exponentiation, then this last results in the two following powers: $m_1^{\alpha_i^a} \cdot m_2^{\beta_i^a} \cdot m^{\delta_i^a} \bmod \widetilde{M}$ and $m_1^{\alpha_i^b} \cdot m_2^{\beta_i^b} \cdot m^{\delta_i^b} \bmod \widetilde{M}$ where $m_1 = m^{a_i} \bmod M$ and $m_2 = m^{b_i} \bmod M$. Therefore, the error is not detected if and only if we have $m_1^{\alpha_i^a + \alpha_i^b} \cdot m_2^{\beta_i^a + \beta_i^b} \cdot m^{\delta_i^a + \delta_i^b} \bmod \widetilde{M} = 1$.

In the uniform fault model, the faulty modulus $\widetilde{M}$ is uniformly distributed over $[0, 2^l[$ where l denotes the bit-length of M. Therefore, the probability of non-detection $\mathcal{P}$ is close to $\mathrm{P}(u_1 \bmod u_2 = 1)$ where u_1 and u_2 are uniform (and independent) random variables over $[0, 2^l[$. This probability equals $2^{-l} \sum_{i=1}^{2^l - 1}(1/i)$ which is strictly lower than 2^{-80} for every $l \geq 86$. The probability of non-detection $\mathcal{P}$ is hence negligible in our context.

5.2 Safe-Error Attacks

As recalled in Sect. 2.2, safe-error attacks divide into two categories: C-safe-error attacks [41] and M-safe-error attacks [40,24].

To prevent C-safe-error attacks one must ensure that no dummy operation is conditionally performed depending on the secret key. Our secure exponentiation does not perform any dummy operation and is hence secure against C-safe-error attacks. When the chain is computed on-the-fly, it must be done in an atomic way in order to thwart simple side channel analysis (see Sect. 6.1). The atomic version of the chain computation algorithm (see Appendix B) makes use of dummy operations and is hence vulnerable to C-safe-error attacks. In that case, these can be thwarted by using the exponent blinding countermeasure (see Sect. 6.2).

To prevent M-safe-error attacks one can either randomize the exponent (using for instance the exponent blinding) or randomize the indices of the registers that are addressed by some exponent bits (or chain bits in our context). When the chain is pre-computed, the exponent cannot be randomized and the *registers indices randomization* introduced hereafter shall be used. The principle is to randomly chose the registers to store the different variables among the different used registers. For instance, in Algorithm 1, a random bit r is picked up so that the registers R_0 and R_1 are switched if r equals 1. In the description of Algorithm 1 this amounts to replace R_γ by $R_{\gamma \oplus r}$. In this way, a M-safe-error attack will imply a faulty output once out of two, independently of the performed operation. This simple countermeasure thwarts the attacks recently published in [24].

6   Toward Side Channel Analysis Resistance

In this section, we address the resistance of our exponentiation algorithm against the two main kinds of side channel analysis (SCA): *simple SCA* and *differential SCA*.

6.1   Simple Side Channel Analysis

Simple SCA [26] exploits the fact that the operation flow of a cryptographic algorithm may depend on the secret key. Different operations may induce different patterns in the side channel leakage which provides secret information to any attacker able to eavesdrop this leakage. To thwart simple SCA, an algorithm must be *atomic* [13], namely, it must have the same operation flow whatever the secret key.

The chain computation algorithm (Algorithm 2) and the double exponentiation algorithm (Algorithm 1) may be vulnerable to simple SCA. To circumvent this weakness, we provide atomic versions of these algorithms in Appendix B.

6.2   Differential Side Channel Analysis

Differential SCA [26] exploits the fact that the side channel leakage reveals information about some key-dependent intermediate variables of the computation. Since its first publication, several improvements of differential SCA have been proposed, especially to attack modular exponentiation [1,17,20,30]. In order to thwart differential SCA, one usually makes use of randomization techniques. The message randomization as well as the modulus randomization are usual countermeasures that can be straightforwardly combined with our method. The exponent is usually randomized using the blinding technique that consists in performing the exponentiation to the power $d' = d + r \cdot \varphi(N)$ for a small random number r [27,30,15]. This technique cannot be straightforwardly applied while using our secure exponentiation algorithm since we have $d' > \varphi(N)$ for every $r > 0$. Therefore, we propose the following simple adaptation: in Step 1 of Algorithm 3, the exponent a is set to $d + r_1 \cdot \varphi(N)$ and the exponent b is set to $r_2 \cdot \varphi(N) - d$ where r_1 and r_2 are two small random numbers with $r_2 \geq r_1 + 2$. Then the rest of the secure exponentiation algorithm does not change. Since $m^{d+r_1 \cdot \varphi(N)} \equiv m^d \bmod N$, the desired signature is computed and since $m^{d+r_1 \cdot \varphi(N)} \cdot m^{r_2 \cdot \varphi(N)-d} \equiv 1 \bmod N$, the final check is correctly carried out.

Remark 9. If the chain ω is pre-computed, the exponent blinding cannot be used. In that case, another kind of randomization (message, modulus) shall be used. However, these do not prevent a differential SCA targeting the chain itself (as for instance the SEMD attack of [30] or the address-bit DPA [20]). To deal with this issue, we suggest to use a Boolean masking such as proposed in [21].

7   Complexity Analysis

In this section we analyze the time complexity and the memory complexity of our proposal. In the sequel, we shall denote by l the bit-length of the exponentiation inputs. Namely for a straightforward RSA we have $l = \lceil \log N \rceil$ and for a RSA-CRT we have $l = \lceil \log N/2 \rceil$.

7.1   Time Complexity

Our secure exponentiation is mainly composed of the chain computation and the double exponentiation. The chain computation loop is shorter than the exponentiation loop and it involves simple operations (*e.g.* substraction, division by 2) whose time complexities are negligible compared to a modular multiplication. Therefore, the time complexity of our proposal mainly depends on the number of multiplications performed by the double exponentiation algorithm (all the more so as the chain may be pre-computed). We shall denote this number by m and we shall define the *multiplications-per-bit ratio* as the coefficient θ satisfying $m = \theta l$.

Some practical values for the expectation and the standard deviation of θ are given in Table 7.1 that were obtained by simulations. For $l \in \{512, \cdots, 1024\}$, the expected multiplications-per-bit ratio is around 1.65. Compared to the classical *square-and-multiply* algorithm, our exponentiation hence requires 10% more multiplications, implying a 10% overhead in average, which is a fair cost for fault analysis resistance. Moreover, the time complexity of our exponentiation is steadier than the one of the square-and-multiply since the standard deviation $\sigma(\theta)$ is lower than $1/5$ and decreasing for $l \geq 512$ while, for the square-and-multiply algorithm, it is constant to $1/4$.

7.2   Memory Complexity

Our double exponentiation algorithm requires three l-bit registers to store the message and the pair of powers. If the chain ω is computed on-the-fly, it requires an additional buffer is necessary to store it.

Table 1. Expectation and standard deviation of the double exponentiation multiplications-per-bit ratio

	$l = 512$	$l = 640$	$l = 768$	$l = 896$	$l = 1024$
$E(\theta)$	1.65	1.66	1.66	1.66	1.66
$\sigma(\theta)$	0.020	0.017	0.017	0.016	0.014

Table 2. Standard deviation of the chain bit-length

	$l = 512$	$l = 640$	$l = 768$	$l = 896$	$l = 1024$
$\sigma\,(n^*)$	$0.015\,l$	$0.013\,l$	$0.011\,l$	$0.010\,l$	$0.010\,l$

We performed simulations to derive the practical values of the expectation and the standard deviation of the chain length n^*. For the expectation, we obtained $\mathrm{E}\,(n^*) \approx 2.03\,l$ for $l \in \{512, \cdots, 1024\}$. For the standard deviation, the obtained values are summarized in Table 2. Approximating the distribution of n^* by a Gaussian, we get $\mathrm{P}\,(n^* > \mathrm{E}\,(n^*) + k\sigma\,(n^*)) = \left(1 - \mathrm{erf}\big(k/\sqrt{2}\big)\right)/2$ where $\mathrm{erf}(\cdot)$ denotes the error function. For $k = 10$ and for $l \in \{512, \cdots, 1024\}$, this probability is lower than 2^{-80}. Consequently, for $l \in \{512, \cdots, 1024\}$, the probability to have $n^* > 2.2\,l$ is negligible in practice, hence ω can be stored in a $(2.2\,l)$-bit buffer.

On the whole, our secure exponentiation requires $5.2\,l$ bits of memory when the chain is computed on-the-fly and it requires $3\,l$ bits of memory when the chain is pre-computed.

For our secure RSA-CRT (see Algorithm 4), the peak of memory consumption is reached in the second exponentiation while s_p and c_p must be kept in memory. This makes a total memory consumption of $7.2\,l$ bits with on-the-fly chain computation and of $5\,l$ bits with pre-computed chain.

7.3   Comparison with Previous Solutions

We analyze hereafter the complexity of previous countermeasures in the literature. As explained in Sect. 2, these can be divided in two categories: the extended modulus based countermeasures and the self-secure exponentiations.

Extended Modulus Based Countermeasures. The time complexity of an extended modulus based countermeasure (such as the Shamir's trick or the Vigilant Scheme) is around the complexity of the main exponentiation loop(s) since the additional computations are negligible. However, such countermeasures are not free in terms of timing since the use of an extended modulus slows down the exponentiation. In fact, the time complexity of a modular multiplication can be written as $l^2 t_0$ where t_0 denotes a constant time that depends on the device architecture. Denoting by k the bit-length of the modulus extension, an extended modulus exponentiation has a time complexity of $m(l + k)^2 t_0$ while a normal exponentiation has a time complexity of $ml^2 t_0$. Besides, the modulus extension implies an increase of the exponentiation execution time by a factor $(1 + k/l)^2$. As an illustration, Table 7.3 gives several values of the induced overhead according to the modulus length and to the extension length. For instance, an RSA 1024 implemented in CRT ($l = 512$) with extended modulus providing a fair level of security ($k = 64$) is about 27% slower than an unprotected one. This time overhead is sizeable; in particular it is significantly greater than the 10% overhead induced by our countermeasure. However, extended modulus based countermeasures enables the use of exponentiation algorithms faster that

Table 3. Time overhead (in %) for an extended modulus based modular exponentiation

	$l = 512$	$l = 768$	$l = 1024$
$k = 32$ (low security)	13	9	6
$k = 64$ (fair security)	27	17	13
$k = 80$ (strong security)	34	22	16

the square-and-multiply such as the q-ary or the sliding windows methods (see for instance [29]). Roughly, a q-ary exponentiation has a multiplications-per-bit ratio of $1 + (2^q - 1)/(q2^q)$ which is lower than or equal to 1.5, but it has a higher memory complexity since it requires $2^{q-1} + 1$ registers. The use of a sliding window allows to slightly improve the time complexity of a q-ary method [28].

The memory complexity of an exponentiation with modulus extension is of $n_r(l + k)$ where n_r denotes the number of registers required by the exponentiation algorithm. For an RSA-CRT, the memory complexity depends on the used countermeasure. For the Vigilant Scheme, the memory consumption peak occurs during the second exponentiation while the values S'_p, i_{qr}, r, R_3 and R_4 must hold in memory (see [38]). This results in a memory consumption of $n_r(l + k) + (l + k) + 3.5\,k = (n_r + 1) \cdot l + (n_r + 4.5) \cdot k$ bits.

Remark 10. We do not detail the memory complexity of the other extended modulus based countermeasures since, for most of them, it is close to the memory complexity of the Vigilant Scheme.

Previous self-secure exponentiations. The Giraud Scheme and the Boscher *et al.* Scheme both have a multiplications-per-bit ratio constant to 2. This implies an average time overhead of 33% compared to the square-and-multiply algorithm and of 21% compared to our exponentiation. However, both of these schemes do not require additional computations contrary to the extended modulus based countermeasures or to our scheme when the chain is computed on-the-fly. Although these additional computations are theoretically negligible, they may induce an overhead for a practical implementation depending on the device architecture.

In terms of memory, we shall focus on the Giraud Scheme since it is less consuming than the Boscher *et al.* Scheme. The secure exponentiation requires two l-bit registers. For the RSA-CRT, the peak of memory consumption is reached during the two recombinations. For instance, the first recombination requires (at least) $3l$ bits of memory while m, S_p and S_q must hold in memory (see [18]) which makes a total complexity of $7l$ bits.

Comparison with our solution. Table 4 provides a comparison between the Giraud Scheme, the Vigilant Scheme and ours for an RSA 1024 with CRT (*i.e.* $l = 512$). For the Vigilant Scheme, we assume a modulus extension of $\{64, 80\}$ bits and a q-ary sliding window exponentiation for $q = 1, 2$ or 3 [29]. The results given in Table 4 shows that our countermeasure is currently one of the most competitive solution to thwart fault analysis for an RSA 1024 with CRT.

Table 4. Memory and time complexities of different fault analysis countermeasures for an RSA 1024 with CRT

Countermeasure	Time ($10^6 \cdot t_0$)	Memory (Kb)
Vigilant [38] ($q = 1$)	$\{511, 484\}$	$\{2.4, \ 2.3\}$
Vigilant [38] ($q = 2$)	$\{468, 444\}$	$\{2.6, \ 2.5\}$
Vigilant [38] ($q = 3$)	$\{440, 417\}$	$\{3.7, \ 3.6\}$
Giraud [18]	537	3.5
This paper	443	2.5 (+1.1)

Remark 11. The time complexity for the Vigilant Scheme with sliding widow is computed as follows. A q-ary exponentiation performs an average of $l \cdot \left(1 + (2^q - 1)/(q2^q)\right)$ multiplications [29] and the use of a sliding window yields an improvement of about 5% for $l = 512$ [28]. Therefore, the time complexity of one exponentiation is estimated to $0.95 \cdot (l + k)^2 t_0 \cdot l \cdot \left(1 + (2^q - 1)/(q2^q)\right)$. Concerning the memory complexity, the sliding window method requires a total of $n_r = 2^{q-1} + 1$ registers.

8   Conclusion

In this paper, we have described a new countermeasure to protect exponentiation and RSA against fault analysis. The core idea of our method is to introduce redundancy in the computation by performing a double exponentiation. To do so, we proposed a double exponentiation algorithm that is based on the computation of an addition chain. We analyzed the security of our solution *vs* fault analysis and we showed how it can be protected against side channel analysis. We also studied the time and memory complexities of our countermeasure which showed that it offers an efficient alternative to the existing schemes. A direction for further research would be to investigate more efficient double exponentiation algorithms and time-memory tradeoffs.

Acknowledgements

I would like to thank Jean-Sébastien Coron, Emmanuelle Dottax, Christophe Giraud, Gilles Piret and Emmanuel Prouff for helpful comments. I am also especially grateful to one of the anonymous reviewers for valuable suggestions.

References

1. Amiel, F., Feix, B., Villegas, K.: Power analysis for secret recovering and reverse engineering of public key algorithms. In: Adams, C., Miri, A., Wiener, M. (eds.) SAC 2007. LNCS, vol. 4876, pp. 110–125. Springer, Heidelberg (2007)
2. Aumüller, C., Bier, P., Fischer, W., Hofreiter, P., Seifert, J.-P.: Fault attacks on RSA with CRT: Concrete results and practical countermeasures. In: Kaliski Jr., B.S., Koç, Ç.K., Paar, C. (eds.) CHES 2002. LNCS, vol. 2523, pp. 260–275. Springer, Heidelberg (2003)

3. Bao, F., Deng, R., Han, Y., Jeng, A., Narasimhalu, A.D., Ngair, T.-H.: Breaking Public Key Cryptosystems an Tamper Resistance Devices in the Presence of Transient Fault. In: Christianson, B., Lomas, M. (eds.) Security Protocols 1997. LNCS, vol. 1361, pp. 115–124. Springer, Heidelberg (1998)

4. Berzati, A., Canovas, C., Goubin, L.: (In)security Against Fault Injection Attacks for CRT-RSA Implementations. In: Breveglieri, L., Gueron, S., Koren, I., Naccache, D., Seifert, J.-P. (eds.) FDTC 2008, pp. 101–107. IEEE Computer Society, Los Alamitos (2008)

5. Berzati, A., Canovas, C., Goubin, L.: Perturbating RSA public keys: An improved attack. In: Oswald, E., Rohatgi, P. (eds.) CHES 2008. LNCS, vol. 5154, pp. 380–395. Springer, Heidelberg (2008)

6. Biham, E., Shamir, A.: Differential fault analysis of secret key cryptosystems. In: Kaliski Jr., B.S. (ed.) CRYPTO 1997. LNCS, vol. 1294, pp. 513–525. Springer, Heidelberg (1997)

7. Blömer, J., Otto, M., Seifert, J.-P.: A New RSA-CRT Algorithm Secure against Bellcore Attacks. In: Jajodia, S., Atluri, V., Jaeger, T. (eds.) CCS 2003, pp. 311–320. ACM Press, New York (2003)

8. Boneh, D., DeMillo, R.A., Lipton, R.J.: On the importance of checking cryptographic protocols for faults. In: Fumy, W. (ed.) EUROCRYPT 1997. LNCS, vol. 1233, pp. 37–51. Springer, Heidelberg (1997)

9. Boreale, M.: Attacking right-to-left modular exponentiation with timely random faults. In: Breveglieri, L., Koren, I., Naccache, D., Seifert, J.-P. (eds.) FDTC 2006. LNCS, vol. 4236, pp. 24–35. Springer, Heidelberg (2006)

10. Boscher, A., Naciri, R., Prouff, E.: CRT RSA algorithm protected against fault attacks. In: Sauveron, D., Markantonakis, K., Bilas, A., Quisquater, J.-J. (eds.) WISTP 2007. LNCS, vol. 4462, pp. 229–243. Springer, Heidelberg (2007)

11. Bos, J., Coster, M.: Addition chain heuristics. In: Brassard, G. (ed.) CRYPTO 1989. LNCS, vol. 435, pp. 400–407. Springer, Heidelberg (1990)

12. Brier, É., Chevallier-Mames, B., Ciet, M., Clavier, C.: Why one should also secure RSA public key elements. In: Goubin, L., Matsui, M. (eds.) CHES 2006. LNCS, vol. 4249, pp. 324–338. Springer, Heidelberg (2006)

13. Chevallier-Mames, B., Ciet, M., Joye, M.: Low-cost Solutions for Preventing Simple Side-Channel Analysis: Side-Channel Atomicity. IEEE Transactions on Computers 53(6), 760–768 (2004)

14. Ciet, M., Joye, M.: Practical Fault Countermeasures for Chinese Remaindering Based RSA. In: Breveglieri, L., Koren, I. (eds.) FDTC 2005, pp. 124–132 (2005)

15. Coron, J.-S.: Resistance against differential power analysis for elliptic curve cryptosystems. In: Koç, Ç.K., Paar, C. (eds.) CHES 1999. LNCS, vol. 1717, pp. 292–302. Springer, Heidelberg (1999)

16. Dottax, E., Giraud, C., Rivain, M., Sierra, Y.: On Second-Order Fault Analysis Resistance for CRT-RSA Implementations. Cryptology ePrint Archive, Report 2009/24 (2009), http://eprint.iacr.org/2009/024

17. Fouque, P.-A., Valette, F.: The Doubling Attack: Why Upwards is Better than Downwards. In: Walter, C.D., Koç, Ç.K., Paar, C. (eds.) CHES 2003. LNCS, vol. 2779, pp. 269–280. Springer, Heidelberg (2003)

18. Giraud, C.: An RSA Implementation Resistant to Fault Attacks and to Simple Power Analysis. IEEE Transactions on Computers 55(9), 1116–1120 (2006)

19. Gordon, D.M.: A Survey of Fast Exponentiation Methods. J. Algorithms 27(1), 129–146 (1998)

20. Itoh, K., Izu, T., Takenak, M.: Address-bit Differential Power Analysis of Cryptographic Schemes OK-ECDH and OK-ECDSA. In: Kaliski Jr., B.S., Koç, Ç.K., Paar, C. (eds.) CHES 2002. LNCS, vol. 2523, pp. 129–143. Springer, Heidelberg (2003)

21. Itoh, K., Izu, T., Takenaka, M.: A Practical Countermeasure against Address-Bit Differential Power Analysis. In: Walter, C.D., Koç, Ç.K., Paar, C. (eds.) CHES 2003. LNCS, vol. 2779, pp. 382–396. Springer, Heidelberg (2003)

22. Joye, M., Lenstra, A., Quisquater, J.-J.: Chinese Remaindering Based Cryptosystems in the Presence of Faults. Journal of Cryptology 12(4), 241–245 (1999)

23. Kim, C.H., Quisquater, J.-J.: Fault Attacks for CRT Based RSA: New Attacks, New Results, and New Countermeasures. In: Sauveron, D., Markantonakis, K., Bilas, A., Quisquater, J.-J. (eds.) WISTP 2007. LNCS, vol. 4462, pp. 215–228. Springer, Heidelberg (2007)

24. Kim, C.H., Shin, J.H., Quisquater, J.-J., Lee, P.J.: Safe-error attack on SPA-FA resistant exponentiations using a HW modular multiplier. In: Nam, K.-H., Rhee, G. (eds.) ICISC 2007. LNCS, vol. 4817, pp. 273–281. Springer, Heidelberg (2007)

25. Knuth, D.: The Art of Computer Programming, 3rd edn. Addison-Wesley, Reading (1988)

26. Kocher, P., Jaffe, J., Jun, B.: Differential Power Analysis. In: Wiener, M. (ed.) CRYPTO 1999. LNCS, vol. 1666, pp. 388–397. Springer, Heidelberg (1999)

27. Kocher, P.: Timing attacks on implementations of diffie-hellman, RSA, DSS, and other systems. In: Koblitz, N. (ed.) CRYPTO 1996. LNCS, vol. 1109, pp. 104–113. Springer, Heidelberg (1996)

28. Koç, Ç.: Analysis of the Sliding Window Techniques for Exponentiation. Computer & Mathematics with applications 30(10), 17–24 (1995)

29. Menezes, A., van Oorschot, P., Vanstone, S.: Handbook of Applied Cryptography. CRC Press, Boca Raton (1997)

30. Messerges, T., Dabbish, E., Sloan, R.: Power Analysis Attacks of Modular Exponentiation in Smartcard. In: Koç, Ç.K., Paar, C. (eds.) CHES 1999. LNCS, vol. 1717, pp. 144–157. Springer, Heidelberg (1999)

31. Mitrinovic, D.S., Sándor, J., Crstici, B.: Handbook of Number Theory. Springer, Heidelberg (1995)

32. Möller, B.: Algorithms for multi-exponentiation. In: Vaudenay, S., Youssef, A.M. (eds.) SAC 2001. LNCS, vol. 2259, pp. 165–180. Springer, Heidelberg (2001)

33. Rivest, R., Shamir, A., Adleman, L.: A Method for Obtaining Digital Signatures and Public-Key Cryptosystems. Communications of the ACM 21(2), 120–126 (1978)

34. Schmidt, J., Herbst, C.: A Practical Fault Attack on Square and Multiply. In: Breveglieri, L., Gueron, S., Koren, I., Naccache, D., Seifert, J.-P. (eds.) FDTC 2008, pp. 53–58. IEEE Computer Society, Los Alamitos (2008)

35. Seifert, J.-P.: On Authenticated Computing and RSA-based Authentication. In: Atluri, V., Meadows, C., Juels, A. (eds.) ACM CCS 2005, pp. 122–127. ACM Press, New York (2005)

36. Shamir, A.: Improved Method and Apparatus for Protecting Public Key Schemes from Timing and Fault Attacks. Publication number: WO9852319 (November 1998)

37. Sun Microsystems. Application Programming Interface – Java Card^{TM} Plateform, Version 2.2.2 (March 2006),
http://java.sun.com/products/javacard/specs.html

38. Vigilant, D.: RSA with CRT: A new cost-effective solution to thwart fault attacks. In: Oswald, E., Rohatgi, P. (eds.) CHES 2008. LNCS, vol. 5154, pp. 130–145. Springer, Heidelberg (2008)
39. Wagner, D.: Cryptanalysis of a Provable Secure CRT-RSA Algorithm. In: Pfitzmann, B., Liu, P. (eds.) CCS 2004, pp. 82–91. ACM Press, New York (2004)
40. Yen, S.-M., Joye, M.: Checking Before Output Not Be Enough Against Fault-Based Cryptanalysis. IEEE Transactions on Computers 49(9), 967–970 (2000)
41. Yen, S.-M., Kim, S.-J., Lim, S.-G., Moon, S.-J.: A countermeasure against one physical cryptanalysis may benefit another attack. In: Kim, K.-c. (ed.) ICISC 2001. LNCS, vol. 2288, pp. 414–427. Springer, Heidelberg (2002)
42. Yen, S.-M., Kim, S.-J., Lim, S.-G., Moon, S.-J.: RSA Speedup with Residue Number System Immune against Hardware Fault Cryptanalysis. IEEE Transactions on Computers 52(4), 461–472 (2003)

A   Proof of Lemma 1

Proof. By the law of total probability, we have:

$$P\left(\mathrm{ord}_M(m)|u\right) = \sum_{\lambda \in \mathcal{D}(\varphi(M))} P\left(\lambda|u\right) P\left(\mathrm{ord}_M(m) = \lambda\right) , \tag{6}$$

where $\mathcal{D}$ is the function mapping a natural integer to the set of its divisors. On the one hand, the probability $P\left(\lambda|u\right)$ equals $1/\lambda$. On the other hand, for every $\lambda \in \mathcal{D}(\varphi(M))$, there are $\varphi(\lambda)$ elements of order λ in $\mathbb{Z}_M^*$ which leads to $P\left(\mathrm{ord}_M(m) = \lambda\right) = \varphi(\lambda)/\varphi(M)$. On the whole, (6) can be rewritten as:

$$P\left(\mathrm{ord}_M(m)|u\right) = \frac{1}{\varphi(M)} \sum_{\lambda \in \mathcal{D}(\varphi(M))} \frac{\varphi(\lambda)}{\lambda} . \tag{7}$$

Since $\varphi(\lambda)/\lambda$ is strictly lower than or equal to 1, we have $P\left(\mathrm{ord}_M(m)|u\right) \leq d(\varphi(M))/\varphi(M)$ where $d(\cdot)$ denotes the divisor function (*i.e.* the function that maps a natural integer to the quantity of its distinct divisors). It is well known that the divisor function satisfies $d(x) < 2\sqrt{x}$ for every x [31] which implies $P\left(\mathrm{ord}_M(m)|u\right) < 2/\sqrt{\varphi(M)}$. Since we have $\varphi(M) > n^{2/3}$ for every $M > 30$ [31], we get (5).   ◇

B   Atomic Algorithms

Looking at the chain computation algorithm, we observe that the main operations (namely operations on large registers) performed at each loop iteration are a division by two and possibly a substraction (depending on the value of τ_i). To render the algorithm atomic both operations must be performed at each loop iteration. The following algorithm describes the atomic version of the chain computation. It makes use of three registers: R_0, R_1 and R_2 which are used to store the values of α_i and β_i as well as a temporary value. It also uses three indices $i_\alpha, i_\beta, i_{tmp} \in \{0, 1, 2\}$ such that α_i is stored in R_{i_α}, β_i is stored in R_{i_β} and the temporary value is stored in $R_{i_{tmp}}$.

Algorithm 5. Atomic double addition chain computation

INPUT: A pair of natural integer (a, b) s.t. $a \leq b$

OUTPUT: The chain $\omega(a, b)$

---

1. $R_{i_\alpha} \leftarrow a;\ R_{i_\beta} \leftarrow b;\ j \leftarrow n^*$
2. **while** $(R_{i_\alpha}, R_{i_\beta}) \neq (0, 1)$ **do**
3. $\quad R_{i_{tmp}} \leftarrow R_{i_\beta} - R_{i_\alpha}$
4. $\quad v \leftarrow R_{i_\beta} \bmod 2$
5. $\quad R_{i_\beta} \leftarrow R_{i_\beta}/2$
6. $\quad t \leftarrow (R_{i_\beta} > R_{i_\alpha})$
7. $\quad \omega_{j-1} \leftarrow t;\ \omega_j \leftarrow t \vee v$
8. $\quad (i_\alpha, i_\beta, i_{tmp}) \leftarrow \big(t \wedge (i_{tmp}, i_\alpha, i_\beta)\big) \vee \big((t \oplus 1) \wedge (i_\alpha, i_\beta, i_{tmp})\big)$
9. $\quad i \leftarrow (t \wedge i_\alpha) \vee \big((t \oplus 1) \wedge i_{tmp}\big)$
10. $\quad R_i \leftarrow 2 \cdot R_i + v$
11. $\quad j \leftarrow j - 1 - (t \oplus 1)$
12. **end while**
13. **return** ω

Notations. In Step 6, the notation $t \leftarrow (R_{i_\beta} > R_{i_\alpha})$ is used to denote the operation that compares the two values in R_{i_β} and R_{i_α} and that returns the binary value t satisfying $t = 1$ if $R_{i_\beta} > R_{i_\alpha}$ and $t = 0$ otherwise. In Steps 8 and 9, the logical AND is extended to the $\{0, 1\} \times \{0, 1\}^n \to \{0, 1\}^n$ operator performing a logical AND between the left argument and each coordinate of the right argument.

Looking at Algorithm 5, we see that, at each loop iteration, the Boolean values t and v represent the values of τ_i and ν_i. One can verify that if $t = 0$ then these values are stored in (ω_{j-1}, ω_j) and j is decremented by two while if $t = 1$ then t is stored in ω_j and j is decremented by 1. Moreover, if $t = 0$ then Steps 8 and 9 have no effect while if $t = 1$ then Step 8 ensures that the indices of the different registers are permuted so that (α_i, β_i) is correctly updated and Step 9 ensures that the value $\beta_i/2$ stored in R_{i_α} is putted back to β_i.

Although Algorithm 5 requires three l-bit registers and a $(2.2\ l)$-bit buffer to store ω (see Sect. 7), its memory consumption can be reduced to $4.2\ l$ bits using the following trick. During the computation of the $1.2\ l$ high order bits of ω, the l low order bits allocated for ω are used as one of the three necessary l-bit registers. Once the $1.2\ l$ high order bits of ω have been computed, the intermediate values α_i and β_i have a bit-length lower than $l/2$. Therefore, the three registers can be allocated on less than $2l$ bits and the low order part of the buffer for ω can be freed.

The following algorithm describes the atomic version of the double modular exponentiation. It makes use of two registers $R_{(0,0)}$ and $R_{(0,1)}$ that are used to store the intermediate results m^{a_i} and m^{b_i} and one more register $R_{(1,0)}$ to store m. It makes also use of two Boolean variables γ and μ. The Boolean γ indicates that m^{a_i} is stored in $R_{(0,\gamma \oplus 1)}$ and that m^{b_i} is stored in $R_{(0,\gamma)}$. And the Boolean μ indicates wether the next modular multiplication is a multiplication by m ($\mu = 0$) or not ($\mu = 1$).

Algorithm 6. Atomic double modular exponentiation

INPUT: An element $m \in \mathbb{Z}_N$, a chain $\omega(a,b)$ s.t. $a \leq b$, a modulus N
OUTPUT: The pair of modular power $(m^a \bmod N, m^b \bmod N)$

---

1. $R_{(0,0)} \leftarrow 1$; $R_{(0,1)} \leftarrow m$; $R_{(1,0)} \leftarrow m$
2. $\gamma \leftarrow 1$; $\mu \leftarrow 1$; $i \leftarrow 0$
3. **while** $i < n$ **do**
4. $\quad t \leftarrow \omega_j \wedge \mu$; $v \leftarrow \omega_{j+1} \wedge \mu$
5. $\quad R_{(0,\gamma \oplus t)} \leftarrow R_{(0,\gamma \oplus t)} \cdot R_{((\mu \oplus 1), \gamma \wedge \mu)} \bmod N$
6. $\quad \mu \leftarrow (t \oplus 1) \wedge v$; $\gamma \leftarrow \gamma \oplus t$
7. $\quad i \leftarrow i + \mu + \mu \wedge t$
8. **end while**
9. **return** $(R_{\gamma \oplus 1}, R_{\gamma})$

While $\mu = 1$, the Boolean t is evaluated to τ_i and, if $\tau_i = 1$, the Boolean v is evaluated to ν_i. Then, while $t = 1$ or $v = 0$ each loop iteration corresponds to a step performing one single multiplication which is done in Step 5. If $t = 0$ and $\nu = 1$, the step must perform two multiplications: $R_{(0,\gamma)}$ by $R_{(0,\gamma)}$ and $R_{(0,\gamma)}$ by $R_{(1,0)}$. The first one is performed in Step 5 afterward the Boolean μ is evaluated to 0 thus indicating that the next loop must perform the multiplication by $R_{(1,0)}$. In that case, i is not incremented and the next loop iteration performs the desired multiplication before evaluating μ to 1 and normally carrying on the computation.

Author Index

Printing: Mercedes-Druck, Berlin
Binding: Stein+Lehmann, Berlin